Quantity	U.S. Customary Unit	SI Equivalent
Acceleration	ft/s²	0.3048 m/s²
	in./s²	0.0254 m/s²
Area	ft²	0.0929 m²
	in²	645.2 mm²
Energy	ft · lb	1.356 J
Force	kip	4.448 kN
	lb	4.448 N
	oz	0.2780 N
Impulse	lb · s	4.448 N · s
Length	ft	0.3048 m
	in.	25.40 mm
	mi	1.609 km
Mass	oz mass	28.35 g
	lb mass	0.4536 kg
	slug	14.59 kg
	ton	907.2 kg
Moment of a force	lb · ft	1.356 N · m
	lb · in.	0.1130 N · m
Moment of inertia		
Of an area	in⁴	0.4162×10^6 mm⁴
Of a mass	lb · ft · s²	1.356 kg · m²
Momentum	lb · s	4.448 kg · m/s
Power	ft · lb/s	1.356 W
	hp	745.7 W
Pressure or stress	lb/ft²	47.88 Pa
	lb/in² (psi)	6.895 kPa
Velocity	ft/s	0.3048 m/s
	in./s	0.0254 m/s
	mi/h (mph)	0.4470 m/s
	mi/h (mph)	1.609 km/h
Volume, solids	ft³	0.02832 m³
	in³	16.39 cm³
Liquids	gal	3.785 l
	qt	0.9464 l
Work	ft · lb	1.356 J

Vector
Mechanics
for
Engineers STATICS
Third Edition

Ferdinand P. Beer

Professor and Chairman
Department of Mechanical Engineering and Mechanics
Lehigh University

E. Russell Johnston, Jr.

Professor and Head
Department of Civil Engineering
University of Connecticut

McGraw-Hill Book Company

New York
St. Louis
San Francisco
Auckland
Bogotá
Düsseldorf
Johannesburg
London
Madrid
Mexico
Montreal
New Delhi
Panama
Paris
São Paulo
Singapore
Sydney
Tokyo
Toronto

Vector Mechanics for Engineers
STATICS

3 4 5 6 7 8 9 0   DODO   7 8 3 2 1 0 9 8 7

This book was set in Laurel by York Graphic Services, Inc. The editors were B. J. Clark and J. W. Maisel; the designer was Merrill Haber; the production supervisor was Thomas J. LoPinto. The drawings were done by Felix Cooper.
R. R. Donnelley & Sons Company was printer and binder.

Library of Congress Cataloging in Publication Data

Beer, Ferdinand Pierre, date
  Vector mechanics for engineers.

  CONTENTS: [1] Statics.
  1. Mechanics. Applied. 2. Vector analysis.
3. Mechanics, Applied—Problems, exercises, etc.
I. Johnston, Elwood Russell, date joint author.
II. Title.
TA350.B3552 1977   531'.01'51563   76-54914
ISBN 0-07-004278-0

Contents

3

RIGID BODIES: EQUIVALENT SYSTEMS OF FORCES 59

4

EQUILIBRIUM OF RIGID BODIES 122

EQUILIBRIUM IN TWO DIMENSIONS 124

EQUILIBRIUM IN THREE DIMENSIONS 150

5
DISTRIBUTED FORCES: CENTROIDS AND CENTERS OF GRAVITY   166

6
ANALYSIS OF STRUCTURES   213

7
FORCES IN BEAMS AND CABLES   266

Preface

The main objective of a first course in mechanics should be to develop in the engineering student the ability to analyze any problem in a simple and logical manner and to apply to its solution a few, well-understood basic principles. It is hoped that this text, designed for the first course in statics offered in the sophomore year, and the volume that follows, *Vector Mechanics for Engineers: Dynamics*, will help the instructor achieve this goal.†

Vector algebra is introduced early in the text and used in the presentation and the discussion of the fundamental principles of mechanics. Vector methods are also used to solve many problems, particularly three-dimensional problems where their application results in a simpler and more concise solution. The emphasis in this text, however, remains on the correct understanding of the principles of mechanics and on their application to the solution of engineering problems, and vector algebra is presented chiefly as a convenient tool.‡

One of the characteristics of the approach used in these volumes is that the mechanics of *particles* has been clearly separated from the mechanics of *rigid bodies*. This approach makes it possible to consider simple practical applications at an early stage and to postpone the introduction of more difficult concepts. In this volume, for example, the statics of particles is treated first (Chap. 2); after the rules of addition and subtraction of vectors have been introduced, the principle of equilibrium of a particle is immediately applied to practical situations involving only concurrent forces. The statics of rigid bodies is considered in Chaps. 3 and 4. In Chap. 3, the vector and scalar products of two vectors are introduced and used to define the moment of a force about a point and about an axis. The presentation of these new concepts is followed by a thorough and rigorous discussion of equivalent systems of forces leading, in Chap. 4, to many practical applications involving the equilibrium of rigid bodies

---

† Both texts are also available in a single volume, *Vector Mechanics for Engineers: Statics and Dynamics*, third edition.

‡ In a parallel text, *Mechanics for Engineers: Statics*, third edition, the use of vector algebra is limited to the addition and subtraction of vectors.

under general force systems. In the volume on dynamics, the same division is observed. The basic concepts of force, mass, and acceleration, of work and energy, and of impulse and momentum are introduced and first applied to problems involving only particles. Thus the student may familiarize himself with the three basic methods used in dynamics and learn their respective advantages before facing the difficulties associated with the motion of rigid bodies.

Since this text is designed for a first course in statics, new concepts have been presented in simple terms and every step explained in detail. On the other hand, by discussing the broader aspects of the problems considered, a definite maturity of approach has been achieved. For example, the concepts of partial constraints and of static indeterminacy are introduced early in the text and used throughout.

The fact that mechanics is essentially a *deductive* science based on a few fundamental principles has been stressed. Derivations have been presented in their logical sequence and with all the rigor warranted at this level. However, the learning process being largely *inductive*, simple applications have been considered first. Thus the statics of particles precedes the statics of rigid bodies, and problems involving internal forces are postponed until Chap. 6. Also, in Chap. 4, equilibrium problems involving only coplanar forces are considered first and solved by ordinary algebra, while problems involving three-dimensional forces and requiring the full use of vector algebra are discussed in the second part of the chapter.

Free-body diagrams are introduced early, and their importance is emphasized throughout the text. Color has been used to distinguish forces from other elements of the free-body diagrams. This makes it easier for the students to identify the forces acting on a given particle or rigid body and to follow the discussion of sample problems and other examples given in the text. Free-body diagrams are used not only to solve equilibrium problems but also to express the equivalence of two systems of forces or, more generally, of two systems of vectors. This approach is particularly useful as a preparation for the study of the dynamics of rigid bodies. As will be shown in the volume on dynamics, by placing the emphasis on "free-body-diagram equations" rather than on the standard algebraic equations of motion, a more intuitive and more complete understanding of the fundamental principles of dynamics may be achieved.

Because of the current trend among American engineers to adopt the international system of units (SI metric units), the SI units most frequently used in mechanics have been introduced in Chap. 1 of this edition and are used throughout the text. Half the sample problems and problems to be assigned have been stated in these units, while the other half retain U.S. customary units. The authors believe that this approach will best serve the needs of the students, who will be entering the engineering profession during the period of transition from one system of units to the other. It also should be recognized that the passage from one system to the other entails more than the use of conversion factors. Since the SI system of units is an absolute system based on the units of time, length, and mass, whereas the U.S. customary system is a gravitational system based on the units of time, length, and force, different approaches are required for the solution of many problems. For example, when SI units are used, a body is generally specified by its mass expressed in kilograms; in most problems of statics it will be necessary to determine the weight of the body in newtons, and an additional calculation will be required for this purpose. On the other hand, when U.S. customary units are used, a body is specified by its weight in pounds and, in dynamics problems, an additional calculation will be required to determine its mass in slugs (or $lb \cdot sec^2/ft$). The authors, therefore, believe that problems assignments should include both types of units. A sufficient number of problems, however, have been provided so that if so desired, two complete sets of assignments may be selected from problems stated in SI units only and two others from problems stated in U.S. customary units. Since the answers to all even-numbered problems stated in U.S. customary units have been given in both systems of units, teachers who wish to give special instruction to their students in the conversion of units may assign these problems and ask their students to use SI units in their solutions.

A large number of optional sections have been included. These sections are indicated by asterisks and may thus easily be distinguished from those which form the core of the basic statics course. They may be omitted without prejudice to the understanding of the rest of the text. Among the topics covered in these additional sections are applications to hydrostatics, shear and bending-moment diagrams for beams, equilibrium of cables, products of inertia and Mohr's circle, mass products of inertia and principal axes of inertia for three-dimensional bodies, and the method of virtual work. The sections on beams are especially useful when the course in statics is immediately followed by a course in mechanics of materials, while the sections on the inertia properties of three-dimensional bodies are primarily intended for the students who will later study in dynamics the motion of rigid bodies in three dimensions.

The material presented in the text and most of the problems require no previous mathematical knowledge beyond algebra, trigonometry, and elementary calculus, and all the elements of vector algebra necessary to the understanding of the text have been carefully presented in Chaps. 2 and 3. In general, a greater emphasis has been placed on the correct understanding of the basic mathematical concepts involved than on the nimble manipulation of mathematical formulas. In this connection, it should be mentioned that the determination of the centroids of composite areas precedes the calculation of centroids by integration, thus making it possible to establish the concept of moment of area firmly before introducing the use of integration.

In several parts of the text the presentation of numerical solutions has been modified to take into account the universal use of calculators by engineering students. Instructions on the proper use of calculators for the solution of typical statics problems have been substituted in Chap. 2 for the notes on the use of the slide rule which were contained in the previous editions of *Vector Mechanics for Engineers.*

The text has been divided into units, each consisting of one or several theory sections, one or several sample problems, and a large number of problems to be assigned. Each unit corresponds to a well-defined topic and generally may be covered in one lesson. In a number of cases, however, the instructor will find it desirable to devote more than one lesson to a given topic. The sample problems have been set up in much the same form that a student will use in solving the assigned problems. They thus serve the double purpose of amplifying the text and demonstrating the type of neat and orderly work that the student should cultivate in his own solutions. Most of the problems to be assigned are of a practical nature and should appeal to the engineering student. They are primarily designed, however, to illustrate the material presented in the text and to help the student understand the basic principles of mechanics. The problems have been grouped according to the portions of material they illustrate and have been arranged in order of increasing difficulty. Problems requiring special attention have been indicated by asterisks. Answers to all even-numbered problems are given at the end of the book.

The authors wish to acknowledge gratefully the many helpful comments and suggestions offered by the users of the previous editions of *Mechanics for Engineers* and of *Vector Mechanics for Engineers.*

FERDINAND P. BEER
E. RUSSELL JOHNSTON, JR.

List of Symbols

a Constant; radius; distance

A, B, C, . . .   Reactions at supports and connections

$A, B, C, \ldots$ Points

A Area

b Width; distance

c Constant

C Centroid

d Distance

e Base of natural logarithms

F   Force; friction force

g Acceleration of gravity

G Center of gravity; constant of gravitation

h Height; sag of cable

i, j, k   Unit vectors along coordinate axes

$I, I_x, \ldots$ Moment of inertia

$\bar{I}$ Centroidal moment of inertia

J Polar moment of inertia

k Spring constant

k_x, k_y, k_O Radius of gyration

$\bar{k}$ Centroidal radius of gyration

l Length

L Length; span

m Mass

M   Couple; moment

$\mathbf{M}_O$ Moment about point O

$\mathbf{M}_O^R$ Moment resultant about point O

M Magnitude of couple or moment; mass of earth

M_{OL} Moment about axis OL

N   Normal component of reaction

O Origin of coordinates

p Pressure

P   Force; vector

$P_{xy}, \ldots$ Product of inertia

Q   Force; vector

r   Position vector

r Radius; distance; polar coordinate

R   Resultant force; resultant vector; reaction

R Radius of earth

s Position vector

s Length of arc; length of cable

S Force; vector

t Thickness

$\mathbf{T}$ Force

T Tension

u Rectangular coordinate

U Work

v Rectangular coordinate

$\mathbf{V}$ Vector product; shearing force

V Volume; potential energy; shear

w Load per unit length

$\mathbf{W}, W$ Weight; load

x, y, z Rectangular coordinates; distances

$\bar{x}, \bar{y}, \bar{z}$ Rectangular coordinates of centroid or center of gravity

α, β, γ Angles

γ Specific weight

δ Elongation

$\delta\mathbf{r}$ Virtual displacement

δU Virtual work

$\boldsymbol{\lambda}$ Unit vector along a line

η Efficiency

θ Angular coordinate; angle; polar coordinate

μ Coefficient of friction

ρ Density

ϕ Angle of friction; angle

Introduction

1.1. What Is Mechanics? Mechanics may be defined as that science which describes and predicts the conditions of rest or motion of bodies under the action of forces. It is divided into three parts: mechanics of *rigid bodies*, mechanics of *deformable bodies*, and mechanics of *fluids*.

The mechanics of rigid bodies is subdivided into *statics* and *dynamics*, the former dealing with bodies at rest, the latter with bodies in motion. In this part of the study of mechanics, bodies are assumed to be perfectly rigid. Actual structures and machines, however, are never absolutely rigid and deform under the loads to which they are subjected. But these deformations are usually small and do not appreciably affect the conditions of equilibrium or motion of the structure under consideration. They are important, though, as far as the resistance of the structure to failure is concerned and are studied in mechanics of materials, which is a part of the mechanics of deformable bodies. The third division of mechanics, the mechanics of fluids, is subdivided into the study of *incompressible fluids* and of *compressible fluids*. An important subdivision of the study of incompressible fluids is *hydraulics*, which deals with problems involving liquids.

Mechanics is a physical science, since it deals with the study of physical phenomena. However, some associate mechanics with mathematics, while many consider it as an engineering subject. Both these views are justified in part. Mechanics is the foundation of most engineering sciences and is an indispensable prerequisite to their study. However, it does not have the *empiricism* found in some engineering sciences, i.e., it does not rely on experience or observation alone; by its rigor and the emphasis it places on deductive reasoning it resembles mathematics. But, again, it is not an *abstract* or even a *pure* science; mechanics is an *applied* science. The purpose of mechanics is to explain and predict physical phenomena and thus to lay the foundations for engineering applications.

1.2. Fundamental Concepts and Principles. Although the study of mechanics goes back to the time of Aristotle (384–322 B.C.) and Archimedes (287–212 B.C.), one has to wait until Newton (1642–1727) to find a satisfactory formulation of its fundamental principles. These principles were later expressed in a modified form by D'Alembert, Lagrange, and Hamilton. Their validity remained unchallenged, however, until Einstein formulated his *theory of relativity* (1905). While its limitations have now been recognized, *newtonian mechanics* still remains the basis of today's engineering sciences.

The basic concepts used in mechanics are *space, time, mass,* and *force.* These concepts cannot be truly defined; they should be accepted on the basis of our intuition and experience and used as a mental frame of reference for our study of mechanics.

The concept of *space* is associated with the notion of the position of a point *P.* The position of *P* may be defined by three lengths measured from a certain reference point, or *origin,* in three given directions. These lengths are known as the *coordinates* of *P.*

To define an event, it is not sufficient to indicate its position in space. The *time* of the event should also be given.

The concept of *mass* is used to characterize and compare bodies on the basis of certain fundamental mechanical experiments. Two bodies of the same mass, for example, will be attracted by the earth in the same manner; they will also offer the same resistance to a change in translational motion.

A *force* represents the action of one body on another. It may be exerted by actual contact or at a distance, as in the case of gravitational forces and magnetic forces. A force is characterized by its *point of application,* its *magnitude,* and its *direction;* a force is represented by a *vector* (Sec. 2.2).

In newtonian mechanics, space, time, and mass are absolute concepts, independent of each other. (This is not true in *relativistic mechanics,* where the time of an event depends upon its position, and where the mass of a body varies with its velocity.) On the other hand, the concept of force is not independent of the other three. Indeed, one of the fundamental principles of newtonian mechanics listed below indicates that the resultant force acting on a body is related to the mass of the body and to the manner in which its velocity varies with time.

We shall study the conditions of rest or motion of particles and rigid bodies in terms of the four basic concepts we have introduced. By *particle* we mean a very small amount of matter which may be assumed to occupy a single point in space. A *rigid body* is a combination of a large number of particles occupying fixed positions with respect to each other. The study of the mechanics of particles is obviously a prerequisite to that of rigid bodies. Besides, the results obtained for a particle may be used directly in a large number of problems dealing with the conditions of rest or motion of actual bodies.

The study of elementary mechanics rests on six fundamental principles based on experimental evidence:

The Parallelogram Law for the Addition of Forces. This states that two forces acting on a particle may be replaced by a single force, called their *resultant*, obtained by drawing the diagonal of the parallelogram which has sides equal to the given forces (Sec. 2.1).

The Principle of Transmissibility. This states that the conditions of equilibrium or of motion of a rigid body will remain unchanged if a force acting at a given point of the rigid body is replaced by a force of the same magnitude and same direction, but acting at a different point, provided that the two forces have the same line of action (Sec. 3.2).

Newton's Three Fundamental Laws. Formulated by Sir Isaac Newton in the latter part of the seventeenth century, these laws may be stated as follows:

FIRST LAW. If the resultant force acting on a particle is zero, the particle will remain at rest (if originally at rest) or will move with constant speed in a straight line (if originally in motion) (Sec. 2.9).

SECOND LAW. If the resultant force acting on a particle is not zero, the particle will have an acceleration proportional to the magnitude of the resultant and in the direction of this resultant force.

As we shall see in Sec. 12.1, this law may be stated as

$$\mathbf{F} = m\mathbf{a} \qquad (1.1)$$

where $\mathbf{F}$, m, and $\mathbf{a}$ represent, respectively, the resultant force acting on the particle, the mass of the particle, and the acceleration of the particle, expressed in a consistent system of units.

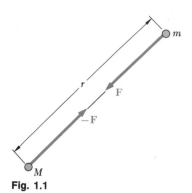

Fig. 1.1

THIRD LAW. The forces of action and reaction between bodies in contact have the same magnitude, same line of action, and opposite sense (Sec. 6.1).

Newton's Law of Gravitation. This states that two particles of mass M and m are mutually attracted with equal and opposite forces $\mathbf{F}$ and $-\mathbf{F}$ (Fig. 1.1) of magnitude F given by the formula

$$F = G\frac{Mm}{r^2} \tag{1.2}$$

where r = distance between the two particles
 G = universal constant called the *constant of gravitation*

Newton's law of gravitation introduces the idea of an action exerted at a distance and extends the range of application of Newton's third law: the action $\mathbf{F}$ and the reaction $-\mathbf{F}$ in Fig. 1.1 are equal and opposite, and they have the same line of action.

A particular case of great importance is that of the attraction of the earth on a particle located on its surface. The force $\mathbf{F}$ exerted by the earth on the particle is then defined as the *weight* $\mathbf{W}$ of the particle. Taking M equal to the mass of the earth, m equal to the mass of the particle, and r equal to the radius R of the earth, and introducing the constant

$$g = \frac{GM}{R^2} \tag{1.3}$$

the magnitude W of the weight of a particle of mass m may be expressed as†

$$W = mg \tag{1.4}$$

The value of R in formula (1.3) depends upon the elevation of the point considered; it also depends upon its latitude, since the earth is not truly spherical. The value of g therefore varies with the position of the point considered. As long as the point actually remains on the surface of the earth, it is sufficiently accurate in most engineering computations to assume that g equals 9.81 m/s^2 or 32.2 ft/s^2.

The principles we have just listed will be introduced in the course of our study of mechanics as they are needed. The study of the statics of particles, carried out in Chap. 2, will be based on the parallelogram law of addition and on Newton's first law alone. The principle of transmissibility will be introduced in Chap. 3 as we begin the study of the statics of rigid bodies, and

† A more accurate definition of the weight $\mathbf{W}$ should take into account the rotation of the earth.

Newton's third law in Chap. 6 as we analyze the forces exerted on each other by the various members forming a structure. It should be noted that the four aforementioned principles underlie the entire study of the statics of particles, rigid bodies, and systems of rigid bodies. In the study of dynamics, Newton's second law and Newton's law of gravitation will be introduced. It will then be shown that Newton's first law is a particular case of Newton's second law (Sec. 12.1) and that the principle of transmissibility could be derived from the other principles and thus eliminated (Sec. 16.5). In the meantime, however, Newton's first and third laws, the parallelogram law of addition, and the principle of transmissibility will provide us with the necessary and sufficient foundation for the study of the entire field of statics.

As noted earlier, the six fundamental principles listed above are based on experimental evidence. Except for Newton's first law and the principle of transmissibility, they are independent principles which cannot be derived mathematically from each other or from any other elementary physical principle. On these principles rests most of the intricate structure of newtonian mechanics. For more than two centuries a tremendous number of problems dealing with the conditions of rest and motion of rigid bodies, deformable bodies, and fluids have been solved by applying these fundamental principles. Many of the solutions obtained could be checked experimentally, thus providing a further verification of the principles from which they were derived. It is only recently that Newton's mechanics was found at fault, in the study of the motion of atoms and in the study of the motion of certain planets, where it must be supplemented by the theory of relativity. But on the human or engineering scale, where velocities are small compared with the velocity of light, Newton's mechanics has yet to be disproved.

1.3. Systems of Units. With the four fundamental concepts introduced in the preceding section are associated the so-called *kinetic units*, i.e., the units of *length, time, mass,* and *force*. These units cannot be chosen independently if Eq. (1.1) is to be satisfied. Three of the units may be defined arbitrarily; they are then referred to as *base units*. The fourth unit, however, must be chosen in accordance with Eq. (1.1) and is referred to as a *derived unit*. Kinetic units selected in that way are said to form a *consistent system of units*.

International System of Units (SI Units†). In this system, which will be in universal use after the United States has com-

---

† SI stands for *Système International d'Unités* (French).

pleted its current conversion, the base units are the units of length, mass, and time, and are called, respectively, the *meter* (m), the *kilogram* (kg), and the *second* (s). All three are arbitrarily defined. The second, which is supposed to represent the 1/86,400 part of the mean solar day, is actually defined as the duration of 9,192,631,770 cycles of the radiation associated with a specified transition of the cesium atom. The meter, originally intended to represent one-ten-millionth of the distance from the equator to the pole, is now defined as 1,650,763.73 wavelengths of the orange-red line of krypton 86. The kilogram, which is approximately equal to the mass of 0.001 m³ of water, is actually defined as the mass of a platinum standard kept at the International Bureau of Weights and Measures at Sèvres, near Paris, France. The unit of force is a derived unit. It is called the *newton* (N) and is defined as the force which gives an acceleration of 1 m/s² to a mass of 1 kg (Fig. 1.2). From Eq. (1.1) we write

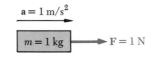

Fig. 1.2

$$1\ N = (1\ kg)(1\ m/s^2) = 1\ kg \cdot m/s^2 \qquad (1.5)$$

The SI units are said to form an *absolute* system of units. This means that the three base units chosen are independent of the location where measurements are made. The meter, the kilogram, and the second may be used anywhere on the earth; they may even be used on another planet. They will always have the same significance.

Like any other force, the *weight* of a body should be expressed in newtons. From Eq. (1.4) it follows that the weight of a body of mass 1 kg (Fig. 1.3) is

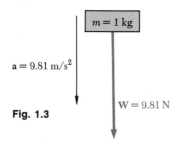

Fig. 1.3

$$
\begin{aligned}
W &= mg \\
&= (1\ kg)(9.81\ m/s^2) \\
&= 9.81\ N
\end{aligned}
$$

Multiples and submultiples of the fundamental SI units may be obtained through the use of the prefixes defined in Table 1.1. The multiples and submultiples of the units of length, mass, and force most frequently used in engineering are, respectively, the *kilometer* (km) and the *millimeter* (mm); the *megagram*† (Mg) and the *gram* (g); and the *kilonewton* (kN). According to Table 1.1, we have

$$
\begin{array}{ll}
1\ km = 1000\ m & 1\ mm = 0.001\ m \\
1\ Mg = 1000\ kg & 1\ g = 0.001\ kg \\
\multicolumn{2}{c}{1\ kN = 1000\ N}
\end{array}
$$

The conversion of these units into meters, kilograms, and newtons, respectively, can be effected by simply moving the decimal point three places to the right or to the left. For example, to

† Also known as a *metric ton*.

Table 1.1 SI Prefixes

Multiplication Factor	Prefix†	Symbol
$1\,000\,000\,000\,000 = 10^{12}$	tera	T
$1\,000\,000\,000 = 10^{9}$	giga	G
$1\,000\,000 = 10^{6}$	mega	M
$1\,000 = 10^{3}$	kilo	k
$100 = 10^{2}$	hecto‡	h
$10 = 10^{1}$	deka‡	da
$0.1 = 10^{-1}$	deci‡	d
$0.01 = 10^{-2}$	centi‡	c
$0.001 = 10^{-3}$	milli	m
$0.000\,001 = 10^{-6}$	micro	μ
$0.000\,000\,001 = 10^{-9}$	nano	n
$0.000\,000\,000\,001 = 10^{-12}$	pico	p
$0.000\,000\,000\,000\,001 = 10^{-15}$	femto	f
$0.000\,000\,000\,000\,000\,001 = 10^{-18}$	atto	a

† The first syllable of every prefix is accented so that the prefix will retain its identity. Thus, the preferred pronunciation of kilometer places the accent on the first syllable, not the second.

‡ The use of these prefixes should be avoided, except for the measurement of areas and volumes and for the nontechnical use of centimeter, as for body and clothing measurements.

convert 3.82 km into meters, one moves the decimal point three places to the right:

$$3.82 \text{ km} = 3820 \text{ m}$$

Similarly, 47.2 mm is converted into meters by moving the decimal point three places to the left:

$$47.2 \text{ mm} = 0.0472 \text{ m}$$

Using the scientific notation, one may also write

$$3.82 \text{ km} = 3.82 \times 10^{3} \text{ m}$$
$$47.2 \text{ mm} = 47.2 \times 10^{-3} \text{ m}$$

The multiples of the unit of time are the *minute* (min) and the *hour* (h). Since

$$1 \text{ min} = 60 \text{ s} \quad \text{and} \quad 1 \text{ h} = 60 \text{ min} = 3600 \text{ s}$$

these multiples cannot be converted as readily as the others.

By using the appropriate multiple or submultiple of a given unit, one may avoid writing very large or very small numbers. For example, one usually writes 427.2 km, rather than 427 200 m, and 2.16 mm, rather than 0.002 16 m.†

† It should be noted that, when more than four digits are used on either side of the decimal point to express a quantity in SI units—as in 427 200 m or 0.002 16 m—spaces, never commas, should be used to separate the digits into groups of three. This is to avoid confusion with the comma which is used in many countries in place of a decimal point.

Units of Area and Volume. The unit of area is the *square meter* (m²), which represents the area of a square of side 1 m; the unit of volume is the *cubic meter* (m³), equal to the volume of a cube of side 1 m. In order to avoid exceedingly small or large numerical values in the computation of areas and volumes, one uses systems of subunits obtained by respectively squaring and cubing, not only the millimeter, but also two intermediate submultiples of the meter, namely, the *decimeter* (dm) and the *centimeter* (cm). Since, by definition,

$$1 \text{ dm} = 0.1 \text{ m} = 10^{-1} \text{ m}$$
$$1 \text{ cm} = 0.01 \text{ m} = 10^{-2} \text{ m}$$
$$1 \text{ mm} = 0.001 \text{ m} = 10^{-3} \text{ m}$$

the submultiples of the unit of area are

$$1 \text{ dm}^2 = (1 \text{ dm})^2 = (10^{-1} \text{ m})^2 = 10^{-2} \text{ m}^2$$
$$1 \text{ cm}^2 = (1 \text{ cm})^2 = (10^{-2} \text{ m})^2 = 10^{-4} \text{ m}^2$$
$$1 \text{ mm}^2 = (1 \text{ mm})^2 = (10^{-3} \text{ m})^2 = 10^{-6} \text{ m}^2$$

and the submultiples of the unit of volume are

$$1 \text{ dm}^3 = (1 \text{ dm})^3 = (10^{-1} \text{ m})^3 = 10^{-3} \text{ m}^3$$
$$1 \text{ cm}^3 = (1 \text{ cm})^3 = (10^{-2} \text{ m})^3 = 10^{-6} \text{ m}^3$$
$$1 \text{ mm}^3 = (1 \text{ mm})^3 = (10^{-3} \text{ m})^3 = 10^{-9} \text{ m}^3$$

It should be noted that when the volume of a liquid is being measured, the cubic decimeter (dm³) is usually referred to as a *liter* (l).

Other derived SI units used to measure the moment of a force, the work of a force, etc., are shown in Table 1.2. While these units will be introduced in later chapters as they are needed, we should note an important rule at this time: When a derived unit is obtained by dividing a base unit by another base unit, a prefix may be used in the numerator of the derived unit but not in its denominator. For example, the constant k of a spring which stretches 20 mm under a load of 100 N will be expressed as

$$k = \frac{100 \text{ N}}{20 \text{ mm}} = \frac{100 \text{ N}}{0.020 \text{ m}} = 5000 \text{ N/m} \quad \text{or} \quad k = 5 \text{ kN/m}$$

but never as $k = 5 \text{ N/mm}$.

U.S. Customary Units. Most practicing American engineers still commonly use a system in which the base units are the units of length, force, and time. These units are, respectively, the *foot* (ft), the *pound* (lb), and the *second* (s). The second is the same as the corresponding SI unit. The foot is defined as 0.3048 m. The pound is defined as the *weight* of a platinum standard, called the *standard pound* and kept at the National Bureau of Standards in Washington, the mass of which is 0.453 592 43 kg. Since the weight of a body depends upon the gravitational attraction of the earth, which varies with location, it is specified that the standard pound should be placed at sea level and at the latitude of 45° to properly define a force of 1 lb. Clearly the U.S. customary units do not form an absolute system of units. Because of their dependence upon the gravitational attraction of the earth, they form a *gravitational* system of units.

While the standard pound also serves as the unit of mass in commercial transactions in the United States, it cannot be so used in engineering computations since such a unit would not be consistent with the base units defined in the preceding paragraph. Indeed, when acted upon by a force of 1 lb, that is, when subjected to its own weight, the standard pound receives the acceleration of gravity, $g = 32.2$ ft/s^2 (Fig. 1.4), not the unit acceleration required by Eq. (1.1). The unit of mass consistent with the foot, the pound, and the second is the mass which receives an acceleration of 1 ft/s^2 when a force of 1 lb is applied

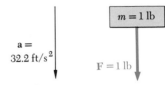

$$a = 32.2 \text{ ft/s}^2$$

$$m = 1 \text{ lb}$$

$$F = 1 \text{ lb}$$

Fig. 1.4

Table 1.2 Principal SI Units Used in Mechanics

Quantity	Unit	Symbol	Formula
Acceleration	Meter per second squared	. . .	m/s^2
Angle	Radian	rad	†
Angular acceleration	Radian per second squared	. . .	rad/s^2
Angular velocity	Radian per second	. . .	rad/s
Area	Square meter	. . .	m^2
Density	Kilogram per cubic meter	. . .	kg/m^3
Energy	Joule	J	N · m
Force	Newton	N	kg · m/s^2
Frequency	Hertz	Hz	s^{-1}
Impulse	Newton-second	. . .	kg · m/s
Length	Meter	m	‡
Mass	Kilogram	kg	‡
Moment of a force	Newton-meter	. . .	N · m
Power	Watt	W	J/s
Pressure	Pascal	Pa	N/m^2
Stress	Pascal	Pa	N/m^2
Time	Second	s	‡
Velocity	Meter per second	. . .	m/s
Volume, solids	Cubic meter	. . .	m^3
Liquids	Liter	l	10^{-3} m^3
Work	Joule	J	N · m

†Supplementary unit (1 revolution = 2π rad = 360°).
‡Base unit.

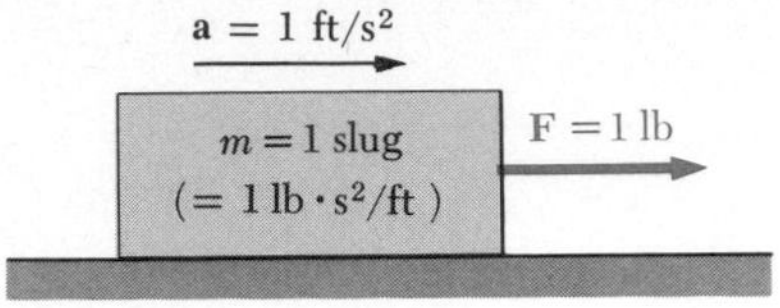

$a = 1\ \text{ft/s}^2$

$m = 1\ \text{slug}$
$(= 1\ \text{lb} \cdot \text{s}^2/\text{ft})$

$F = 1\ \text{lb}$

Fig. 1.5

to it (Fig. 1.5). This unit, sometimes called a *slug*, can be derived from the equation $F = ma$ after substituting 1 lb and 1 ft/s² for F and a, respectively. We write

$$F = ma \qquad 1\ \text{lb} = (1\ \text{slug})(1\ \text{ft/s}^2)$$

and obtain

$$1\ \text{slug} = \frac{1\ \text{lb}}{1\ \text{ft/s}^2} = 1\ \text{lb} \cdot \text{s}^2/\text{ft} \qquad (1.6)$$

Comparing Figs. 1.4 and 1.5, we conclude that the slug is a mass 32.2 times larger than the mass of the standard pound.

The fact that in the U.S. customary system of units bodies are characterized by their weight in pounds, rather than by their mass in slugs, will be a convenience in the study of statics, where we shall constantly deal with weights and other forces and only seldom with masses. However, in the study of dynamics, where forces, masses, and accelerations are involved, we shall have to express the mass m in slugs of a body of which the weight W has been given in pounds. Recalling Eq. (1.4), we shall write

$$m = \frac{W}{g} \qquad (1.7)$$

where g is the acceleration of gravity ($g = 32.2\ \text{ft/s}^2$).

Other U.S. customary units frequently encountered in engineering problems are the *mile* (mi), equal to 5280 ft; the *inch* (in.), equal to $\frac{1}{12}$ ft; and the *kilopound* (kip), equal to a force of 1000 lb. The *ton* is often used to represent a mass of 2000 lb but, like the pound, must be converted into slugs in engineering computations.

The conversion into feet, pounds, and seconds of quantities expressed in other U.S. customary units is generally more involved and requires greater attention than the corresponding operation in SI units. If, for example, the magnitude of a velocity is given as $v = 30\ \text{mi/h}$, we shall proceed as follows to convert it to ft/s. First we write

$$v = 30\frac{\text{mi}}{\text{h}}$$

Since we want to get rid of the unit miles and introduce instead the unit feet, we should multiply the right-hand member of the equation by an expression containing miles in the denominator and feet in the numerator. But, since we do not want to change the value of the right-hand member, the expression used should have a value equal to unity. The quotient (5280 ft)/(1 mi) is such an expression. Operating in a similar way to transform the unit hour into seconds, we write

$$v = \left(30\frac{\text{mi}}{\text{h}}\right)\left(\frac{5280 \text{ ft}}{1 \text{ mi}}\right)\left(\frac{1 \text{ h}}{3600 \text{ s}}\right)$$

Carrying out the numerical computations and canceling out units which appear both in the numerator and the denominator, we obtain

$$v = 44\frac{\text{ft}}{\text{s}} = 44 \text{ ft/s}$$

1.4. Conversion from One System of Units to Another. There are many instances when an engineer wishes to convert into SI units a numerical result obtained in U.S. customary units or vice versa. Because the unit of time is the same in both systems, only two kinetic base units need be converted. Thus, since all other kinetic units can be derived from these base units, only two conversion factors need be memorized.†

Units of Length. By definition the U.S. customary unit of length is

$$1 \text{ ft} = 0.3048 \text{ m} \tag{1.8}$$

It follows that

$$1 \text{ mi} = 5280 \text{ ft} = 5280(0.3048 \text{ m}) = 1609 \text{ m}$$

or

$$1 \text{ mi} = 1.609 \text{ km} \tag{1.9}$$

Also

$$1 \text{ in.} = \tfrac{1}{12} \text{ ft} = \tfrac{1}{12}(0.3048 \text{ m}) = 0.0254 \text{ m}$$

or

$$1 \text{ in.} = 25.4 \text{ mm} \tag{1.10}$$

† While all conversion factors are given in this section with four significant figures, it should be noted that such an accuracy is seldom needed in engineering computations (Sec. 1.6). Therefore, for most calculations the given values may be rounded off as follows:

$$1 \text{ ft} = 0.305 \text{ m} \qquad 1 \text{ lb} = 4.45 \text{ N} \qquad 1 \text{ slug} = 14.6 \text{ kg}$$

Units of Force. Recalling that the U.S. customary unit of force (pound) is defined as the weight of the standard pound (of mass 0.4536 kg) at sea level and at the latitude of 45° (where $g = 9.807$ m/s²), and using Eq. (1.4), we write

$$W = mg$$
$$1 \text{ lb} = (0.4536 \text{ kg})(9.807 \text{ m/s}^2) = 4.448 \text{ kg} \cdot \text{m/s}^2$$

or, recalling (1.5),

$$1 \text{ lb} = 4.448 \text{ N} \tag{1.11}$$

Units of Mass. The U.S. customary unit of mass (slug) is a derived unit. Thus, using (1.6), (1.8), and (1.11), we write

$$1 \text{ slug} = 1 \text{ lb} \cdot \text{s}^2/\text{ft} = \frac{1 \text{ lb}}{1 \text{ ft/s}^2} = \frac{4.448 \text{ N}}{0.3048 \text{ m/s}^2} = 14.59 \text{ N} \cdot \text{s}^2/\text{m}$$

and, recalling (1.5),

$$1 \text{ slug} = 1 \text{ lb} \cdot \text{s}^2/\text{ft} = 14.59 \text{ kg} \tag{1.12}$$

Although it cannot be used as a consistent unit of mass, we recall that the mass of the standard pound is, by definition,

$$1 \text{ pound mass} = 0.4536 \text{ kg} \tag{1.13}$$

This constant may be used to determine the *mass* in SI units (kilograms) of a body which has been characterized by its *weight* in U.S. customary units (pounds).

To convert into SI units a derived U.S. customary unit, one simply multiplies or divides the appropriate conversion factors. For example, to convert into SI units the moment of a force which was found to be $M = 46$ lb · in., we use formulas (1.10) and (1.11) and write

$$M = 46 \text{ lb} \cdot \text{in.} = 46(4.45 \text{ N})(25.4 \text{ mm})$$
$$= 5200 \text{ N} \cdot \text{mm} = 5.20 \text{ N} \cdot \text{m}$$

The conversion factors given in this section may also be used to convert into U.S. customary units a numerical result obtained in SI units. For example, if the moment of a force was found to be $M = 40$ N · m, we write, following the procedure used in the last paragraph of Sec. 1.3,

$$M = 40 \text{ N} \cdot \text{m} = (40 \text{ N} \cdot \text{m})\left(\frac{1 \text{ lb}}{4.45 \text{ N}}\right)\left(\frac{1 \text{ ft}}{0.305 \text{ m}}\right)$$

Carrying out the numerical computations and canceling out units which appear both in the numerator and the denominator, we obtain

$$M = 29.5 \text{ lb} \cdot \text{ft}$$

The U.S. customary units most frequently used in mechanics are listed in Table 1.3 with their SI equivalents.

Table 1.3 U.S. Customary Units and Their SI Equivalents

Quantity	U.S. Customary Unit	SI Equivalent
Acceleration	ft/s^2	0.3048 m/s^2
	in./s^2	0.0254 m/s^2
Area	ft^2	0.0929 m^2
	in^2	645.2 mm^2
Energy	$\text{ft} \cdot \text{lb}$	1.356 J
Force	kip	4.448 kN
	lb	4.448 N
	oz	0.2780 N
Impulse	$\text{lb} \cdot \text{s}$	$4.448 \text{ N} \cdot \text{s}$
Length	ft	0.3048 m
	in.	25.40 mm
	mi	1.609 km
Mass	oz mass	28.35 g
	lb mass	0.4536 kg
	slug	14.59 kg
	ton	907.2 kg
Moment of a force	$\text{lb} \cdot \text{ft}$	$1.356 \text{ N} \cdot \text{m}$
	$\text{lb} \cdot \text{in.}$	$0.1130 \text{ N} \cdot \text{m}$
Moment of inertia		
Of an area	in^4	$0.4162 \times 10^6 \text{ mm}^4$
Of a mass	$\text{lb} \cdot \text{ft} \cdot \text{s}^2$	$1.356 \text{ kg} \cdot \text{m}^2$
Momentum	$\text{lb} \cdot \text{s}$	$4.448 \text{ kg} \cdot \text{m/s}$
Power	$\text{ft} \cdot \text{lb/s}$	1.356 W
	hp	745.7 W
Pressure or stress	lb/ft^2	47.88 Pa
	lb/in^2 (psi)	6.895 kPa
Velocity	ft/s	0.3048 m/s
	in./s	0.0254 m/s
	mi/h (mph)	0.4470 m/s
	mi/h (mph)	1.609 km/h
Volume, solids	ft^3	0.02832 m^3
	in^3	16.39 cm^3
Liquids	gal	3.785 l
	qt	0.9464 l
Work	$\text{ft} \cdot \text{lb}$	1.356 J

1.5. Method of Problem Solution.

The student should approach a problem in mechanics as he would approach an actual engineering situation. By drawing on his own experience and on his intuition, he will find it easier to understand and formulate the problem. Once the problem has been clearly stated, however, there is no place in its solution for the student's particular fancy. *The solution must be based on the six fundamental principles stated in Sec. 1.2 or on theorems derived from them.* Every step taken must be justified on that basis. Strict rules must be followed, which lead to the solution in an almost automatic fashion, leaving no room for the student's intuition or "feeling." After an answer has been obtained, it should be checked. Here again, the student may call upon his common sense and personal experience. If not completely satisfied with the result obtained, he should carefully check his formulation of the problem, the validity of the methods used for its solution, and the accuracy of his computations.

The *statement* of a problem should be clear and precise. It should contain the given data and indicate what information is required. A neat drawing showing all quantities involved should be included. Separate diagrams should be drawn for all bodies involved, indicating clearly the forces acting on each body. These diagrams are known as *free-body diagrams* and are described in detail in Secs. 2.10 and 4.2.

The *fundamental principles* of mechanics listed in Sec. 1.2 *will be used to write equations* expressing the conditions of rest or motion of the bodies considered. Each equation should be clearly related to one of the free-body diagrams. The student will then proceed to solve the problem, observing strictly the usual rules of algebra and recording neatly the various steps taken.

After the answer has been obtained, it should be *carefully checked*. Mistakes in *reasoning* may often be detected by checking the units. For example, to determine the moment of a force of 50 N about a point 0.60 m from its line of action, we would have written (Sec. 3.11)

$$M = Fd = (50 \text{ N})(0.60 \text{ m}) = 30 \text{ N} \cdot \text{m}$$

The unit $\text{N} \cdot \text{m}$ obtained by multiplying newtons by meters is the correct unit for the moment of a force; if another unit had been obtained, we would have known that some mistake had been made.

Errors in *computation* will usually be found by substituting the numerical values obtained into an equation which has not yet been used and verifying that the equation is satisfied. The importance of correct computations in engineering cannot be overemphasized.

1.6. Numerical Accuracy. The accuracy of the solution of a problem depends upon two items: (1) the accuracy of the given data; (2) the accuracy of the computations performed.

The solution cannot be more accurate than the less accurate of these two items. For example, if the loading of a bridge is known to be 75,000 lb with a possible error of 100 lb either way, the relative error which measures the degree of accuracy of the data is

$$\frac{100 \text{ lb}}{75,000 \text{ lb}} = 0.0013 = 0.13 \text{ percent}$$

It would then be meaningless, in computing the reaction at one of the bridge supports, to record it as 14,322 lb. The accuracy of the solution cannot be greater than 0.13 percent, no matter how accurate the computations are, and the possible error in the answer may be as large as $(0.13/100)(14,322 \text{ lb}) \approx 20$ lb. The solution should be properly recorded as $14,320 \pm 20$ lb.

In engineering problems, the data are seldom known with an accuracy greater than 0.2 percent. It is therefore unnecessary to carry out computations with a greater accuracy. For this reason, all calculations in this text shall be carried out with an accuracy of 0.2 percent. A practical rule to accomplish this is to use 4 figures to record numbers beginning with a "1" and 3 figures in all other cases. Unless otherwise indicated, the data given in a problem should be assumed known with a comparable degree of accuracy. A force of 40 lb, for example, should be read 40.0 lb, and a force of 15 lb should be read 15.00 lb.

Pocket electronic calculators are widely used by practicing engineers and engineering students. The speed and accuracy of these calculators facilitates the numerical computations in the solution of many problems. However, the student should not record more significant figures than can be justified, merely because they are easily obtained. As noted above, an accuracy greater than 0.2 percent is seldom necessary or meaningful in the solution of practical engineering problems.

2 Statics of Particles

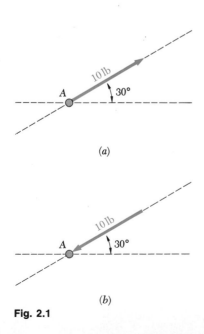

(a)

(b)

Fig. 2.1

FORCES IN A PLANE

2.1. Force on a Particle. Resultant of Two Forces. A force represents the action of one body on another. It is characterized by its *point of application*, its *magnitude*, and its *direction*. In this chapter we shall study the effect of forces on particles. The use of the word "particles" does not imply that we shall restrict our study to that of small corpuscles. It means that the size and shape of the bodies under consideration will not affect the solution of the problems treated in this chapter and that all the forces acting on a given body will be assumed to have the same point of application. Each force will thus be completely defined by its magnitude and direction.

The magnitude of a force is characterized by a certain number of units. As indicated in Chap. 1, the SI units used by engineers to measure the magnitude of a force are the newton (N) and its multiple the kilonewton (kN), equal to 1000 N, while the U.S. customary units used for the same purpose are the pound (lb) and its multiple the kilopound (kip, or k), equal to 1000 lb. The direction of a force is defined by the *line of action* and the *sense* of the force. The line of action is the infinite straight line along which the force acts; it is characterized by the angle it forms with some fixed axis (Fig. 2.1). The force itself is represented by a segment of that line; through the use of an appropriate scale, the length of this segment may be chosen to represent the magnitude of the force. Finally, the sense of the force should be indicated by an arrowhead. It is important in defining a force to indicate its sense. Two forces, such as those shown in Fig. 2.1*a* and *b*, having the same magnitude and the same line of action but different sense, will have directly opposite effects on a particle.

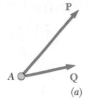

(a)

Experimental evidence shows that two forces **P** and **Q** acting on a particle A (Fig. 2.2*a*) may be replaced by a single force **R** which has the same effect on the particle (Fig. 2.2*c*). This force is called the *resultant* of the forces **P** and **Q** and may be obtained, as shown in Fig. 2.2*b*, by constructing a parallelogram, using **P** and **Q** as two sides of the parallelogram. *The diagonal that passes through A represents the resultant.* This is known as the *parallelogram law* for the addition of two forces. This law is based on experimental evidence; it cannot be proved or derived mathematically.

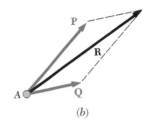

(b)

2.2. Vectors. It appears from the above that forces do not obey the rules of addition defined in ordinary arithmetic or algebra. For example, two forces acting at a right angle to each other, one of 4 lb and the other of 3 lb, add up to a force of 5 lb, *not* to a force of 7 lb. Forces are not the only expressions which follow the parallelogram law of addition. As we shall see later, *displacements, velocities, accelerations, momenta* are other examples of physical quantities possessing magnitude and direction and which are added according to the parallelogram law. All these quantities may be represented mathematically by *vectors,* while those physical quantities which do not have direction, such as *volume, mass,* or *energy,* are represented by ordinary numbers or *scalars.*

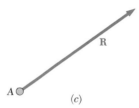

(c)

Fig. 2.2

Vectors are defined as *mathematical expressions possessing magnitude and direction, which add according to the parallelogram law.* Vectors are represented by arrows in the illustrations and will be distinguished from scalar quantities in this text through the use of boldface type (**P**). In longhand writing, a vector may be characterized by drawing a short arrow above the letter used to represent it ($\vec{P}$), or by underlining the letter ($\underline{P}$). The last method is gaining wider acceptance since it can also be used on a typewriter. The magnitude of a vector defines the length of the arrow used to represent the vector. In this text, italic type will be used to denote the magnitude of a vector. Thus, the magnitude of the vector **P** will be referred to as *P.*

A vector used to represent a force acting on a given particle has a well-defined point of application, namely, the particle itself. Such a vector is said to be a *fixed,* or *bound,* vector and cannot be moved without modifying the conditions of the problem. Other physical quantities, however, such as couples (see Chap. 3), are represented by vectors which may be freely moved

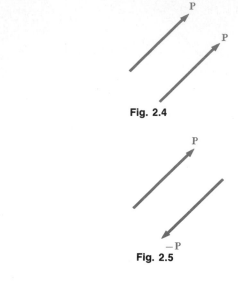

Fig. 2.4

Fig. 2.5

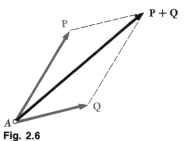

Fig. 2.6

in space; these vectors are called *free* vectors. Still other physical quantities, such as forces acting on a rigid body (see Chap. 3), are represented by vectors which may be moved, or slid, along their line of action; they are known as *sliding* vectors.†

Two vectors which have the same magnitude and the same direction are said to be *equal*, whether or not they also have the same point of application (Fig. 2.4); equal vectors may be denoted by the same letter.

The *negative vector* of a given vector **P** is defined as a vector having the same magnitude as **P** and a direction opposite to that of **P** (Fig. 2.5); the negative of the vector **P** is denoted by −**P**. The vectors **P** and −**P** are commonly referred to as *equal and opposite* vectors. Clearly, we have

$$\mathbf{P} + (-\mathbf{P}) = 0$$

2.3. Addition of Vectors. We saw in the preceding section that, by definition, vectors add according to the parallelogram law. Thus the sum of two vectors **P** and **Q** is obtained by attaching the two vectors to the same point A and constructing a parallelogram, using **P** and **Q** as two sides of the parallelogram (Fig. 2.6). The diagonal that passes through A represents the sum of the vectors **P** and **Q**, and this sum is denoted by **P** + **Q.** The fact that the sign + is used to denote both vector and scalar addition should not cause any confusion if vector and scalar quantities are always carefully distinguished. Thus, we should note that the magnitude of the vector **P** + **Q** is *not,* in general, equal to the sum $P + Q$ of the magnitudes of the vectors **P** and **Q.**

† Some expressions have magnitude and direction, but do not add according to the parallelogram law. While these expressions may be represented by arrows, they *cannot* be considered as vectors.

A group of such expressions are the finite rotations of a rigid body. Place a closed book on a table in front of you, so that it lies in the usual fashion, with its front cover up and its binding to the left. Now rotate it through 180° about an axis parallel to the binding (Fig. 2.3a); this rotation may be represented by an arrow of length equal to 180 units and oriented as shown. Picking up the book as it lies in its new position, rotate it now through 180° about a

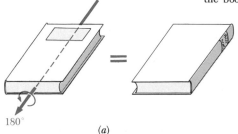

(a)

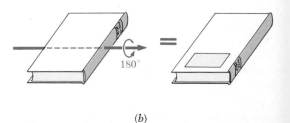

(b)

Fig. 2.3 Finite rotations of a rigid body.

Since the parallelogram constructed on the vectors **P** and **Q** does not depend upon the order in which **P** and **Q** are selected, we conclude that the addition of two vectors is *commutative*, and we write

$$\mathbf{P} + \mathbf{Q} = \mathbf{Q} + \mathbf{P} \qquad (2.1)$$

From the parallelogram law, we can derive an alternate method for determining the sum of two vectors. This method, known as the *triangle rule*, is derived as follows: Consider Fig. 2.6, where the sum of the vectors **P** and **Q** has been determined by the parallelogram law. Since the side of the parallelogram opposite **Q** is equal to **Q** in magnitude and direction, we could draw only half of the parallelogram (Fig. 2.7*a*). The sum of the two vectors may thus be found by *arranging* **P** *and* **Q** *in tip-to-tail fashion and then connecting the tail of* **P** *with the tip of* **Q**. In Fig. 2.7*b*, the other half of the parallelogram is considered, and the same result is obtained. This confirms the fact that vector addition is commutative.

The *subtraction* of a vector is defined as the addition of the corresponding negative vector. Thus, the vector **P** − **Q** representing the difference between the vectors **P** and **Q** is obtained by adding to **P** the negative vector −**Q** (Fig. 2.8). We write

$$\mathbf{P} - \mathbf{Q} = \mathbf{P} + (-\mathbf{Q}) \qquad (2.2)$$

Here again we should observe that, while the same sign is used to denote both vector and scalar subtraction, confusion will be avoided if care is taken to distinguish between vector and scalar quantities.

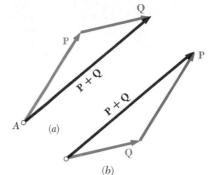

Fig. 2.7

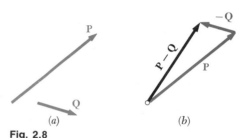

Fig. 2.8

horizontal axis perpendicular to the binding (Fig. 2.3*b*); this second rotation may be represented by an arrow 180 units long and oriented as shown. But the book could have been placed in this final position through a single 180° rotation about a vertical axis (Fig. 2.3*c*). We conclude that the sum of the two 180° rotations represented by arrows directed respectively along the *z* and *x* axes is a 180° rotation represented by an arrow directed along the *y* axis (Fig. 2.3*d*). Clearly, the finite rotations of a rigid body *do not* obey the parallelogram law of addition; therefore they *cannot* be represented by vectors.

We shall now consider the *sum of three or more vectors*. The sum of three vectors **P, Q,** and **S** will, *by definition*, be obtained by first adding the vectors **P** and **Q**, and then adding the vector **S** to the vector **P + Q.** We thus write

$$\mathbf{P} + \mathbf{Q} + \mathbf{S} = (\mathbf{P} + \mathbf{Q}) + \mathbf{S} \qquad (2.3)$$

Similarly, the sum of four vectors will be obtained by adding the fourth vector to the sum of the first three. It follows that the sum of any number of vectors may be obtained by applying repeatedly the parallelogram law to successive pairs of vectors until all the given vectors are replaced by a single vector.

If the given vectors are *coplanar*, i.e., if they are contained in the same plane, their sum may be easily obtained graphically. In that case, the repeated application of the triangle rule will be preferred to the application of the parallelogram law. In Fig. 2.9 the sum of three vectors **P, Q,** and **S** was obtained in that manner. The triangle rule was first applied to obtain the sum **P + Q** of the vectors **P** and **Q;** it was applied again to obtain the sum of the vectors **P + Q** and **S.** The determination of the vector **P + Q,** however, could have been omitted and the sum of the three vectors could have been obtained directly, as shown in Fig. 2.10, by *arranging the given vectors in tip-to-tail fashion and connecting the tail of the first vector with the tip of the last one.* This is known as the *polygon rule* for the addition of vectors.

We observe that the result obtained would have been unchanged if, as shown in Fig. 2.11, the vectors **Q** and **S** had been replaced by their sum **Q + S.** We may thus write

$$\mathbf{P} + \mathbf{Q} + \mathbf{S} = (\mathbf{P} + \mathbf{Q}) + \mathbf{S} = \mathbf{P} + (\mathbf{Q} + \mathbf{S}) \qquad (2.4)$$

which expresses the fact that vector addition is *associative*. Recalling that vector addition has also been shown, in the case of two vectors, to be commutative, we write

$$\mathbf{P} + \mathbf{Q} + \mathbf{S} = (\mathbf{P} + \mathbf{Q}) + \mathbf{S} = \mathbf{S} + (\mathbf{P} + \mathbf{Q})$$
$$= \mathbf{S} + (\mathbf{Q} + \mathbf{P}) = \mathbf{S} + \mathbf{Q} + \mathbf{P} \qquad (2.5)$$

This expression, as well as others which may be obtained in the same way, shows that the order in which several vectors are added together is immaterial (Fig. 2.12).

Product of a Scalar and a Vector. Since it is convenient to denote the sum **P + P** by 2**P,** the sum **P + P + P** by 3**P,** and, in general, the sum of *n* equal vectors **P** by the product *n***P,** we shall define the product *n***P** of a positive integer *n* and a vector **P** as a vector having the same direction as **P** and the magnitude *nP*. Extending this definition to include all scalars, and recalling the

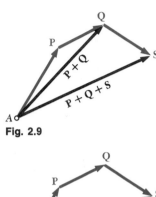

Fig. 2.9

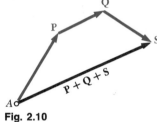

Fig. 2.10

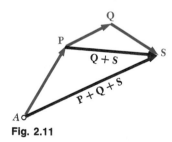

Fig. 2.11

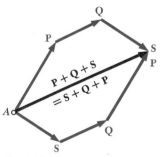

Fig. 2.12

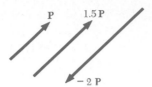

Fig. 2.13

definition of a negative vector given in Sec. 2.2, we define the product $k\mathbf{P}$ of a scalar k and a vector $\mathbf{P}$ as a vector having the same direction as $\mathbf{P}$ (if k is positive), or a direction opposite to that of $\mathbf{P}$ (if k is negative), and a magnitude equal to the product of P and of the absolute value of k (Fig. 2.13).

2.4. Resultant of Several Concurrent Forces. Consider a particle A acted upon by several coplanar forces, i.e., by several forces contained in the same plane (Fig. 2.14a). Since the forces considered here all pass through A, they are also said to be *concurrent*. The vectors representing the forces acting on A may be added by the polygon rule (Fig. 2.14b). Since the use of the polygon rule is equivalent to the repeated application of

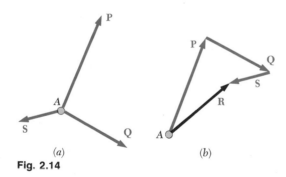

(a) (b)

Fig. 2.14

the parallelogram law, the vector $\mathbf{R}$ thus obtained represents the resultant of the given concurrent forces, i.e., the single force which has the same effect on the particle A as the given forces. As indicated above, the order in which the vectors $\mathbf{P}$, $\mathbf{Q}$, and $\mathbf{S}$ representing the given forces are added together is immaterial.

2.5. Resolution of a Force into Components. We have seen that two or more forces acting on a particle may be replaced by a single force which has the same effect on the particle. Conversely, a single force $\mathbf{F}$ acting on a particle may be replaced by two or more forces which, together, have the same effect on the particle. These forces are called the *components* of the original force $\mathbf{F}$, and the process of substituting them for $\mathbf{F}$ is called *resolving the force* $\mathbf{F}$ *into components*.

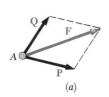
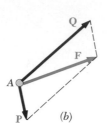
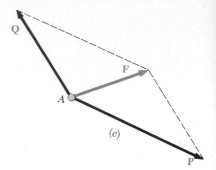

(a) *(b)* *(c)*

Fig. 2.15

Clearly, for each force **F** there exist an infinite number of possible sets of components. Sets of *two components* **P** *and* **Q** are the most important as far as practical applications are concerned. But, even then, the number of ways in which a given force **F** may be resolved into two components is unlimited (Fig. 2.15). Two cases are of particular interest:

1. *One of the Two Components,* **P,** *Is Known.* The second component, **Q,** is obtained by applying the triangle rule and joining the tip of **P** to the tip of **F** (Fig. 2.16); the magnitude and direction of **Q** are determined graphically or by trigonometry. Once **Q** has been determined, both components **P** and **Q** should be applied at *A*.

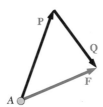

Fig. 2.16

2. *The Line of Action of Each Component Is Known.* The magnitude and sense of the components are obtained by applying the parallelogram law and drawing lines, through the tip of **F,** parallel to the given lines of action (Fig. 2.17). This process leads to two well-defined components, **P** and **Q,** which may be determined graphically or by applying the law of sines.

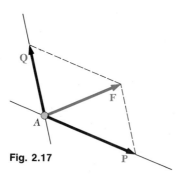

Fig. 2.17

Many other cases may be encountered; for example, the direction of one component may be known while the magnitude of the other component is to be as small as possible (see Sample Prob. 2.2). In all cases the appropriate triangle or parallelogram is drawn, which satisfies the given conditions.

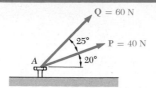

SAMPLE PROBLEM 2.1

The two forces **P** and **Q** act on a bolt A. Determine their resultant.

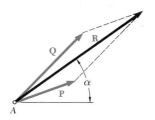

Graphical Solution. A parallelogram with sides equal to **P** and **Q** is drawn to scale. The magnitude and direction of the resultant are measured and found to be

$$R = 98 \text{ N} \qquad \alpha = 35° \qquad R = 98 \text{ N} \measuredangle 35° \qquad \blacktriangleleft$$

The triangle rule may also be used. Forces **P** and **Q** are drawn in tip-to-tail fashion. Again the magnitude and direction of the resultant are measured.

$$R = 98 \text{ N} \qquad \alpha = 35° \qquad R = 98 \text{ N} \measuredangle 35° \qquad \blacktriangleleft$$

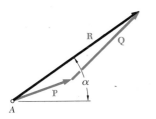

Trigonometric Solution. The triangle rule is again used; two sides and the included angle are known. We apply the law of cosines.

$$R^2 = P^2 + Q^2 - 2PQ \cos B$$
$$R^2 = (40 \text{ N})^2 + (60 \text{ N})^2 - 2(40 \text{ N})(60 \text{ N}) \cos 155°$$
$$R = 97.7 \text{ N}$$

Now, applying the law of sines, we write

$$\frac{\sin A}{Q} = \frac{\sin B}{R} \qquad \frac{\sin A}{60 \text{ N}} = \frac{\sin 155°}{97.7 \text{ N}} \qquad (1)$$

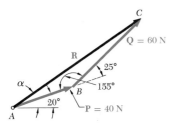

Solving Eq. (1) for $\sin A$, we have

$$\sin A = \frac{(60 \text{ N}) \sin 155°}{97.7 \text{ N}}$$

Using a calculator, we first compute the quotient, then its arc sine, and obtain

$$A = 15.0° \qquad \alpha = 20° + A = 35.0°$$
$$R = 97.7 \text{ N} \measuredangle 35.0° \qquad \blacktriangleleft$$

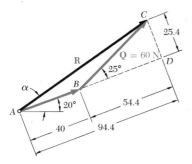

Alternate Trigonometric Solution. We construct the right triangle BCD and compute

$$CD = (60 \text{ N}) \sin 25° = 25.4 \text{ N}$$
$$BD = (60 \text{ N}) \cos 25° = 54.4 \text{ N}$$

Then, using triangle ACD, we obtain

$$\tan A = \frac{25.4 \text{ N}}{94.4 \text{ N}} \qquad A = 15.0°$$
$$R = 25.4/\sin A \qquad R = 97.7 \text{ N}$$

Again,

$$\alpha = 20° + A = 35.0° \qquad R = 97.7 \text{ N} \measuredangle 35.0° \qquad \blacktriangleleft$$

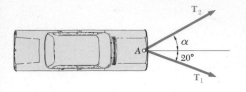

SAMPLE PROBLEM 2.2

A disabled automobile is pulled by means of two ropes as shown. If the resultant of the two forces exerted by the ropes is a 300-lb force parallel to the axis of the automobile, find (*a*) the tension in each of the ropes, knowing that $\alpha = 30°$, (*b*) the value of α such that the tension in rope *2* is minimum.

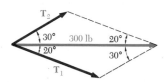

a. **Tension for** $\alpha = 30°$. *Graphical Solution.* The parallelogram law is used; the diagonal (resultant) is known to be equal to 300 lb and to be directed to the right. The sides are drawn parallel to the ropes. If the drawing is done to scale, we measure

$$T_1 = 196 \text{ lb} \qquad T_2 = 134 \text{ lb} \quad \blacktriangleleft$$

Trigonometric Solution. The triangle rule may be used. We note that the triangle shown represents half of the parallelogram shown above. Using the law of sines, we write

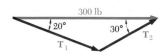

$$\frac{T_1}{\sin 30°} = \frac{T_2}{\sin 20°} = \frac{300 \text{ lb}}{\sin 130°}$$

With a calculator, we first compute and store the value of the last quotient. Multiplying this value successively by $\sin 30°$ and $\sin 20°$, we obtain

$$T_1 = 195.8 \text{ lb} \qquad T_2 = 133.9 \text{ lb} \quad \blacktriangleleft$$

b. **Value of** α **for Minimum** T_2. To determine the value of α such that the tension in rope *2* is minimum, the triangle rule is again used. In the sketch shown, line *1-1* is the known direction of $\mathbf{T}_1$. Several possible directions of $\mathbf{T}_2$ are shown by the lines *2-2*. We note that the minimum value of T_2 occurs when $\mathbf{T}_1$ and $\mathbf{T}_2$ are perpendicular. The minimum value of T_2 is

$$T_2 = (300 \text{ lb}) \sin 20° = 102.6 \text{ lb}$$

Corresponding values of T_1 and α are

$$T_1 = (300 \text{ lb}) \cos 20° = 282 \text{ lb}$$
$$\alpha = 90° - 20° \qquad\qquad \alpha = 70° \quad \blacktriangleleft$$

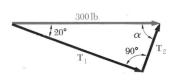

PROBLEMS†

2.1 and 2.2 Determine graphically the magnitude and direction of the resultant of the two forces shown, using in each problem (*a*) the parallelogram law, (*b*) the triangle rule.

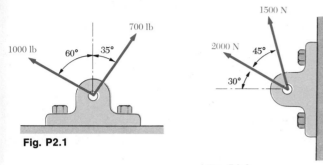

Fig. P2.1

Fig. P2.2

2.3 Two structural members *B* and *C* are riveted to the bracket *A*. Knowing that the tension in member *B* is 2500 lb and that the tension in *C* is 2000 lb, determine graphically the magnitude and direction of the resultant force acting on the bracket.

2.4 Two structural members *B* and *C* are riveted to the bracket *A*. Knowing that the tension in member *B* is 6 kN and that the tension in *C* is 10 kN, determine graphically the magnitude and direction of the resultant force acting on the bracket.

2.5 The force **F** of magnitude 100 lb is to be resolved into two components along the lines *a-a* and *b-b*. Determine by trigonometry the angle α, knowing that the component of **F** along the line *a-a* is to be 70 lb.

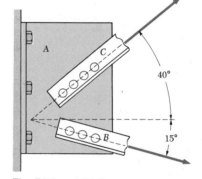

Fig. P2.3 and P2.4

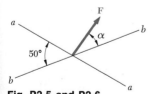

Fig. P2.5 and P2.6

2.6 The force **F** of magnitude 800 N is to be resolved into two components along the lines *a-a* and *b-b*. Determine by trigonometry the angle α, knowing that the component of **F** along the line *b-b* is to be 120 N.

† Answers to all even-numbered problems are given at the end of the book.

2.7 Knowing that $\alpha = 30°$, determine the magnitude of the force **P** so that the resultant force exerted on the cylinder is vertical. What is the corresponding magnitude of the resultant?

2.8 The barge B is pulled by two tugboats A and C. The tension in cable AB is 4000 lb, and the resultant of the two forces applied at B is directed along the axis of the barge. Determine by trigonometry (*a*) the tension in cable BC, (*b*) the magnitude of the resultant of the two forces applied at B.

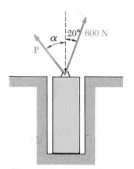

Fig. P2.7 and P2.9

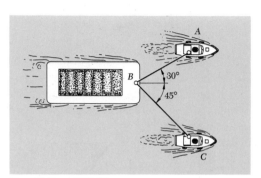

Fig. P2.8 and P2.10

2.9 A cylinder is to be lifted by two cables. Knowing that the tension in one cable is 600 N, determine the magnitude and direction of the force **P** so that the resultant is a vertical force of 900 N.

2.10 The barge B is pulled by two tugboats A and C. At a given instant the tension in cable AB is 4500 lb and the tension in cable BC is 2000 lb. Determine by trigonometry the magnitude and direction of the resultant of the two forces applied at B at that instant.

2.11 Solve Prob. 2.3 by trigonometry.

2.12 Determine by trigonometry the magnitude and direction of the force **P** so that the resultant of **P** and the 900-N force is a vertical force of 2700 N directed downward.

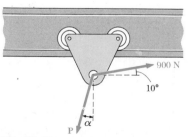

Fig. P2.12

2.13 If the resultant of the two forces acting on the cylinder of Prob. 2.7 is to be vertical, find (*a*) the value of α for which the magnitude of **P** is minimum, (*b*) the corresponding magnitude of **P**.

2.6. Rectangular Components of a Force. Unit Vectors.†

In many problems it will be found desirable to resolve a force into two components which are perpendicular to each other. In Fig. 2.18, the force **F** has been resolved into a component $\mathbf{F}_x$ along the x axis and a component $\mathbf{F}_y$ along the y axis. The parallelogram drawn to obtain the two components is a *rectangle*, and $\mathbf{F}_x$ and $\mathbf{F}_y$ are called *rectangular components*.

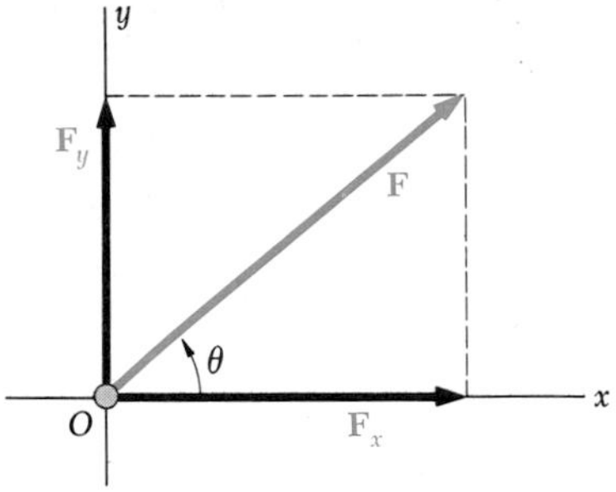

The x and y axes are usually chosen horizontal and vertical, respectively, as in Fig. 2.18; they may, however, be chosen in any two perpendicular directions, as shown in Fig. 2.19. In determining the rectangular components of a force, the student should think of the construction lines shown in Figs. 2.18 and 2.19 as being *parallel* to the x and y axes, rather than *perpendicular* to these axes. This practice will help avoid mistakes in determining *oblique* components as in Sec. 2.5.

We shall, at this point, introduce two vectors of magnitude 1, directed respectively along the positive x and y axes. These vectors are called *unit vectors* and are denoted by **i** and **j**, respectively (Fig. 2.20). Recalling the definition of the product of a scalar and a vector given in Sec. 2.3, we note that the rectangular components $\mathbf{F}_x$ and $\mathbf{F}_y$ of a force **F** may be obtained by multiplying respectively the unit vectors **i** and **j** by appropriate scalars (Fig. 2.21). We write

$$\mathbf{F}_x = F_x\mathbf{i} \qquad \mathbf{F}_y = F_y\mathbf{j} \tag{2.6}$$

and

$$\mathbf{F} = F_x\mathbf{i} + F_y\mathbf{j} \tag{2.7}$$

While the scalars F_x and F_y may be positive or negative, depending upon the sense of $\mathbf{F}_x$ and of $\mathbf{F}_y$, their absolute values are respectively equal to the magnitudes of the component forces $\mathbf{F}_x$ and $\mathbf{F}_y$. The scalars F_x and F_y are called the *scalar components* of the force **F**, while the actual component forces $\mathbf{F}_x$ and $\mathbf{F}_y$ should be referred to as the *vector components* of **F**. However, when

† The properties established in Secs. 2.6 and 2.7 may be readily extended to the rectangular components of any vector quantity.

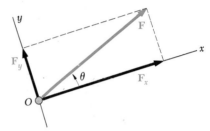

Fig. 2.18

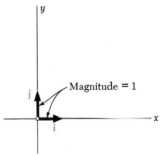

Fig. 2.19

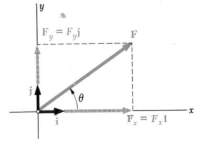

Fig. 2.20

Fig. 2.21

there exists no possibility of confusion, the vector as well as the scalar components of **F** may be referred to simply as the *components* of **F**. We note that the scalar component F_x is positive when the vector component $\mathbf{F}_x$ has the same sense as the unit vector **i** (i.e., the same sense as the positive x axis) and negative when $\mathbf{F}_x$ has the opposite sense. A similar conclusion may be drawn regarding the sign of the scalar component F_y.

Denoting by F the magnitude of the force **F** and by θ the angle between **F** and the x axis, measured counterclockwise from the positive x axis (Fig. 2.21), we may express the scalar components of **F** as follows:

$$F_x = F \cos \theta \qquad F_y = F \sin \theta \qquad (2.8)$$

We note that the relations obtained hold for any value of the angle θ from $0°$ to $360°$, and that they define the signs as well as the absolute values of the scalar components F_x and F_y.

Example 1. A force of 800 N is exerted on a bolt A as shown in Fig. 2.22a. Determine the horizontal and vertical components of the force.

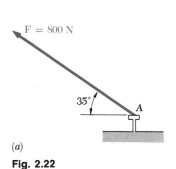

(a)

(b)

Fig. 2.22

In order to obtain the correct sign for the scalar components F_x and F_y, the value $\theta = 180° - 35° = 145°$ should be substituted for θ in the relations (2.8). However, it will be found more practical to determine by inspection the signs of F_x and of F_y (Fig. 2.22b) and to use the trigonometric functions of the angle $\alpha = 35°$. We write therefore

$$F_x = -F \cos \alpha = -(800 \text{ N}) \cos 35° = -655 \text{ N}$$
$$F_y = +F \sin \alpha = +(800 \text{ N}) \sin 35° = +459 \text{ N}$$

The vector components of **F** are thus

$$\mathbf{F}_x = -(655 \text{ N})\mathbf{i} \qquad \mathbf{F}_y = +(459 \text{ N})\mathbf{j}$$

and we may write **F** in the form

$$\mathbf{F} = -(655 \text{ N})\mathbf{i} + (459 \text{ N})\mathbf{j}$$

Example 2. A man pulls with a force of 300 N on a rope attached to a building, as shown in Fig. 2.23*a*. What are the horizontal and vertical components of the force exerted by the rope at point *A*?

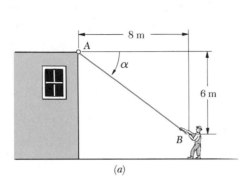

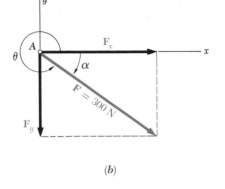

(*a*)   (*b*)

Fig. 2.23

It is seen from Fig. 2.23*b* that

$$F_x = +(300 \text{ N}) \cos \alpha \qquad F_y = -(300 \text{ N}) \sin \alpha$$

Observing that $AB = 10$ m, we find from Fig. 2.23*a*

$$\cos \alpha = \frac{8 \text{ m}}{AB} = \frac{8 \text{ m}}{10 \text{ m}} = \frac{4}{5} \qquad \sin \alpha = \frac{6 \text{ m}}{AB} = \frac{6 \text{ m}}{10 \text{ m}} = \frac{3}{5}$$

We thus obtain

$$F_x = +(300 \text{ N})\tfrac{4}{5} = +240 \text{ N} \qquad F_y = -(300 \text{ N})\tfrac{3}{5} = -180 \text{ N}$$

and write

$$\mathbf{F} = (240 \text{ N})\mathbf{i} - (180 \text{ N})\mathbf{j}$$

When a force **F** is defined by its rectangular components F_x and F_y (see Fig. 2.21), the angle θ defining its direction can be obtained by writing

$$\tan \theta = \frac{F_y}{F_x} \qquad\qquad (2.9)$$

The magnitude F of the force may be obtained by applying the Pythagorean theorem and writing

$$F = \sqrt{F_x^2 + F_y^2} \qquad\qquad (2.10)$$

However, once θ has been found, it is usually easier to determine the magnitude of the force by the process of solving one of the formulas (2.8) for F.

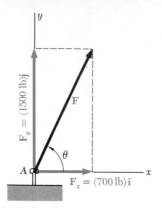

Fig. 2.24

Example 3. A force $\mathbf{F} = (700 \text{ lb})\mathbf{i} + (1500 \text{ lb})\mathbf{j}$ is applied to a bolt A. Determine the magnitude of the force and the angle θ it forms with the horizontal.

First we draw a diagram showing the two rectangular components of the force and the angle θ (Fig. 2.24). From Eq. (2.9), we write

$$\tan \theta = \frac{F_y}{F_x} = \frac{1500 \text{ lb}}{700 \text{ lb}}$$

Using a calculator†, we enter 1500 lb and divide by 700 lb; computing the arc tangent of the quotient, we obtain $\theta = 65.0°$. Solving the second of Eqs. (2.8) for F, we have

$$F = \frac{F_y}{\sin \theta} = \frac{1500 \text{ lb}}{\sin 65.0°} = 1655 \text{ lb}$$

The last calculation is facilitated if the value of F_y is stored when originally entered; it may then be recalled to be divided by $\sin \theta$.

2.7. Addition of Forces by Summing *x* and *y* Components. It was seen in Sec. 2.1 that forces should be added according to the parallelogram law. From this law, two other methods, more readily applicable to the *graphical* solution of problems, were derived in Secs. 2.3 and 2.4: The triangle rule for the addition of two forces and the polygon rule for the addition of three or more forces. It was also seen that the force triangle used to define the resultant of two forces could be used to obtain a *trigonometric* solution.

When three or more forces are to be added, no practical trigonometric solution may be obtained from the force polygon which defines the resultant of the forces. In this case, an *analytic* solution of the problem may be obtained by resolving each force into two rectangular components. Consider, for instance, three forces $\mathbf{P}$, $\mathbf{Q}$, and $\mathbf{S}$ acting on a particle A (Fig. 2.25a). Their resultant $\mathbf{R}$ is defined by the relation

$$\mathbf{R} = \mathbf{P} + \mathbf{Q} + \mathbf{S} \tag{2.11}$$

---

†It is assumed that the calculator used has keys for the computation of trigonometric and inverse trigonometric functions. Some calculators also have keys for the direct conversion of rectangular coordinates into polar coordinates, and vice versa. Such calculators eliminate the need for the computation of trigonometric functions in Examples 1, 2, and 3 and in problems of the same type.

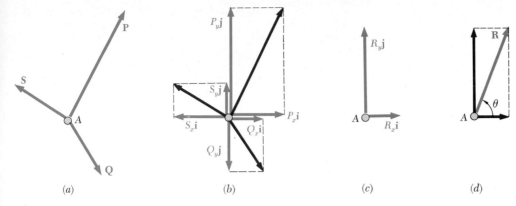

(a) (b) (c) (d)

Fig. 2.25

Resolving each force into its rectangular components, we write

$$R_x\mathbf{i} + R_y\mathbf{j} = P_x\mathbf{i} + P_y\mathbf{j} + Q_x\mathbf{i} + Q_y\mathbf{j} + S_x\mathbf{i} + S_y\mathbf{j}$$
$$= (P_x + Q_x + S_x)\mathbf{i} + (P_y + Q_y + S_y)\mathbf{j}$$

from which it follows that

$$R_x = P_x + Q_x + S_x \qquad R_y = P_y + Q_y + S_y \qquad (2.12)$$

or, for short,

$$R_x = \Sigma F_x \qquad R_y = \Sigma F_y \qquad (2.13)$$

We thus conclude that *the scalar components R_x and R_y of the resultant $\mathbf{R}$ of several forces acting on a particle are obtained by adding algebraically the corresponding scalar components of the given forces.*†

In practice, the determination of the resultant $\mathbf{R}$ is carried out in three steps as illustrated in Fig. 2.25. First, the given forces shown in Fig. 2.25a are resolved into their x and y components (Fig. 2.25b). Adding these components, we obtain the x and y components of $\mathbf{R}$ (Fig. 2.25c). Finally, the resultant $\mathbf{R} = R_x\mathbf{i} + R_y\mathbf{j}$ is determined by applying the parallelogram law (Fig. 2.25d). The procedure just described will be carried out most efficiently if the computations are arranged in a table. While it is the only practical analytic method for adding three or more forces, it is also often preferred to the trigonometric solution in the case of the addition of two forces.

---

† Clearly, this result also applies to the addition of other vector quantities, such as velocities, accelerations, or momenta.

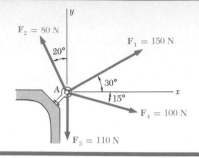

Four forces act on bolt A as shown. Determine the resultant of the forces on the bolt.

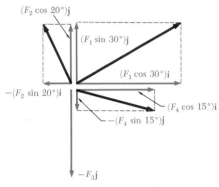

Solution. The x and y components of each force are determined by trigonometry as shown and are entered in the table below. According to the convention adopted in Sec. 2.6, the scalar number representing a force component is positive if the force component has the same sense as the corresponding coordinate axis. Thus, x components acting to the right and y components acting upward are represented by positive numbers.

Force	Magnitude, N	x Component, N	y Component, N
F_1	150	+129.9	+75.0
F_2	80	−27.4	+75.2
F_3	110	0	−110.0
F_4	100	+96.6	−25.9
		$R_x = +199.1$	$R_y = +14.3$

Thus, the resultant **R** of the four forces is

$$\mathbf{R} = R_x\mathbf{i} + R_y\mathbf{j} \qquad \mathbf{R} = (199.1\ \text{N})\mathbf{i} + (14.3\ \text{N})\mathbf{j} \quad \blacktriangleleft$$

The magnitude and direction of the resultant may now be determined. From the triangle shown, we have

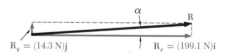

$$\tan \alpha = \frac{R_y}{R_x} = \frac{14.3\ \text{N}}{199.1\ \text{N}} \qquad \alpha = 4.1°$$

$$R = \frac{14.3\ \text{N}}{\sin \alpha} = 199.6\ \text{N} \qquad R = 199.6\ \text{N} \measuredangle 4.1° \quad \blacktriangleleft$$

With a calculator, the last computation may be facilitated if the value of R_y is stored when originally entered; it may then be recalled to be divided by $\sin \alpha$. (Also see the footnote on p. 30.)

PROBLEMS

2.14 A force of 2.5 kN is applied to a cable attached to the bracket. What are the horizontal and vertical components of this force?

2.15 The hydraulic cylinder *GE* exerts on member *DF* a force **P** directed along line *GE*. Knowing that **P** must have a 600-N component perpendicular to member *DF*, determine the magnitude of **P** and of its component parallel to *DF*.

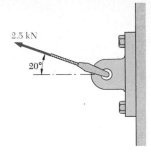

Fig. P2.14

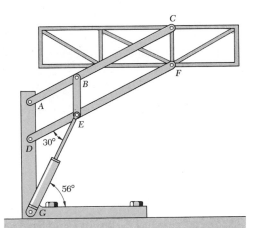

Fig. P2.15

2.16 Determine the *x* and *y* components of each of the forces shown.

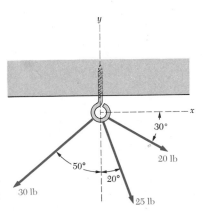

Fig. P2.16 and P2.17

2.17 Determine separately the components of the 20-lb force and of the 30-lb force in directions parallel and perpendicular to the line of action of the 25-lb force.

2.18 The tension in the support wire *AB* is 650 N. Determine the horizontal and vertical components of the force acting on the pin at *A*.

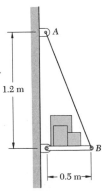

Fig. P2.18

2.19 Determine the x and y components of each of the forces shown.

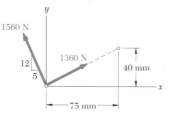

Fig. P2.19

2.20 and 2.21 The x and y components of a force **F** are as shown. Determine the magnitude and direction of the force **F**.

Fig. P2.20

Fig. P2.21 350 N

2.22 Determine the resultant of the three forces of Prob. 2.16.

2.23 Determine the resultant of the two forces of Prob. 2.19.

2.24 Using x and y components, solve Prob. 2.2.

2.25 Using x and y components, solve Prob. 2.3.

2.26 Two cables which have known tensions are attached at point B. A third cable AB is used as a guy wire and is also attached at B. Determine the required tension in AB so that the resultant of the forces exerted by the three cables will be vertical.

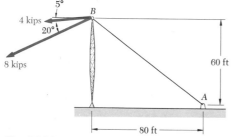

Fig. P2.26

2.27 A collar which may slide on a vertical rod is subjected to the three forces shown. The direction of the force **F** may be varied. If possible, determine the direction of the force **F** so that the resultant of the three forces is horizontal, knowing that the magnitude of **F** is (a) 480 lb, (b) 280 lb.

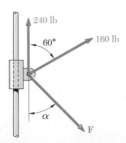

Fig. P2.27

2.28 The directions of the 300-N forces may vary, but the angle between the forces is always 40°. Determine the value of α for which the resultant of the forces acting at A is directed parallel to the plane b-b.

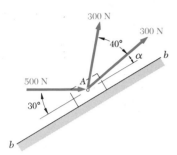

Fig. P2.28

2.29 Show that the forces $\mathbf{P} = P_x\mathbf{i} + P_y\mathbf{j}$ and $\mathbf{Q} = Q_x\mathbf{i} + Q_y\mathbf{j}$ are perpendicular to each other if and only if $P_xQ_x + P_yQ_y = 0$.

2.8. Equilibrium of a Particle. In the preceding sections, we discussed the methods for determining the resultant of several forces acting on a particle. Although this has not occurred in any of the problems considered so far, it is quite possible for the resultant to be zero. In such a case, the net effect of the given forces is zero, and the particle is said to be in equilibrium. We thus have the following definition: *When the resultant of all the forces acting on a particle is zero, the particle is in equilibrium.*

A particle which is acted upon by two forces will be in equilibrium if the two forces have the same magnitude, same line of action, and opposite sense. The resultant of the two forces is then zero. Such a case is shown in Fig. 2.26.

Another case of equilibrium of a particle is represented in Fig. 2.27, where four forces are shown acting on A. In Fig. 2.28, the resultant of the given forces is determined by the polygon rule. Starting from point O with $\mathbf{F}_1$ and arranging the forces in tip-to-tail fashion, we find that the tip of $\mathbf{F}_4$ coincides with the starting point O. Thus the resultant $\mathbf{R}$ of the given system of forces is zero, and the particle is in equilibrium.

The closed polygon drawn in Fig. 2.28 provides a *graphical* expression of the equilibrium of A. To express *algebraically* the conditions for the equilibrium of a particle, we write

$$\mathbf{R} = \Sigma\mathbf{F} = 0 \qquad (2.14)$$

Fig. 2.26

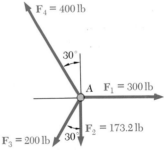

Fig. 2.27

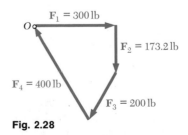

Fig. 2.28

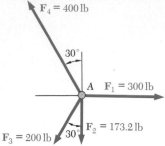

Fig. 2.27 (repeated)

Resolving each force **F** into rectangular components, we have

$$\Sigma(F_x\mathbf{i} + F_y\mathbf{j}) = 0 \qquad \text{or} \qquad (\Sigma F_x)\mathbf{i} + (\Sigma F_y)\mathbf{j} = 0$$

We conclude that the necessary and sufficient conditions for the equilibrium of a particle are

$$\Sigma F_x = 0 \qquad \Sigma F_y = 0 \qquad\qquad (2.15)$$

Returning to the particle shown in Fig. 2.27 we check that the equilibrium conditions are satisfied. We write

$$\Sigma F_x = 300\text{ lb} - (200\text{ lb})\sin 30° - (400\text{ lb})\sin 30°$$
$$= 300\text{ lb} - 100\text{ lb} - 200\text{ lb} = 0$$
$$\Sigma F_y = -173.2\text{ lb} - (200\text{ lb})\cos 30° + (400\text{ lb})\cos 30°$$
$$= -173.2\text{ lb} - 173.2\text{ lb} + 346.4\text{ lb} = 0$$

2.9. Newton's First Law of Motion. In the latter part of the seventeenth century, Sir Isaac Newton formulated three fundamental laws upon which the science of mechanics is based. The first of these laws can be stated as follows:

If the resultant force acting on a particle is zero, the particle will remain at rest (if originally at rest) or will move with constant speed in a straight line (if originally in motion).

From this law and from the definition of equilibrium given in Sec. 2.8, it is seen that a particle in equilibrium either is at rest or is moving in a straight line with constant speed. In the following section, various problems concerning the equilibrium of a particle will be considered.

2.10. Problems Involving the Equilibrium of a Particle. Free-Body Diagram. In practice, a problem in engineering mechanics is derived from an actual physical situation. A sketch showing the physical conditions of the problem is known as a *space diagram*.

The methods of analysis discussed in the preceding sections apply to a system of forces acting on a particle. A large number of problems involving actual structures, however, may be reduced to problems concerning the equilibrium of a particle. This is done by choosing a significant particle and drawing a separate diagram showing this particle and all the forces acting on it. Such a diagram is called a *free-body diagram*.

As an example, consider the 75-kg crate shown in the space diagram of Fig. 2.29a. This crate was lying between two buildings, and it is now being lifted onto a truck, which will remove it. The crate is supported by a vertical cable, which is joined at A to two ropes which pass over pulleys attached to the buildings at B and C. It is desired to determine the tension in each of the ropes AB and AC.

In order to solve this problem, a free-body diagram must be drawn, showing a particle in equilibrium. Since we are interested in the rope tensions, the free-body diagram should include at least one of these tensions and, if possible, both tensions. Point A is seen to be a good free body for this problem. The free-body diagram of point A is shown in Fig. 2.29b. It shows point A and the forces exerted on A by the vertical cable and the two ropes. The force exerted by the cable is directed downward and is equal to the weight W of the crate. Recalling Eq. (1.4), we write

$$W = mg = (75 \text{ kg})(9.81 \text{ m/s}^2) = 736 \text{ N}$$

and indicate this value in the free-body diagram. The forces exerted by the two ropes are not known. Since they are respectively equal in magnitude to the tension in rope AB and rope AC, we denote them by $\mathbf{T}_{AB}$ and $\mathbf{T}_{AC}$ and draw them away from A in the directions shown in the space diagram. No other detail is included in the free-body diagram.

Since point A is in equilibrium, the three forces acting on it must form a closed triangle when drawn in tip-to-tail fashion. This *force triangle* has been drawn in Fig. 2.29c. The values T_{AB} and T_{AC} of the tension in the ropes may be found graphically if the triangle is drawn to scale, or they may be found by trigonometry. If the latter method of solution is chosen, we use the law of sines and write

$$\frac{T_{AB}}{\sin 60°} = \frac{T_{AC}}{\sin 40°} = \frac{736 \text{ N}}{\sin 80°}$$

$$T_{AB} = 647 \text{ N} \qquad T_{AC} = 480 \text{ N}$$

When a particle is in *equilibrium under three forces*, the problem may always be solved by drawing a force triangle. When a particle is in *equilibrium under more than three forces*, the problem may be solved graphically by drawing a force polygon. If an analytic solution is desired, the *equations of equilibrium* given in Sec. 2.8 should be solved:

$$\Sigma F_x = 0 \qquad \Sigma F_y = 0 \qquad (2.15)$$

These equations may be solved for no more than *two unknowns*; similarly, the force triangle used in the case of equilibrium under three forces may be solved for two unknowns.

The more common types of problems are those where the two unknowns represent (1) the two components (or the magnitude and direction) of a single force, (2) the magnitude of two forces each of known direction. Problems involving the determination of the maximum or minimum value of the magnitude of a force are also encountered (see Probs. 2.34, 2.35, and 2.36).

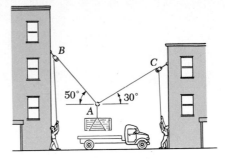

(a) Space diagram

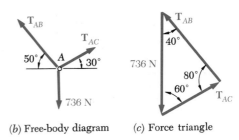

(b) Free-body diagram   (c) Force triangle

Fig. 2.29

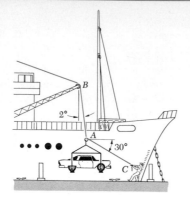

SAMPLE PROBLEM 2.4

In a ship-unloading operation, a 3500-lb automobile is supported by a cable. A rope is tied to the cable at A and pulled in order to center the automobile over its intended position. The angle between the cable and the vertical is 2°, while the angle between the rope and the horizontal is 30°. What is the tension in the rope?

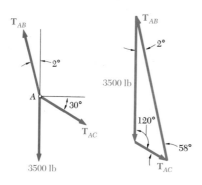

Solution. Point A is chosen as a free body, and the complete free-body diagram is drawn. T_{AB} is the tension in the cable AB, and T_{AC} is the tension in the rope.

Equilibrium Condition. Since only three forces act on the free body, we draw a force triangle to express that it is in equilibrium. Using the law of sines, we write

$$\frac{T_{AB}}{\sin 120°} = \frac{T_{AC}}{\sin 2°} = \frac{3500 \text{ lb}}{\sin 58°}$$

With a calculator, we first compute and store the value of the last quotient. Multiplying this value successively by sin 120° and sin 2°, we obtain

$$T_{AB} = 3570 \text{ lb} \qquad T_{AC} = 144 \text{ lb} \quad \blacktriangleleft$$

SAMPLE PROBLEM 2.5

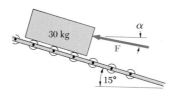

Determine the magnitude and direction of the smallest force $\mathbf{F}$ which will maintain the package shown in equilibrium. Note that the force exerted by the rollers on the package is perpendicular to the incline.

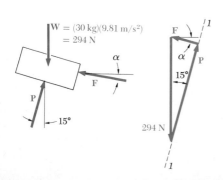

Solution. We choose the package as a free body, assuming that it may be treated as a particle. We draw the corresponding free-body diagram.

Equilibrium Condition. Since only three forces act on the free body, we draw a force triangle to express that it is in equilibrium. Line *1-1* represents the known direction of $\mathbf{P}$. In order to obtain the minimum value of the force $\mathbf{F}$, we choose the direction of $\mathbf{F}$ perpendicular to that of $\mathbf{P}$. From the geometry of the triangle obtained, we find

$$F = (294 \text{ N}) \sin 15° = 76.1 \text{ N} \qquad \alpha = 15°$$
$$F = 76.1 \text{ N} \angle 15° \quad \blacktriangleleft$$

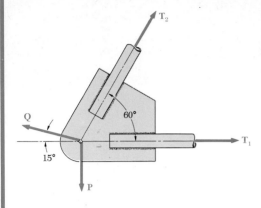

Two forces **P** and **Q** of magnitude $P = 1000$ lb and $Q = 1200$ lb are applied to the aircraft connection shown. Knowing that the connection is in equilibrium, determine the tensions T_1 and T_2.

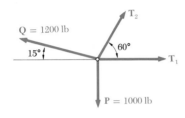

Solution. The connection is considered a particle and taken as a free body. It is acted upon by four forces directed as shown. Each force is resolved into its x and y components.

$$\mathbf{P} = -(1000 \text{ lb})\mathbf{j}$$
$$\mathbf{Q} = -(1200 \text{ lb}) \cos 15°\mathbf{i} + (1200 \text{ lb}) \sin 15°\mathbf{j}$$
$$= -(1159 \text{ lb})\mathbf{i} + (311 \text{ lb})\mathbf{j}$$
$$\mathbf{T}_1 = T_1\mathbf{i}$$
$$\mathbf{T}_2 = T_2 \cos 60°\mathbf{i} + T_2 \sin 60°\mathbf{j}$$
$$= 0.500\,T_2\mathbf{i} + 0.866\,T_2\mathbf{j}$$

Equilibrium Condition. Since the connection is in equilibrium, the resultant of the forces must be zero. Thus

$$\mathbf{R} = \mathbf{P} + \mathbf{Q} + \mathbf{T}_1 + \mathbf{T}_2 = 0$$

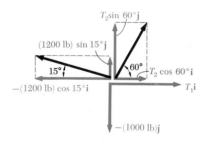

Substituting for **P**, **Q**, $\mathbf{T}_1$, and $\mathbf{T}_2$ the expressions obtained above, and factoring the unit vectors **i** and **j**, we have

$$(-1159 \text{ lb} + T_1 + 0.500\,T_2)\mathbf{i} + (-1000 \text{ lb} + 311 \text{ lb} + 0.866\,T_2)\mathbf{j} = 0$$

This equation will be satisfied if, and only if, the coefficients of **i** and **j** are equal to zero. We thus obtain the following two equilibrium equations, which express, respectively, that the sum of the x components and the sum of the y components of the given forces must be zero.

$$(\Sigma F_x = 0\text{:}) \qquad -1159 \text{ lb} + T_1 + 0.500\,T_2 = 0$$
$$(\Sigma F_y = 0\text{:}) \qquad -1000 \text{ lb} + 311 \text{ lb} + 0.866\,T_2 = 0$$

Solving these equations, we find

$$T_1 = 761 \text{ lb} \qquad T_2 = 796 \text{ lb} \quad \blacktriangleleft$$

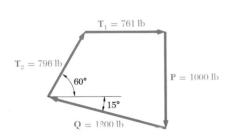

In drawing the free-body diagram, we assumed a sense for each unknown force. A positive sign in the answer indicates that the assumed sense is correct. The complete force polygon may be drawn to check the results.

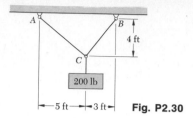

Fig. P2.30

PROBLEMS

2.30 through 2.33 Two cables are tied together at C and loaded as shown. Determine the tension in AC and BC.

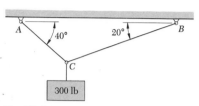

Fig. P2.31

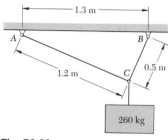

Fig. P2.32

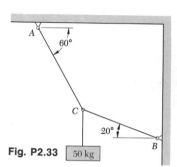

Fig. P2.33

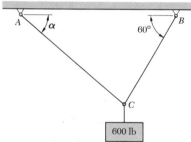

Fig. P2.34 and P2.35

2.34 A 600-lb block is supported by the two cables AC and BC. (a) For what value of α is the tension in cable AC minimum? (b) What are the corresponding values of the tension in cables AC and BC?

2.35 A 600-lb block is supported by the two cables AC and BC. Determine (a) the value of α for which the larger of the cable tensions is as small as possible, (b) the corresponding values of the tension in cables AC and BC.

2.36 Two ropes are tied together at C. If the maximum permissible tension in each rope is 2.5 kN, what is the maximum force **F** that may be applied? In what direction must this maximum force act?

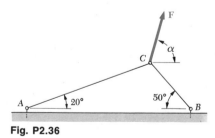

Fig. P2.36

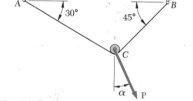

Fig. P2.37

2.37 The force **P** is applied to a small wheel which rolls on the cable ACB. Knowing that the tension in both parts of the cable is 750 N, determine the magnitude and direction of **P**.

2.38 Two forces **A** and **B** of magnitude $A = 5000\,\text{N}$ and $B = 2500\,\text{N}$ are applied to the connection shown. Knowing that the connection is in equilibrium, determine the magnitudes of the forces **C** and **D**.

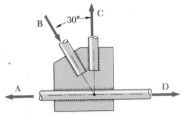

Fig. P2.38 and P2.39

2.39 Two forces **C** and **D** of magnitude $C = 800\,\text{lb}$ and $D = 1500\,\text{lb}$ are applied to the connection shown. Knowing that the connection is in equilibrium, determine the magnitudes of the forces **A** and **B**.

2.40 Knowing that $P = 100\,\text{lb}$, determine the tension in cables AC and BC.

2.41 Determine the range of values of **P** for which both cables remain taut.

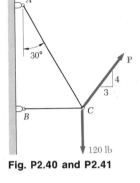

Fig. P2.40 and P2.41

2.42 In order to lift a 1200-lb crate, a cable sling 14 ft 4 in. long is wrapped around the crate and hung to the crane cable EF. Determine the tension in the cable sling in each of the two cases shown.

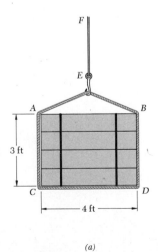

(a)

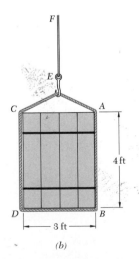

(b)

Fig. P2.42

2.43 A 200-kg crate is supported by several rope-and-pulley arrangements as shown. Determine for each arrangement the tension in the rope. (The tension in the rope is the same on each side of a simple pulley. This can be proved by the methods of Chap. 4.)

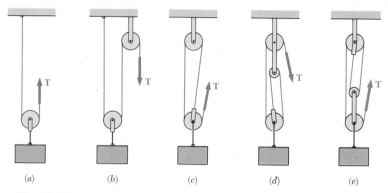

(a) (b) (c) (d) (e)

Fig. P2.43

2.44 Solve parts *b* and *d* of Prob. 2.43 assuming that the free end of the rope is attached to the crate.

2.45 Express the weight *W* required to maintain equilibrium in terms of *P*, *d*, and *h*.

2.46 If in the diagram shown $W = 80$ lb, $P = 10$ lb, and $d = 20$ in., determine the value of *h* consistent with equilibrium.

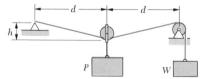

Fig. P2.45 and P2.46

2.47 The collar *A* may slide freely on the horizontal smooth rod. The spring attached to the collar has a constant of 10 lb/in. and is undeformed when the collar is directly below support *B*. Determine the magnitude of the force **P** required to maintain equilibrium when (*a*) $c = 9$ in., (*b*) $c = 16$ in.

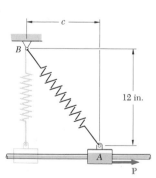

Fig. P2.47

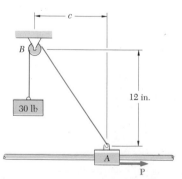

Fig. P2.48

2.48 The collar *A* may slide freely on the horizontal smooth rod. Determine the magnitude of the force **P** required to maintain equilibrium when (*a*) $c = 9$ in., (*b*) $c = 16$ in.

*2.49 A 150-kg block is attached to a small pulley which may roll on the cable *ACB*. The pulley and load are held in the position shown by a second cable *DE* which is parallel to the portion *CB* of the main cable. Determine (*a*) the tension in cable *ACB*, (*b*) the tension in cable *DE*. Neglect the radius of the pulleys and the weight of the cables.

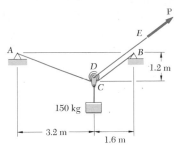

Fig. P2.49

FORCES IN SPACE

2.11. Rectangular Components of a Force in Space.

The problems considered in the first part of this chapter involved only two dimensions; they could be formulated and solved in a single plane. In this section and in the remaining sections of the chapter, we shall discuss problems involving the three dimensions of space.

Consider a force **F** acting at the origin *O* of the system of rectangular coordinates *x, y, z*. To define the direction of **F**, we may draw the vertical plane *OBAC* containing **F** and shown in Fig. 2.30*a*. This plane passes through the vertical *y* axis; its orientation is defined by the angle ϕ it forms with the *xy* plane,

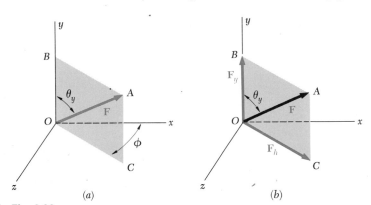

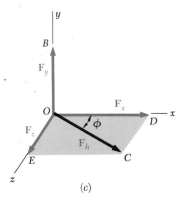

(a)   (b)   (c)

Fig. 2.30

while the direction of **F** within the plane is defined by the angle θ_y that **F** forms with the *y* axis. The force **F** may be resolved into a vertical component $\mathbf{F}_y$ and a horizontal component $\mathbf{F}_h$; this operation, shown in Fig. 2.30*b*, is carried out inside plane *OBAC* according to the rules developed in the first part of the chapter. The corresponding scalar components are

$$F_y = F \cos \theta_y \qquad F_h = F \sin \theta_y \qquad (2.16)$$

But $\mathbf{F}_h$ may be resolved into two rectangular components $\mathbf{F}_x$ and $\mathbf{F}_z$ along the *x* and *z* axes, respectively. This operation, shown in Fig. 2.30*c*, is carried out inside the *xz* plane. We obtain the following expressions for the corresponding scalar components:

$$F_x = F_h \cos \phi = F \sin \theta_y \cos \phi \qquad (2.17)$$
$$F_z = F_h \sin \phi = F \sin \theta_y \sin \phi$$

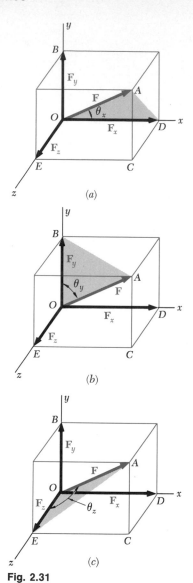

Fig. 2.31

The given force $\mathbf{F}$ has thus been resolved into three rectangular vector components $\mathbf{F}_x$, $\mathbf{F}_y$, $\mathbf{F}_z$, directed along the three coordinate axes.

Applying the Pythagorean theorem to the triangles OAB and OCD of Fig. 2.30, we write

$$F^2 = (OA)^2 = (OB)^2 + (BA)^2 = F_y^2 + F_h^2$$
$$F_h^2 = (OC)^2 = (OD)^2 + (DC)^2 = F_x^2 + F_z^2$$

Eliminating F_h^2 from these two equations and solving for F, we obtain the following relation between the magnitude of $\mathbf{F}$ and its rectangular scalar components:

$$F = \sqrt{F_x^2 + F_y^2 + F_z^2} \tag{2.18}$$

The relationship existing between the force $\mathbf{F}$ and its three components $\mathbf{F}_x$, $\mathbf{F}_y$, $\mathbf{F}_z$ is more easily visualized if a "box" having $\mathbf{F}_x$, $\mathbf{F}_y$, $\mathbf{F}_z$ for edges is drawn as shown in Fig. 2.31. The force $\mathbf{F}$ is then represented by the diagonal OA of this box. Fig. 2.31b shows the right triangle OAB used to derive the first of the formulas (2.16): $F_y = F \cos \theta_y$. In Fig. 2.31a and c, two other right triangles have also been drawn: OAD and OAE. These triangles are seen to occupy in the box positions comparable with that of triangle OAB. Denoting by θ_x and θ_z, respectively, the angles that $\mathbf{F}$ forms with the x and z axes, we may derive two formulas similar to $F_y = F \cos \theta_y$. We thus write

$$F_x = F \cos \theta_x \qquad F_y = F \cos \theta_y \qquad F_z = F \cos \theta_z \tag{2.19}$$

The three angles θ_x, θ_y, θ_z define the direction of the force $\mathbf{F}$; they are more commonly used for this purpose than the angles θ_y and ϕ introduced at the beginning of this section. The cosines of θ_x, θ_y, θ_z are known as the direction cosines of the force $\mathbf{F}$.

Introducing the unit vectors $\mathbf{i}$, $\mathbf{j}$, and $\mathbf{k}$, directed respectively along the x, y, and z axes (Fig. 2.32), we may express $\mathbf{F}$ in the form

$$\mathbf{F} = F_x\mathbf{i} + F_y\mathbf{j} + F_z\mathbf{k} \tag{2.20}$$

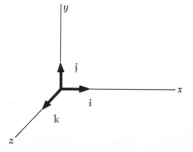

Fig. 2.32

where the scalar components F_x, F_y, F_z are defined by the relations (2.19).

Example 1. A force of 500 N forms angles of 60, 45, and 120°, respectively, with the x, y, and z axes. Find the components F_x, F_y, and F_z of the force.

Substituting $F = 500$ N, $\theta_x = 60°$, $\theta_y = 45°$, $\theta_z = 120°$ into formulas (2.19), we write

$$F_x = (500 \text{ N}) \cos 60° = +250 \text{ N}$$
$$F_y = (500 \text{ N}) \cos 45° = +354 \text{ N}$$
$$F_z = (500 \text{ N}) \cos 120° = -250 \text{ N}$$

Carrying into Eq. (2.20) the values obtained for the scalar components of **F**, we have

$$\mathbf{F} = (250 \text{ N})\mathbf{i} + (354 \text{ N})\mathbf{j} - (250 \text{ N})\mathbf{k}$$

As in the case of two-dimensional problems, the plus sign indicates that the component has the same sense as the corresponding axis, and the minus sign that it has the opposite sense.

The angle a force **F** forms with an axis should be measured from the positive side of the axis and will always be comprised between 0 and 180°. An angle θ_x smaller than 90° (acute) indicates that **F** (assumed attached at O) is on the same side of the yz plane as the positive x axis; $\cos \theta_x$ and F_x will then be positive. An angle θ_x larger than 90° (obtuse) would indicate that **F** is on the other side of the yz plane; $\cos \theta_x$ and F_x would then be negative. In Example 1 the angles θ_x and θ_y are acute, while θ_z is obtuse: consequently, F_x and F_y are positive, while F_z is negative.

Substituting into (2.20) the expressions obtained for F_x, F_y, F_z in (2.19), we write

$$\mathbf{F} = F(\cos \theta_x \mathbf{i} + \cos \theta_y \mathbf{j} + \cos \theta_z \mathbf{k}) \qquad (2.21)$$

which shows that the force **F** may be expressed as the product of the scalar F and of the vector

$$\boldsymbol{\lambda} = \cos \theta_x \mathbf{i} + \cos \theta_y \mathbf{j} + \cos \theta_z \mathbf{k} \qquad (2.22)$$

Clearly, the vector $\boldsymbol{\lambda}$ is a vector of magnitude equal to 1 and of the same direction as **F** (Fig. 2.33). We shall refer to $\boldsymbol{\lambda}$ as the *unit vector* along the line of action of **F**. It follows from (2.22) that the components of the unit vector $\boldsymbol{\lambda}$ are respectively equal to the direction cosines of the line of action of **F**:

$$\lambda_x = \cos \theta_x \qquad \lambda_y = \cos \theta_y \qquad \lambda_z = \cos \theta_z \qquad (2.23)$$

We should observe that the values of the three angles θ_x, θ_y, θ_z are not independent. Expressing that the sum of the squares of

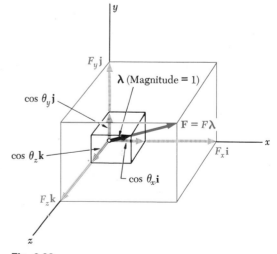

Fig. 2.33

the components of $\boldsymbol{\lambda}$ is equal to the square of its magnitude, we write

$$\lambda_x^2 + \lambda_y^2 + \lambda_z^2 = 1$$

or, substituting for λ_x, λ_y, λ_z from (2.23),

$$\cos^2 \theta_x + \cos^2 \theta_y + \cos^2 \theta_z = 1 \qquad (2.24)$$

In Example 1, for instance, once the values $\theta_x = 60°$ and $\theta_y = 45°$ have been selected, the value of θ_z *must* be equal to $60°$ or $120°$ in order to satisfy identity (2.24).

When the components F_x, F_y, F_z of a force $\mathbf{F}$ are given, the magnitude F of the force is obtained from (2.18).† The relations (2.19) may then be solved for the direction cosines,

$$\cos \theta_x = \frac{F_x}{F} \qquad \cos \theta_y = \frac{F_y}{F} \qquad \cos \theta_z = \frac{F_z}{F} \qquad (2.25)$$

and the angles θ_x, θ_y, θ_z characterizing the direction of $\mathbf{F}$ may be found.

Example 2. A force $\mathbf{F}$ has the components $F_x = 20$ lb, $F_y = -30$ lb, $F_z = 60$ lb. Determine its magnitude F and the angles θ_x, θ_y, θ_z it forms with the axes of coordinates.

From formula (2.18) we obtain†

$$\begin{aligned} F &= \sqrt{F_x^2 + F_y^2 + F_z^2} \\ &= \sqrt{(20\ \text{lb})^2 + (-30\ \text{lb})^2 + (60\ \text{lb})^2} \\ &= \sqrt{4900}\ \text{lb} = 70\ \text{lb} \end{aligned}$$

Substituting the values of the components and magnitude of $\mathbf{F}$ into Eqs. (2.25), we write

$$\cos \theta_x = \frac{F_x}{F} = \frac{20\ \text{lb}}{70\ \text{lb}} \qquad \cos \theta_y = \frac{F_y}{F} = \frac{-30\ \text{lb}}{70\ \text{lb}} \qquad \cos \theta_z = \frac{F_z}{F} = \frac{60\ \text{lb}}{70\ \text{lb}}$$

Calculating successively each quotient and its arc cosine, we obtain

$$\theta_x = 73.4° \qquad \theta_y = 115.4° \qquad \theta_z = 31.0°$$

The computations indicated may easily be carried out with a calculator.

---

† With a calculator programmed to convert rectangular coordinates into polar coordinates, the following procedure will be found more expeditious for computing F: First determine F_h from its two rectangular components F_x and F_z (Fig. 2.30c), then determine F from its two rectangular components F_h and F_y (Fig. 2.30b). The actual order in which the three components F_x, F_y, F_z are entered is immaterial.

2.12. Force Defined by Its Magnitude and Two Points on Its Line of Action.

In many applications, the direction of a force $\mathbf{F}$ is defined by the coordinates of two points, $M(x_1, y_1, z_1)$ and $N(x_2, y_2, z_2)$, located on its line of action (Fig. 2.34). Consider the vector $\overrightarrow{MN}$ joining M and N and of the same

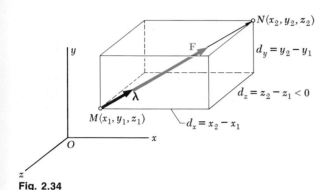

Fig. 2.34

sense as $\mathbf{F}$. Denoting its scalar components by d_x, d_y, d_z, respectively, we write

$$\overrightarrow{MN} = d_x\mathbf{i} + d_y\mathbf{j} + d_z\mathbf{k} \qquad (2.26)$$

The unit vector $\boldsymbol{\lambda}$ along the line of action of $\mathbf{F}$ (i.e., along the line MN) may be obtained by dividing the vector $\overrightarrow{MN}$ by its magnitude MN. Substituting for $\overrightarrow{MN}$ from (2.26) and observing that MN is equal to the distance d from M to N, we write

$$\boldsymbol{\lambda} = \frac{\overrightarrow{MN}}{MN} = \frac{1}{d}(d_x\mathbf{i} + d_y\mathbf{j} + d_z\mathbf{k}) \qquad (2.27)$$

Recalling that $\mathbf{F}$ is equal to the product of F and $\boldsymbol{\lambda}$, we have

$$\mathbf{F} = F\boldsymbol{\lambda} = \frac{F}{d}(d_x\mathbf{i} + d_y\mathbf{j} + d_z\mathbf{k}) \qquad (2.28)$$

from which it follows that the scalar components of $\mathbf{F}$ are, respectively,

$$F_x = \frac{Fd_x}{d} \qquad F_y = \frac{Fd_y}{d} \qquad F_z = \frac{Fd_z}{d} \qquad (2.29)$$

The relations (2.29) considerably simplify the determination of the components of a force $\mathbf{F}$ of given magnitude F when the line of action of $\mathbf{F}$ is defined by two points M and N. Subtracting the coordinates of M from those of N, we first determine the components of the vector $\overrightarrow{MN}$ and the distance d from M to N:

$$d_x = x_2 - x_1 \qquad d_y = y_2 - y_1 \qquad d_z = z_2 - z_1$$
$$d = \sqrt{d_x^2 + d_y^2 + d_z^2}$$

Substituting for F and for d_x, d_y, d_z, and d into the relations (2.29), we obtain the components F_x, F_y, F_z of the force.

The angles θ_x, θ_y, θ_z that $\mathbf{F}$ forms with the coordinate axes may then be obtained from Eqs. (2.25). Comparing Eqs. (2.22) and (2.27), we may also write

$$\cos \theta_x = \frac{d_x}{d} \qquad \cos \theta_y = \frac{d_y}{d} \qquad \cos \theta_z = \frac{d_z}{d} \qquad (2.30)$$

and determine the angles θ_x, θ_y, θ_z directly from the components and magnitude of the vector MN.

2.13. Addition of Concurrent Forces in Space. We shall determine the resultant $\mathbf{R}$ of two or more forces in space by summing their rectangular components. Graphical or trigonometric methods are generally not practical in the case of forces in space.

The method followed here is similar to that used in Sec. 2.7 with coplanar forces. Setting

$$\mathbf{R} = \Sigma \mathbf{F}$$

we resolve each force into its rectangular components and write

$$R_x\mathbf{i} + R_y\mathbf{j} + R_z\mathbf{k} = \Sigma(F_x\mathbf{i} + F_y\mathbf{j} + F_z\mathbf{k})$$
$$= (\Sigma F_x)\mathbf{i} + (\Sigma F_y)\mathbf{j} + (\Sigma F_z)\mathbf{k}$$

from which it follows that

$$R_x = \Sigma F_x \qquad R_y = \Sigma F_y \qquad R_z = \Sigma F_z \qquad (2.31)$$

The magnitude of the resultant and the angles θ_x, θ_y, θ_z it forms with the axes of coordinates are obtained by the method of Sec. 2.11. We write

$$R = \sqrt{R_x^2 + R_y^2 + R_z^2} \qquad (2.32)$$

$$\cos \theta_x = \frac{R_x}{R} \qquad \cos \theta_y = \frac{R_y}{R} \qquad \cos \theta_z = \frac{R_z}{R} \qquad (2.33)$$

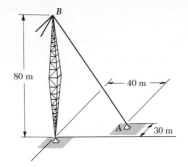

80 m

B

40 m

A

30 m

A tower guy wire is anchored by means of a bolt at A. The tension in the wire is 2500 N. Determine (a) the components F_x, F_y, F_z of the force acting on the bolt, (b) the angles θ_x, θ_y, θ_z defining the direction of the force.

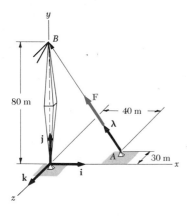

a. **Components of the Force.** The line of action of the force acting on the bolt passes through A and B, and the force is directed from A to B. The components of the vector $\overrightarrow{AB}$, which has the same direction as the force, are

$$d_x = -40 \text{ m} \qquad d_y = +80 \text{ m} \qquad d_z = +30 \text{ m}$$

The total distance from A to B is

$$AB = d = \sqrt{d_x^2 + d_y^2 + d_z^2} = 94.3 \text{ m}$$

Denoting by $\mathbf{i}$, $\mathbf{j}$, $\mathbf{k}$ the unit vectors along the coordinate axes, we have

$$\overrightarrow{AB} = -(40 \text{ m})\mathbf{i} + (80 \text{ m})\mathbf{j} + (30 \text{ m})\mathbf{k}$$

Introducing the unit vector $\boldsymbol{\lambda} = \overrightarrow{AB}/AB$, we write

$$\mathbf{F} = F\boldsymbol{\lambda} = F\frac{\overrightarrow{AB}}{AB} = \frac{2500 \text{ N}}{94.3 \text{ m}}\overrightarrow{AB}$$

Substituting the expression found for AB, we obtain

$$\mathbf{F} = \frac{2500 \text{ N}}{94.3 \text{ m}}[-(40 \text{ m})\mathbf{i} + (80 \text{ m})\mathbf{j} + (30 \text{ m})\mathbf{k}]$$

$$\mathbf{F} = -(1060 \text{ N})\mathbf{i} + (2120 \text{ N})\mathbf{j} + (795 \text{ N})\mathbf{k}$$

The components of $\mathbf{F}$, therefore, are

$$F_x = -1060 \text{ N} \qquad F_y = +2120 \text{ N} \qquad F_z = +795 \text{ N} \quad \blacktriangleleft$$

b. **Direction of the Force.** Using Eqs. (2.25), we write

$$\cos \theta_x = \frac{F_x}{F} = \frac{-1060 \text{ N}}{2500 \text{ N}} \qquad \cos \theta_y = \frac{F_y}{F} = \frac{+2120 \text{ N}}{2500 \text{ N}}$$

$$\cos \theta_z = \frac{F_z}{F} = \frac{+795 \text{ N}}{2500 \text{ N}}$$

Calculating successively each quotient and its arc cosine, we obtain

$$\theta_x = 115.1° \qquad \theta_y = 32.0° \qquad \theta_z = 71.5° \quad \blacktriangleleft$$

(*Note.* This result could have been obtained by using the components and magnitude of the vector $\overrightarrow{AB}$ rather than those of the force $\mathbf{F}$.)

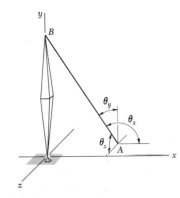

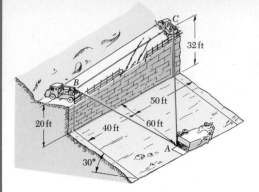

SAMPLE PROBLEM 2.8

In order to move a wrecked truck, two cables are attached to the truck at A and pulled by winches B and C as shown. Determine the resultant of the forces exerted on the truck by the two cables, knowing that the tension is 2000 lb in cable AB and 1500 lb in cable AC.

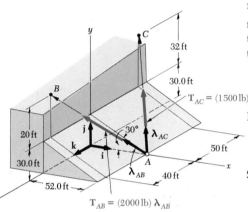

Solution. The force exerted by each cable on the truck will be resolved into x, y, and z components. We first determine the components and magnitudes of the vectors $\overrightarrow{AB}$ and $\overrightarrow{AC}$, measuring them from the truck toward each winch. Denoting by $\mathbf{i}$, $\mathbf{j}$, $\mathbf{k}$ the unit vectors along the coordinate axes, we write

$$\overrightarrow{AB} = -(52\ \text{ft})\mathbf{i} + (50\ \text{ft})\mathbf{j} + (40\ \text{ft})\mathbf{k} \qquad AB = 82.5\ \text{ft}$$
$$\overrightarrow{AC} = -(52\ \text{ft})\mathbf{i} + (62\ \text{ft})\mathbf{j} - (50\ \text{ft})\mathbf{k} \qquad AC = 95.1\ \text{ft}$$

Denoting by $\boldsymbol{\lambda}_{AB}$ the unit vector along AB, we have

$$\mathbf{T}_{AB} = T_{AB}\boldsymbol{\lambda}_{AB} = T_{AB}\frac{\overrightarrow{AB}}{AB} = \frac{2000\ \text{lb}}{82.5\ \text{ft}}\overrightarrow{AB}$$

Substituting the expression found for $\overrightarrow{AB}$, we obtain

$$\mathbf{T}_{AB} = \frac{2000\ \text{lb}}{82.5\ \text{ft}}[-(52\ \text{ft})\mathbf{i} + (50\ \text{ft})\mathbf{j} + (40\ \text{ft})\mathbf{k}]$$
$$\mathbf{T}_{AB} = -(1260\ \text{lb})\mathbf{i} + (1212\ \text{lb})\mathbf{j} + (970\ \text{lb})\mathbf{k}$$

Denoting by $\boldsymbol{\lambda}_{AC}$ the unit vector along AC, we obtain in a similar way

$$\mathbf{T}_{AC} = T_{AC}\boldsymbol{\lambda}_{AC} = T_{AC}\frac{\overrightarrow{AC}}{AC} = \frac{1500\ \text{lb}}{95.1\ \text{lb}}\overrightarrow{AC}$$
$$\mathbf{T}_{AC} = -(820\ \text{lb})\mathbf{i} + (978\ \text{lb})\mathbf{j} - (788\ \text{lb})\mathbf{k}$$

The resultant $\mathbf{R}$ of the forces exerted by the two cables is

$$\mathbf{R} = \mathbf{T}_{AB} + \mathbf{T}_{AC} = -(2080\ \text{lb})\mathbf{i} + (2190\ \text{lb})\mathbf{j} + (182\ \text{lb})\mathbf{k}$$

The magnitude and direction of the resultant are now determined.

$$R = \sqrt{R_x^2 + R_y^2 + R_z^2} = \sqrt{(-2080)^2 + (2190)^2 + (182)^2}$$
$$R = 3030\ \text{lb} \quad \blacktriangleleft$$

From Eqs. (2.33) we obtain

$$\cos\theta_x = \frac{R_x}{R} = \frac{-2080\ \text{lb}}{3030\ \text{lb}} \qquad \cos\theta_y = \frac{R_y}{R} = \frac{+2190\ \text{lb}}{3030\ \text{lb}}$$
$$\cos\theta_z = \frac{R_z}{R} = \frac{+182\ \text{lb}}{3030\ \text{lb}}$$

Calculating successively each quotient and its arc cosine, we have

$$\theta_x = 133.4° \qquad \theta_y = 43.7° \qquad \theta_z = 86.6° \quad \blacktriangleleft$$

PROBLEMS

2.50 Determine (*a*) the *x*, *y*, and *z* components of the 250-N force, (*b*) the angles θ_x, θ_y, and θ_z that the force forms with the coordinate axes.

2.51 Determine (*a*) the *x*, *y*, and *z* components of the 300-N force, (*b*) the angles θ_x, θ_y, and θ_z that the force forms with the coordinate axes.

2.52 The angle between the guy wire *AB* and the mast is 20°. Knowing that the tension in *AB* is 300 lb, determine (*a*) the *x*, *y*, and *z* components of the force exerted on the boat at *B*, (*b*) the angles θ_x, θ_y, and θ_z defining the direction of the force exerted at *B*.

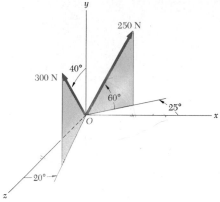

Fig. P2.50 and P2.51

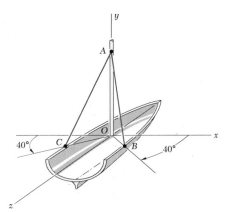

Fig. P2.52 and P2.53

2.53 The angle between the guy wire *AC* and the mast is 20°. Knowing that the tension in *AC* is 300 lb, determine (*a*) the *x*, *y*, and *z* components of the force exerted on the boat at *C*, (*b*) the angles θ_x, θ_y, and θ_z defining the direction of the force exerted at *C*.

2.54 A 250-lb force acts at the origin in a direction defined by the angles $\theta_x = 65°$ and $\theta_y = 40°$. It is also known that the *z* component of the force is positive. Determine the value of θ_z and the components of the force.

2.55 A force acts at the origin in a direction defined by the angles $\theta_y = 120°$ and $\theta_z = 75°$. It is known that the *x* component of the force is +40 N. Determine the magnitude of the force and the value of θ_x.

2.56 Determine the magnitude and direction of the force $\mathbf{F} = (700 \text{ N})\mathbf{i} - (820 \text{ N})\mathbf{j} + (960 \text{ N})\mathbf{k}$.

2.57 Determine the magnitude and direction of the force $\mathbf{F} = -(240 \text{ lb})\mathbf{i} - (320 \text{ lb})\mathbf{j} + (600 \text{ lb})\mathbf{k}$.

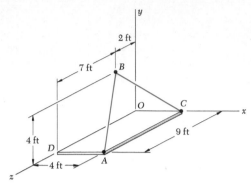

Fig. P2.58, P2.59, and P2.63

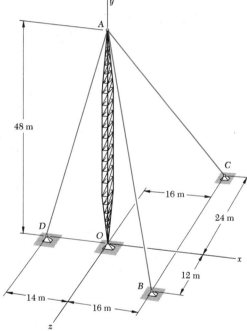

Fig. P2.64 and P2.65

2.58 Knowing that the tension in cable AB is 450 lb, determine the components of the force exerted on the plate at A.

2.59 Knowing that the tension in cable BC is 225 lb, determine the components of the force exerted on the plate at C.

2.60 Determine the angles θ_x, θ_y, and θ_z which define the direction of the force exerted on point A in Prob. 2.58.

2.61 Determine the angles θ_x, θ_y, and θ_z which define the direction of the force exerted on point C in Prob. 2.59.

2.62 Determine the two possible values of θ_y for a force $\mathbf{F}$, (a) if the force forms equal angles with the positive x, y, and z axes, (b) if the force forms equal angles with the positive y and z axes and an angle of 45° with the positive x axis.

2.63 The tension in each of the cables AB and BC is 1350 lb. Determine the components of the resultant of the forces exerted on point B.

2.64 Knowing that the tension in AB is 39 kN, determine the required values of the tension in AC and AD so that the resultant of the three forces applied at A is vertical.

2.65 Knowing that the tension in AC is 28 kN, determine the required values of the tension in AB and AD so that the resultant of the three forces applied at A is vertical.

2.14. Equilibrium of a Particle in Space. According to the definition given in Sec. 2.8, a particle A is in equilibrium if the resultant of all the forces acting on A is zero. The components R_x, R_y, R_z of the resultant are given by the relations (2.31); expressing that the components of the resultant are zero, we write

$$\Sigma F_x = 0 \qquad \Sigma F_y = 0 \qquad \Sigma F_z = 0 \qquad (2.34)$$

Equations (2.34) represent the necessary and sufficient conditions for the equilibrium of a particle in space. They may be used to solve problems dealing with the equilibrium of a particle involving no more than three unknowns.

To solve such problems, we first should draw a free-body diagram showing the particle in equilibrium and *all* the forces acting on it. We may then write the equations of equilibrium (2.34) and solve them for three unknowns. In the more common types of problems, these unknowns will represent (1) the three components of a single force or (2) the magnitude of three forces each of known direction.

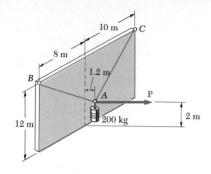

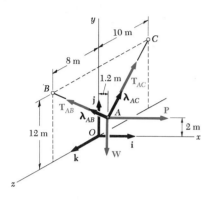

SAMPLE PROBLEM 2.9

A 200-kg cylinder is hung by means of two cables AB and AC, which are attached to the top of a vertical wall. A horizontal force $\mathbf{P}$ perpendicular to the wall holds the cylinder in the position shown. Determine the magnitude of $\mathbf{P}$ and the tension in each cable.

Solution. Point A is chosen as a free body; this point is subjected to four forces, three of which are of unknown magnitude.

Introducing the unit vectors $\mathbf{i}$, $\mathbf{j}$, $\mathbf{k}$, we resolve each force into rectangular components.

$$
\begin{aligned}
\mathbf{P} &= P\mathbf{i} \\
\mathbf{W} &= -mg\mathbf{j} = -(200 \text{ kg})(9.81 \text{ m/s}^2)\mathbf{j} = -(1962 \text{ N})\mathbf{j}
\end{aligned} \tag{1}
$$

In the case of $\mathbf{T}_{AB}$ and $\mathbf{T}_{AC}$, it is necessary first to determine the components and magnitudes of the vectors $\overrightarrow{AB}$ and $\overrightarrow{AC}$. Denoting by $\boldsymbol{\lambda}_{AB}$ the unit vector along AB, we write

$$\overrightarrow{AB} = -(1.2 \text{ m})\mathbf{i} + (10 \text{ m})\mathbf{j} + (8 \text{ m})\mathbf{k} \qquad AB = 12.86 \text{ m}$$

$$\boldsymbol{\lambda}_{AB} = \frac{\overrightarrow{AB}}{12.86 \text{ m}} = -0.0933\mathbf{i} + 0.778\mathbf{j} + 0.622\mathbf{k}$$

$$\mathbf{T}_{AB} = T_{AB}\boldsymbol{\lambda}_{AB} = -0.0933T_{AB}\mathbf{i} + 0.778T_{AB}\mathbf{j} + 0.622T_{AB}\mathbf{k} \tag{2}$$

Denoting by $\boldsymbol{\lambda}_{AC}$ the unit vector along AC, we write in a similar way

$$\overrightarrow{AC} = -(1.2 \text{ m})\mathbf{i} + (10 \text{ m})\mathbf{j} - (10 \text{ m})\mathbf{k} \qquad AC = 14.19 \text{ m}$$

$$\boldsymbol{\lambda}_{AC} = \frac{\overrightarrow{AC}}{14.19 \text{ m}} = -0.0846\mathbf{i} + 0.705\mathbf{j} - 0.705\mathbf{k}$$

$$\mathbf{T}_{AC} = T_{AC}\boldsymbol{\lambda}_{AC} = -0.0846T_{AC}\mathbf{i} + 0.705T_{AC}\mathbf{j} - 0.705T_{AC}\mathbf{k} \tag{3}$$

Equilibrium Condition. Since A is in equilibrium, we must have

$$\Sigma\mathbf{F} = 0: \qquad\qquad \mathbf{T}_{AB} + \mathbf{T}_{AC} + \mathbf{P} + \mathbf{W} = 0$$

or, substituting from (1), (2), (3) for the forces and factoring $\mathbf{i}$, $\mathbf{j}$, $\mathbf{k}$,

$$
\begin{aligned}
&(-0.0933T_{AB} - 0.0846T_{AC} + P)\mathbf{i} \\
&\quad + (0.778T_{AB} + 0.705T_{AC} - 1962 \text{ N})\mathbf{j} + (0.622T_{AB} - 0.705T_{AC})\mathbf{k} = 0
\end{aligned}
$$

Setting the coefficients of $\mathbf{i}$, $\mathbf{j}$, $\mathbf{k}$ equal to zero, we write three scalar equations, which express that the sums of the x, y, and z components of the forces are respectively equal to zero.

$$
\begin{aligned}
(\Sigma F_x = 0:) &\qquad -0.0933T_{AB} - 0.0846T_{AC} + P = 0 \\
(\Sigma F_y = 0:) &\qquad +0.778T_{AB} + 0.705T_{AC} - 1962 \text{ N} = 0 \\
(\Sigma F_z = 0:) &\qquad +0.622T_{AB} - 0.705T_{AC} = 0
\end{aligned}
$$

Solving these equations, we obtain

$$P = 235 \text{ N} \qquad T_{AB} = 1401 \text{ N} \qquad T_{AC} = 1236 \text{ N} \quad \blacktriangleleft$$

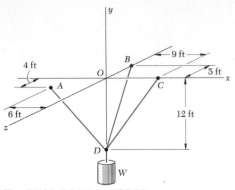

Fig. P2.66, P2.67, and P2.68

PROBLEMS

2.66 A load W is supported by three cables as shown. Determine the value of W, knowing that the tension in cable BD is 975 lb.

2.67 A load W is supported by three cables as shown. Determine the value of W, knowing that the tension in cable CD is 300 lb.

2.68 A load W of magnitude 555 lb is supported by three cables as shown. Determine the tension in each cable.

2.69 Three cables are connected at D, where a 15.60-kN force is applied as shown. Determine the tension in each cable.

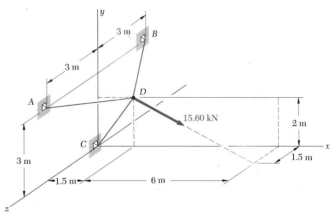

Fig. P2.69 and P2.70

2.70 In addition to the 15.60-kN force shown, a force $\mathbf{P}$ is applied at D in a direction parallel to the y axis. Determine the required magnitude and sense of $\mathbf{P}$ if the tension in cable CD is to be zero.

2.71 A triangular plate of weight 18 lb is supported by three wires as shown. Determine the tension in each wire.

2.72 A 15-kg cylinder is supported by three wires as shown. Determine the tension in each wire when $h = 60$ mm.

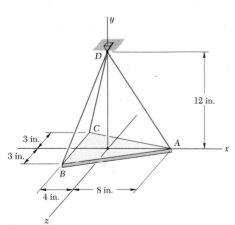

Fig. P2.71

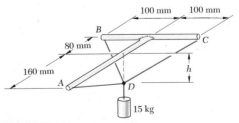

Fig. P2.72

2.73 Solve Prob. 2.72 for the case when $h = 20$ mm. (Explain why this arrangement results in higher tensions.)

2.74 Collar A weighs 5.6 lb and may slide freely on a smooth vertical rod; it is connected to collar B by wire AB. Knowing that the length of wire AB is 18 in., determine the tension in the wire when (*a*) $c = 2$ in., (*b*) $c = 8$ in.

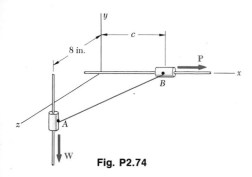

Fig. P2.74

2.75 Solve Prob. 2.74 when (*a*) $c = 14$ in., (*b*) $c = 16$ in.

2.76 The 30-kg crate is held on the incline by the wire AB and by the horizontal force $\mathbf{P}$ which is directed parallel to the z axis. Since the crate is mounted on casters, the force exerted by the incline on the crate is perpendicular to the incline. Determine the magnitude of $\mathbf{P}$ and the tension in wire AB.

2.77 Three cables are connected at D, where an upward force of 30 kN is applied. Determine the tension in each cable.

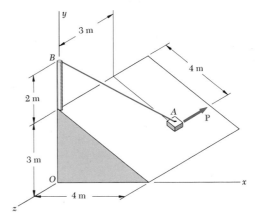

Fig. P2.76

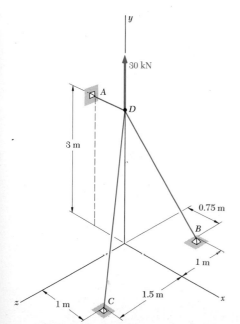

Fig. P2.77

2.78 Two wires are attached to the top of pole *CD*. It is known that the force exerted by the pole is vertical and that the 2500-N force applied to point *C* is horizontal. If the 2500-N force is parallel to the *z* axis ($\alpha = 90°$), determine the tension in each cable.

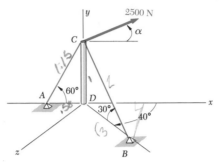

Fig. P2.78

*2.79 In Prob. 2.78, determine (*a*) the value of the angle α for which the tension in cable *AC* is maximum, (*b*) the corresponding tension in each cable.

REVIEW PROBLEMS

2.80 Determine the range of values of *P* for which the resultant of the three forces applied at *A* does not exceed 2400 N.

2.81 Knowing that the magnitude of the force **P** is 4000 N, determine the resultant of the three forces applied at *A*.

2.82 A 12-ft length of steel pipe weighing 600 lb is lifted by a crane cable *CD*. Determine the tension in the cable sling *ACB*, knowing that the length of the sling is (*a*) 15 ft, (*b*) 20 ft.

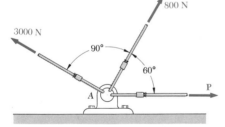

Fig. P2.80 and P2.81

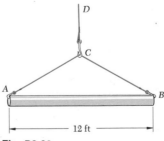

Fig. P2.82

2.83 Three cables are joined at D where two forces $\mathbf{P} = (800\text{ lb})\mathbf{i}$ and $\mathbf{Q} = 0$ are applied. Determine the tension in each cable.

2.84 Three cables are joined at D where two forces $\mathbf{P} = (700\text{ lb})\mathbf{i}$ and $\mathbf{Q} = (300\text{ lb})\mathbf{k}$ are applied. Determine the tension in each cable.

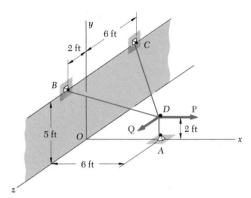

Fig. P2.83, P2.84, and P2.85

2.85 Determine the angles θ_x, θ_y, and θ_z for the force exerted (a) at B by cable BD, (b) at C by cable CD.

2.86 Determine the tension in each cable when $F = 800\text{ N}$ and $\alpha = 90°$.

*2.87 The maximum permissible tension in each of the cables AC and CB is 1500 N. For any given value of α, there is a maximum force $\mathbf{F}$ which may be applied without exceeding the permissible tension in either of the ropes. Considering only cases in which both ropes remain taut, determine the magnitude of the maximum force $\mathbf{F}$ as a function of the angle α.

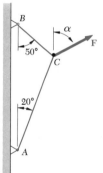

Fig. P2.86 and P2.87

2.88 A 100-kg crate is to be supported by the rope-and-pulley arrangement shown. Determine the required magnitude and direction of the force $\mathbf{T}$.

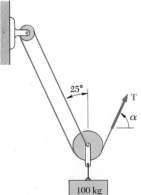

Fig. P2.88

2.89 A 2-kN force acts at the origin in a direction defined by the angles $\theta_y = 46.0°$ and $\theta_z = 80.0°$. It is also known that the x component of the force is positive. Determine the value of θ_x and the components of the force.

2.90 The uniform circular ring weighs 60 lb and is 18 in. in diameter. It is supported by three wires, each of length 15 in. If $\alpha = 150°$, $\beta = 120°$, and $\gamma = 90°$, determine the tension in each wire.

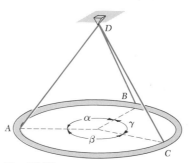

Fig. P2.90

2.91 A painter's scaffold can be supported in the two ways illustrated. If the maximum permissible tension in rope ABC is 100 lb and the maximum permissible tension in rope BD is 150 lb, what is the greatest load that can be carried by each arrangement? (Neglect the effect of the horizontal distance from the building to the scaffold.)

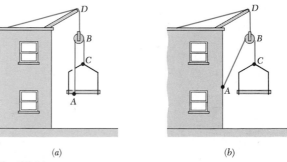

(a)   (b)

Fig. P2.91

Rigid Bodies: Equivalent Systems of Forces

3.1. Rigid Bodies. External and Internal Forces.
In the preceding chapter it was assumed that each of the bodies considered could be treated as a single particle. Such a view, however, is not always possible, and a body, in general, should be treated as a combination of a large number of particles. The size of the body will have to be taken into consideration, as well as the fact that forces will act on different particles and thus will have different points of application.

Most of the bodies considered in elementary mechanics are assumed to be *rigid*, a *rigid body* being defined as one which does not deform. Actual structures and machines, however, are never absolutely rigid and deform under the loads to which they are subjected. But these deformations are usually small and do not appreciably affect the conditions of equilibrium or motion of the structure under consideration. They are important, though, as far as the resistance of the structure to failure is concerned, and are considered in the study of mechanics of materials.

Forces acting on rigid bodies may be separated into two groups: (1) *external forces;* (2) *internal forces.*

1. The *external forces* represent the action of other bodies on the rigid body under consideration. They are entirely responsible for the external behavior of the rigid body. They will either cause it to move or assure that it remains at rest. We shall be concerned only with external forces in this chapter and in Chaps. 4 and 5.
2. The *internal forces* are the forces which hold together the particles forming the rigid body. If the rigid body is structurally composed of several parts, the forces holding the component parts together are also defined as internal forces. Internal forces will be considered in Chaps. 6 and 7.

Fig. 3.1

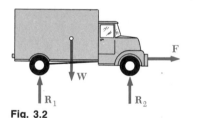

Fig. 3.2

As an example of external forces, we shall consider the forces acting on a disabled truck that men are pulling forward by means of a rope attached to the front bumper (Fig. 3.1). The external forces acting on the truck are shown in a *free-body diagram* (Fig. 3.2). Let us first consider the *weight* of the truck. Although it embodies the effect of the earth's pull on each of the particles forming the truck, the weight may be represented by the single force **W**. The *point of application* of this force, i.e., the point at which the force acts, is defined as the *center of gravity* of the truck. It will be seen in Chap. 5 how centers of gravity may be determined. The weight **W** tends to make the truck move vertically downward. In fact, it would actually cause the truck to move downward, i.e., to fall, if it were not for the presence of the ground. The ground opposes the downward motion of the truck by means of the reactions **R**$_1$ and **R**$_2$. These forces are exerted *by* the ground *on* the truck and must therefore be included among the external forces acting on the truck.

The men pulling on the rope exert the force **F.** The point of application of **F** is on the front bumper. The force **F** tends to make the truck move forward in a straight line and does actually make it move, since no external force opposes this motion. (Rolling resistance has been neglected here for simplicity.) This forward motion of the truck, during which all straight lines remain parallel to themselves (the floor of the truck remains horizontal, and its walls remain vertical), is known as a *translation*. Other forces might cause the truck to move differently. For example, the force exerted by a jack placed under the front axle would cause the truck to pivot about its rear axle. Such a motion is a *rotation*. It may be concluded, therefore, that each of the *external forces* acting on a *rigid body* is capable, if unopposed, of imparting to the rigid body a motion of translation or rotation, or both.

3.2. Principle of Transmissibility. Equivalent Forces.

The *principle of transmissibility* states that the conditions of equilibrium or of motion of a rigid body will remain unchanged if a force **F** acting at a given point of the rigid body is replaced by a force **F′** of the same magnitude and same direction, but acting at a different point, *provided that the two forces have the same line of action* (Fig. 3.3). The two forces **F** and **F′** have the same effect on the rigid body and are said to be *equivalent*. This principle, which states in fact that the action of a force may be *transmitted* along its line of action, is based on experimental evidence. It *cannot* be derived from the properties established so far in this text and must therefore be accepted as an experimental law. However, as we shall see in Sec. 16.5, the principle of transmissibility may be derived from the study of the dynamics of rigid bodies, but this study requires the introduction of all Newton's three laws and of a number of other concepts as well. We shall therefore base our study of the statics of rigid bodies on the three principles introduced so far, i.e., the parallelogram law of addition, Newton's first law, and the principle of transmissibility.

It was indicated in Chap. 2 that the forces acting on a particle could be represented by vectors. These vectors had a well-defined point of application, namely, the particle itself, and were therefore fixed, or bound, vectors. In the case of forces acting on a rigid body, however, the point of application of the force does not matter, as long as the line of action remains unchanged. Thus forces acting on a rigid body must be represented by a different kind of vectors, known as *sliding vectors*, since these vectors may be allowed to slide along their line of action. We should note that all the properties which will be derived in the following sections for the forces acting on a rigid body will be valid more generally for any system of sliding vectors. In order to keep our presentation more intuitive, however, we shall carry it out in terms of physical forces rather than in terms of mathematical sliding vectors.

Returning to the example of the truck, we first observe that the line of action of the force **F** is a horizontal line passing through both the front and the rear bumpers of the truck (Fig. 3.4). Using the principle of transmissibility, we may therefore replace **F** by an *equivalent force* **F′** acting on the rear bumper.

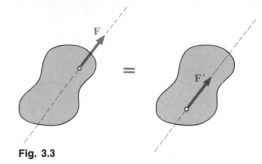

Fig. 3.3

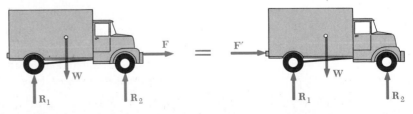

Fig. 3.4

In other words, the conditions of motion are unaffected, and all the other external forces acting on the truck ($\mathbf{W}$, $\mathbf{R}_1$, $\mathbf{R}_2$) remain unchanged if the men push on the rear bumper instead of pulling on the front bumper.

The principle of transmissibility and the concept of equivalent forces have limitations, however. Consider, for example, a short bar AB acted upon by equal and opposite axial forces $\mathbf{P}_1$ and $\mathbf{P}_2$, as shown in Fig. 3.5a. According to the principle of transmissibility, the force $\mathbf{P}_2$ may be replaced by a force $\mathbf{P}_2'$ having the same magnitude, same direction, and same line of action, but acting at A instead of B (Fig. 3.5b). The forces $\mathbf{P}_1$ and $\mathbf{P}_2'$ acting on the same particle may be added according to the rules of Chap. 2, and, being equal and opposite, their sum is found equal to zero. The original system of forces shown in Fig. 3.5a is thus equivalent to no force at all (Fig. 3.5c) from the point of view of the external behavior of the bar.

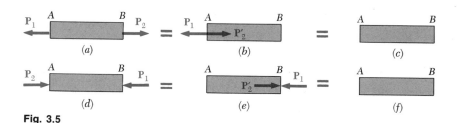

Fig. 3.5

Consider now the two equal and opposite forces $\mathbf{P}_1$ and $\mathbf{P}_2$ acting on the bar AB as shown in Fig. 3.5d. The force $\mathbf{P}_2$ may be replaced by a force $\mathbf{P}_2'$ having the same magnitude, same direction, and same line of action, but acting at B instead of A (Fig. 3.5e). The forces $\mathbf{P}_1$ and $\mathbf{P}_2'$ may then be added, and their sum is found again to be zero (Fig. 3.5f). From the point of view of the mechanics of rigid bodies, the systems shown in Fig. 3.5a and d are thus equivalent. But the *internal forces* and *deformations* produced by the two systems are clearly different. The bar of Fig. 3.5a is in *tension* and, if not absolutely rigid, will increase in length slightly; the bar of Fig. 3.5d is in *compression* and, if not absolutely rigid, will decrease in length slightly. Thus, while the principle of transmissibility may be used freely to determine the conditions of motion or equilibrium of rigid bodies and to compute the external forces acting on these bodies, it should be avoided, or at least used with care, in determining internal forces and deformations.

$V = P \times Q$

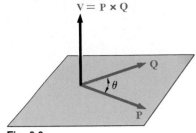

Fig. 3.6

3.3. Vector Product of Two Vectors.
In order to gain a better understanding of the effect of a force on a rigid body, we shall introduce a new concept, the concept of *moment of a force about a point.* This concept will be more clearly understood, and we shall be able to apply it more effectively, if we first add to the mathematical tools at our disposal by defining the *vector product* of two vectors.

The vector product of two vectors **P** and **Q** is defined as the vector **V** which satisfies the following conditions:

1. The line of action of **V** is perpendicular to the plane containing **P** and **Q** (Fig. 3.6).
2. The magnitude of **V** is the product of the magnitudes of **P** and **Q** and of the sine of the angle θ formed by **P** and **Q** (the measure of which will always be 180° or less); we thus have

$$V = PQ \sin \theta \qquad (3.1)$$

3. The sense of **V** is such that a man located at the tip of **V** will observe as counterclockwise the rotation through θ which brings the vector **P** in line with the vector **Q**; note that if **P** and **Q** do not have a common point of application, they should first be redrawn from the same point. The three vectors **P, Q,** and **V**—taken in that order—are said to form a *right-handed triad.*†

As stated above, the vector **V** satisfying these three conditions (which define it uniquely) is referred to as the vector product of **P** and **Q**; it is represented by the mathematical expression

$$V = P \times Q \qquad (3.2)$$

Because of the notation used, the vector product of two vectors **P** and **Q** is also referred to as the *cross product* of **P** and **Q**.

It follows from Eq. (3.1) that, when two vectors **P** and **Q** have either the same direction or opposite directions, their vector product is zero. In the general case when the angle θ formed by the two vectors is neither 0° nor 180°, Eq. (3.1) may be given a simple geometric interpretation: The magnitude V of the vector product of **P** and **Q** measures the area of the parallelogram which has **P** and **Q** for sides (Fig. 3.7). The vector product **P** × **Q** will therefore remain unchanged if we replace **Q** by a vector **Q′** coplanar with **P** and **Q** and such that the line joining the tips of

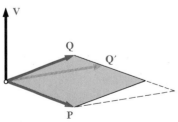

Fig. 3.7

† We should note that the $x, y,$ and z axes used in Chap. 2 form a right-handed system of orthogonal axes and that the unit vectors **i, j, k** defined in Sec. 2.11 form a right-handed orthogonal triad.

Q and **Q′** is parallel to **P**. We write

$$\mathbf{V} = \mathbf{P} \times \mathbf{Q} = \mathbf{P} \times \mathbf{Q}' \qquad (3.3)$$

From the third condition used to define the vector product **V** of **P** and **Q**, namely, the condition stating that **P**, **Q**, and **V** must form a right-handed triad, it follows that vector products *are not commutative*, i.e., **Q** × **P** is not equal to **P** × **Q**. Indeed, we may easily check that **Q** × **P** is represented by the vector −**V** equal and opposite to **V**. We thus write

$$\mathbf{Q} \times \mathbf{P} = -(\mathbf{P} \times \mathbf{Q}) \qquad (3.4)$$

Example. Let us compute the vector product **V** = **P** × **Q** of the vector **P** of magnitude 6 lying in the *zx* plane at an angle of 30° with the *x* axis, and of the vector **Q** of magnitude 4 lying along the *x* axis (Fig. 3.8).

It follows immediately from the definition of the vector product that the vector **V** must lie along the *y* axis, have the magnitude

$$V = PQ \sin \theta = (6)(4) \sin 30° = 12$$

and be directed upward.

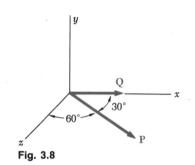

Fig. 3.8

We saw that the commutative property does not apply to vector products. We may wonder whether the *distributive* property holds, i.e., whether the relation

$$\mathbf{P} \times (\mathbf{Q}_1 + \mathbf{Q}_2) = \mathbf{P} \times \mathbf{Q}_1 + \mathbf{P} \times \mathbf{Q}_2 \qquad (3.5)$$

is valid. The answer is *yes*. Many readers are probably willing to accept without formal proof an answer which they intuitively feel is correct. However, since the entire structure of vector algebra and of statics depends upon the relation (3.5), we shall take time out to derive it.

We may, without any loss of generality, assume that **P** is directed along the *y* axis (Fig. 3.9*a*). Denoting by **Q** the sum of **Q**₁ and **Q**₂, we drop perpendiculars from the tips of **Q**, **Q**₁, and **Q**₂ onto the *zx* plane, defining in this way the vectors **Q′**, **Q′**₁, and **Q′**₂. These vectors will be referred to, respectively, as the *projections* of **Q**, **Q**₁, and **Q**₂ on the *zx* plane. Recalling the property expressed by Eq. (3.3), we note that the left-hand member of Eq. (3.5) may be replaced by **P** × **Q′** and that, similarly, the vector products **P** × **Q**₁ and **P** × **Q**₂ may respectively be replaced by **P** × **Q′**₁ and **P** × **Q′**₂. Thus, the relation to be proved may be written in the form

$$\mathbf{P} \times \mathbf{Q}' = \mathbf{P} \times \mathbf{Q}_1' + \mathbf{P} \times \mathbf{Q}_2' \qquad (3.5')$$

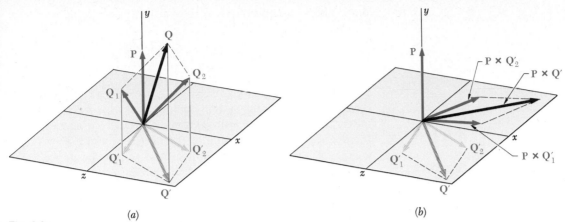

Fig. 3.9

We now observe that $\mathbf{P} \times \mathbf{Q}'$ may be obtained from $\mathbf{Q}'$ by multiplying this vector by the scalar P and rotating it counterclockwise through 90° in the zx plane (Fig. 3.9b); the other two vector products in (3.5') may be obtained in the same manner from $\mathbf{Q}'_1$ and $\mathbf{Q}'_2$, respectively. Now, since the projection of a parallelogram onto an arbitrary plane is a parallelogram, the projection $\mathbf{Q}'$ of the sum $\mathbf{Q}$ of $\mathbf{Q}_1$ and $\mathbf{Q}_2$ must be the sum of the projections $\mathbf{Q}'_1$ and $\mathbf{Q}'_2$ of $\mathbf{Q}_1$ and $\mathbf{Q}_2$ on the same plane (Fig. 3.9a). This relation between the vectors $\mathbf{Q}'$, $\mathbf{Q}'_1$, and $\mathbf{Q}'_2$ will still hold after the three vectors have been multiplied by the scalar P and rotated through 90° (Fig. 3.9b). Thus the relation (3.5') has been proved and we can now be sure that the distributive property holds for vector products.

The third property, the associative property, does not apply to vector products; we have in general

$$(\mathbf{P} \times \mathbf{Q}) \times \mathbf{S} \neq \mathbf{P} \times (\mathbf{Q} \times \mathbf{S}) \tag{3.6}$$

3.4. Vector Products Expressed in Terms of Rectangular Components.

We shall now determine the vector product of any two of the unit vectors, $\mathbf{i}, \mathbf{j}, \mathbf{k}$, defined in Chap. 2. Consider first the product $\mathbf{i} \times \mathbf{j}$ (Fig. 3.10a). Since both vectors have a magnitude equal to one and since they are at a right angle to each other, their vector product will also be a unit vector. This unit vector must be $\mathbf{k}$, since the vectors $\mathbf{i}, \mathbf{j}, \mathbf{k}$ are mutually perpendicular and form a right-handed triad. On the other hand, the product $\mathbf{j} \times \mathbf{i}$ will be equal to $-\mathbf{k}$ since the 90° rotation which brings $\mathbf{j}$ into $\mathbf{i}$ is observed as counterclockwise by a man located at the tip of $-\mathbf{k}$ (Fig. 3.10b). Finally it should be observed that the vector product of a unit vector by itself, such as $\mathbf{i} \times \mathbf{i}$, is equal to zero, since both vectors have the same direc-

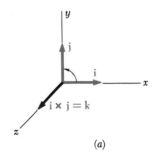

(a)

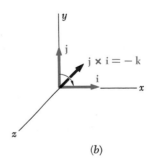

(b)

Fig. 3.10

tion. The vector products of the various possible pairs of unit vectors are

$$
\begin{array}{lll}
\mathbf{i} \times \mathbf{i} = 0 & \mathbf{j} \times \mathbf{i} = -\mathbf{k} & \mathbf{k} \times \mathbf{i} = \mathbf{j} \\
\mathbf{i} \times \mathbf{j} = \mathbf{k} & \mathbf{j} \times \mathbf{j} = 0 & \mathbf{k} \times \mathbf{j} = -\mathbf{i} \\
\mathbf{i} \times \mathbf{k} = -\mathbf{j} & \mathbf{j} \times \mathbf{k} = \mathbf{i} & \mathbf{k} \times \mathbf{k} = 0
\end{array}
\qquad (3.7)
$$

By arranging in a circle and in counterclockwise order the three letters representing the unit vectors (Fig. 3.11), we may simplify the determination of the sign of the vector product of two unit vectors: The product of two unit vectors will be positive if they follow each other in counterclockwise order, and negative otherwise.

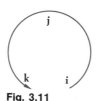

Fig. 3.11

We may now easily express the vector product $\mathbf{V}$ of two given vectors $\mathbf{P}$ and $\mathbf{Q}$ in terms of the rectangular components of these vectors. Resolving $\mathbf{P}$ and $\mathbf{Q}$ into components, we first write

$$
\mathbf{V} = \mathbf{P} \times \mathbf{Q} = (P_x\mathbf{i} + P_y\mathbf{j} + P_z\mathbf{k}) \times (Q_x\mathbf{i} + Q_y\mathbf{j} + Q_z\mathbf{k})
$$

Making use of the distributive property, we express $\mathbf{V}$ as the sum of vector products such as $P_x\mathbf{i} \times Q_y\mathbf{j}$. Observing that each of the expressions obtained is equal to the vector product of two unit vectors, such as $\mathbf{i} \times \mathbf{j}$, multiplied by the product of two scalars, such as P_xQ_y, and recalling the identities (3.7), we obtain, after factoring $\mathbf{i}$, $\mathbf{j}$, and $\mathbf{k}$,

$$
\mathbf{V} = (P_yQ_z - P_zQ_y)\mathbf{i} + (P_zQ_x - P_xQ_z)\mathbf{j} + (P_xQ_y - P_yQ_x)\mathbf{k} \quad (3.8)
$$

The rectangular components of the vector product $\mathbf{V}$ are thus found to be

$$
\begin{aligned}
V_x &= P_yQ_z - P_zQ_y \\
V_y &= P_zQ_x - P_xQ_z \\
V_z &= P_xQ_y - P_yQ_x
\end{aligned}
\qquad (3.9)
$$

Returning to Eq. (3.8), we observe that its right-hand member represents the expansion of a determinant. The vector product $\mathbf{V}$ may thus be expressed in the following form, more easily memorized:†

$$
\mathbf{V} = \begin{vmatrix} \mathbf{i} & \mathbf{j} & \mathbf{k} \\ P_x & P_y & P_z \\ Q_x & Q_y & Q_z \end{vmatrix}
\qquad (3.10)
$$

† Any determinant consisting of three rows and three columns may be evaluated by repeating the first and second columns and forming products along each diagonal line. The sum of the products obtained along the colored lines is then subtracted from the sum of the products obtained along the black lines.

3.5. Moment of a Force about a Point.

Let us now consider a force **F** acting on a rigid body (Fig. 3.12). As we know, the force **F** is represented by a vector which defines its magnitude and direction. However, the effect of the force on the rigid body depends also upon its point of application A. The position of A may be conveniently defined by the vector **r** which joins the fixed reference point O with A; this vector is known as the *position vector* of A.† The position vector **r** and the force **F** define the plane shown in Fig. 3.12.

We shall define the *moment of* **F** *about* O as the vector product of **r** and **F**:

$$\mathbf{M}_O = \mathbf{r} \times \mathbf{F} \tag{3.11}$$

According to the definition of the vector product given in Sec. 3.3, the moment $\mathbf{M}_O$ must be perpendicular to the plane containing O and the force **F**. The sense of $\mathbf{M}_O$ is defined by the sense of the rotation which would bring the vector **r** in line with the vector **F**; this rotation should be observed as *counterclockwise* by an observer located at the tip of $\mathbf{M}_O$. Another way of defining the sense of $\mathbf{M}_O$ is furnished by the *right-hand rule:* Close your right hand and hold it so that your fingers are curled in the sense of the rotation that **F** would impart to the rigid body about a fixed axis directed along the line of action of $\mathbf{M}_O$; your thumb will indicate the sense of the moment $\mathbf{M}_O$.

Finally, denoting by θ the angle between the lines of action of the position vector **r** and the force **F**, we find that the magnitude of the moment of **F** about O is

$$M_O = rF \sin \theta = Fd \tag{3.12}$$

where d represents the perpendicular distance from O to the line of action of **F**. Since the tendency of a force **F** to make a rigid body rotate about a fixed axis perpendicular to the force depends upon the distance of **F** from that axis, as well as upon the magnitude of **F**, we note that *the magnitude of* $\mathbf{M}_O$ *measures the tendency of the force* **F** *to make the rigid body rotate about a fixed axis directed along* $\mathbf{M}_O$.

In the SI system of units, where a force is expressed in newtons (N) and a distance in meters (m), the moment of a force will be

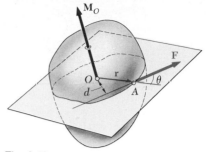

Fig. 3.12

---

† We may easily verify that position vectors obey the law of vector addition and, thus, are truly vectors. Consider, for example, the position vectors **r** and **r'** of A with respect to two reference points O and O', and the position vector **s** of O with respect to O' (Fig. 3.40a, Sec. 3.15). We check that the position vector $\mathbf{r}' = \overrightarrow{O'A}$ may be obtained from the position vectors $\mathbf{s} = \overrightarrow{O'O}$ and $\mathbf{r} = \overrightarrow{OA}$ by applying the triangle rule for the addition of vectors.

expressed in newton-meters (N · m). In the U.S. customary system of units, where a force is expressed in pounds and a distance in feet or inches, the moment of a force will be expressed in lb · ft or lb · in.

We may observe that the moment M_O of a force about a point, while it depends upon the magnitude, the line of action, and the sense of the force, does *not* depend upon the actual position of the point of application of the force along its line of action. Conversely, the moment M_O of a force F does not characterize the position of the point of application of F.

However, as we shall see presently, the moment M_O of a force F of given magnitude and direction *completely defines the line of action of* F. Indeed, the line of action of F must lie in a plane through O perpendicular to the moment M_O; its distance d from O must be equal to the quotient M_O/F of the magnitudes of M_O and F; and the sense of M_O determines whether the line of action of F is to be drawn on one side or the other of the point O.

We recall from Sec. 3.2 that the principle of transmissibility states that two forces F and F' are equivalent (i.e., have the same effect on a rigid body) if they have the same magnitude, same direction, and same line of action. This principle may now be restated as follows: *Two forces F and F' are equivalent if, and only if, they are equal* (i.e., have the same magnitude and same direction) *and have equal moments about a given point O*. The necessary and sufficient condition for two forces F and F' to be equivalent is thus

$$F = F' \qquad \text{and} \qquad M_O = M'_O \tag{3.13}$$

We should observe that it follows from this statement that if the relations (3.13) hold for a given point O, they will hold for any other point.

Problems Involving Only Two Dimensions. Many applications deal with two-dimensional structures, i.e., structures which have length and breadth, but only negligible depth, and which are subjected to forces contained in the plane of the structure. Two-dimensional structures and the forces acting on them may be readily represented on a sheet of paper or on a blackboard. Their analysis is therefore considerably simpler than that of three-dimensional structures and forces.

Consider, for example, a rigid slab acted upon by a force F (Fig. 3.13). The moment of F about a point O chosen in the plane of the figure is represented by a vector M_O perpendicular to that plane and of magnitude Fd. In the case of Fig. 3.13a the vector M_O points *out* of the paper, while in the case of Fig. 3.13b it points *into* the paper. As we look at the figure, we observe the action of **F** in the first case as counterclockwise, and in the second case as clockwise. Therefore, it is natural to refer to the sense of the moment of **F** about O in Fig. 3.13a as counterclockwise $\circlearrowleft$, and in Fig. 3.13b as clockwise $\circlearrowright$.

Since the moment of a force **F** acting in the plane of the figure must be perpendicular to that plane, we need only specify the *magnitude* and the *sense* of the moment of **F** about O. This may be done by assigning to the magnitude M_O of the moment a positive or negative sign, according to whether the vector $\mathbf{M}_O$ points out of or into the paper.

3.6. Varignon's Theorem. The distributive property of vector products may be used to determine the moment of the resultant of several *concurrent forces.* If several forces $\mathbf{F}_1, \mathbf{F}_2, \ldots$ are applied at the same point A (Fig. 3.14), and if we denote by **r** the position vector of A, it follows immediately from formula (3.5) that

$$\mathbf{r} \times (\mathbf{F}_1 + \mathbf{F}_2 + \cdots) = \mathbf{r} \times \mathbf{F}_1 + \mathbf{r} \times \mathbf{F}_2 + \cdots \quad (3.14)$$

In words, *the moment about a given point O of the resultant of several concurrent forces is equal to the sum of the moments of the various forces about the same point O.* This property was originally established by the French mathematician Varignon (1654–1722), long before the introduction of vector algebra, and is known as *Varignon's theorem.*

The relation (3.14) makes it possible to replace the direct determination of the moment of a force **F** by the determination of the moments of two or more component forces. As we shall see in the next section, **F** will generally be resolved into components parallel to the coordinate axes. However, it may be found more expeditious in some instances to resolve **F** into components which are not parallel to the coordinate axes (see Sample Prob. 3.3).

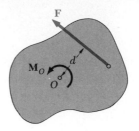

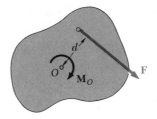

(a) $M_O = + Fd$

(b) $M_O = - Fd$

Fig. 3.13

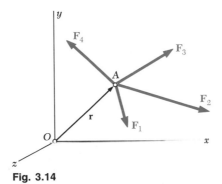

Fig. 3.14

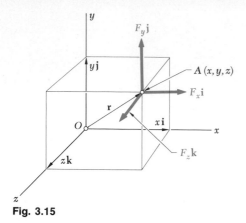

Fig. 3.15

3.7. Rectangular Components of the Moment of a Force.

In general, the determination of the moment of a force in space will be considerably simplified if the force and the position vector of its point of application are resolved into rectangular x, y, and z components. Consider, for example, the moment $\mathbf{M}_O$ about O of a force $\mathbf{F}$ of components F_x, F_y, and F_z, applied at a point A of coordinates x, y, and z (Fig. 3.15). Observing that the components of the position vector $\mathbf{r}$ are respectively equal to the coordinates x, y, and z of the point A, we write

$$\mathbf{r} = x\mathbf{i} + y\mathbf{j} + z\mathbf{k} \tag{3.15}$$
$$\mathbf{F} = F_x\mathbf{i} + F_y\mathbf{j} + F_z\mathbf{k} \tag{3.16}$$

Substituting for $\mathbf{r}$ and $\mathbf{F}$ from (3.15) and (3.16) into

$$\mathbf{M}_O = \mathbf{r} \times \mathbf{F} \tag{3.11}$$

and recalling the results obtained in Sec. 3.4, we write the moment $\mathbf{M}_O$ of $\mathbf{F}$ about O in the form

$$\mathbf{M}_O = M_x\mathbf{i} + M_y\mathbf{j} + M_z\mathbf{k} \tag{3.17}$$

where the components M_x, M_y, and M_z are defined by the relations

$$\begin{aligned} M_x &= yF_z - zF_y \\ M_y &= zF_x - xF_z \\ M_z &= xF_y - yF_x \end{aligned} \tag{3.18}$$

As we shall see in Sec. 3.10, the scalar components M_x, M_y, and M_z of the moment $\mathbf{M}_O$ measure the tendency of the force $\mathbf{F}$ to impart to a rigid body a motion of rotation about the x, y, and z axes, respectively. Substituting from (3.18) into (3.17), we may also write $\mathbf{M}_O$ in the form of the determinant

$$\mathbf{M}_O = \begin{vmatrix} \mathbf{i} & \mathbf{j} & \mathbf{k} \\ x & y & z \\ F_x & F_y & F_z \end{vmatrix} \tag{3.19}$$

To compute the moment $\mathbf{M}_B$ about an arbitrary point B of a force $\mathbf{F}$ applied at A (Fig. 3.16), we must use the vector $\Delta\mathbf{r} = \mathbf{r}_A - \mathbf{r}_B$ instead of the vector $\mathbf{r}$. We write

$$\mathbf{M}_B = \Delta\mathbf{r} \times \mathbf{F} = (\mathbf{r}_A - \mathbf{r}_B) \times \mathbf{F} \tag{3.20}$$

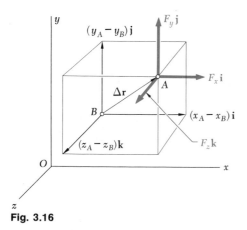

Fig. 3.16

or, using the determinant form,

$$M_B = \begin{vmatrix} i & j & k \\ \Delta x & \Delta y & \Delta z \\ F_x & F_y & F_z \end{vmatrix} \quad (3.21)$$

where Δx, Δy, Δz are the components of the vector Δr joining A and B:

$$\Delta x = x_A - x_B \qquad \Delta y = y_A - y_B \qquad \Delta z = z_A - z_B$$

In the case of *problems involving only two dimensions*, the force F may be assumed to lie in the xy plane (Fig. 3.17).

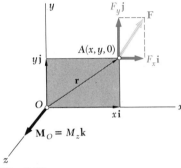

Fig. 3.17

Carrying $z = 0$ and $F_z = 0$ into the relations (3.19), we obtain

$$M_O = (xF_y - yF_x)k$$

We check that the moment of F about O is perpendicular to the plane of the figure and that it is completely defined by the scalar

$$M_O = M_z = xF_y - yF_x \quad (3.22)$$

As noted earlier, a positive value for M_O indicates that the vector M_O points out of the paper (the force F tends to rotate the body counterclockwise about O), and a negative value that the vector M_O points into the paper (the force F tends to rotate the body clockwise about O).

To compute the moment about $B(x_B, y_B)$ of a force lying in the xy plane and applied at $A(x_A, y_A)$ (Fig. 3.18), we carry $\Delta z = 0$ and $F_z = 0$ in the relations (3.21) and check that the vector M_B is perpendicular to the xy plane and is defined in magnitude and sense by the scalar

$$M_B = (x_A - x_B)F_y - (y_A - y_B)F_x \quad (3.23)$$

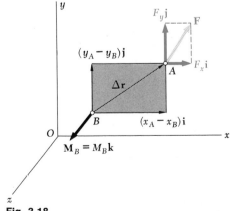

Fig. 3.18

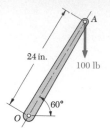

SAMPLE PROBLEM 3.1

A 100-lb vertical force is applied to the end of a lever which is attached to a shaft at O. Determine (a) the moment of the 100-lb force about O; (b) the magnitude of the horizontal force applied at A which creates the same moment about O; (c) the smallest force applied at A which creates the same moment about O; (d) how far from the shaft a 240-lb vertical force must act to create the same moment about O; (e) whether any one of the forces obtained in parts b, c, and d is equivalent to the original force.

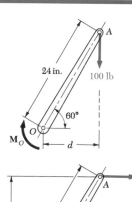

a. **Moment about O.** The perpendicular distance from O to the line of action of the 100-lb force is

$$d = (24 \text{ in.}) \cos 60° = 12 \text{ in.}$$

The magnitude of the moment about O of the 100-lb force is

$$M_O = Fd = (100 \text{ lb})(12 \text{ in.}) = 1200 \text{ lb} \cdot \text{in.}$$

Since the force tends to rotate the lever clockwise about O, the moment will be represented by a vector $\mathbf{M}_O$ perpendicular to the plane of the figure and pointing *into* the paper. We express this fact by writing

$$\mathbf{M}_O = 1200 \text{ lb} \cdot \text{in.} \; \downarrow \quad \blacktriangleleft$$

b. **Horizontal Force.** In this case, we have

$$d = (24 \text{ in.}) \sin 60° = 20.8 \text{ in.}$$

Since the moment about O must be 1200 lb · in., we write

$$M_O = Fd \qquad 1200 \text{ lb} \cdot \text{in.} = F(20.8 \text{ in.})$$
$$F = 57.7 \text{ lb} \qquad\qquad \mathbf{F} = 57.7 \text{ lb} \rightarrow \quad \blacktriangleleft$$

c. **Smallest Force.** Since $M_O = Fd$, the smallest value of F occurs when d is maximum. We choose the force perpendicular to OA and find $d = 24$ in.; thus

$$M_O = Fd \qquad 1200 \text{ lb} \cdot \text{in.} = F(24 \text{ in.})$$
$$F = 50 \text{ lb} \qquad\qquad \mathbf{F} = 50 \text{ lb} \; \diagdown 30° \quad \blacktriangleleft$$

d. **240-lb Vertical Force.** In this case $M_O = Fd$ yields

$$1200 \text{ lb} \cdot \text{in.} = (240 \text{ lb})d \qquad d = 5 \text{ in.}$$
but $\qquad\qquad OB \cos 60° = d \qquad\qquad OB = 10 \text{ in.} \quad \blacktriangleleft$

e. None of the forces considered in parts b, c, and d is equivalent to the original 100-lb force. Although they have the same moment about O, they have different x and y components. In other words, although each force tends to rotate the shaft in the same manner, each causes the lever to pull on the shaft in a different way.

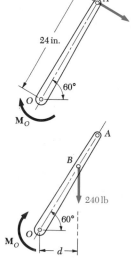

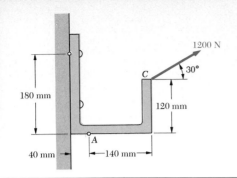

SAMPLE PROBLEM 3.2

A force of 1200 N acts on a bracket as shown. Determine the moment $\mathbf{M}_A$ of the force about A.

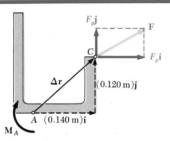

Solution. The moment $\mathbf{M}_A$ is obtained by forming the vector product

$$\mathbf{M}_A = \Delta\mathbf{r} \times \mathbf{F}$$

where $\Delta\mathbf{r}$ is the vector joining A and C. Resolving $\Delta\mathbf{r}$ and $\mathbf{F}$ into rectangular components, we write

$$\Delta\mathbf{r} = \Delta x\mathbf{i} + \Delta y\mathbf{j} = (0.140 \text{ m})\mathbf{i} + (0.120 \text{ m})\mathbf{j}$$
$$\mathbf{F} = F_x\mathbf{i} + F_y\mathbf{j} = (1200 \text{ N}) \cos 30°\mathbf{i} + (1200 \text{ N}) \sin 30°\mathbf{j}$$
$$= (1039 \text{ N})\mathbf{i} + (600 \text{ N})\mathbf{j}$$

Recalling the relations (3.7) for the cross products of unit vectors, we obtain

$$\mathbf{M}_A = \Delta\mathbf{r} \times \mathbf{F} = [(0.140 \text{ m})\mathbf{i} + (0.120 \text{ m})\mathbf{j}] \times [(1039 \text{ N})\mathbf{i} + (600 \text{ N})\mathbf{j}]$$
$$= (84.0 \text{ N} \cdot \text{m})\mathbf{k} - (124.7 \text{ N} \cdot \text{m})\mathbf{k}$$
$$\mathbf{M}_A = -(40.7 \text{ N} \cdot \text{m})\mathbf{k} \qquad\qquad \mathbf{M}_A = 40.7 \text{ N} \cdot \text{m} \downarrow \quad \blacktriangleleft$$

The moment $\mathbf{M}_A$ is a vector perpendicular to the plane of the figure and pointing *into* the paper.

SAMPLE PROBLEM 3.3

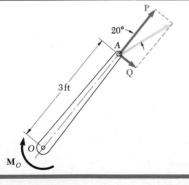

A 30-lb force acts on the end of the 3-ft lever as shown. Determine the moment of the force about O.

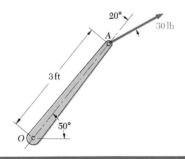

Solution. The force is replaced by two components, one component $\mathbf{P}$ in the direction of OA and one component $\mathbf{Q}$ perpendicular to OA. Since O is on the line of action of $\mathbf{P}$, the moment of $\mathbf{P}$ about O is zero and the moment of the 30-lb force reduces to the moment of $\mathbf{Q}$, which is clockwise and, thus, represented by a negative scalar.

$$Q = (30 \text{ lb}) \sin 20° = 10.26 \text{ lb}$$
$$M_O = -Q(3 \text{ ft}) = -(10.26 \text{ lb})(3 \text{ ft}) = -30.8 \text{ lb} \cdot \text{ft}$$

Since the value obtained for the scalar M_O is negative, the moment $\mathbf{M}_O$ points *into* the paper. We write

$$M_O = 30.8 \text{ lb} \cdot \text{ft} \downarrow \quad \blacktriangleleft$$

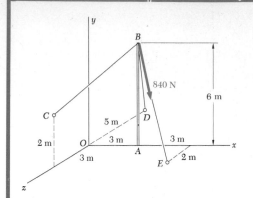

A pole AB, 6 m long, is held by three guy wires as shown. Determine the moment about C of the force exerted by wire BE on point B. The tension T in wire BE is known to be 840 N.

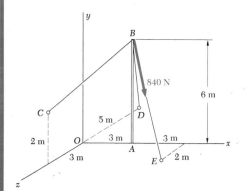

Solution. The moment $\mathbf{M}_C$ about C of the force exerted by wire BE on point B is obtained by forming the vector product

$$\mathbf{M}_C = \Delta \mathbf{r} \times \mathbf{F} \tag{1}$$

where $\Delta \mathbf{r}$ is the vector $\overrightarrow{CB}$ joining points C and B,

$$\Delta \mathbf{r} = \overrightarrow{CB} = (3 \text{ m})\mathbf{i} + (4 \text{ m})\mathbf{j} - (3 \text{ m})\mathbf{k} \tag{2}$$

and where $\mathbf{F}$ is the 840-N force directed along BE. Introducing the unit vector $\boldsymbol{\lambda} = \overrightarrow{BE}/BE$, we write

$$\mathbf{F} = F\boldsymbol{\lambda} = (840 \text{ N})\frac{\overrightarrow{BE}}{BE} \tag{3}$$

Resolving the vector $\overrightarrow{BE}$ into rectangular components, we have

$$\overrightarrow{BE} = (3 \text{ m})\mathbf{i} - (6 \text{ m})\mathbf{j} + (2 \text{ m})\mathbf{k} \qquad BE = 7 \text{ m}$$

Substituting into (3), we obtain

$$\mathbf{F} = \frac{840 \text{ N}}{7 \text{ m}}[(3 \text{ m})\mathbf{i} - (6 \text{ m})\mathbf{j} + (2 \text{ m})\mathbf{k}]$$

$$\mathbf{F} = (360 \text{ N})\mathbf{i} - (720 \text{ N})\mathbf{j} + (240 \text{ N})\mathbf{k} \tag{4}$$

Substituting for $\Delta \mathbf{r}$ and $\mathbf{F}$ from (2) and (4) into (1), and recalling the relations (3.7), we obtain

$$\begin{aligned}
\mathbf{M}_C = \Delta \mathbf{r} \times \mathbf{F} &= (3\mathbf{i} + 4\mathbf{j} - 3\mathbf{k}) \times (360\mathbf{i} - 720\mathbf{j} + 240\mathbf{k}) \\
&= (3)(-720)\mathbf{k} + (3)(240)(-\mathbf{j}) + (4)(360)(-\mathbf{k}) \\
&\quad + (4)(240)\mathbf{i} + (-3)(360)\mathbf{j} + (-3)(-720)(-\mathbf{i}) \\
\mathbf{M}_C &= -(1200 \text{ N} \cdot \text{m})\mathbf{i} - (1800 \text{ N} \cdot \text{m})\mathbf{j} - (3600 \text{ N} \cdot \text{m})\mathbf{k} \quad \blacktriangleleft
\end{aligned}$$

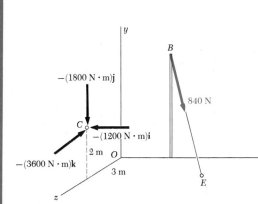

Alternate Solution. As indicated in Sec. 3.7, the moment $\mathbf{M}_C$ may be expressed in the form of a determinant. Substituting for Δx, Δy, Δz the differences between the coordinates of B and C, and for F_x, F_y, F_z the values obtained above for the components of the 840-N force, we have

$$\mathbf{M}_C = \begin{vmatrix} \mathbf{i} & \mathbf{j} & \mathbf{k} \\ \Delta x & \Delta y & \Delta z \\ F_x & F_y & F_z \end{vmatrix} = \begin{vmatrix} \mathbf{i} & \mathbf{j} & \mathbf{k} \\ 3 & 4 & -3 \\ 360 & -720 & 240 \end{vmatrix}$$

$$\mathbf{M}_C = -(1200 \text{ N} \cdot \text{m})\mathbf{i} - (1800 \text{ N} \cdot \text{m})\mathbf{j} - (3600 \text{ N} \cdot \text{m})\mathbf{k} \quad \blacktriangleleft$$

PROBLEMS

3.1 A 150-N force is applied to the control lever at *A*. Knowing that the distance *AB* is 250 mm, determine the moment of the force about *B* when α is 50°.

3.2 Knowing that the distance *AB* is 250 mm, determine the maximum moment about *B* which can be caused by the 150-N force. In what direction should the force act?

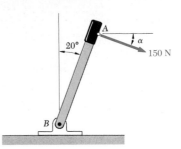

Fig. P3.1 and P3.2

3.3 A 450-N force is applied at *A* as shown. Determine (a) the moment of the 450-N force about *D*, (b) the smallest force applied at *B* which creates the same moment about *D*.

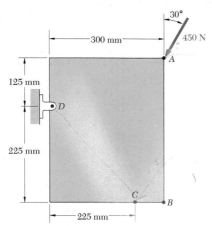

Fig. P3.3 and P3.4

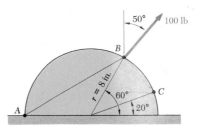

Fig. P3.5 and P3.8

3.4 A 450-N force is applied at *A*. Determine (a) the moment of the 450-N force about *D*, (b) the magnitude and sense of the horizontal force applied at *C* which creates the same moment about *D*, (c) the smallest force applied at *C* which creates the same moment about *D*.

3.5 and 3.6 Compute the moment of the 100-lb force about *A*, (a) by using the definition of the moment of a force, (b) by resolving the force into horizontal and vertical components, (c) by resolving the force into components along *AB* and in the direction perpendicular to *AB*.

3.7 and 3.8 Determine the moment of the 100-lb force about *C*.

3.9 In Prob. 3.7, determine the perpendicular distance from *C* to the line of action of the 100-lb force.

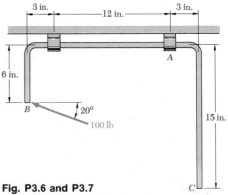

Fig. P3.6 and P3.7

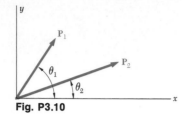

Fig. P3.10

3.10 Form the vector product $\mathbf{P}_1 \times \mathbf{P}_2$ and use the result obtained to prove the identity $\sin(\theta_1 - \theta_2) = \sin\theta_1 \cos\theta_2 - \cos\theta_1 \sin\theta_2$.

3.11 A force $\mathbf{F} = F_x\mathbf{i} + F_y\mathbf{j}$ acts at a point of coordinates x and y. Derive an expression for the perpendicular distance d from the line of action of $\mathbf{F}$ to the origin O of the system of coordinates.

3.12 The line of action of a force $\mathbf{P}$ passes through the two points $A\,(x_1, y_1)$ and $B\,(x_2, y_2)$. If the force is directed from A to B, determine the moment of the force about the origin.

3.13 Determine the moment about the origin O of the force $\mathbf{F} = -2\mathbf{i} - 3\mathbf{j} + 5\mathbf{k}$ which acts at a point A. Assume that the position vector of A is (a) $\mathbf{r} = \mathbf{i} + \mathbf{j} + \mathbf{k}$, (b) $\mathbf{r} = 4\mathbf{i} + 6\mathbf{j} - 10\mathbf{k}$, (c) $\mathbf{r} = 4\mathbf{i} + 3\mathbf{j} - 5\mathbf{k}$.

3.14 Determine the moment about the origin O of the force $\mathbf{F} = 4\mathbf{i} + 10\mathbf{j} + 6\mathbf{k}$ which acts at a point A. Assume that the position vector of A is (a) $\mathbf{r} = 2\mathbf{i} - 3\mathbf{j} + 4\mathbf{k}$, (b) $\mathbf{r} = 2\mathbf{i} + 6\mathbf{j} + 3\mathbf{k}$, and (c) $\mathbf{r} = 2\mathbf{i} + 5\mathbf{j} + 6\mathbf{k}$.

3.15 A precast-concrete wall section is temporarily held by cables as shown. Knowing that the tension in cable BC is 900 lb, determine the moment about the origin of coordinates O of the force exerted on the wall section at C.

3.16 Knowing that the tension in cable AB is 700 lb, determine the moment about the origin of coordinates O of the force exerted on the wall section at A.

3.17 A force $\mathbf{P}$ of magnitude 360 N is applied at point B as shown. Determine the moment of $\mathbf{P}$ about (a) the origin of coordinates O, (b) point D.

3.18 A force $\mathbf{Q}$ of magnitude 450 N is applied at point C as shown. Determine the moment of $\mathbf{Q}$ about (a) the origin of coordinates O, (b) point D.

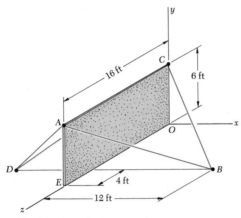

Fig. P3.15 and P3.16

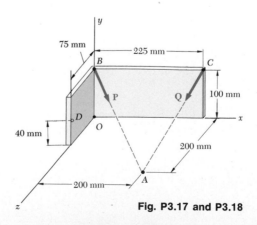

Fig. P3.17 and P3.18

3.19 The line of action of the force **P** of magnitude 420 lb passes through the two points A and B as shown. Compute the moment of **P** about O using the position vector (a) of point A, (b) of point B.

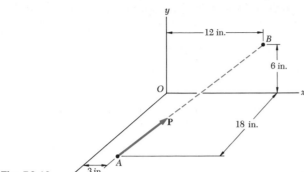

Fig. P3.19

3.20 A force **P** of magnitude 200 N acts along the diagonal BC of the bent plate shown. Determine the moment of **P** about point E.

3.21 In Prob. 3.20, determine the perpendicular distance from the line of action of **P** to point E.

3.22 In Prob. 3.19, determine the perpendicular distance from the line of action of **P** to the origin O.

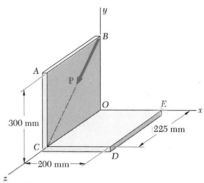

Fig. P3.20

3.23 A single force **F** acts at a point A of coordinates $x = y = z = a$. Show that $M_x + M_y + M_z = 0$; i.e., show that the *algebraic* sum of the rectangular components of the moment of **F** about O is zero.

3.24 In Sample Prob. 3.4, determine the perpendicular distance from the line BE to the point C.

3.8. Scalar Product of Two Vectors.

We shall now expand our knowledge of vector algebra and introduce the *scalar product* of two vectors.

The scalar product of two vectors **P** and **Q** is defined as the product of the magnitudes of **P** and **Q** and of the cosine of the angle θ formed by **P** and **Q** (Fig. 3.19). The scalar product of **P** and **Q** is denoted by **P · Q**. We write therefore

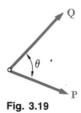

Fig. 3.19

$$\mathbf{P} \cdot \mathbf{Q} = PQ \cos \theta \qquad (3.24)$$

Note that the expression just defined is not a vector, but a *scalar*, which explains the name *scalar product*; because of the notation used, **P · Q** is also referred to as the *dot product* of the vectors **P** and **Q**.

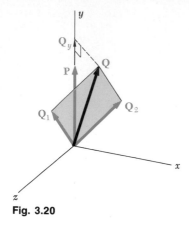

Fig. 3.20

It follows from its very definition that the scalar product of two vectors is *commutative*, i.e., that

$$\mathbf{P} \cdot \mathbf{Q} = \mathbf{Q} \cdot \mathbf{P} \tag{3.25}$$

To prove that the scalar product is also *distributive*, we must prove the relation

$$\mathbf{P} \cdot (\mathbf{Q}_1 + \mathbf{Q}_2) = \mathbf{P} \cdot \mathbf{Q}_1 + \mathbf{P} \cdot \mathbf{Q}_2 \tag{3.26}$$

We may, without any loss of generality, assume that $\mathbf{P}$ is directed along the y axis (Fig. 3.20). Denoting by $\mathbf{Q}$ the sum of $\mathbf{Q}_1$ and $\mathbf{Q}_2$, and by θ_y the angle $\mathbf{Q}$ forms with the y axis, we express the left-hand member of (3.26) as follows:

$$\mathbf{P} \cdot (\mathbf{Q}_1 + \mathbf{Q}_2) = \mathbf{P} \cdot \mathbf{Q} = PQ \cos \theta_y = PQ_y \tag{3.27}$$

where Q_y is the y component of $\mathbf{Q}$. We may, in a similar way, express the right-hand member of (3.26) as

$$\mathbf{P} \cdot \mathbf{Q}_1 + \mathbf{P} \cdot \mathbf{Q}_2 = P(Q_1)_y + P(Q_2)_y \tag{3.28}$$

Since $\mathbf{Q}$ is the sum of $\mathbf{Q}_1$ and $\mathbf{Q}_2$, its y component must be equal to the sum of the y components of $\mathbf{Q}_1$ and $\mathbf{Q}_2$. Thus, the expressions obtained in (3.27) and (3.28) are equal and the relation (3.26) has been proved.

As far as the third property—the associative property—is concerned, we note that this property cannot apply to scalar products. Indeed, $(\mathbf{P} \cdot \mathbf{Q}) \cdot \mathbf{S}$ has no meaning, since $\mathbf{P} \cdot \mathbf{Q}$ is not a vector, but a scalar.

We shall now express the scalar product of two vectors $\mathbf{P}$ and $\mathbf{Q}$ in terms of their rectangular components. Resolving $\mathbf{P}$ and $\mathbf{Q}$ into components, we first write

$$\mathbf{P} \cdot \mathbf{Q} = (P_x \mathbf{i} + P_y \mathbf{j} + P_z \mathbf{k}) \cdot (Q_x \mathbf{i} + Q_y \mathbf{j} + Q_z \mathbf{k})$$

Making use of the distributive property, we express $\mathbf{P} \cdot \mathbf{Q}$ as the sum of scalar products such as $P_x \mathbf{i} \cdot Q_x \mathbf{i}$ and $P_x \mathbf{i} \cdot Q_y \mathbf{j}$. But we may easily check from the definition of the scalar product that the scalar products of the unit vectors are either zero or one.

$$\begin{array}{lll} \mathbf{i} \cdot \mathbf{i} = 1 & \mathbf{j} \cdot \mathbf{j} = 1 & \mathbf{k} \cdot \mathbf{k} = 1 \\ \mathbf{i} \cdot \mathbf{j} = 0 & \mathbf{j} \cdot \mathbf{k} = 0 & \mathbf{k} \cdot \mathbf{i} = 0 \end{array} \tag{3.29}$$

Thus, the expression obtained for $\mathbf{P} \cdot \mathbf{Q}$ reduces to

$$\mathbf{P} \cdot \mathbf{Q} = P_x Q_x + P_y Q_y + P_z Q_z \tag{3.30}$$

In the particular case when $\mathbf{P}$ and $\mathbf{Q}$ are equal, we check that

$$\mathbf{P} \cdot \mathbf{P} = P_x^2 + P_y^2 + P_z^2 = P^2 \tag{3.31}$$

Applications

1. *Angle formed by two given vectors.* Let two vectors be given in terms of their components.

$$\mathbf{P} = P_x\mathbf{i} + P_y\mathbf{j} + P_z\mathbf{k}$$
$$\mathbf{Q} = Q_x\mathbf{i} + Q_y\mathbf{j} + Q_z\mathbf{k}$$

To determine the angle formed by the two vectors we shall equate the expressions obtained in (3.24) and (3.30) for their scalar product:

$$PQ \cos\theta = P_xQ_x + P_yQ_y + P_zQ_z$$

Solving for $\cos\theta$, we write

$$\cos\theta = \frac{P_xQ_x + P_yQ_y + P_zQ_z}{PQ} \qquad (3.32)$$

2. *Projection of a vector on a given axis.* Consider a vector **P** forming an angle θ with an axis, or directed line, OL (Fig. 3.21). The *projection of* **P** *on the axis* OL is defined as the scalar

$$P_{OL} = P\cos\theta \qquad (3.33)$$

We note that the projection P_{OL} is equal in absolute value to the length of the segment OA; it will be positive if OA has the same sense as the axis OL, i.e., if θ is acute, and negative otherwise. If **P** and OL are at a right angle, the projection of **P** on OL is zero.

Consider now a vector **Q** directed along OL and of the same sense as OL (Fig. 3.22). The scalar product of **P** and **Q** may be expressed as

$$\mathbf{P}\cdot\mathbf{Q} = PQ\cos\theta = P_{OL}Q \qquad (3.34)$$

from which it follows that

$$P_{OL} = \frac{\mathbf{P}\cdot\mathbf{Q}}{Q} = \frac{P_xQ_x + P_yQ_y + P_zQ_z}{Q} \qquad (3.35)$$

In the particular case when the vector selected along OL is the unit vector $\boldsymbol{\lambda}$ (Fig. 3.23), we write

$$P_{OL} = \mathbf{P}\cdot\boldsymbol{\lambda} \qquad (3.36)$$

Resolving **P** and $\boldsymbol{\lambda}$ into rectangular components, and recalling from Sec. 2.11 that the components of $\boldsymbol{\lambda}$ along the coordinate axes are respectively equal to the direction cosines of OL, we express the projection of **P** on OL as

$$P_{OL} = P_x\cos\theta_x + P_y\cos\theta_y + P_z\cos\theta_z \qquad (3.37)$$

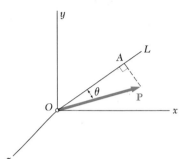

Fig. 3.21

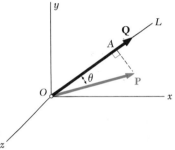

Fig. 3.22

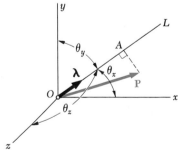

Fig. 3.23

where θ_x, θ_y, and θ_z denote the angles that the axis OL forms with the coordinate axes.

3.9. Mixed Triple Product of Three Vectors. We define the *mixed triple product* of the three vectors **S**, **P**, and **Q** as the scalar expression

$$\mathbf{S} \cdot (\mathbf{P} \times \mathbf{Q}) \qquad (3.38)$$

obtained by forming the scalar product of **S** with the vector product of **P** and **Q**.†

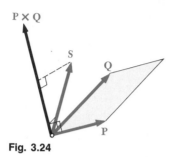

A simple geometrical interpretation may be given for the mixed triple product of **S**, **P**, and **Q** (Fig. 3.24). We first recall from Sec. 3.3 that the vector **P** × **Q** is perpendicular to the plane containing **P** and **Q**, and that its magnitude is equal to the area of the parallelogram constructed on **P** and **Q**. On the other hand, Eq. (3.34) indicates that the scalar product of **S** and **P** × **Q** may be obtained by multiplying the magnitude of **P** × **Q** (i.e., the area of the parallelogram built on **P** and **Q**) by the projection of **S** on the vector **P** × **Q** (i.e., by the projection of **S** on the normal to the plane containing the parallelogram). The mixed triple product is thus equal, in absolute value, to the volume of the parallelepiped having the vectors **S**, **P**, and **Q** for sides (Fig. 3.25). We may check that the sign of the mixed triple product will be positive if **S**, **P**, and **Q** form a right-handed triad, and negative if they form a left-handed triad [i.e., **S** · (**P** × **Q**) will be negative if the rotation which brings **P** into line with **Q** is observed as clockwise from the tip of **S**]. The mixed triple product will be zero if **S**, **P**, and **Q** are coplanar.

Fig. 3.24

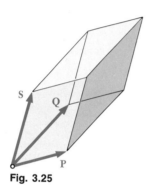

Fig. 3.25

Since the parallelepiped defined in the preceding paragraph is independent of the order in which the three vectors are taken, the six mixed triple products which may be formed with **S**, **P**, and **Q** will all have the same absolute value, although not the same sign. We check that

$$\mathbf{S} \cdot (\mathbf{P} \times \mathbf{Q}) = \mathbf{P} \cdot (\mathbf{Q} \times \mathbf{S}) = \mathbf{Q} \cdot (\mathbf{S} \times \mathbf{P})$$
$$= -\mathbf{S} \cdot (\mathbf{Q} \times \mathbf{P}) = -\mathbf{P} \cdot (\mathbf{S} \times \mathbf{Q}) = -\mathbf{Q} \cdot (\mathbf{P} \times \mathbf{S}) \quad (3.39)$$

Arranging in a circle and in counterclockwise order the letters representing the three vectors (Fig. 3.26), we note that the sign of the mixed triple product is conserved if the vectors are permuted in such a way that they are still read in counterclockwise order. Such a permutation is said to be a *circular permutation*. It also follows from (3.39) that the mixed triple product of **S**, **P**, and **Q** may be defined equally well as **S** · (**P** × **Q**) or (**S** × **P**) · **Q**.

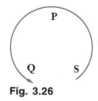

Fig. 3.26

---

† Another kind of triple product will be introduced later (Chap. 15): the *vector triple product* **S** × (**P** × **Q**).

We shall now express the mixed triple product of the vectors **S**, **P**, and **Q** in terms of the rectangular components of these vectors. Denoting **P** × **Q** by **V**, and using formula (3.30) to express the scalar product of **S** and **V**, we write

$$\mathbf{S} \cdot (\mathbf{P} \times \mathbf{Q}) = \mathbf{S} \cdot \mathbf{V} = S_x V_x + S_y V_y + S_z V_z$$

Substituting from the relations (3.9) for the components of **V**, we obtain

$$\mathbf{S} \cdot (\mathbf{P} \times \mathbf{Q}) = S_x(P_y Q_z - P_z Q_y) + S_y(P_z Q_x - P_x Q_z) \\ + S_z(P_x Q_y - P_y Q_x) \quad (3.40)$$

This expression may be written in a more compact form if we observe that it represents the expansion of a determinant:

$$\mathbf{S} \cdot (\mathbf{P} \times \mathbf{Q}) = \begin{vmatrix} S_x & S_y & S_z \\ P_x & P_y & P_z \\ Q_x & Q_y & Q_z \end{vmatrix} \quad (3.41)$$

By applying the rules governing the permutation of rows in a determinant, we could easily verify the relations (3.39) which were derived earlier from geometrical considerations.

3.10. Moment of a Force about a Given Axis. Now that we have further increased our knowledge of vector algebra, we shall introduce a new concept, the concept of *moment of a force about an axis*. Consider again a force **F** acting on a rigid body and the moment $\mathbf{M}_O$ of that force about O (Fig. 3.27). Let OL be an axis through O; *we define the moment M_{OL} of **F** about OL as the projection OC of the moment $\mathbf{M}_O$ on the axis OL.* Denoting by $\boldsymbol{\lambda}$ the unit vector along OL, and recalling the expressions (3.36) and (3.11) obtained earlier for the projection of a vector on a given axis and for the moment $\mathbf{M}_O$ of a force **F**, we write

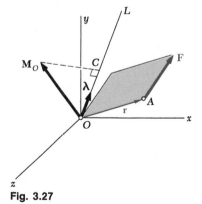

Fig. 3.27

$$M_{OL} = \boldsymbol{\lambda} \cdot \mathbf{M}_O = \boldsymbol{\lambda} \cdot (\mathbf{r} \times \mathbf{F}) \quad (3.42)$$

which shows that the moment M_{OL} of **F** about the axis OL is the scalar obtained by forming the mixed triple product of $\boldsymbol{\lambda}$, **r**, and **F**. Expressing M_{OL} in the form of a determinant, we write

$$M_{OL} = \begin{vmatrix} \lambda_x & \lambda_y & \lambda_z \\ x & y & z \\ F_x & F_y & F_z \end{vmatrix} \quad (3.43)$$

where λ_x, λ_y, λ_z = direction cosines of axis OL
x, y, z = coordinates of point of application of $\mathbf{F}$
F_x, F_y, F_z = components of force $\mathbf{F}$

The physical significance of the moment M_{OL} of a force $\mathbf{F}$ about a fixed axis OL becomes more apparent if we resolve $\mathbf{F}$ into two rectangular components $\mathbf{F}_1$ and $\mathbf{F}_2$, with $\mathbf{F}_1$ parallel to OL and $\mathbf{F}_2$ lying in a plane P perpendicular to OL (Fig. 3.28). Resolving $\mathbf{r}$ similarly into two components $\mathbf{r}_1$ and $\mathbf{r}_2$, and substituting for $\mathbf{F}$ and $\mathbf{r}$ into (3.42), we write

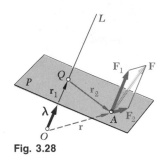

Fig. 3.28

$$M_{OL} = \boldsymbol{\lambda} \cdot [(\mathbf{r}_1 + \mathbf{r}_2) \times (\mathbf{F}_1 + \mathbf{F}_2)]$$
$$= \boldsymbol{\lambda} \cdot (\mathbf{r}_1 \times \mathbf{F}_1) + \boldsymbol{\lambda} \cdot (\mathbf{r}_1 \times \mathbf{F}_2) + \boldsymbol{\lambda} \cdot (\mathbf{r}_2 \times \mathbf{F}_1) + \boldsymbol{\lambda} \cdot (\mathbf{r}_2 \times \mathbf{F}_2)$$

Noting that all mixed triple products, except the last one, are equal to zero, since they involve vectors which are coplanar when drawn from a common origin (Sec. 3.9), we have

$$M_{OL} = \boldsymbol{\lambda} \cdot (\mathbf{r}_2 \times \mathbf{F}_2) \tag{3.44}$$

The vector product $\mathbf{r}_2 \times \mathbf{F}_2$ is perpendicular to the plane P and represents the moment of the component $\mathbf{F}_2$ of $\mathbf{F}$ about the point Q where OL intersects P. Therefore, the scalar M_{OL}, which will be positive if $\mathbf{r}_2 \times \mathbf{F}_2$ and OL have the same sense, and negative otherwise, measures the tendency of $\mathbf{F}_2$ to make the rigid body rotate about the fixed axis OL. Since the other component $\mathbf{F}_1$ of $\mathbf{F}$ does not tend to make the body rotate about OL, we conclude that *the moment M_{OL} of $\mathbf{F}$ about OL measures the tendency of the force $\mathbf{F}$ to impart to the rigid body a motion of rotation about the fixed axis OL.*

It follows from the definition of the moment of a force about an axis that the moment of $\mathbf{F}$ about a coordinate axis is equal to the component of $\mathbf{M}_O$ along that axis. Substituting successively each of the unit vectors $\mathbf{i}$, $\mathbf{j}$, and $\mathbf{k}$ for $\boldsymbol{\lambda}$ in (3.42), we check that the expressions thus obtained for the *moments of $\mathbf{F}$ about the coordinate axes* are respectively equal to the expressions obtained in Sec. 3.7 for the components of the moment $\mathbf{M}_O$ of $\mathbf{F}$ about O.

$$\begin{aligned} M_x &= yF_z - zF_y \\ M_y &= zF_x - xF_z \\ M_z &= xF_y - yF_x \end{aligned} \tag{3.18}$$

We observe that, just as the components F_x, F_y, and F_z of a force $\mathbf{F}$ acting on a rigid body measure, respectively, the tendency of $\mathbf{F}$ to move the rigid body in the x, y, and z directions, the moments M_x, M_y, and M_z of $\mathbf{F}$ about the coordinate axes measure the tendency of $\mathbf{F}$ to impart to the rigid body a motion of rotation about the x, y, and z axes, respectively.

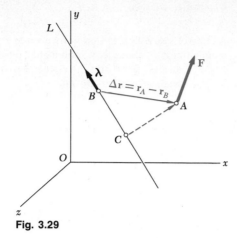

Fig. 3.29

More generally, the moment of a force $\mathbf{F}$ applied at A about an axis which does not pass through the origin is obtained by choosing an arbitrary point B on the axis (Fig. 3.29) and determining the projection on the axis BL of the moment $\mathbf{M}_B$ of $\mathbf{F}$ about B. We write

$$M_{BL} = \boldsymbol{\lambda} \cdot \mathbf{M}_B = \boldsymbol{\lambda} \cdot (\Delta\mathbf{r} \times \mathbf{F}) \qquad (3.45)$$

where $\Delta\mathbf{r} = \mathbf{r}_A - \mathbf{r}_B$ represents the vector joining B and A. Expressing M_{BL} in the form of a determinant, we have

$$M_{BL} = \begin{vmatrix} \lambda_x & \lambda_y & \lambda_z \\ \Delta_x & \Delta_y & \Delta_z \\ F_x & F_y & F_z \end{vmatrix} \qquad (3.46)$$

where $\lambda_x, \lambda_y, \lambda_z$ = direction cosines of axis BL
$\Delta x = x_A - x_B, \Delta y = y_A - y_B, \Delta z = z_A - z_B$
F_x, F_y, F_z = components of force $\mathbf{F}$
It should be noted that the result obtained is independent of the choice of the point B on the given axis. Indeed, denoting by M_{CL} the result obtained with a different point C, we have

$$M_{CL} = \boldsymbol{\lambda} \cdot [(\mathbf{r}_A - \mathbf{r}_C) \times \mathbf{F}]$$
$$= \boldsymbol{\lambda} \cdot [(\mathbf{r}_A - \mathbf{r}_B) \times \mathbf{F}] + \boldsymbol{\lambda} \cdot [(\mathbf{r}_B - \mathbf{r}_C) \times \mathbf{F}]$$

But, since the vectors $\boldsymbol{\lambda}$ and $\mathbf{r}_B - \mathbf{r}_C$ lie in the same line, the volume of the parallelepiped having the vectors $\boldsymbol{\lambda}, \mathbf{r}_B - \mathbf{r}_C$, and $\mathbf{F}$ for sides is zero, as is the mixed triple product of these three vectors (Sec. 3.9). The expression obtained for M_{CL} thus reduces to its first term, which is the expression used earlier to define M_{BL}.

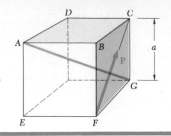

A cube of side a is acted upon by a force $\mathbf{P}$ as shown. Determine the moment of $\mathbf{P}$ (a) about A, (b) about the edge AB, (c) about the diagonal AG of the cube. (d) Using the result of part c, determine the perpendicular distance from AG to FC.

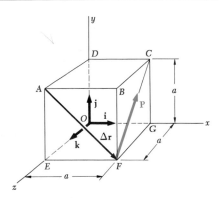

a. **Moment about A.** Choosing x, y, and z axes as shown, we resolve into rectangular components the force $\mathbf{P}$ and the vector $\Delta\mathbf{r} = \overrightarrow{AF}$ which joins A and the point of application F of $\mathbf{P}$.

$$\Delta\mathbf{r} = a\mathbf{i} - a\mathbf{j} = a(\mathbf{i} - \mathbf{j})$$
$$\mathbf{P} = (P/\sqrt{2})\mathbf{j} - (P/\sqrt{2})\mathbf{k} = (P/\sqrt{2})(\mathbf{j} - \mathbf{k})$$

The moment of $\mathbf{P}$ about A is

$$\mathbf{M}_A = \Delta\mathbf{r} \times \mathbf{P} = a(\mathbf{i} - \mathbf{j}) \times (P/\sqrt{2})(\mathbf{j} - \mathbf{k})$$
$$\mathbf{M}_A = (aP/\sqrt{2})(\mathbf{i} + \mathbf{j} + \mathbf{k}) \quad \blacktriangleleft$$

b. **Moment about AB.** Projecting $\mathbf{M}_A$ on AB, we write

$$M_{AB} = \mathbf{i} \cdot \mathbf{M}_A = \mathbf{i} \cdot (aP/\sqrt{2})(\mathbf{i} + \mathbf{j} + \mathbf{k}) \qquad M_{AB} = aP/\sqrt{2} \quad \blacktriangleleft$$

We verify that, since AB is parallel to the x axis, M_{AB} is also the x component of the moment $\mathbf{M}_A$.

c. **Moment about Diagonal AG.** The moment of $\mathbf{P}$ about AG is obtained by projecting $\mathbf{M}_A$ on AG. Denoting by $\boldsymbol{\lambda}$ the unit vector along AG, we note that

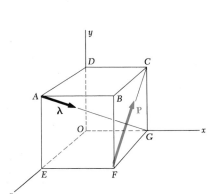

$$\boldsymbol{\lambda} = \frac{\overrightarrow{AG}}{AG} = \frac{a\mathbf{i} - a\mathbf{j} - a\mathbf{k}}{a\sqrt{3}} = (1/\sqrt{3})(\mathbf{i} - \mathbf{j} - \mathbf{k})$$

and write

$$M_{AG} = \boldsymbol{\lambda} \cdot \mathbf{M}_A = (1/\sqrt{3})(\mathbf{i} - \mathbf{j} - \mathbf{k}) \cdot (aP/\sqrt{2})(\mathbf{i} + \mathbf{j} + \mathbf{k})$$
$$M_{AG} = (aP/\sqrt{6})(1 - 1 - 1) \qquad M_{AG} = -aP/\sqrt{6} \quad \blacktriangleleft$$

Alternate Method. The moment of $\mathbf{P}$ about AG may also be expressed in the form of a determinant:

$$M_{AG} = \begin{vmatrix} \lambda_x & \lambda_y & \lambda_z \\ \Delta x & \Delta y & \Delta z \\ F_x & F_y & F_z \end{vmatrix} = \begin{vmatrix} 1/\sqrt{3} & -1/\sqrt{3} & -1/\sqrt{3} \\ a & -a & 0 \\ 0 & P/\sqrt{2} & -P/\sqrt{2} \end{vmatrix} = -aP/\sqrt{6}$$

d. **Perpendicular Distance from AG to FC.** We first observe that $\mathbf{P}$ is perpendicular to the diagonal AG. This may be checked by forming the scalar product $\mathbf{P} \cdot \boldsymbol{\lambda}$ and verifying that it is zero:

$$\mathbf{P} \cdot \boldsymbol{\lambda} = (P/\sqrt{2})(\mathbf{j} - \mathbf{k}) \cdot (1/\sqrt{3})(\mathbf{i} - \mathbf{j} - \mathbf{k}) = (P/\sqrt{6})(0 - 1 + 1) = 0$$

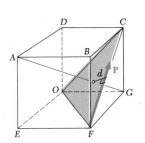

The moment M_{AG} may then be expressed as $-Pd$, where d is the perpendicular distance from AG to FC. (The negative sign is used since the rotation imparted to the cube by $\mathbf{P}$ appears as clockwise to an observer at G.) Recalling the value found for M_{AG} in part c,

$$M_{AG} = -Pd = -aP/\sqrt{6} \qquad d = a/\sqrt{6} \quad \blacktriangleleft$$

PROBLEMS

3.25 Form the scalar product $\mathbf{P}_1 \cdot \mathbf{P}_2$ and use the result obtained to prove the identity $\cos(\theta_1 - \theta_2) = \cos\theta_1 \cos\theta_2 + \sin\theta_1 \sin\theta_2$.

3.26 Given the vectors $\mathbf{P} = 2\mathbf{i} + \mathbf{j} + 2\mathbf{k}$, $\mathbf{Q} = 3\mathbf{i} + 4\mathbf{j} - 5\mathbf{k}$, and $\mathbf{S} = -4\mathbf{i} + \mathbf{j} - 2\mathbf{k}$, compute the scalar products $\mathbf{P} \cdot \mathbf{Q}$, $\mathbf{P} \cdot \mathbf{S}$, and $\mathbf{Q} \cdot \mathbf{S}$.

3.27 Several cables are attached to the top of the tower at A. Determine the angle formed by cables AB and AC.

3.28 Determine the angle formed by cables AD and AB.

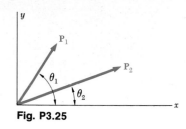

Fig. P3.25

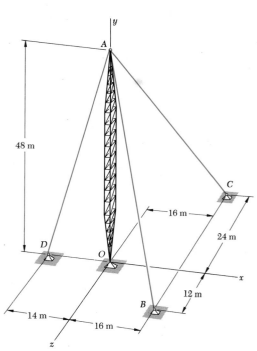

Fig. P3.27 and P3.28

3.29 The rod ABC consists of a straight segment AB and a circular portion BC. Knowing that $\alpha = 30°$ and that the tension in wire AD is 300 N, determine (a) the angle formed by wires AC and AD, (b) the projection on AC of the force exerted by wire AD at point A.

3.30 The rod ABC consists of a straight segment AB and a circular portion BC. Knowing that $\alpha = 45°$ and that the tension in wire AC is 450 N, determine (a) the angle formed by wires AC and AD, (b) the projection on AD of the force exerted by wire AC at point A.

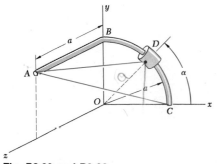

Fig. P3.29 and P3.30

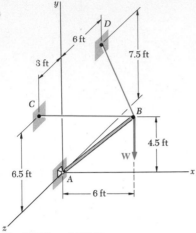

Fig. P3.31 and P3.32

3.31 Knowing that the tension in cable *BC* is 280 lb, determine (*a*) the angle between cable *BC* and the boom *AB*, (*b*) the projection on *AB* of the force exerted by cable *BC* at point *B*.

3.32 Knowing that the tension in cable *BD* is 180 lb, determine (*a*) the angle between cable *BD* and the boom *AB*, (*b*) the projection on *AB* of the force exerted by cable *BD* at point *B*.

3.33 Given the vectors $\mathbf{P} = \mathbf{i} + 2\mathbf{j} + 3\mathbf{k}$, $\mathbf{Q} = \mathbf{i} + 2\mathbf{j}$, and $\mathbf{S} = \mathbf{i}$, compute $\mathbf{P} \cdot (\mathbf{Q} \times \mathbf{S})$, $(\mathbf{P} \times \mathbf{Q}) \cdot \mathbf{S}$, and $(\mathbf{S} \times \mathbf{Q}) \cdot \mathbf{P}$.

3.34 Given the vectors $\mathbf{P} = 3\mathbf{i} + 2\mathbf{j} + \mathbf{k}$, $\mathbf{Q} = 5\mathbf{i} + \mathbf{j} - 2\mathbf{k}$, and $\mathbf{S} = \mathbf{i} + 3\mathbf{j} + S_z\mathbf{k}$, determine the value of S_z for which the three vectors are coplanar.

3.35 The jib crane is oriented so that the boom *DA* is parallel to the *x* axis. At the instant shown the tension in cable *AB* is 13 kN. Determine the moment about each of the coordinate axes of the force exerted on *A* by the cable *AB*.

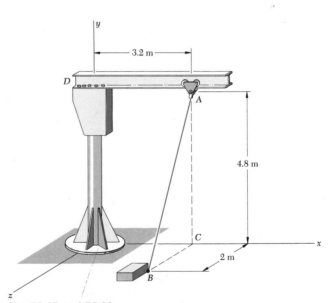

Fig. P3.35 and P3.36

3.36 The jib crane is oriented so that the boom *DA* is parallel to the *x* axis. Determine the maximum permissible tension in the cable *AB* if the absolute values of the moments about the coordinate axes of the force exerted on *A* must be as follows: $|M_x| \leq 10$ kN · m, $|M_y| \leq 6$ kN · m, $|M_z| \leq 16$ kN · m.

3.37 A vertical force **P** of magnitude 60 lb is applied to the crank at A. Knowing that $\theta = 75°$, determine the moment of **P** about each of the coordinate axes.

3.38 Determine the magnitude of the force **P** and the angle θ, knowing that the moments of **P** about the x and z axes are, respectively, $M_x = -160$ lb · in. and $M_z = -240$ lb · in.

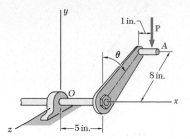

Fig. P3.37 and P3.38

3.39 A force **F** is applied at point C of the pipe $ABCD$. Determine the components F_y and F_z of the force, knowing that $F_x = +100$ N and that its moments about the x and y axes are, respectively, $M_x = +75$ N · m and $M_y = -140$ N · m.

3.40 A force **F** of unknown magnitude and direction is applied at point C of the pipe $ABCD$. Determine the moment M_y of **F** about the y axis, knowing that $M_x = +150$ N · m and $M_z = +90$ N · m.

3.41 The force $\mathbf{F} = (100\ \text{N})\mathbf{i} + (150\ \text{N})\mathbf{j} + (300\ \text{N})\mathbf{k}$ is applied at point C of the pipe $ABCD$. Determine the moment of **F** about (a) a line joining points B and E, (b) a line joining points A and D.

3.42 Two rods are welded together to form a T-shaped lever which is acted upon by a 130-lb force as shown. Determine the moment of the force about rod AB.

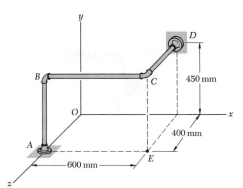

Fig. P3.39, P3.40, and P3.41

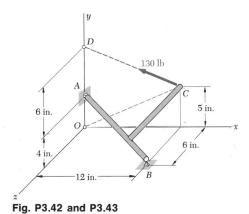

Fig. P3.42 and P3.43

3.43 Solve Prob. 3.42 assuming that the 130-lb force is directed along the line CO.

3.44 The rectangular plate $ABCD$ is held by hinges along its edge AD and by the wire BE. Knowing that the tension in the wire is 546 N, determine the moment about AD of the force exerted by the wire at point B.

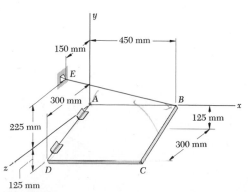

Fig. P3.44

3.45 A regular tetrahedron has six edges each of length *a*. A single force **P** is directed along one edge. Determine the moment of **P** about each of the other five edges.

3.46 Two forces $\mathbf{F}_1$ and $\mathbf{F}_2$ in space have the same magnitude *F*. Prove that the moment of $\mathbf{F}_1$ about the line of action of $\mathbf{F}_2$ is equal to the moment of $\mathbf{F}_2$ about the line of action of $\mathbf{F}_1$.

∗**3.47** Use the result obtained in Prob. 3.42 to determine the perpendicular distance between the lines *AB* and *CD*.

∗**3.48** Use the result obtained in Prob. 3.44 to determine the perpendicular distance between the lines *AD* and *BE*.

3.11. Moment of a Couple. *Two forces **F** and −**F**, having the same magnitude, parallel lines of action, and opposite sense are said to form a couple* (Fig. 3.30). Clearly, the sum of the components of the two forces in any direction is zero. The sum of the moments of the two forces about a given point, however, is not zero. While the two forces will not translate the body on which they act, they will tend to make it rotate.

Denoting by $\mathbf{r}_A$ and $\mathbf{r}_B$, respectively, the position vectors of the points of application of **F** and −**F** (Fig. 3.31), we find that the sum of the moments of the two forces about *O* is

$$\mathbf{r}_A \times \mathbf{F} + \mathbf{r}_B \times (-\mathbf{F}) = (\mathbf{r}_A - \mathbf{r}_B) \times \mathbf{F}$$

Setting $\mathbf{r}_A - \mathbf{r}_B = \mathbf{r}$, where **r** is the vector joining the points of application of the two forces, we conclude that the sum of the moments of **F** and −**F** about *O* is represented by the vector

$$\mathbf{M} = \mathbf{r} \times \mathbf{F} \tag{3.47}$$

The vector **M** is called the *moment of the couple;* it is a vector perpendicular to the plane containing the two forces and its magnitude is

$$M = rF \sin \theta = Fd \tag{3.48}$$

where *d* is the perpendicular distance between the lines of action of **F** and −**F**. The sense of **M** is defined by the right-hand rule.

Since the vector **r** in (3.47) is independent of the choice of the origin *O* of the coordinate axes, we note that the same result would have been obtained if the moments of **F** and −**F** had been computed about a different point *O′*. Thus, the moment **M** of a couple is a *free vector* (Sec. 2.2) which may be applied at any point (Fig. 3.32).

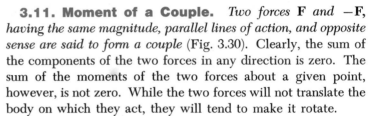

Fig. 3.30

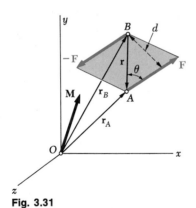

Fig. 3.31

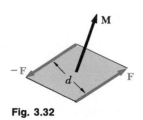

Fig. 3.32

From the definition of the moment of a couple, it also follows that two couples, one consisting of the forces $\mathbf{F}_1$ and $-\mathbf{F}_1$, the other of the forces $\mathbf{F}_2$ and $-\mathbf{F}_2$ (Fig. 3.33), will have equal moments if

$$F_1 d_1 = F_2 d_2 \tag{3.49}$$

and if the two couples lie in parallel planes (or in the same plane) and have the same sense.

3.12. Equivalent Couples. Consider the three couples shown in Fig. 3.34, which are made to act successively on the same rectangular box. As seen in the preceding section, the only motion a couple may impart to a rigid body is a rotation. Since each of the three couples shown has the same moment $\mathbf{M}$ (same direction and same magnitude $M = 120$ lb · in.), we may expect the three couples to have the same effect on the box.

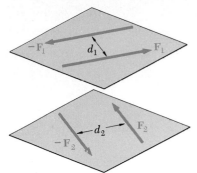

Fig. 3.33

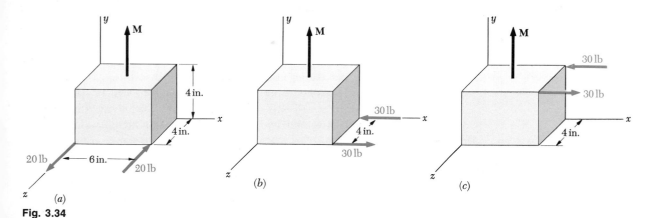

Fig. 3.34

As reasonable as this conclusion may appear, we should not accept it hastily. While intuitive feeling is of great help in the study of mechanics, it should not be accepted as a substitute for logical reasoning. Before stating that two systems (or groups) of forces have the same effect on a rigid body, we should prove that fact on the basis of the experimental evidence introduced so far. This evidence consists of the parallelogram law for the addition of two forces (Sec. 2.1) and of the principle of transmissibility (Sec. 3.2). Therefore, we shall state that *two systems of forces are equivalent* (i.e., they have the same effect on a rigid body) *if we can transform one of them into the other by means of one or several of the following operations:* (1) replacing two forces acting on the same particle by their resultant; (2) resolving a force into two components; (3) canceling two equal and opposite forces acting on the same particle; (4) attaching to the same

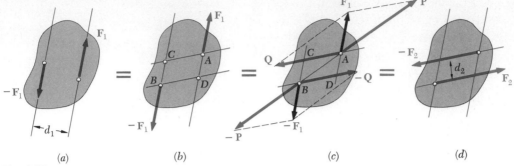

Fig. 3.35

particle two equal and opposite forces; (5) moving a force along its line of action. Each of these operations is easily justified on the basis of the parallelogram law or the principle of transmissibility.

Let us now prove that *two couples having the same moment* **M** *are equivalent.* First, we shall consider two couples contained in the same plane, and we shall assume that this plane coincides with the plane of the figure (Fig. 3.35). The first couple consists of the forces $\mathbf{F}_1$ and $-\mathbf{F}_1$, of magnitude F_1 and at a distance d_1 from each other (Fig. 3.35a), and the second couple of the forces $\mathbf{F}_2$ and $-\mathbf{F}_2$, of magnitude F_2 and at a distance d_2 from each other (Fig. 3.35d). Since the two couples have the same moment **M** perpendicular to the plane of the figure, they must have the same sense (assumed here counterclockwise) and the relation

$$F_1 d_1 = F_2 d_2 \tag{3.49}$$

must be satisfied. To prove that they are equivalent, we shall show that the first couple may be transformed into the second by means of the operations listed above.

Denoting by A, B, C, D the points of intersection of the lines of action of the two couples, we first slide the forces $\mathbf{F}_1$ and $-\mathbf{F}_1$ until they are attached, respectively, at A and B, as shown in Fig. 3.35b. The force $\mathbf{F}_1$ is then resolved into a component $\mathbf{P}$ along line AB and a component $\mathbf{Q}$ along AC (Fig. 3.35c); similarly, the force $-\mathbf{F}_1$ is resolved into $-\mathbf{P}$ along AB and $-\mathbf{Q}$ along BD. The forces $\mathbf{P}$ and $-\mathbf{P}$ have the same magnitude, same line of action, and opposite sense; they may be moved along their common line of action until they are applied at the same point and then canceled. Thus the couple formed by $\mathbf{F}_1$ and $-\mathbf{F}_1$ reduces to a couple consisting of $\mathbf{Q}$ and $-\mathbf{Q}$.

We shall now show that the forces $\mathbf{Q}$ and $-\mathbf{Q}$ are respectively equal to the forces $-\mathbf{F}_2$ and $\mathbf{F}_2$. The moment of the couple formed by $\mathbf{Q}$ and $-\mathbf{Q}$ may be obtained by computing the moment of $\mathbf{Q}$ about B; similarly, the moment of the couple formed by $\mathbf{F}_1$ and $-\mathbf{F}_1$ is the moment of $\mathbf{F}_1$ about B. But, by Varignon's theorem, the moment of $\mathbf{F}_1$ is equal to the sum of the moments of its components $\mathbf{P}$ and $\mathbf{Q}$. Since the moment of $\mathbf{P}$ about B is zero, the moment of the couple formed by $\mathbf{Q}$ and $-\mathbf{Q}$ must be equal to the moment of the couple formed by $\mathbf{F}_1$ and $-\mathbf{F}_1$. Recalling (3.49), we write

$$Qd_2 = F_1d_1 = F_2d_2 \qquad \text{and} \qquad Q = F_2$$

Thus the forces $\mathbf{Q}$ and $-\mathbf{Q}$ are respectively equal to the forces $-\mathbf{F}_2$ and $\mathbf{F}_2$, and the couple of Fig. 3.35a is equivalent to the couple of Fig. 3.35d.

Next we shall consider two couples contained in parallel planes P_1 and P_2 and prove that they are equivalent if they have the same moment. In view of the foregoing we may assume that the couples consist of forces of the same magnitude F acting along parallel lines (Fig. 3.36a and d). We propose to show that the couple contained in plane P_1 may be transformed into the couple contained in plane P_2 by means of the standard operations listed above.

Let us consider the two planes defined respectively by the lines of action of $\mathbf{F}_1$ and $-\mathbf{F}_2$, and of $-\mathbf{F}_1$ and $\mathbf{F}_2$ (Fig. 3.36b). At a point on their line of intersection we attach two forces $\mathbf{F}_3$ and $-\mathbf{F}_3$, respectively equal to $\mathbf{F}_1$ and $-\mathbf{F}_1$. The couple formed by $\mathbf{F}_1$ and $-\mathbf{F}_3$ may be replaced by a couple consisting of $\mathbf{F}_3$ and $-\mathbf{F}_2$ (Fig. 3.36c), since both couples have clearly the same moment and are contained in the same plane. Similarly, the couple formed by $-\mathbf{F}_1$ and $\mathbf{F}_3$ may be replaced by a couple consisting of $-\mathbf{F}_3$ and $\mathbf{F}_2$. Canceling the two equal and opposite forces $\mathbf{F}_3$ and $-\mathbf{F}_3$, we obtain the desired couple in plane P_2 (Fig. 3.36d). Thus, we conclude that two couples having the same moment $\mathbf{M}$ are equivalent, whether they are contained in the same plane or in parallel planes.

The property we have just established is very important for the correct understanding of the mechanics of rigid bodies. It indicates that, when a couple acts on a rigid body, it does not matter where the two forces forming the couple act, or what magnitude and direction they have. The only thing which counts is the *moment* of the couple (magnitude and direction). Couples with the same moment will have the same effect on the rigid body.

3.13. Addition of Couples. Consider two intersecting planes P_1 and P_2 and two couples acting respectively in P_1 and P_2. We may, without any loss of generality, assume that the couple in P_1 consists of two forces $\mathbf{F}_1$ and $-\mathbf{F}_1$ perpendicular to the line

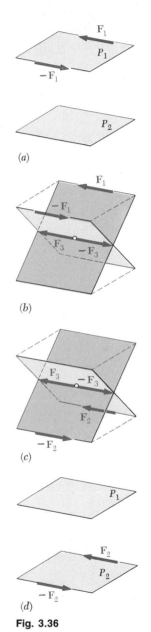

(a)

(b)

(c)

(d)

Fig. 3.36

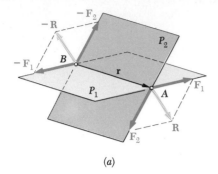

(a)

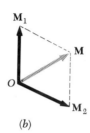

(b)

Fig. 3.37

of intersection of the two planes and acting respectively at A and B (Fig. 3.37a). Similarly, we assume that the couple in P_2 consists of two forces $\mathbf{F}_2$ and $-\mathbf{F}_2$ perpendicular to AB and acting respectively at A and B. It is clear that the resultant $\mathbf{R}$ of $\mathbf{F}_1$ and $\mathbf{F}_2$ and the resultant $-\mathbf{R}$ of $-\mathbf{F}_1$ and $-\mathbf{F}_2$ form a couple. Denoting by $\mathbf{r}$ the vector joining B to A, and recalling the definition of the moment of a couple (Sec. 3.11), we express the moment $\mathbf{M}$ of the resulting couple as follows:

$$\mathbf{M} = \mathbf{r} \times \mathbf{R} = \mathbf{r} \times (\mathbf{F}_1 + \mathbf{F}_2)$$

and, by Varignon's theorem,

$$\mathbf{M} = \mathbf{r} \times \mathbf{F}_1 + \mathbf{r} \times \mathbf{F}_2$$

But the first term in the expression obtained represents the moment $\mathbf{M}_1$ of the couple in P_1, and the second term the moment $\mathbf{M}_2$ of the couple in P_2. We have

$$\mathbf{M} = \mathbf{M}_1 + \mathbf{M}_2 \qquad (3.50)$$

and we conclude that the sum of two couples of moments $\mathbf{M}_1$ and $\mathbf{M}_2$ is a couple of moment $\mathbf{M}$ equal to the vector sum of $\mathbf{M}_1$ and $\mathbf{M}_2$ (Fig. 3.37b).

3.14. Couples May Be Represented by Vectors. As we saw in Sec. 3.12, couples which have the same moment, whether they act in the same plane or in parallel planes, are equivalent. There is therefore no need to draw the actual forces forming a given couple in order to define its effect on a rigid body (Fig. 3.38a). It is sufficient to draw an arrow equal in magnitude and direction to the moment $\mathbf{M}$ of the couple (Fig. 3.38b). On the other hand, we saw in Sec. 3.13 that the sum of two couples is itself a couple and that the moment $\mathbf{M}$ of the resultant couple may be obtained by forming the vector sum of the moments $\mathbf{M}_1$ and $\mathbf{M}_2$ of the given couples. Thus, couples obey the law of addition of vectors, and the arrow used in Fig. 3.38b to represent the couple defined in Fig. 3.38a may truly be considered a vector.

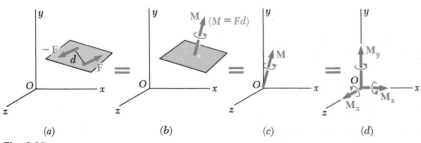

(a)   (b)   (c)   (d)

Fig. 3.38

The vector representing a couple is called a *couple vector*. Note that, in Fig. 3.38, a colored arrow is used to distinguish the couple vector, *which represents the couple itself*, from the vector representing the *moment* of the couple, and that the symbol ∫ is added to avoid any confusion with vectors representing forces. A couple vector, like the moment of a couple, is a free vector. Its point of application, therefore, may be chosen at the origin of the system of coordinates, if so desired (Fig. 3.38c). Furthermore, the couple vector **M** may be resolved into component vectors **M**$_x$, **M**$_y$, and **M**$_z$, directed along the axes of coordinates (Fig. 3.38d) and representing couples acting, respectively, in the *yz*, *zx*, and *xy* planes.

3.15. Resolution of a Given Force into a Force at O and a Couple.

Consider a force **F** acting on a rigid body at a point *A* defined by the position vector **r** (Fig. 3.39a). Suppose that for some reason we would rather have the force act at point *O*. We know that we can move **F** along its line of action (principle of transmissibility); but we cannot move it to a point *O* away from the original line of action without modifying the action of **F** on the rigid body.

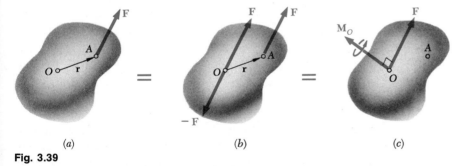

(a)	(b)	(c)

Fig. 3.39

We may, however, attach two forces at point *O*, one equal to **F** and the other equal to −**F**, without modifying the action of the original force on the rigid body (Fig. 3.39b). As a result of this transformation, a force **F** is now applied at *O*; the other two forces form a couple of moment **M**$_O$ = **r** × **F**. Thus, *any force* **F** *acting on a rigid body may be moved to an arbitrary point O, provided that a couple is added, of moment equal to the moment of* **F** *about O*. The couple tends to impart to the rigid body the same motion of rotation about *O* that the force **F** tended to produce before it was transferred to *O*. The couple is represented by a couple vector **M**$_O$ perpendicular to the plane containing **r** and **F**. Since **M**$_O$ is a free vector, it may be applied anywhere; for convenience, however, the couple vector is usually attached at *O*, together with **F**, and the combination obtained is referred to as a *force-couple system* (Fig. 3.39c).

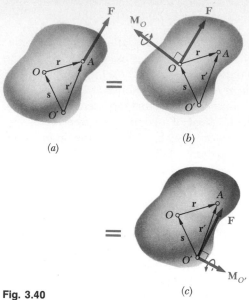

Fig. 3.40

(a)

(b)

(c)

If the force **F** had been moved from A to a different point O' (Fig. 3.40a and c), the moment $\mathbf{M}_{O'} = \mathbf{r}' \times \mathbf{F}$ of **F** about O' should have been computed, and a new force-couple system, consisting of **F** and of the couple vector $\mathbf{M}_{O'}$, would have been attached at O'. The relation existing between the moments of **F** about O and O' is obtained by writing

$$\mathbf{M}_{O'} = \mathbf{r}' \times \mathbf{F} = (\mathbf{r} + \mathbf{s}) \times \mathbf{F} = \mathbf{r} \times \mathbf{F} + \mathbf{s} \times \mathbf{F}$$

$$\mathbf{M}_{O'} = \mathbf{M}_O + \mathbf{s} \times \mathbf{F} \tag{3.51}$$

where **s** is the vector joining O' to O. Thus, the moment $\mathbf{M}_{O'}$ of **F** about O' is obtained by adding to the moment $\mathbf{M}_O$ of **F** about O the vector product $\mathbf{s} \times \mathbf{F}$ representing the moment about O' of the force **F** applied at O.

This result could also have been established by observing that, in order to transfer to O' the force-couple system attached at O (Fig. 3.40b and c), the couple vector $\mathbf{M}_O$ may be freely moved to O'; to move the force **F** from O to O', however, it is necessary to add to **F** a couple vector $\mathbf{s} \times \mathbf{F}$ representing the moment about O' of the force **F** applied at O. Thus, the couple vector $\mathbf{M}_{O'}$ must be the sum of $\mathbf{M}_O$ and $\mathbf{s} \times \mathbf{F}$.

As noted above, the force-couple system obtained by transferring a force **F** from a point A to a point O consists of **F** and of a couple vector $\mathbf{M}_O = \mathbf{r} \times \mathbf{F}$ perpendicular to **F**. Conversely, any force-couple system consisting of a force **F** and of a couple vector $\mathbf{M}_O$ which are *mutually perpendicular* may be replaced by a single equivalent force. This is done by moving the force **F** in the plane perpendicular to $\mathbf{M}_O$ until its moment about O becomes equal to the couple vector $\mathbf{M}_O$ to be eliminated.

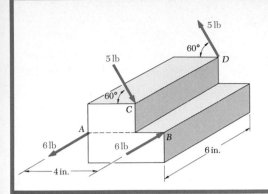

SAMPLE PROBLEM 3.6

Two couples act on a block as shown. Replace these two couples with a single equivalent couple.

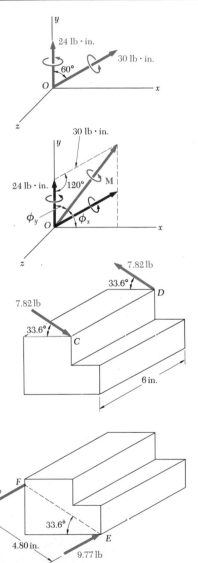

Solution. Each of the given couples is represented by a couple vector which is perpendicular to the plane of the couple and of a magnitude equal to the moment of the couple. The sense of each vector is obtained by applying the right-hand rule, and for convenience both couple vectors are attached at the origin.

The single couple equivalent to the two given couples will be represented by the resultant of the two couple vectors. The magnitude of the resultant couple vector **M** is obtained from the law of cosines.

$$M^2 = (24)^2 + (30)^2 - (2)(24)(30) \cos 120°$$
$$M = 46.9 \text{ lb} \cdot \text{in.}$$

The angle ϕ_y that the resultant couple vector forms with the vertical is obtained from the law of sines.

$$\frac{\sin \phi_y}{30 \text{ lb} \cdot \text{in.}} = \frac{\sin 120°}{46.9 \text{ lb} \cdot \text{in.}} \qquad \phi_y = 33.6°$$

The angle that the couple vector forms with the x axis is

$$\phi_x = 90° - 33.6° = 56.4°$$

and the angle it forms with the z axis is $\phi_z = 90°$. Thus the resultant couple vector **M** is defined by

$$M = 46.9 \text{ lb} \cdot \text{in.} \qquad \phi_x = 56.4° \qquad \phi_y = 33.6° \qquad \phi_z = 90° \quad \blacktriangleleft$$

The single couple equivalent to the two original couples is a couple of magnitude $M = 46.9$ lb · in., acting in a plane parallel to the z axis and forming an angle of 33.6° with the horizontal plane. This couple may be formed in many ways, for example, by two 7.82-lb forces acting at corners C and D in the directions shown, or by two 9.77-lb forces parallel to the z axis and applied at E and F as shown.

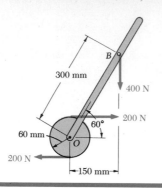

SAMPLE PROBLEM 3.7

Replace the couple and force shown by an equivalent single force applied to the lever. Determine the distance from the shaft to the point of application of this equivalent force.

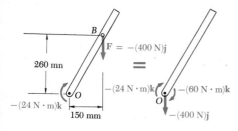

Solution. First the given force and couple are replaced by an equivalent force-couple system at O. We move the force $\mathbf{F} = -(400\ \text{N})\mathbf{j}$ to O and at the same time add a couple of moment $\mathbf{M}_O$ equal to the moment about O of the force in its original position.

$$\mathbf{M}_O = \overrightarrow{OB} \times \mathbf{F} = [(0.150\ \text{m})\mathbf{i} + (0.260\ \text{m})\mathbf{j}] \times (-400\ \text{N})\mathbf{j}$$
$$= -(60\ \text{N} \cdot \text{m})\mathbf{k}$$

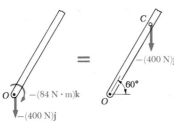

This couple is added to the couple of moment $-(24\ \text{N} \cdot \text{m})\mathbf{k}$ formed by the two 200-N forces, and a couple of moment $-(84\ \text{N} \cdot \text{m})\mathbf{k}$ is obtained. This last couple may be eliminated by applying $\mathbf{F}$ at a point C chosen in such a way that

$$-(84\ \text{N} \cdot \text{m})\mathbf{k} = \overrightarrow{OC} \times \mathbf{F}$$
$$= [(OC)\cos 60°\mathbf{i} + (OC)\sin 60°\mathbf{j}] \times (-400\ \text{N})\mathbf{j}$$
$$= -(OC)\cos 60°(400\ \text{N})\mathbf{k}$$

We conclude that

$$(OC)\cos 60° = 0.210\ \text{m} = 210\ \text{mm} \qquad OC = 420\ \text{mm} \quad \blacktriangleleft$$

Alternate Solution. Since the effect of a couple does not depend on its location, the couple of moment $-(24\ \text{N} \cdot \text{m})\mathbf{k}$ may be moved to B; we thus obtain a force-couple system at B. The couple may now be eliminated by applying $\mathbf{F}$ at a point C chosen in such a way that

$$-(24\ \text{N} \cdot \text{m})\mathbf{k} = BC \times \mathbf{F}$$
$$= -(BC)\cos 60°(400\ \text{N})\mathbf{k}$$

We conclude that

$$(BC)\cos 60° = 0.060\ \text{m} = 60\ \text{mm} \qquad BC = 120\ \text{mm}$$
$$OC = OB + BC = 300\ \text{mm} + 120\ \text{mm} \qquad OC = 420\ \text{mm} \quad \blacktriangleleft$$

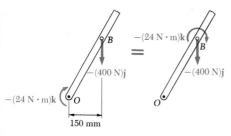

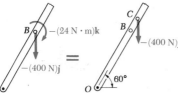

PROBLEMS

3.49 The two couples shown are applied to a 120- by 160-mm plate. Knowing that $P_1 = P_2 = 150$ N and $Q_1 = Q_2 = 200$ N, prove that their sum is zero (*a*) by adding their moments, (*b*) by combining $\mathbf{P}_1$ and $\mathbf{Q}_1$ into their resultant $\mathbf{R}_1$, combining $\mathbf{P}_2$ and $\mathbf{Q}_2$ into their resultant $\mathbf{R}_2$, and then showing that $\mathbf{R}_1$ and $\mathbf{R}_2$ are equal and opposite and have the same line of action.

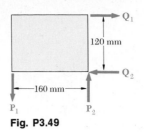

Fig. P3.49

3.50 A couple formed by two 975-N forces is applied to the pulley assembly shown. Determine an equivalent couple which is formed by (*a*) vertical forces acting at A and C, (*b*) the smallest possible forces acting at B and D, (*c*) the smallest possible forces which can be attached to the assembly.

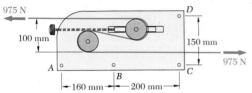

Fig. P3.50

3.51 Four 1-in.-diameter pegs are attached to a board as shown. Two strings are passed around the pegs and pulled with forces of magnitude $P = 20$ lb and $Q = 35$ lb. Determine the resultant couple acting on the board.

3.52 A multiple-drilling machine is used to drill simultaneously six holes in the steel plate shown. Each drill exerts a clockwise couple of magnitude 40 lb · in. on the plate. Determine an equivalent couple formed by the smallest possible forces acting (*a*) at A and C, (*b*) at A and D, (*c*) on the plate.

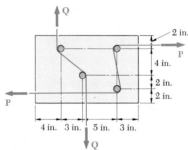

Fig. P3.51

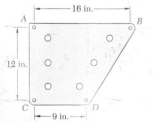

Fig. P3.52

3.53 The axles and drive shaft of an automobile are acted upon by the three couples shown. Replace, these three couples by a single equivalent couple.

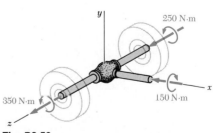

Fig. P3.53

3.54 Determine the components of a single couple equivalent to the two couples shown. Check the result obtained by adding the moments of the individual forces about the coordinate axes.

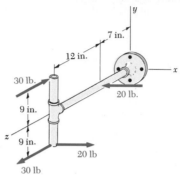

Fig. P3.54

3.55 The couple vectors M_1 and M_2 represent couples which are contained in the planes ABC and ACD, respectively. Assuming that $M_1 = M_2 = M$, determine a single couple equivalent to the two given couples.

3.56 Three shafts are connected to a gearbox as shown. Shaft A is horizontal and shafts B and C lie in the vertical yz plane. Determine the components of the resultant couple exerted on the gearbox.

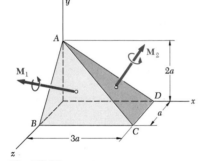

Fig. P3.55

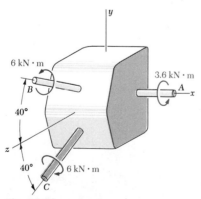

Fig. P3.56

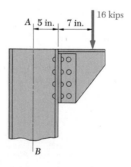

Fig. P3.57

3.57 A crane column supports a 16-kip load as shown. Reduce the load to an axial force along AB and a couple.

3.58 A 400-N force is applied to a bent plate as shown. Determine an equivalent force-couple system (a) at A, (b) at B.

3.59 Knowing that $\alpha = 60°$, replace the force and couple shown by a single force applied at a point located (a) on line AB, (b) on line CD. In each case determine the distance from the center O to the point of application of the force.

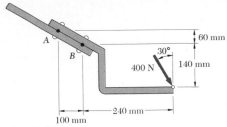

Fig. P3.58

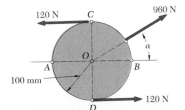

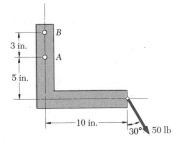

Fig. P3.59 and P3.60

3.60 The force and couple shown are to be replaced by an equivalent single force. Determine the required value of α so that the line of action of the single equivalent force will pass through point B.

3.61 A 50-lb force is applied to a corner plate as shown. Determine (a) an equivalent force-couple system at A, (b) two horizontal forces at A and B which form a couple equivalent to the couple found in part a.

Fig. P3.61 and P3.62

3.62 A 50-lb force is applied to a corner plate as shown. Determine (a) an equivalent force-couple system at B, (b) two horizontal forces at A and B which form a couple equivalent to the couple found in part a.

3.63 The force **P** has a magnitude of 300 N and is applied in a direction perpendicular to the handle ($\beta = 0$). Assuming $\alpha = 0$, replace the force **P** by (a) an equivalent force-couple system at B, (b) an equivalent system formed by two parallel forces at B and C.

3.64 Replace the force **P** by an equivalent system formed by two parallel forces at B and C. Show (a) that these forces are parallel to the force **P**, (b) that the magnitude of these forces is independent of both α and β.

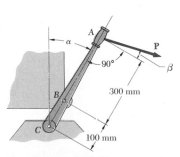

Fig. P3.63 and P3.64

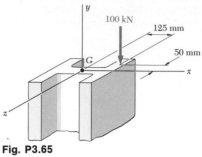

Fig. P3.65

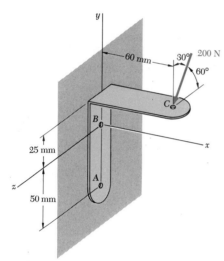

Fig. P3.66

3.65 A 100-kN load is applied eccentrically on a column. Determine the components of the force and couple at G which are equivalent to the 100-kN load.

3.66 A 200-N force is applied as shown on the bracket ABC. Determine the components of the force and couple at A which are equivalent to this force.

3.67 A precast-concrete wall section is temporarily held by cables as shown. The tension in cable AB is 700 lb. Replace the force exerted on the wall section at A by a force-couple system located (a) at the origin of coordinates O, (b) at point E.

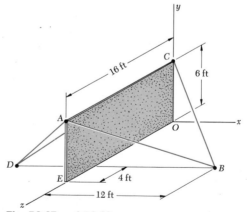

Fig. P3.67 and P3.68

3.68 A precast-concrete wall section is temporarily held by cables as shown. The tension in cable BC is 900 lb. Replace the force exerted on the wall section at C by a force-couple system located (a) at the origin of coordinates O, (b) at point E.

3.69   Five separate force-couple systems act at the corners of a rectangular box as shown. Find two force-couple systems which are equivalent.

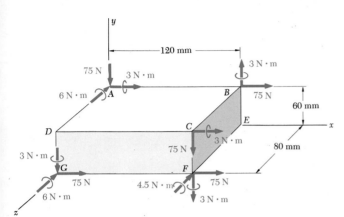

Fig. P3.69 and P3.70

3.70   Determine which of the force-couple systems shown is equivalent to a force-couple system located at the origin O and consisting of $\mathbf{F} = -(75 \text{ N})\mathbf{j}$ and $\mathbf{M}_O = (9 \text{ N} \cdot \text{m})\mathbf{i} - (9 \text{ N} \cdot \text{m})\mathbf{k}$.

3.71   Determine the dimension a for which the force-couple system shown may be replaced by a single equivalent force. If $F = 40$ lb and $M = 200$ lb·in., determine the point where the line of action of the single equivalent force intersects the yz plane.

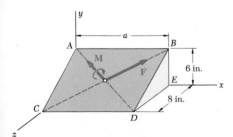

Fig. P3.71 and P3.72

3.72   A force $\mathbf{F}$ of magnitude 40 lb and a couple $\mathbf{M}$ of magnitude 200 lb·in. act on a wedge as shown. Knowing that $a = 12$ in., determine an equivalent force-couple system at point E.

3.16. Reduction of a System of Forces to One Force and One Couple.

Consider a system of forces F_1, F_2, F_3, etc., acting on a rigid body at the points A_1, A_2, A_3, etc., defined by the position vectors r_1, r_2, r_3, etc. (Fig. 3.41a). As seen in the preceding section, F_1 may be moved from A_1 to a given point O if a couple of moment M_1 equal to the moment $r_1 \times F_1$ of F_1 about O is added to the original system of forces. Repeating this procedure with F_2, F_3, etc., we obtain the system shown in Fig. 3.41b, consisting of forces acting at O and of couples.

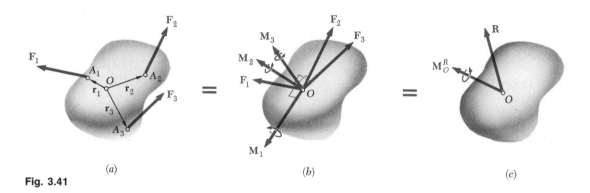

(a)   (b)   (c)

Fig. 3.41

Since the forces are now concurrent, they may be added vectorially and replaced by their resultant R. Similarly, the couple vectors M_1, M_2, M_3, etc., may be added vectorially and replaced by a single couple vector M_O^R. Any system of forces, however complex, may thus be reduced to an *equivalent force-couple system acting at a given point O* (Fig. 3.41c). We should note that, while each of the couple vectors M_1, M_2, M_3, etc., in Fig. 3.41b is perpendicular to the corresponding force, the resultant force R and the resultant couple vector M_O^R in Fig. 3.41c will not, in general, be perpendicular to each other.

The equivalent force-couple system is defined by the equations

$$R = \Sigma F \qquad M_O^R = \Sigma M_O = \Sigma(r \times F) \qquad (3.52)$$

which express that the force R is obtained by adding all the forces of the system, while the moment M_O^R of the couple, called *moment resultant* of the system, is obtained by adding the moments about O of all the forces of the system.

Once a given system of forces has been reduced to a force and a couple at a point O, it may easily be reduced to a force and a couple at another point O'. While the resultant force R will remain unchanged, the new couple vector $M_{O'}^R$ will be equal to the sum of the couple vector $\mathbf{M}_O^R$ and of the moment about O' of the force $\mathbf{R}$ attached at O (Fig. 3.42). We have

$$\mathbf{M}_{O'}^R = \mathbf{M}_O^R + \mathbf{s} \times \mathbf{R} \qquad (3.53)$$

In practice, the reduction of a given system of forces to a single force $\mathbf{R}$ at O and a couple $\mathbf{M}_O^R$ will be carried out in terms of components. Resolving each position vector $\mathbf{r}$ and each force $\mathbf{F}$ of the system into rectangular components, we write

$$\mathbf{r} = x\mathbf{i} + y\mathbf{j} + z\mathbf{k} \qquad (3.54)$$
$$\mathbf{F} = F_x\mathbf{i} + F_y\mathbf{j} + F_z\mathbf{k} \qquad (3.55)$$

Substituting for $\mathbf{r}$ and $\mathbf{F}$ into (3.52) and factoring the unit vectors $\mathbf{i}$, $\mathbf{j}$, $\mathbf{k}$, we obtain $\mathbf{R}$ and $\mathbf{M}_O^R$ in the form

$$\mathbf{R} = R_x\mathbf{i} + R_y\mathbf{j} + R_z\mathbf{k} \qquad \mathbf{M}_O^R = M_x^R\mathbf{i} + M_y^R\mathbf{j} + M_z^R\mathbf{k} \qquad (3.56)$$

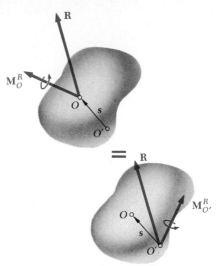

Fig. 3.42

The components R_x, R_y, R_z represent, respectively, the sums of the x, y, and z components of the given forces and measure the tendency of the system to impart to the rigid body a motion of translation in the x, y, or z direction. Similarly, the components M_x^R, M_y^R, M_z^R represent, respectively, the sums of the moments of the given forces about the x, y, and z axes and measure the tendency of the system to impart to the rigid body a motion of rotation about the x, y, or z axis.

If the magnitude and direction of the force $\mathbf{R}$ are desired, they may be obtained from the components R_x, R_y, R_z by means of the relations (2.18) and (2.19) of Sec. 2.11; similar computations will yield the magnitude and direction of the couple vector $\mathbf{M}_O^R$.

3.17. Equivalent Systems of Forces. We have seen in the preceding section that any system of forces acting on a rigid body may be reduced to a force-couple system at a given point O. This equivalent force-couple system characterizes completely the effect of the given system on the rigid body. *Two systems of forces are equivalent, therefore, if they may be reduced to the same force-couple system at a given point O.* Recalling that the force-couple system at O is defined by the relations (3.52), we state: *Two systems of forces $\mathbf{F}_1$, $\mathbf{F}_2$, $\mathbf{F}_3$, etc., and $\mathbf{F}_1'$, $\mathbf{F}_2'$, $\mathbf{F}_3'$, etc., are equivalent if, and only if, the sums of the forces and the sums of the moments about a given point O of the forces of the two systems are, respectively, equal.* Expressed mathematically, the necessary and sufficient conditions for the two systems of forces to be equivalent are

$$\Sigma \mathbf{F} = \Sigma \mathbf{F}' \quad \text{and} \quad \Sigma \mathbf{M}_O = \Sigma \mathbf{M}'_O \qquad (3.57)$$

Note that, to prove that two systems of forces are equivalent, the second of the relations (3.57) needs to be established with respect to *only one point O.* It will hold, however, with respect to *any point* if the two systems are equivalent.

Resolving the forces and moments in (3.57) into their rectangular components, we may express the necessary and sufficient conditions for the equivalence of two systems of forces acting on a rigid body as follows:

$$\begin{aligned} \Sigma F_x = \Sigma F'_x \qquad \Sigma F_y = \Sigma F'_y \qquad \Sigma F_z = \Sigma F'_z \\ \Sigma M_x = \Sigma M'_x \qquad \Sigma M_y = \Sigma M'_y \qquad \Sigma M_z = \Sigma M'_z \end{aligned} \quad (3.58)$$

These equations have a simple physical significance. They express that two systems of forces are equivalent if they tend to impart to the rigid body (1) the same translation in the x, y, and z directions, respectively, and (2) the same rotation about the x, y, and z axes, respectively.

3.18. Equipollent Systems of Vectors. When two systems of vectors satisfy Eqs. (3.57) or (3.58), i.e., when their resultants and their moment resultants about an arbitrary point O are respectively equal, the two systems are said to be *equipollent.* The result established in the preceding section may thus be restated as follows: *If two systems of forces acting on a rigid body are equipollent, they are also equivalent.*

It is important to note that this statement does not apply to *any* system of vectors. Consider for example a system of forces acting on a set of independent particles which do *not* form a rigid body. A different system of forces acting on the same particles may happen to be equipollent to the first one, i.e., it may have the same resultant and the same moment resultant. Yet, since different forces will now act on the various particles, their effects on these particles will be different; the two systems of forces, while equipollent, are *not equivalent.*

Summarizing, we can state that two equipollent systems of forces acting on a rigid body are equivalent. Similarly, two equipollent systems of sliding vectors (see Sec. 3.2) are also equivalent. However, any other two equipollent systems of vectors will not in general be equivalent.

3.19. Further Reduction of a System of Forces. We saw in Sec. 3.16 that any given system of forces acting on a rigid body may be reduced to an equivalent force-couple system at O, consisting of a force $\mathbf{R}$ equal to the sum of the forces of the system, and of a couple vector $\mathbf{M}_O^R$ equal to the moment resultant of the system.

When $\mathbf{R} = 0$, the force-couple system reduces to the couple vector $\mathbf{M}_O^R$. The given system of forces may then be reduced to a single couple, called the *resultant couple* of the system.

We shall now investigate the conditions under which a given system of forces may be reduced to a single force. It follows from Sec. 3.15 that the force-couple system at O may be replaced by a single force $\mathbf{R}$ acting along a new line of action if $\mathbf{R}$ and $\mathbf{M}_O^R$ are mutually perpendicular. The systems of forces which may be reduced to a single force, or *resultant,* are therefore the systems for which the force $\mathbf{R}$ and the couple vector $\mathbf{M}_O^R$ are mutually perpendicular. While this condition *is generally not satisfied* by systems of forces in space, it *will be satisfied* by systems consisting (1) of concurrent forces, (2) of coplanar forces, or (3) of parallel forces. We shall discuss these cases separately.

1. *Concurrent forces* are applied at the same point and may therefore be added directly into their resultant $\mathbf{R}$. Thus, they always reduce to a single force. Concurrent forces have been discussed in detail in Chap. 2.

2. *Coplanar forces* act in the same plane, which we shall assume here to be the plane of the figure (Fig. 3.43*a*). The sum $\mathbf{R}$ of the forces of the system will also lie in the plane of the figure, while the moment of each force about O, and thus the moment resultant $\mathbf{M}_O^R$, will be perpendicular to that plane. The force-couple system at O consists therefore of a force $\mathbf{R}$ and a couple vector $\mathbf{M}_O^R$ which are mutually perpendicular (Fig. 3.43*b*).† They may be reduced to a single force $\mathbf{R}$ by moving $\mathbf{R}$ in the plane of the figure until its moment about O becomes equal to $\mathbf{M}_O^R$. The distance from O to the line of action of $\mathbf{R}$ is $d = M_O^R / R$ (Fig. 3.43*c*).

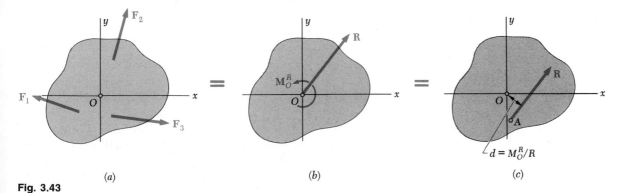

(a)   (b)   (c)

Fig. 3.43

† Since the couple vector $\mathbf{M}_O^R$ is perpendicular to the plane of the figure, it has been represented by the symbol ↻. A counterclockwise couple ↺ corresponds to a vector pointing out of the paper, and a clockwise couple ↻ to a vector pointing into the paper.

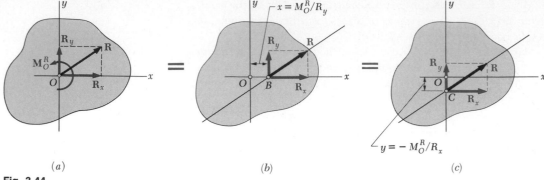

(a)　　　　　　　　　(b)　　　　　　　　　(c)

Fig. 3.44

As noted in Sec. 3.16, the reduction of a system of forces is considerably simplified if the forces are resolved into rectangular components. The force-couple system at O is then characterized by the components (Fig. 3.44a)

$$R_x = \Sigma F_x \qquad R_y = \Sigma F_y \qquad M_z^R = M_O^R = \Sigma M_O \quad (3.59)$$

To reduce the system to a single force $\mathbf{R}$ we shall express that the moment of $\mathbf{R}$ about O must be equal to $\mathbf{M}_O^R$. Denoting by x and y the coordinates of the point of application A of the resultant, and recalling formula (3.22), we write

$$xR_y - yR_x = M_O^R$$

which represents the equation of the line of action of $\mathbf{R}$. We may also determine directly the x and y intercepts of the line of action of the resultant by noting that $\mathbf{M}_O^R$ must be equal to the moment about O of the y component of $\mathbf{R}$ when $\mathbf{R}$ is attached at B (Fig. 3.44b), and to the moment of its x component when $\mathbf{R}$ is attached at C (Fig. 3.44c).

3. *Parallel forces* have parallel lines of action and may or may not have the same sense. Assuming here that the forces are parallel to the y axis (Fig. 3.45a), we note that their sum $\mathbf{R}$ will also be parallel to the y axis. On the other hand, since the moment of a given force must be perpendicular to that force, the moment about O of each force of the system, and thus the moment resultant $\mathbf{M}_O^R$, will lie in the zx plane. The force-couple system at O consists therefore of a force $\mathbf{R}$ and a

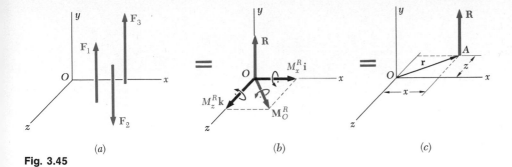

Fig. 3.45

couple vector $\mathbf{M}_O^R$ which are mutually perpendicular (Fig. 3.45b). They may be reduced to a single force $\mathbf{R}$ (Fig. 3.45c) or, if $\mathbf{R} = 0$, to a single couple of moment $\mathbf{M}_O^R$.

In practice, the force-couple system at O will be characterized by the components

$$R_y = \Sigma F_y \qquad M_x^R = \Sigma M_x \qquad M_z^R = \Sigma M_z \qquad (3.60)$$

The reduction of the system to a single force may be carried out by moving $\mathbf{R}$ to a new point of application $A(x,0,z)$ chosen so that the moment of $\mathbf{R}$ about O is equal to $\mathbf{M}_O^R$. We write

$$\mathbf{r} \times \mathbf{R} = \mathbf{M}_O^R$$
$$(x\mathbf{i} + z\mathbf{k}) \times R_y\mathbf{j} = M_x^R\mathbf{i} + M_z^R\mathbf{k}$$

Computing the vector products and equating the coefficients of the corresponding unit vectors in both members of the equation, we obtain two scalar equations which define the coordinates of A:

$$-zR_y = M_x^R \qquad xR_y = M_z^R$$

These equations express that the moments of $\mathbf{R}$ about the x and z axes must, respectively, be equal to M_x^R and M_z^R.

In the general case of a system of forces in space, the force-couple system at O consists of a force $\mathbf{R}$ and a couple vector $\mathbf{M}_O^R$ which are not perpendicular, and neither of which is zero (Fig. 3.46a). Thus, the system of forces *cannot* be reduced to a single force or a single couple. The couple vector, however, may be replaced by two other couple vectors obtained by resolving $\mathbf{M}_O^R$

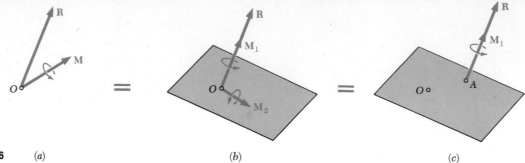

Fig. 3.46   (a)   (b)   (c)

into a component $\mathbf{M}_1$ along $\mathbf{R}$ and a component $\mathbf{M}_2$ in a plane perpendicular to $\mathbf{R}$ (Fig. 3.46b). The couple vector $\mathbf{M}_2$ and the force $\mathbf{R}$ may then be replaced by a single force $\mathbf{R}$ acting along a new line of action. The original system of forces thus reduces to $\mathbf{R}$ and to the couple vector $\mathbf{M}_1$ (Fig. 3.46c), i.e., to $\mathbf{R}$ and a couple acting in the plane perpendicular to $\mathbf{R}$. This particular force-couple system is called a *wrench* because the resulting combination of push and twist is the same that would be caused by an actual wrench. The line of action of $\mathbf{R}$ is known as the *axis of the wrench*, and the ratio M_1/R is called the *pitch* of the wrench.

Recalling the expression (3.35) obtained for the projection of a vector on the line of action of another vector, we note that the projection of $\mathbf{M}_O^R$ on the line of action of $\mathbf{R}$ is

$$M_1 = \frac{\mathbf{R} \cdot \mathbf{M}_O^R}{R} \tag{3.61}$$

Thus†

$$\text{Pitch of wrench} = \frac{M_1}{R} = \frac{\mathbf{R} \cdot \mathbf{M}_O^R}{R^2} \tag{3.62}$$

An important particular case of the reduction of a system of forces to a force-couple system occurs when both the force $\mathbf{R}$ and the couple vector $\mathbf{M}_O^R$ are equal to zero. The system of forces is said to be *equivalent to zero*. Such a system has no effect on the rigid body on which it acts, and the rigid body is said to be in *equilibrium*. This case will be considered in detail in the next chapter.

---

† The expressions obtained for the projection of the couple vector on the line of action of $\mathbf{R}$ and for the pitch of the wrench are independent of the choice of point O. Using the relation (3.53), we check that if a different point O' had been used, the numerator in (3.61) and (3.62) would be

$$\mathbf{R} \cdot \mathbf{M}_{O'}^R = \mathbf{R} \cdot (\mathbf{M}_O^R + \mathbf{s} \times \mathbf{R}) = \mathbf{R} \cdot \mathbf{M}_O^R + \mathbf{R} \cdot (\mathbf{s} \times \mathbf{R})$$

Since the mixed triple product $\mathbf{R} \cdot (\mathbf{s} \times \mathbf{R})$ is identically equal to zero, we have

$$\mathbf{R} \cdot \mathbf{M}_{O'}^R = \mathbf{R} \cdot \mathbf{M}_O^R \tag{3.63}$$

Thus, the scalar product $\mathbf{R} \cdot \mathbf{M}_O^R$ is independent of the choice of point O.

SAMPLE PROBLEM 3.8

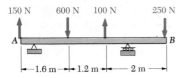

150 N   600 N   100 N   250 N

A   B

|—1.6 m—|—1.2 m—|—— 2 m ——|

A 4.80-m beam is subjected to the forces shown. Reduce the given system of forces to (*a*) an equivalent force-couple system at *A*, (*b*) an equivalent force-couple system at *B*, (*c*) a single force or resultant.

Note. Since the reactions at the supports are not included in the given system of forces, the given system will not maintain the beam in equilibrium.

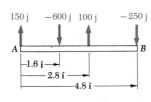

150 j   −600 j 100 j   −250 j

A   B

|—1.6 i—|
|—— 2.8 i ——|
|———— 4.8 i ————|

−(600 N)j

A   B

−(1880 N · m)k

a. **Force-couple System at *A*.** The force-couple system at *A* equivalent to the given system of forces consists of a force **R** and a couple $\mathbf{M}_A^R$ defined as follows:

$$\mathbf{R} = \Sigma\mathbf{F}$$
$$= (150\text{ N})\mathbf{j} - (600\text{ N})\mathbf{j} + (100\text{ N})\mathbf{j} - (250\text{ N})\mathbf{j} = -(600\text{ N})\mathbf{j}$$
$$\mathbf{M}_A^R = \Sigma(\mathbf{r} \times \mathbf{F})$$
$$= (1.6\mathbf{i}) \times (-600\mathbf{j}) + (2.8\mathbf{i}) \times (100\mathbf{j}) + (4.8\mathbf{i}) \times (-250\mathbf{j})$$
$$= -(1880\text{ N} \cdot \text{m})\mathbf{k}$$

The equivalent force-couple system at *A* is thus

$$\mathbf{R} = 600\text{ N} \downarrow \qquad \mathbf{M}_A^R = 1880\text{ N} \cdot \text{m} \; \rlap{)} \quad \blacktriangleleft$$

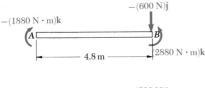

−(1880 N · m)k   −(600 N)j

A   B

|———— 4.8 m ————| (2880 N · m)k

b. **Force-couple System at *B*.** We shall find a force-couple system at *B* equivalent to the force-couple system at *A* determined in part *a*. The force **R** is unchanged, but a new couple $\mathbf{M}_B^R$ must be determined, the moment of which is equal to the moment about *B* of the force-couple system determined in part *a*. Thus, we have

$$\mathbf{M}_B^R = \mathbf{M}_A^R + \overrightarrow{BA} \times \mathbf{R}$$
$$= -(1880\text{ N} \cdot \text{m})\mathbf{k} + (-4.8\text{ m})\mathbf{i} \times (-600\text{ N})\mathbf{j}$$
$$= -(1880\text{ N} \cdot \text{m})\mathbf{k} + (2880\text{ N} \cdot \text{m})\mathbf{k} = +(1000\text{ N} \cdot \text{m})\mathbf{k}$$

The equivalent force-couple system at *B* is thus

$$\mathbf{R} = 600\text{ N} \downarrow \qquad \mathbf{M}_B^R = 1000\text{ N} \cdot \text{m} \; \rlap{)} \quad \blacktriangleleft$$

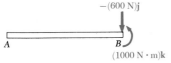

−(600 N)j

A   B

(1000 N · m)k

c. **Single Force or Resultant.** The resultant of the given system of forces is equal to **R** and its point of application must be such that the moment of **R** about *A* is equal to $\mathbf{M}_A^R$. We write

$$\mathbf{r} \times \mathbf{R} = \mathbf{M}_A^R$$
$$x\mathbf{i} \times (-600\text{ N})\mathbf{j} = -(1880\text{ N} \cdot \text{m})\mathbf{k}$$
$$-x(600\text{ N})\mathbf{k} = -(1880\text{ N} \cdot \text{m})\mathbf{k}$$

and conclude that $x = 3.13$ m. Thus, the single force equivalent to the given system is defined as

$$\mathbf{R} = 600\text{ N} \downarrow \qquad x = 3.13\text{ m} \quad \blacktriangleleft$$

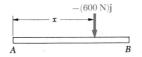

|——— x ———| −(600 N)j

A   B

SAMPLE PROBLEM 3.9

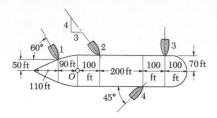

Four tugboats are used to bring an ocean liner to its pier. Each tugboat exerts a 5000-lb force in the direction shown. Determine (a) the equivalent force-couple system at the foremast O, (b) the point on the hull where a single, more powerful tugboat should push to produce the same effect as the original four tugboats.

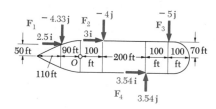

a. Force-couple System at O. Each of the given forces is resolved into components in the diagram shown (kip units are used). The force-couple system at O equivalent to the given system of forces consists of a force $\mathbf{R}$ and a couple $\mathbf{M}_O^R$ defined as follows:

$$\mathbf{R} = \Sigma\mathbf{F}$$
$$= (2.50\mathbf{i} - 4.33\mathbf{j}) + (3.00\mathbf{i} - 4.00\mathbf{j}) + (-5.00\mathbf{j}) + (3.54\mathbf{i} + 3.54\mathbf{j})$$
$$= 9.04\mathbf{i} - 9.79\mathbf{j}$$

$$\mathbf{M}_O^R = \Sigma(\mathbf{r} \times \mathbf{F})$$
$$= (-90\mathbf{i} + 50\mathbf{j}) \times (2.50\mathbf{i} - 4.33\mathbf{j})$$
$$+ (100\mathbf{i} + 70\mathbf{j}) \times (3.00\mathbf{i} - 4.00\mathbf{j})$$
$$+ (400\mathbf{i} + 70\mathbf{j}) \times (-5.00\mathbf{j})$$
$$+ (300\mathbf{i} - 70\mathbf{j}) \times (3.54\mathbf{i} + 3.54\mathbf{j})$$
$$= (390 - 125 - 400 - 210 - 2000 + 1062 + 248)\mathbf{k}$$
$$= -1035\mathbf{k}$$

The equivalent force-couple system at O is thus

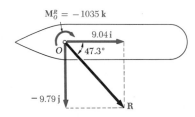

$$\mathbf{R} = (9.04 \text{ kips})\mathbf{i} - (9.79 \text{ kips})\mathbf{j} \qquad \mathbf{M}_O^R = -(1035 \text{ kip} \cdot \text{ft})\mathbf{k}$$
or $\qquad \mathbf{R} = 13.33 \text{ kips} \searrow 47.3° \qquad \mathbf{M}_O^R = 1035 \text{ kip} \cdot \text{ft} \downarrow$ ◀

Remark. Since all the forces are contained in the plane of the figure, we could have expected the sum of their moments to be perpendicular to that plane. Note that the moment of each force component could have been obtained directly from the diagram by forming the product of its magnitude and its perpendicular distance to O, and assigning to this product a positive or a negative sign, depending upon the sense of the moment.

b. Single Tugboat. The force exerted by a single tugboat must be equal to $\mathbf{R}$ and its point of application A must be such that the moment of $\mathbf{R}$ about O is equal to $\mathbf{M}_O^R$. Observing that the position vector of A is

$$\mathbf{r} = x\mathbf{i} + 70\mathbf{j}$$

we write

$$\mathbf{r} \times \mathbf{R} = \mathbf{M}_O^R$$
$$(x\mathbf{i} + 70\mathbf{j}) \times (9.04\mathbf{i} - 9.79\mathbf{j}) = -1035\mathbf{k}$$
$$-x(9.79)\mathbf{k} - 633\mathbf{k} = -1035\mathbf{k}$$
$$x = 41.1 \text{ ft} \quad ◀$$

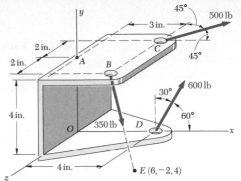

Three cables are attached to a bracket as shown. Replace the forces exerted by the cables with an equivalent force-couple system at A.

Solution. We first determine the vectors $\Delta \mathbf{r}$ joining point A with the points of application of the forces and resolve the forces into rectangular components. Observing that $\mathbf{F}_B = (350 \text{ lb})\boldsymbol{\lambda}_{BE}$ where

$$\boldsymbol{\lambda}_{BE} = \frac{\overrightarrow{BE}}{BE} = \frac{1}{7}(3\mathbf{i} - 6\mathbf{j} + 2\mathbf{k})$$

we have

$$\Delta \mathbf{r}_B = \overrightarrow{AB} = 3\mathbf{i} + 2\mathbf{k} \qquad \mathbf{F}_B = 150\mathbf{i} - 300\mathbf{j} + 100\mathbf{k}$$
$$\Delta \mathbf{r}_C = \overrightarrow{AC} = 3\mathbf{i} - 2\mathbf{k} \qquad \mathbf{F}_C = 354\mathbf{i} \qquad\quad - 354\mathbf{k}$$
$$\Delta \mathbf{r}_D = \overrightarrow{AD} = 4\mathbf{i} - 4\mathbf{j} \qquad \mathbf{F}_D = 300\mathbf{i} + 520\mathbf{j}$$

The force-couple system at A equivalent to the given forces consists of a force $\mathbf{R} = \Sigma\mathbf{F}$ and a couple $\mathbf{M}_A^R = \Sigma(\Delta\mathbf{r} \times \mathbf{F})$. The force $\mathbf{R}$ is readily obtained by adding respectively the x, y, and z components of the forces:

$$\mathbf{R} = \Sigma\mathbf{F} = (804 \text{ lb})\mathbf{i} + (220 \text{ lb})\mathbf{j} - (254 \text{ lb})\mathbf{k} \quad \blacktriangleleft$$

The computation of $\mathbf{M}_A^R$ will be facilitated if we express the moments $\Delta\mathbf{r} \times \mathbf{F}$ in the form of determinants (Sec. 3.7):

$$\Delta \mathbf{r}_B \times \mathbf{F}_B = \begin{vmatrix} \mathbf{i} & \mathbf{j} & \mathbf{k} \\ 3 & 0 & 2 \\ 150 & -300 & 100 \end{vmatrix} = 600\mathbf{i} \qquad -900\mathbf{k}$$

$$\Delta \mathbf{r}_C \times \mathbf{F}_C = \begin{vmatrix} \mathbf{i} & \mathbf{j} & \mathbf{k} \\ 3 & 0 & -2 \\ 354 & 0 & -354 \end{vmatrix} = \qquad 354\mathbf{j}$$

$$\Delta \mathbf{r}_D \times \mathbf{F}_D = \begin{vmatrix} \mathbf{i} & \mathbf{j} & \mathbf{k} \\ 4 & -4 & 0 \\ 300 & 520 & 0 \end{vmatrix} = \qquad 3280\mathbf{k}$$

Adding the expressions obtained, we have

$$\mathbf{M}_A^R = \Sigma(\Delta\mathbf{r} \times \mathbf{F}) = (600 \text{ lb} \cdot \text{in.})\mathbf{i} + (354 \text{ lb} \cdot \text{in.})\mathbf{j} + (2380 \text{ lb} \cdot \text{in.})\mathbf{k} \quad \blacktriangleleft$$

The rectangular components of the force $\mathbf{R}$ and the couple $\mathbf{M}_A^R$ are shown in the adjoining sketch.

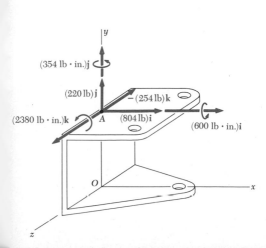

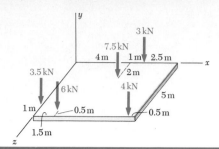

A rectangular slab, 5 by 7.5 m, supports five columns which exert on the slab the forces indicated. Determine the magnitude and point of application of the single force equivalent to the given forces.

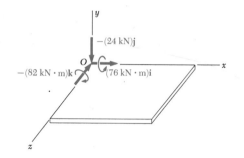

Solution. We shall first reduce the given system of forces to a force-couple system at the origin O of the coordinates. This force-couple system consists of a force $\mathbf{R}$ and a couple $\mathbf{M}_O^R$ defined as follows:

$$\mathbf{R} = \Sigma\mathbf{F} \qquad \mathbf{M}_O^R = \Sigma(\mathbf{r} \times \mathbf{F})$$

The position vectors of the points of application of the various forces are determined and the computations are arranged in tabular form.

r, m	F, kN	$r \times F$, kN·m
$5\mathbf{i}$	$-3\mathbf{j}$	$-15\mathbf{k}$
$4\mathbf{k}$	$-3.5\mathbf{j}$	$14\mathbf{i}$
$7\mathbf{i} + 5\mathbf{k}$	$-4\mathbf{j}$	$20\mathbf{i} - 28\mathbf{k}$
$1.5\mathbf{i} + 4.5\mathbf{k}$	$-6\mathbf{j}$	$27\mathbf{i} - 9\mathbf{k}$
$4\mathbf{i} + 2\mathbf{k}$	$-7.5\mathbf{j}$	$15\mathbf{i} - 30\mathbf{k}$
	$\mathbf{R} = -24\mathbf{j}$	$\mathbf{M}_O^R = 76\mathbf{i} - 82\mathbf{k}$

Since the force $\mathbf{R}$ and the couple vector $\mathbf{M}_O^R$ are mutually perpendicular, the force-couple system obtained may be reduced further to a single force $\mathbf{R}$. The new point of application of $\mathbf{R}$ will be selected in the plane of the slab and in such a way that the moment of $\mathbf{R}$ about O will be equal to $\mathbf{M}_O^R$. Denoting by $\mathbf{r}$ the position vector of the desired point of application, and by x and z its coordinates, we write

$$\mathbf{r} \times \mathbf{R} = \mathbf{M}_O^R$$
$$(x\mathbf{i} + z\mathbf{k}) \times (-24\mathbf{j}) = 76\mathbf{i} - 82\mathbf{k}$$
$$-24x\mathbf{k} + 24z\mathbf{i} = 76\mathbf{i} - 82\mathbf{k}$$

from which it follows that

$$-24x = -82 \qquad 24z = 76$$
$$x = 3.42 \text{ m} \qquad z = 3.17 \text{ m}$$

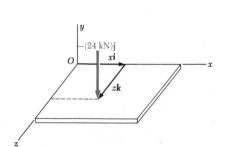

We conclude that the resultant of the given system of forces is

$$\mathbf{R} = 24 \text{ kN} \downarrow \qquad \text{at } x = 3.42 \text{ m}, z = 3.17 \text{ m} \quad \blacktriangleleft$$

PROBLEMS

3.73 A 4-m beam is loaded in the various ways represented in the figure. Find two loadings which are equivalent.

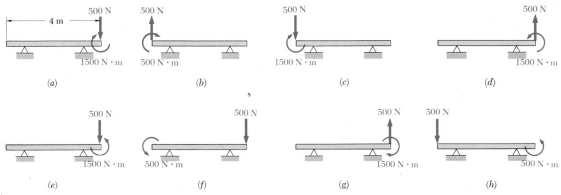

Fig. P3.73

3.74 A 4-m beam is loaded as shown. Determine the loading of Prob. 3.73 which is equivalent to this loading.

3.75 Two parallel forces **P** and **Q** are applied at the ends of a beam AB of length L. Find the distance x from A to the line of action of their resultant. Check the formula obtained by assuming $L = 8$ in. and (a) $P = 10$ lb down, $Q = 30$ lb down; (b) $P = 10$ lb down, $Q = 30$ lb up.

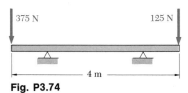

Fig. P3.74

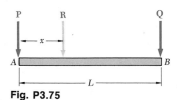

Fig. P3.75

3.76 Determine the distance from point A to the line of action of the resultant of the three forces shown when (a) $a = 1$ m, (b) $a = 1.5$ m, (c) $a = 2.5$ m.

3.77 Solve Prob. 3.76, assuming that the load at C is reduced to 6 kN.

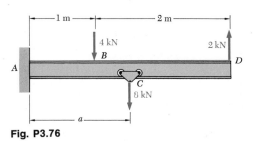

Fig. P3.76

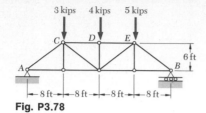

3 kips 4 kips 5 kips

6 ft

8 ft — 8 ft — 8 ft — 8 ft

Fig. P3.78

3.78 For the truss and loading shown, determine the resultant of the loads and the distance from point A to its line of action.

3.79 Four packages are transported at constant speed from A to B by the conveyor. At the instant shown, determine the resultant of the loading and the location of its line of action.

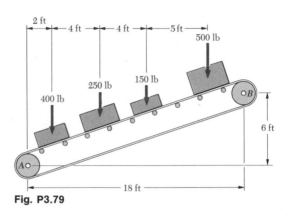

2 ft
4 ft
4 ft
5 ft
500 lb
150 lb
250 lb
400 lb
6 ft
18 ft

Fig. P3.79

3.80 Solve Prob. 3.79, assuming that the 250-lb package has been removed.

3.81 An angle bracket is subjected to the system of forces shown. Find the resultant of the system and the point of intersection of its line of action with (*a*) line AB, (*b*) line BC.

3.82 Find the resultant of the system shown and the point of intersection of its line of action with (*a*) line AC, (*b*) line CD.

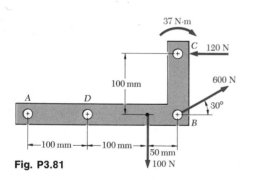

37 N·m

C 120 N

100 mm

600 N

30°

A D

B

100 mm — 100 mm

50 mm

100 N

Fig. P3.81

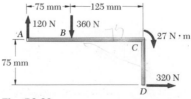

75 mm — 125 mm

120 N   360 N

A B

27 N · m

C

75 mm

320 N

D

Fig. P3.82

3.83 The roof of a building frame is subjected to the wind loading shown. Determine (*a*) the equivalent force-couple system at *D*, (*b*) the resultant of the loading and its line of action.

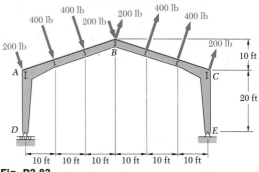

Fig. P3.83

3.84 A bracket is subjected to the system of forces and couples shown. Find the resultant of the system and the point of intersection of its line of action with (*a*) line *AB*, (*b*) line *BC*, (*c*) line *CD*.

3.85 Two cables exert forces of 90 kN each on a truss of weight *W* = 200 kN. Find the resultant force acting on the truss and the point of intersection of its line of action with line *AB*.

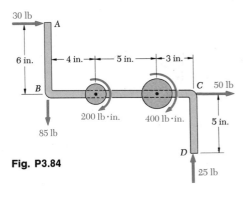

Fig. P3.84

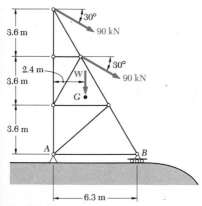

Fig. P3.85

3.86 A force **P** of given magnitude *P* is applied to the edge of a semicircular plate of radius *a* as shown. (*a*) Replace **P** by an equivalent force-couple system at the point *D* obtained by drawing the perpendicular from *B* to the *x* axis. (*b*) Determine the value of *θ* for which the moment of the equivalent force-couple system at *D* is maximum.

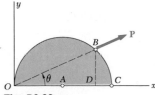

Fig. P3.86

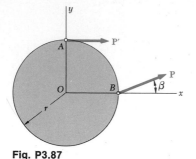

Fig. P3.87

*3.87 Two forces, each of magnitude P, are applied to the edge of a circular disk as shown. Knowing that the line of action of the resultant of the two forces is tangent to the edge of the disk, determine the required value of β.

3.88 The 12-ft boom AB has a fixed end A and is subjected to its 100-lb weight and to the 140-lb force exerted by cable BC. Determine the force and couple at A equivalent to the two given forces.

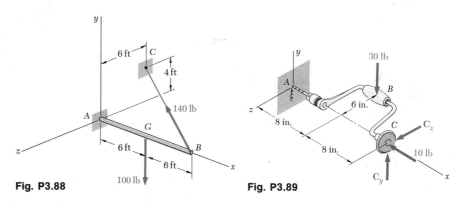

Fig. P3.88

Fig. P3.89

3.89 In drilling a hole in a wall, a man applies a vertical 30-lb force at B on the brace and bit, while pushing at C with a 10-lb force. The brace lies in the horizontal xz plane. (a) Determine the other components of the total force which should be exerted at C if the bit is not to be bent about the y and z axes (i.e., if the system of forces applied on the brace is to have zero moment about both the y and z axes). (b) Reduce the 30-lb force and the total force at C to an equivalent force and couple at A.

3.90 In order to tighten the joint between the tapped faucet A and the pipe AC, a plumber uses two pipe wrenches as shown. By exerting a 250-N force on each wrench, at a distance of 200 mm from the axis of the pipe and in a direction perpendicular to the pipe and to the wrench, he prevents the pipe from rotating, and thus avoids loosening or further tightening the joint between the pipe and the tapped elbow C. Replace the two given forces by an equivalent force-couple system at D and determine whether the plumber's action tends to tighten or loosen the joint between (a) pipe CD and elbow D, (b) elbow D and pipe DE. Assume all threads to be right-handed.

3.91 In Prob. 3.90, replace the two given forces by an equivalent force-couple system at E and determine whether the plumber's action tends to tighten or loosen the joint between (a) pipe DE and elbow E, (b) elbow E and pipe EO. Assume all threads to be right-handed.

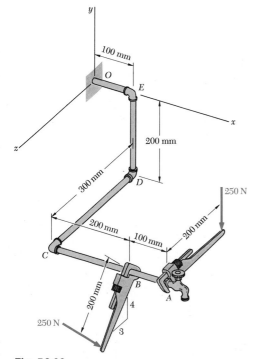

Fig. P3.90

3.92 Four horizontal forces act on a vertical quarter-circular plate of radius 250 mm. Determine the magnitude and point of application of the resultant of the four forces if $P = 40$ N.

3.93 Determine the magnitude of the force **P** for which the resultant of the four forces acts on the rim of the plate.

3.94 A square plate of side a supports four loads as shown. Determine the magnitude and point of application of the resultant of the four loads.

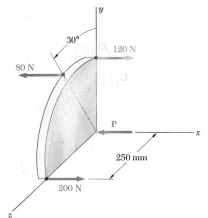

Fig. P3.92 and P3.93

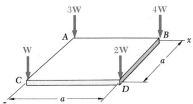

Fig. P3.94 and P3.95

3.95 A square plate of side a supports four loads as shown. Determine the magnitude and point of application of the smallest additional load which must be applied to the plate if the resultant of the five loads is to pass through the center of the plate.

3.96 Reduce the system of forces shown to a wrench. (Specify the axis and pitch of the wrench.)

3.97 Solve Prob. 3.96 if the 40-lb forces are replaced by 20-lb forces.

3.98 The worm-gear speed reducer shown weighs 300 N; the center of gravity is located on the x axis at $x = 200$ mm. Replace the weight and couples shown by a wrench. (Specify the axis and pitch of the wrench.)

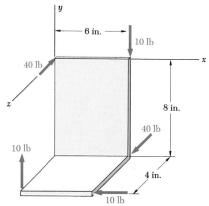

Fig. P3.96

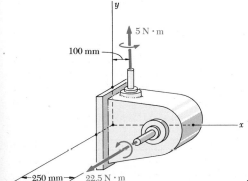

Fig. P3.98

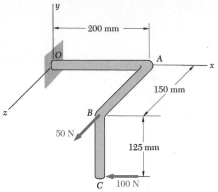

Fig. P3.99

3.99 Replace the two forces shown by (*a*) a force-couple system at the origin, (*b*) a wrench. (Specify the axis and pitch of the wrench.)

3.100 Two forces of magnitude P act along the diagonals of the faces of a cube of side a as shown. Replace the two forces by a system consisting of (*a*) a single force at O and a couple, (*b*) a wrench. (Specify the axis and pitch of the wrench.)

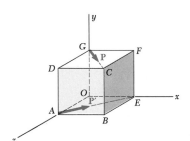

Fig. P3.100

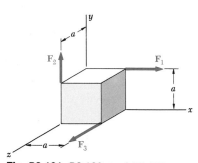

Fig. P3.101, P3.102, and P3.103

3.101 Three forces act on a cube of side a as shown. Determine the magnitude and axis of the wrench equivalent to the forces, if $F_1 = F_2 = P$ and $F_3 = 0$.

3.102 Three forces act on a cube of side a as shown. Determine the magnitude and axis of the wrench equivalent to the three forces, if $F_1 = F_2 = F_3 = P$.

*3.103 Three forces act on a cube of side a as shown. Determine the magnitude of $\mathbf{F}_3$, knowing that $F_1 = F_2 = P$ and that the system may be reduced to a single force $\mathbf{R}$. Also determine the magnitude of $\mathbf{R}$ and its line of action.

3.104 Knowing that $R = 70$ lb and $M = 140$ lb $\cdot$ in., replace the given wrench by a system of two forces chosen in such a way that one force acts at point B and the other force lies in the xz plane.

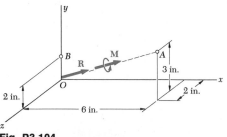

Fig. P3.104

3.105 (a) Reduce the wrench shown to a system consisting of two forces perpendicular to the y axis and applied respectively at A and B. (b) Solve part a assuming $R = 240$ N, $M = 40$ N·m, $a = 125$ mm, and $b = 250$ mm.

3.106 Show that, in general, a wrench may be replaced by two forces chosen in such a way that one force passes through a given point while the other force lies in a given plane.

*__3.107__ Show that a wrench may be replaced by two perpendicular forces, one of which is applied at a given point.

*__3.108__ Show that a wrench may be replaced by two forces, one of which has a prescribed line of action.

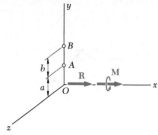

Fig. P3.105

REVIEW PROBLEMS

3.109 In order to move a 173-lb crate, two men push on it while two other men pull on it by means of ropes. The force exerted by man A is 150 lb, and that exerted by man B is 50 lb; both forces are horizontal. Man C pulls with a force equal to 80 lb and man D with a force equal to 120 lb. Both cables form an angle of 30° with the vertical. Determine the resultant of all forces acting on the crate.

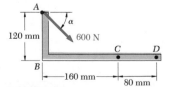

Fig. P3.109

3.110 Three horizontal forces are applied as shown to a machine arm. Determine the resultant of the loads and the point where its line of action intersects AD if the magnitude of **P** is (a) $P = 50$ lb, (b) $P = 500$ lb, (c) $P = 350$ lb.

3.111 A 150- by 300-mm plate is subjected to four loads. Find the resultant of the four loads and the two points at which the line of action of the resultant intersects the edge of the plate.

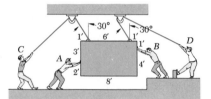

Fig. P3.110

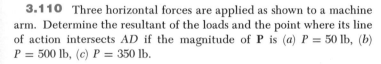

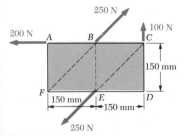

Fig. P3.111

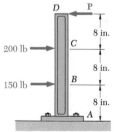

3.112 Knowing that $\alpha = 45°$, replace the 600-N force by (a) an equivalent force-couple system at C, (b) an equivalent system formed by two parallel forces at B and C.

Fig. P3.112

3.113 Four column loads act on a square foundation mat as shown. Determine the magnitude and point of application of the resultant of the four loads.

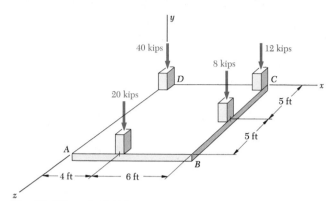

Fig. P3.113 and P3.114

3.114 Determine the magnitude and point of application of the smallest additional load which must be applied to the foundation mat if the resultant of the five loads is to pass through the center of the mat.

3.115 A force **P** of magnitude 700 lb acts along a line joining points C and D as shown. Determine the moment of **P** (*a*) about point A, (*b*) about line AB.

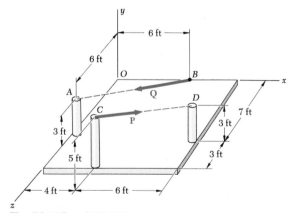

Fig. P3.115 and P3.116

3.116 A force **Q** of magnitude 210 lb acts along a line joining points B and A as shown. Determine the moment of **Q** (*a*) about point D, (*b*) about line DC.

3.117 A gearbox is acted upon by three couples and a horizontal force as shown. Determine (a) the equivalent force-couple system at the origin, (b) the equivalent wrench. (Specify the axis and pitch of the wrench.)

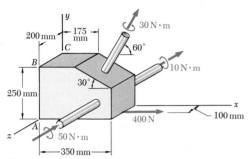

Fig. P3.117

3.118 Three forces act on the plate shown; the horizontal force **P** is parallel to the x axis and the vertical forces **Q** and −**Q** form a couple of moment $M = Qr$. Determine (a) the wrench equivalent to the applied system, (b) the locus of the points of intersection of the axis of the wrench with the yz plane as θ varies.

3.119 Given $P = 400$ N, $Q = 800$ N, and $r = 200$ mm, determine the wrench equivalent to the given system when (a) $θ = 30°$, (b) $θ = 90°$. (In each case specify the axis and pitch of the wrench.)

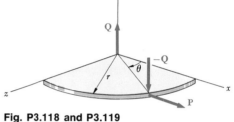

Fig. P3.118 and P3.119

3.120 Two forces are applied to the vertical pole as shown. Determine the components of the force and couple at O equivalent to the two forces.

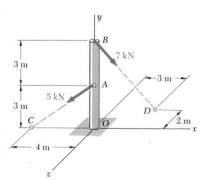

Fig. P3.120

CHAPTER 4 Equilibrium of Rigid Bodies

4.1. Rigid Body in Equilibrium. *A rigid body is said to be in equilibrium when the external forces acting on it form a system of forces equivalent to zero,* i.e., when the external forces may be reduced to no force and no couple. Setting $\mathbf{R}$ and $\mathbf{M}_O^R$ equal to zero in the relations (3.52), we obtain the following necessary and sufficient conditions for the equilibrium of a rigid body:

$$\Sigma \mathbf{F} = 0 \qquad \Sigma \mathbf{M}_O = \Sigma(\mathbf{r} \times \mathbf{F}) = 0 \qquad (4.1)$$

Resolving each force and each moment into its rectangular components, we find that the necessary and sufficient conditions for the equilibrium of a rigid body may also be expressed by the six scalar equations

$$\Sigma F_x = 0 \qquad \Sigma F_y = 0 \qquad \Sigma F_z = 0 \qquad (4.2)$$
$$\Sigma M_x = 0 \qquad \Sigma M_y = 0 \qquad \Sigma M_z = 0 \qquad (4.3)$$

Equations (4.2) express the fact that the components of the external forces in the x, y, and z directions are balanced; Eqs. (4.3) express the fact that the moments of the external forces about the x, y, and z axes are balanced. The system of the external forces, therefore, will impart no motion of translation or rotation to the rigid body considered.

4.2. Free-body Diagram. In solving a problem concerning the equilibrium of a rigid body, it is essential to consider *all* the forces acting on the body; it is equally important to exclude any force which is not directly applied on the body. Omitting a force or adding an extraneous one would destroy the conditions of equilibrium. Therefore, the first step in the solution of the problem should consist in drawing a *free-body diagram* of the rigid body under consideration. Free-body diagrams have already been used on many occasions in Chap. 2. However, in view of their importance to the solution of equilibrium problems, we shall summarize here the various steps which must be followed in drawing a free-body diagram.

First, a clear decision is made regarding the choice of the free body to be used. This body is then detached from the ground and separated from any other body. The contour of the body thus isolated is sketched.

All external forces are then indicated. These forces represent the action exerted *on* the free body *by* the ground and the bodies which have been detached; they should be applied at the various points where the free body was supported by the ground or connected to the other bodies. The *weight* of the free body should also be included among the external forces, since it represents the attraction exerted by the earth on the various particles forming the free body. As will be seen in Chap. 5, the weight should be applied at the center of gravity of the body. When the free body is made of several parts, the forces the various parts exert on each other should *not* be included among the external forces. These forces are internal forces as far as the free body is concerned.

The magnitude and direction of the *known external forces* should be clearly marked on the free-body diagram. Care should be taken to indicate the sense of the force exerted *on* the free body, not that of the force exerted *by* the free body. Known external forces generally include the *weight* of the free body and *forces applied* for a given purpose.

Unknown external forces usually consist of the *reactions*—sometimes called *constraining forces*—through which the ground and other bodies oppose a possible motion of the free body and thus constrain it to remain in the same position. Reactions are exerted at the points where the free body is *supported* or *connected* to other bodies. They will be discussed in detail in Secs. 4.3 and 4.8.

The free-body diagram should also include dimensions, since these may be needed in the computation of moments of forces. Any other detail, however, should be omitted.

EQUILIBRIUM IN TWO DIMENSIONS

4.3. Reactions at Supports and Connections for a Two-dimensional Structure. In the first part of this chapter we shall consider the equilibrium of a two-dimensional structure; i.e., we shall assume that the structure considered and the forces applied to it are contained in the plane of the figure. Clearly, the reactions needed to maintain the structure in the same position will also be contained in the plane of the figure.

The reactions exerted on a two-dimensional structure may be divided into three groups, corresponding to three types of *supports*, or *connections:*

1. *Reactions Equivalent to a Force with Known Line of Action.* Supports and connections causing reactions of this group include *rollers, rockers, frictionless surfaces, short links and cables, collars on frictionless rods,* and *frictionless pins in slots.* Each of these supports and connections can prevent motion in one direction only. They are shown in Fig. 4.1, together with the reaction they produce. Reactions of this group involve *one unknown,* namely, the magnitude of the reaction; this magnitude should be denoted by an appropriate letter. The line of action of the reaction is known and should be indicated clearly in the free-body diagram. The sense of the reaction must be as shown in Fig. 4.1 in the case of a frictionless surface (away from the surface) or of a cable (tension in the direction of the cable). The reaction may be directed either way in the case of double-track rollers, links, collars on rods, and pins in slots. Single-track rollers and rockers are generally assumed to be reversible, and thus the corresponding reactions may also be directed either way.

2. *Reactions Equivalent to a Force of Unknown Direction.* Supports and connections causing reactions of this group include *frictionless pins in fitted holes, hinges,* and *rough surfaces.* They can prevent translation of the free body in all directions, but they cannot prevent the body from rotating about the connection. Reactions of this group involve *two unknowns* and are usually represented by their x and y components. In the case of a rough surface, the component normal to the surface must be directed away from the surface.

3. *Reactions Equivalent to a Force and a Couple.* These reactions are caused by *fixed supports* which oppose any motion of the free body and thus constrain it completely. Fixed supports actually produce forces over the entire surface of contact; these forces, however, form a system which may be reduced to a force and a couple. Reactions of this group

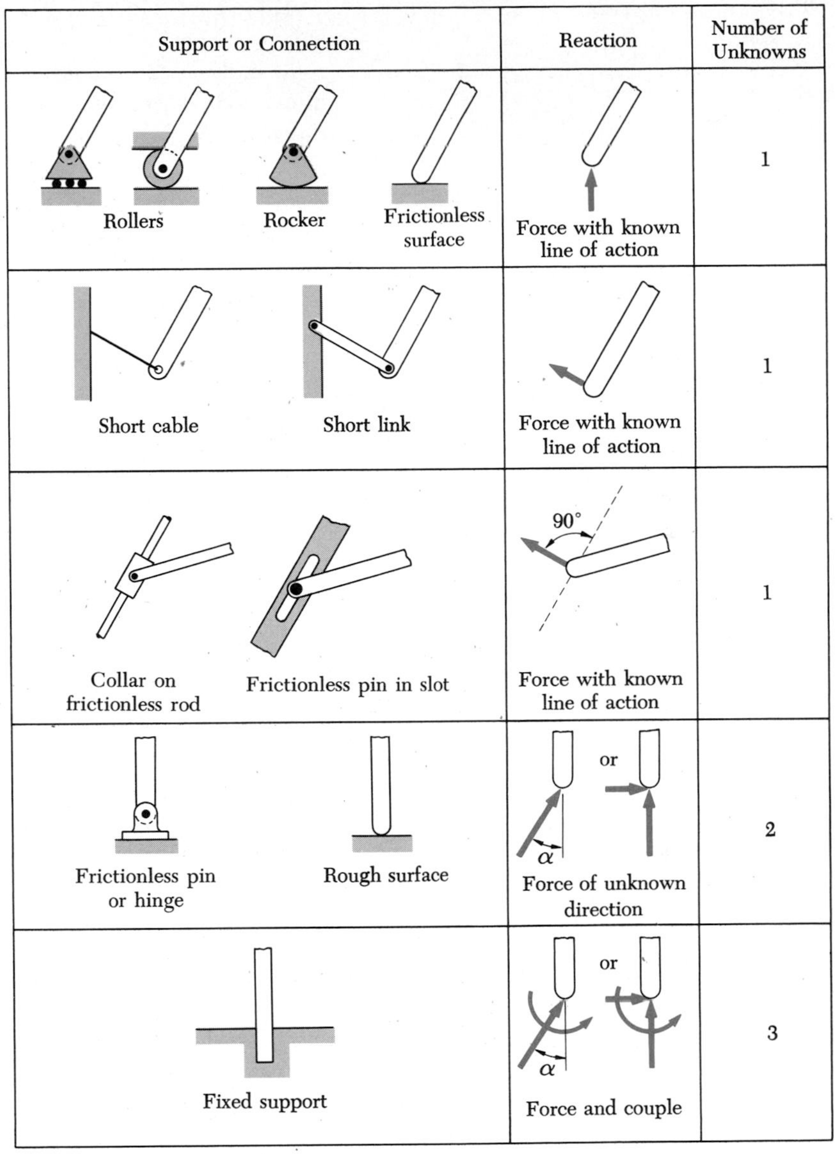

Support or Connection				Reaction	Number of Unknowns
Rollers	Rocker	Frictionless surface		Force with known line of action	1
Short cable		Short link		Force with known line of action	1
Collar on frictionless rod		Frictionless pin in slot		Force with known line of action	1
Frictionless pin or hinge		Rough surface		Force of unknown direction	2
Fixed support				Force and couple	3

Fig. 4.1 Reactions at supports and connections.

involve *three unknowns,* consisting usually of the two components of the force and the moment of the couple.

When the sense of an unknown force or couple is not clearly apparent, no attempt should be made at determining it. Instead, the sense of the force or couple should be arbitrarily assumed; the sign of the answer obtained will indicate whether the assumption is correct or not.

4.4. Equilibrium of a Rigid Body in Two Dimensions.

The conditions stated in Sec. 4.1 for the equilibrium of a rigid body become considerably simpler in the case of a two-dimensional structure. Choosing the x and y axes in the plane of the structure, we have

$$F_z = 0 \qquad M_x = M_y = 0 \qquad M_z = M_O$$

for each of the forces applied to the structure. Thus, the six equations of equilibrium derived in Sec. 4.1 reduce to

$$\Sigma F_x = 0 \qquad \Sigma F_y = 0 \qquad \Sigma M_O = 0 \qquad (4.4)$$

and to three trivial identities $0 = 0$. Since the third of the equations (4.4) must be satisfied regardless of the choice of the origin O, we may write the equations of equilibrium for a two-dimensional structure in the more general form

$$\Sigma F_x = 0 \qquad \Sigma F_y = 0 \cdot \qquad \Sigma M_A = 0 \qquad (4.5)$$

where A is any point in the plane of the structure. The three equations obtained may be solved for no more than *three unknowns*.

We saw in the preceding section that unknown forces usually consist of reactions, and that the number of unknowns corresponding to a given reaction depends upon the type of support or connection causing that reaction. Referring to Sec. 4.3, we check that the equilibrium equations (4.5) may be used to determine the reactions of two rollers and one cable, or of one fixed support, or of one roller and one pin in a fitted hole, etc.

Consider, for instance, the truss shown in Fig. 4.2a, which is subjected to the given forces **P**, **Q**, and **S**. The truss is held in place by a pin at A and a roller at B. The pin prevents point A from moving by exerting on the truss a force which may be resolved into the components $\mathbf{A}_x$ and $\mathbf{A}_y$; the roller keeps the truss from rotating about A by exerting the vertical force **B**. The free-body diagram of the truss is shown in Fig. 4.2b; it includes the reactions $\mathbf{A}_x$, $\mathbf{A}_y$, and **B** as well as the applied forces **P**, **Q**, **S**, and the weight **W** of the truss. Expressing that the sum of the moments about A of all the forces shown in Fig. 4.2b is zero, we write the equation $\Sigma M_A = 0$, which may be solved for the magnitude B since it does not contain A_x or A_y. Expressing, then, that the sum of the x components and the sum of the y components of the forces are zero, we write the equations $\Sigma F_x = 0$ and $\Sigma F_y = 0$, which may be solved for the components A_x and A_y, respectively.

Additional equations could be obtained by expressing that the sum of the moments of the external forces about points other

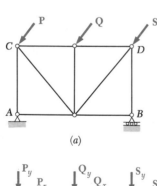

(a)

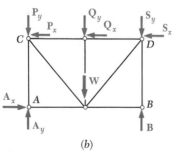

(b)

Fig. 4.2

than A is zero. We could write, for instance, $\Sigma M_B = 0$. Such a statement, however, does not contain any new information, since it has already been established that the system of the forces shown in Fig. 4.2b is equivalent to zero. The additional equation *is not independent* and cannot be used to determine a fourth unknown. It will be useful, however, for checking the solution obtained from the original three equations of equilibrium.

While the three equations of equilibrium cannot be *augmented* by additional equations, any of them may be *replaced* by another equation. Therefore, an alternate system of equations of equilibrium is

$$\Sigma F_x = 0 \qquad \Sigma M_A = 0 \qquad \Sigma M_B = 0 \qquad (4.6)$$

where the line AB is chosen in a direction different from the y direction (Fig. 4.2b). These equations are sufficient conditions for the equilibrium of the truss. The first two equations indicate that the external forces must reduce to a single vertical force at A. Since the third equation requires that the moment of this force be zero about a point B which is not on its line of action, the force must be zero and the rigid body is in equilibrium.

A third possible set of equations of equilibrium is

$$\Sigma M_A = 0 \qquad \Sigma M_B = 0 \qquad \Sigma M_C = 0 \qquad (4.7)$$

where the points A, B, and C are not in a straight line (Fig. (4.2b). The first equation requires that the external forces reduce to a single force at A; the second equation requires that this force pass through B; the third, that it pass through C. Since the points A, B, C are not in a straight line, the force must be zero, and the rigid body is in equilibrium.

The equation $\Sigma M_A = 0$, which expresses that the sum of the moments of the forces about pin A is zero, possesses a more definite physical meaning than either of the other two equations (4.7). These two equations express a similar idea of balance, but with respect to points about which the rigid body is not actually hinged. They are, however, as useful as the first equation, and our choice of equilibrium equations should not be unduly influenced by the physical meaning of these equations. Indeed, it will be desirable in practice to choose equations of equilibrium containing only one unknown, since this eliminates the necessity of solving simultaneous equations. Equations containing only one unknown may be obtained by summing moments about the point of intersection of the lines of action of two unknown forces or, if these forces are parallel, by summing components in a direction perpendicular to their common direction. In the case

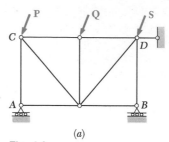

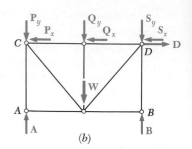

(a)

(b)

Fig. 4.3

of the truss of Fig. 4.3, for example, which is held by rollers at A and B and a short link at D, the reactions at A and B may be eliminated by summing x components. The reactions at A and D will be eliminated by summing moments about C and the reactions at B and D by summing moments about D. The equations obtained are

$$\Sigma F_x = 0 \qquad \Sigma M_C = 0 \qquad \Sigma M_D = 0$$

Each of these equations contains only one unknown.

4.5. Statically Indeterminate Reactions. Partial Constraints. In each of the two examples considered in the preceding section (Figs. 4.2 and 4.3), the types of supports used were such that the rigid body could not possibly move under the given loads or under any other loading conditions. In such cases, the rigid body is said to be *completely constrained*. We also recall that the reactions corresponding to these supports involved *three unknowns* and could be determined by solving the three equations of equilibrium. When such a situation exists, the reactions are said to be *statically determinate*.

Consider now the truss shown in Fig. 4.4a, which is held by pins at A and B. These supports provide more constraints than are necessary to keep the truss from moving under the given loads or under any other loading conditions. We also note from the free-body diagram of Fig. 4.4b that the corresponding reactions involve *four unknowns*. Since, as was pointed out in Sec. 4.4, only three independent equilibrium equations are available, there are *more unknowns than equations*, and all the unknowns cannot be determined. While the equations $\Sigma M_A = 0$ and $\Sigma M_B = 0$ yield the vertical components B_y and A_y, respectively, the equation $\Sigma F_x = 0$ gives only the sum $A_x + B_x$ of the horizontal components of the reactions at A and B. The components A_x and B_x are said to be *statically indeterminate*. They could be determined by considering the deformations produced in the truss by the given loading, but this method is beyond the scope of statics and belongs to the study of mechanics of materials.

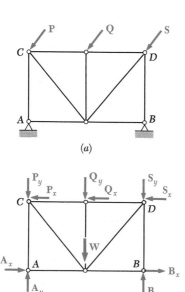

(a)

(b)

Fig. 4.4 Statically indeterminate reactions

The supports used to hold the truss shown in Fig. 4.5*a* consist of rollers at *A* and *B*. Clearly, the constraints provided by these supports are not sufficient to keep the truss from moving. While any vertical motion is prevented, the truss is free to move horizontally. The truss is said to be *partially constrained*.† Turning our attention to Fig. 4.5*b*, we note that the reactions at *A* and *B* involve only *two unknowns*. Since three equations of equilibrium must still be satisfied, there are *fewer unknowns than equations*, and one of the equilibrium equations will not be satisfied. While the equations $\Sigma M_A = 0$ and $\Sigma M_B = 0$ can be satisfied by a proper choice of reactions at *A* and *B*, the equation $\Sigma F_x = 0$ will not be satisfied unless the sum of the horizontal components of the applied forces happens to be zero. We thus check that the equilibrium of the truss of Fig. 4.5 cannot be maintained under general loading conditions.

It appears from the above that, if a rigid body is to be completely constrained and if the reactions at its supports are to be statically determinate, *there must be as many unknowns as there are equations of equilibrium*. When this condition is *not* satisfied, we may be sure that the rigid body is not completely constrained, or that the reactions at its supports are not statically determinate, or both.

We should note, however, that, while *necessary*, the above condition is *not sufficient*. In other words, the fact that the number of unknowns is equal to the number of equations is no guarantee that the body is completely constrained or that the reactions at its supports are statically determinate. Consider, for example, the truss shown in Fig. 4.6*a*, which is held by rollers at *A*, *B*, and *E*. While there are three unknown reactions, **A**, **B**, and **E** (Fig. 4.6*b*), we find that the equation $\Sigma F_x = 0$ will not be satisfied unless the sum of the horizontal components of the applied forces happens to be zero. There is a sufficient number of constraints, but these constraints are not properly arranged, and the truss is free to move horizontally. We say that the truss is *improperly constrained*. Since only two equilibrium equations are left for determining the three unknowns, the reactions will be statically indeterminate. Thus, improper constraints also produce static indeterminacy.

Another example of improper constraints—and of static indeterminacy—is provided by the truss shown in Fig. 4.7. This

---

† Partially constrained bodies are often referred to as *unstable*. However, in order to avoid any confusion between this type of instability, due to insufficient constraints, and the type of instability considered in Chap. 10, which relates to the behavior of a rigid body when its equilibrium is disturbed, we shall restrict the use of the words *stable* and *unstable* to the latter case.

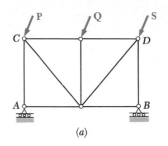

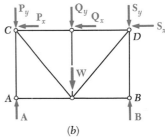

(a)

(b)

Fig. 4.5 Partial constraints

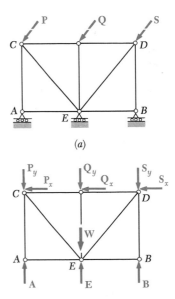

(a)

(b)

Fig. 4.6 Improper constraints

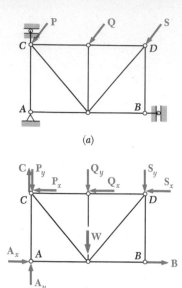

Fig. 4.7 Improper constraints

truss is held by a pin at A and by rollers at B and C, which altogether involve four unknowns. Since only three independent equilibrium equations are available, the reactions at the supports are statically indeterminate. On the other hand, we note that the equation $\Sigma M_A = 0$ cannot be satisfied under general loading conditions, since the lines of action of the reactions $\mathbf{B}$ and $\mathbf{C}$ are required to pass through A. We conclude that the truss can rotate about A and that it is improperly constrained.†

The examples of Figs. 4.6 and 4.7 lead us to conclude that *a rigid body is improperly constrained whenever the supports, even though they may provide a sufficient number of reactions, are arranged in such a way that the reactions must be either concurrent or parallel.‡*

In summary, to be sure that a two-dimensional rigid body is completely constrained and that the reactions at its supports are statically determinate, we should check that the reactions involve three—and only three—unknowns, and that the supports are arranged in such a way that they do not require the reactions to be either concurrent or parallel.

Supports involving statically indeterminate reactions should be used with care in the *design* of structures, and only with a full knowledge of the problems they may cause. On the other hand, the *analysis* of structures possessing statically indeterminate reactions often may be partially carried out by the methods of statics. In the case of the truss of Fig. 4.4, for example, the vertical components of the reactions at A and B were obtained from the equilibrium equations.

For obvious reasons, supports producing partial or improper constraints should be avoided in the design of stationary structures. However, a partially or improperly constrained structure will not necessarily collapse; under particular loading conditions, equilibrium may be maintained. For example, the trusses of Figs. 4.5 and 4.6 will be in equilibrium if the applied forces $\mathbf{P}$, $\mathbf{Q}$, and $\mathbf{S}$ are vertical. Besides, structures which are designed to move *should* be only partially constrained. A railroad car, for instance, would be of little use if it were completely constrained by having its brakes applied permanently.

---

† Rotation of the truss about A requires some "play" in the supports at B and C. In practice such play will always exist. Besides, we may note that if the play is kept small, the displacements of the rollers B and C and, thus, the perpendicular distances from A to the lines of action of the reactions $\mathbf{B}$ and $\mathbf{C}$, will also be small. The equation $\Sigma M_A = 0$ will then require that $\mathbf{B}$ and $\mathbf{C}$ be very large, a situation which may well result in the failure of the supports at B and C.

‡ Because this situation arises from an inadequate arrangement or *geometry* of the supports, it is often referred to as *geometric instability*.

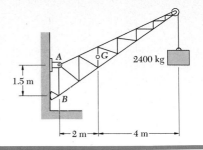

A fixed crane has a mass of 1000 kg and is used to lift a 2400-kg crate. It is held in place by a pin at A and a rocker at B. The center of gravity of the crane is located at G. Determine the components of the reactions at A and B.

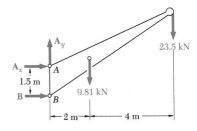

Solution. A free-body diagram of the crane is drawn. Multiplying the masses of the crane and of the crate by $g = 9.81$ m/s^2, we obtain the corresponding weights, that is, 9810 N or 9.81 kN, and 23 500 N or 23.5 kN. The reaction at pin A is a force of unknown direction, represented by its components A_x and A_y. The reaction at the rocker B is perpendicular to the rocker surface; thus it is horizontal. We assume that A_x, A_y, and B act in the directions shown.

Determination of B. We express that the sum of the moments of all external forces about point A is zero. The equation obtained will contain neither A_x nor A_y since the moments of A_x and A_y about A are zero. Multiplying the magnitude of each force by its perpendicular distance from B, we write

$$+\circlearrowleft \Sigma M_A = 0: \qquad +B(1.5 \text{ m}) - (9.81 \text{ kN})(2 \text{ m}) - (23.5 \text{ kN})(6 \text{ m}) = 0$$
$$B = +107.1 \text{ kN} \qquad\qquad \mathbf{B} = 107.1 \text{ kN} \rightarrow \quad \blacktriangleleft$$

Since the result is positive, the reaction is directed as assumed.

Determination of A_x. The magnitude A_x is determined by expressing that the sum of the horizontal components of all external forces is zero.

$$\xrightarrow{+} \Sigma F_x = 0: \qquad A_x + B = 0 \qquad A_x + 107.1 \text{ kN} = 0$$
$$A_x = -107.1 \text{ kN} \qquad\qquad \mathbf{A}_x = 107.1 \text{ kN} \leftarrow \quad \blacktriangleleft$$

Since the result is negative, the sense of $\mathbf{A}_x$ is opposite to that assumed originally.

Determination of A_y. The sum of the vertical components must also equal zero.

$$+\uparrow \Sigma F_y = 0: \qquad A_y - 9.81 \text{ kN} - 23.5 \text{ kN} = 0$$
$$A_y = +33.3 \text{ kN} \qquad\qquad \mathbf{A}_y = 33.3 \text{ kN} \uparrow \quad \blacktriangleleft$$

Adding vectorially the components $\mathbf{A}_x$ and $\mathbf{A}_y$, we find that the reaction at A is 112.2 kN $\measuredangle$17.3°.

Check. The values obtained for the reactions may be checked by recalling that the sum of the moments of all external forces about any point must be zero. For example, considering point B, we write

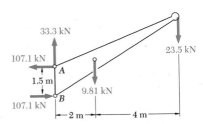

$$+\circlearrowleft \Sigma M_B = -(9.81 \text{ kN})(2 \text{ m}) - (23.5 \text{ kN})(6 \text{ m}) + (107.1 \text{ kN})(1.5 \text{ m}) = 0$$

SAMPLE PROBLEM 4.2

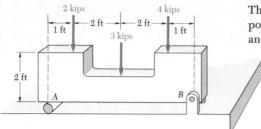

Three loads are applied to a steel plate as shown. The plate is supported by a roller at A and by a pin at B. Determine the reactions at A and B.

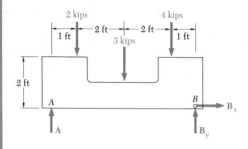

Solution. A free-body diagram of the plate is drawn. The reaction at A is vertical and is denoted by **A**. The reaction at B is represented by components $\mathbf{B}_x$ and $\mathbf{B}_y$. Each component is assumed to act in the direction shown.

Equilibrium Equations. We write the following three equilibrium equations and solve for the reactions indicated:

$$\xrightarrow{+} \Sigma F_x = 0: \qquad\qquad B_x = 0 \qquad\qquad \mathbf{B}_x = 0 \blacktriangleleft$$

$$+\,\mathanswer{}\Sigma M_A = 0:$$
$$-(2\text{ kips})(1\text{ ft}) - (3\text{ kips})(3\text{ ft}) - (4\text{ kips})(5\text{ ft}) + B_y(6\text{ ft}) = 0$$
$$B_y = +5.17\text{ kips} \qquad \mathbf{B}_y = 5.17\text{ kips} \uparrow \blacktriangleleft$$

$$+\,\mathansw{}\Sigma M_B = 0:$$
$$-A(6\text{ ft}) + (2\text{ kips})(5\text{ ft}) + (3\text{ kips})(3\text{ ft}) + (4\text{ kips})(1\text{ ft}) = 0$$
$$A = +3.83\text{ kips} \qquad A = 3.83\text{ kips} \uparrow \blacktriangleleft$$

Check. The results are checked by adding the vertical components of all the external forces.

$$+\uparrow\Sigma F_y = +5.17\text{ kips} + 3.83\text{ kips} - 2\text{ kips} - 3\text{ kips} - 4\text{ kips} = 0$$

Remark. In this problem the reactions at both A and B are vertical; however, these reactions are vertical for different reasons. At A, the plate is supported by a roller; hence the reaction cannot have any horizontal component. At B, the horizontal component of the reaction is zero because it must satisfy the equilibrium equation $\Sigma F_x = 0$ and because none of the other forces acting on the plate has a horizontal component.

We could have noticed at first glance that the reaction at B was vertical and dispensed with the horizontal component $\mathbf{B}_x$. This, however, is a bad practice. In following it, we would run the risk of forgetting the component $\mathbf{B}_x$ when the loading conditions require such a component (i.e., when a horizontal load is included). Also, the component $\mathbf{B}_x$ was found to be zero by using and solving an equilibrium equation, $\Sigma F_x = 0$. By setting $\mathbf{B}_x$ equal to zero immediately, we might not realize that we actually make use of this equation and thus might lose track of the number of equations available for solving the problem.

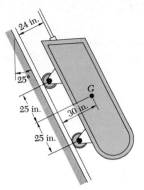

SAMPLE PROBLEM 4.3

A loading car is at rest on a track forming an angle of 25° with the vertical. The gross weight of the car and its load is 5500 lb, and it is applied at a point 30 in. from the track, halfway between the two axles. The car is held by a cable attached 24 in. from the track. Determine the tension in the cable and the reaction at each pair of wheels.

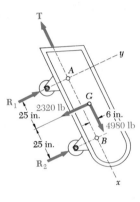

Solution. A free-body diagram of the car is drawn. The reaction at each wheel is perpendicular to the track, and the tension force **T** is parallel to the track. For convenience, we choose the x axis parallel to the track and the y axis perpendicular to the track. The 5500-lb weight is then resolved into x and y components.

$$W_x = +(5500 \text{ lb}) \cos 25° = +4980 \text{ lb}$$
$$W_y = -(5500 \text{ lb}) \sin 25° = -2320 \text{ lb}$$

Equilibrium Equations. We take moments about A to eliminate **T** and $\mathbf{R}_1$ from the computation.

$$+\curvearrowleft \Sigma M_A = 0:$$
$$-(2320 \text{ lb})(25 \text{ in.}) - (4980 \text{ lb})(6 \text{ in.}) + R_2(50 \text{ in.}) = 0$$
$$R_2 = +1758 \text{ lb} \qquad \mathbf{R}_2 = 1758 \text{ lb} \nearrow \quad \blacktriangleleft$$

Now, taking moments about B to eliminate **T** and $\mathbf{R}_2$ from the computation, we write

$$+\curvearrowleft \Sigma M_B = 0: \quad (2320 \text{ lb})(25 \text{ in.}) - (4980 \text{ lb})(6 \text{ in.}) - R_1(50 \text{ in.}) = 0$$
$$R_1 = +562 \text{ lb} \qquad \mathbf{R}_1 = 562 \text{ lb} \nearrow \quad \blacktriangleleft$$

The value of T is found by writing

$$\searrow +\Sigma F_x = 0: \quad +4980 \text{ lb} - T = 0$$
$$T = +4980 \text{ lb} \qquad \mathbf{T} = 4980 \text{ lb} \nwarrow \quad \blacktriangleleft$$

The computed values of the reactions are shown in the adjacent sketch.

Check. The computations are verified by writing

$$\nearrow +\Sigma F_y = +562 \text{ lb} + 1758 \text{ lb} - 2320 \text{ lb} = 0$$

A check could also have been obtained by computing moments about any point except A or B.

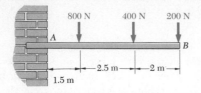

800 N 400 N 200 N

SAMPLE PROBLEM 4.4

A cantilever beam is loaded as shown. The beam is fixed at the left end and free at the right end. Determine the reaction at the fixed end.

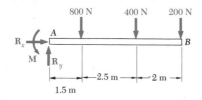

800 N 400 N 200 N

Solution. The portion of the beam which is embedded in the wall is subjected to a large number of forces. These forces, however, are equivalent to a force of components $\mathbf{R}_x$ and $\mathbf{R}_y$ and a couple $\mathbf{M}$.

Equilibrium Equations

$$\xrightarrow{+} \Sigma F_x = 0: \qquad R_x = 0 \qquad\qquad\qquad R_x = 0 \blacktriangleleft$$

$$+\uparrow\Sigma F_y = 0: \qquad R_y - 800\ \text{N} - 400\ \text{N} - 200\ \text{N} = 0$$
$$R_y = +1400\ \text{N} \qquad\qquad R_y = 1400\ \text{N} \uparrow \blacktriangleleft$$

$$+\, \jmath\, \Sigma M_A = 0:$$
$$-(800\ \text{N})(1.5\ \text{m}) - (400\ \text{N})(4\ \text{m}) - (200\ \text{N})(6\ \text{m}) + M = 0$$
$$M = +4000\ \text{N} \cdot \text{m} \qquad M = 4000\ \text{N} \cdot \text{m}\, \jmath \blacktriangleleft$$

The reaction at the fixed end consists of a vertical upward force of 1400 N and of a 4000-N · m counterclockwise couple.

SAMPLE PROBLEM 4.5

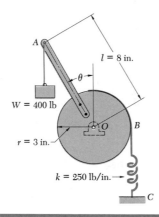

A 400-lb weight is attached to the lever AO as shown. The constant of the spring BC is $k = 250$ lb/in., and the spring is unstretched when $\theta = 0$. Determine the position or positions of equilibrium.

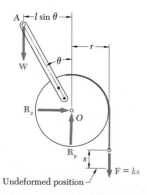

Solution. *Force Exerted by Spring.* Denoting by s the deflection of the spring from its undeformed position, and noting that $s = r\theta$, we write

$$F = ks = kr\theta$$

Equilibrium Equation. Summing the moments of $\mathbf{W}$ and $\mathbf{F}$ about O, we write

$$+\, \jmath\, \Sigma M_O = 0: \qquad Wl\sin\theta - r(kr\theta) = 0 \qquad \sin\theta = \frac{kr^2}{Wl}\theta$$

Substituting the given data, we obtain

$$\sin\theta = \frac{(250\ \text{lb/in.})\,(3\ \text{in.})^2}{(400\ \text{lb})\,(8\ \text{in.})}\theta \qquad \sin\theta = 0.703\theta$$

Solving by trial and error, we find $\qquad \theta = 0 \qquad \theta = 80.3° \blacktriangleleft$

PROBLEMS

4.1 The 40-ft boom *AB* weighs 2 kips; the distance from the axle *A* to the center of gravity *G* of the boom is 20 ft. For the position shown, determine the tension *T* in the cable and the reaction at *A*.

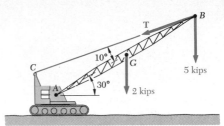

Fig. P4.1

4.2 The 600-kg forklift truck is used to hold the 150-kg crate *C* in the position shown. Determine the reactions (*a*) at each of the two wheels *A* (one wheel on each side of the truck), (*b*) at the single steerable wheel *B*.

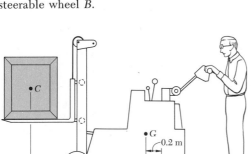

Fig. P4.2

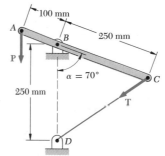

Fig. P4.3

4.3 Knowing that the magnitude of the vertical force **P** is 400 N, determine (*a*) the tension in the cable *CD*, (*b*) the reaction at *B*.

4.4 The ladder *AB*, of length *L* and weight *W*, can be raised by the cable *BC*. Determine the tension *T* required to raise end *B* just off the floor (*a*) in terms of *W* and θ, (*b*) if $h = 8$ ft, $L = 10$ ft, and $W = 35$ lb.

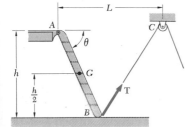

Fig. P4.4

4.5 A block of weight *W* is to be supported by the winch shown. Determine the required magnitude of the force **P** (*a*) in terms of *W*, *r*, *l*, and θ, (*b*) if $W = 100$ lb, $r = 3$ in., $l = 15$ in., and $\theta = 60°$.

4.6 Determine the values of θ for which the winch is in equilibrium when $W = 125$ lb, $P = 50$ lb, $r = 3$ in., and $l = 15$ in.

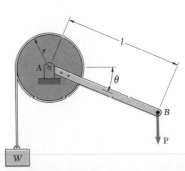

Fig. P4.5 and P4.6

4.7 The required tension in cable AB is 1200 N. Determine (a) the vertical force **P** which must be applied to the pedal, (b) the corresponding reaction at C.

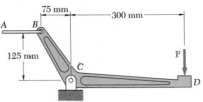

Fig. P4.7 and P4.8

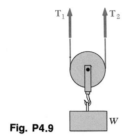

Fig. P4.9

4.8 Determine the maximum tension which may be developed in cable AB if the maximum allowable magnitude of the reaction at C is 2.6 kN.

4.9 A load W is supported by the pulley as shown. Prove that, if the pulley is in equilibrium, the tensions T_1 and T_2 are both equal to $W/2$.

4.10 Determine the reactions at A and B when $\alpha = 60°$.

4.11 Determine the reactions at A and B when $\alpha = 90°$.

4.12 Determine the value of α for which the reaction at B is minimum. What are the corresponding reactions at A and B?

4.13 A light rod, supported by rollers at B, C, and D, is subjected to a 200-lb force applied at A. If $\beta = 0$, determine (a) the reactions at B, C, and D, (b) the rollers which may safely be removed for this loading.

Fig. P4.10, P4.11, and P4.12

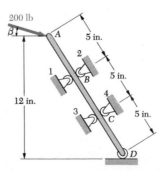

Fig. P4.13

4.14 Solve Prob. 4.13 when the 200-lb force is directed vertically downward, i.e., when $\beta = 90°$.

4.15 A movable bracket is held at rest by a cable attached at C and by frictionless rollers at A and B. For the loading shown, determine the tension in the cable and the reactions at A and B.

4.16 A 90-kg overhead garage door consists of a uniform rectangular panel AC, 2.4 m high, supported by the cable AE attached at the middle of the upper edge of the door and by two sets of frictionless rollers at A and B. Each set consists of two rollers located on either side of the door. The rollers A are free to move in horizontal channels, while the rollers B are guided by vertical channels. If the door is held in the position for which $BD = 1.2$ m, determine (a) the tension in cable AE, (b) the reaction at each of the four rollers.

4.17 In Prob. 4.16, determine the minimum allowable value of the distance BD if the tension in cable AE is not to exceed 5 kN.

4.18 Two external shafts of a gearbox carry torques as shown. Determine the vertical components of the forces which must be exerted by the bolts at A and B to maintain the gearbox in equilibrium.

4.19 Determine the range of values of α for which the semicircular rod can be maintained in equilibrium by the small wheel at D and the rollers at B and C.

4.20 Determine the reactions at B, C, and D (a) if $\alpha = 0$, (b) if $\alpha = 30°$.

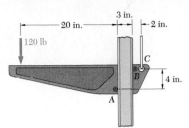

Fig. P4.15

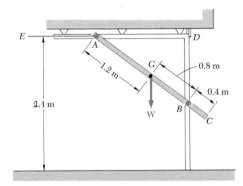

Fig. P4.16 and P4.17

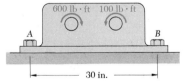

Fig. P4.18

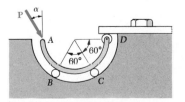

Fig. P4.19 and P4.20

4.21 A couple **M** is applied to a bent rod *AB* which may be supported in four different ways as shown. In each case determine the reactions at the supports.

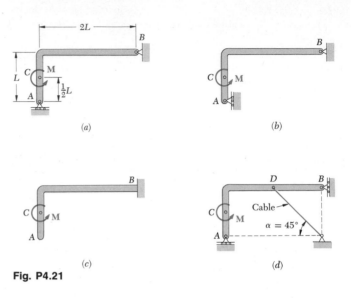

Fig. P4.21

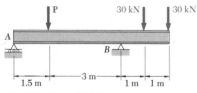

Fig. P4.22 and P4.23

4.22 The maximum allowable value for each of the reactions is 150 kN and the reaction at *A* must be directed upward. Neglecting the weight of the beam, determine the range of values of *P* for which the beam is safe.

4.23 Determine the reactions at *A* and *B* for the beam and loading shown when (*a*) $P = 75$ kN, (*b*) $P = 150$ kN.

4.24 The 10-ft beam *AB* rests upon, but is not attached to, supports at *C* and *D*. Neglecting the weight of the beam, determine the range of values of *P* for which the beam will remain in equilibrium.

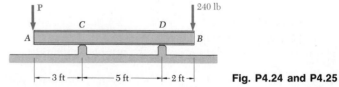

Fig. P4.24 and P4.25

4.25 The 10-ft uniform beam *AB* weighs 100 lb; it rests upon, but is not attached to, supports at *C* and *D*. Determine the range of values of *P* for which the beam will remain in equilibrium.

4.26 A traffic-signal pole may be supported in the three ways shown; in part *c* the tension in cable *BC* is known to be 1950 N. Determine the reactions for each type of support shown.

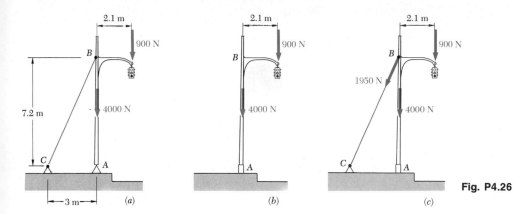

Fig. P4.26

4.27 The mechanism shown is designed to measure the tension in a heavy paper tape used in a manufacturing process. The tape passes over three rollers which are free to rotate about their axles *A*, *B*, and *C*. The bearings of *A* and *C* are fixed in a vertical plate, while the bearing of *B* is located at the end of a bell crank pivoted at *D* and attached to a spring *EF*. As the tension in the tape varies, the bell crank rotates about *D*, and the pointer *E* indicates the appropriate reading on the scale. Determine (*a*) the force exerted at *E* by the spring when the tension in the tape is 150 N, (*b*) the corresponding reaction at *D*.

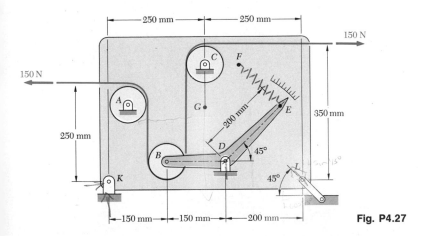

Fig. P4.27

4.28 Knowing that the weight of the mechanism of Prob. 4.27 is 200 N and that its center of gravity *G* is located directly below *C*, determine the reactions at the supports *K* and *L* when the tension in the tape is 150 N.

4.29 The beam *AC* is part of the roof structure of a small build-ing. It is supported at *C* by a riveted connection and at *B* by the cable *BDF*. (*a*) If the tension in the cable is 39 kips, determine the reaction at the riveted connection *C*. (*b*) For what value of the tension would the couple at *C* be zero?

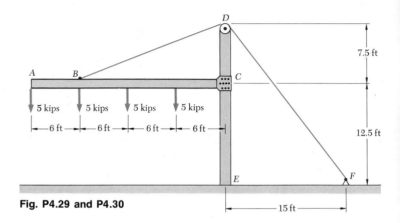

Fig. P4.29 and P4.30

4.30 The frame shown supports part of the roof of a small build-ing. (*a*) If the tension in the cable is 30 kips, determine the reaction at the fixed end *E*. (*b*) For what value of the tension would the couple at *E* be zero?

4.31 In the pivoted motor mount, or Rockwood drive, the weight of the motor is used to maintain tension in the drive belt. When the motor is at rest, the tensions T_1 and T_2 may be assumed equal. The mass of the motor is 90 kg, and the diameter of the drive pulley is 150 mm. Assuming that the mass of the platform *AB* is negligible, determine (*a*) the tension in the belt, (*b*) the reaction at *C* when the motor is at rest.

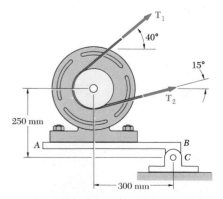

Fig. P4.31

4.32 The horizontal member *ABC* of the rig shown weighs 1000 lb and is supported by a pin *B* and a cable *EADC*. Since the cable passes over pulleys at *A* and *D*, the tension may be assumed to be the same in all portions of the cable. If the rig raises a 3000-lb load at a distance $a = 12$ ft from the vertical member *DF*, determine (*a*) the tension in the cable, (*b*) the horizontal and vertical components of the reaction at *B*.

4.33 For the rig of Prob. 4.32, determine (*a*) the maximum distance *a* from the vertical member *DF* at which a 3000-lb load may be supported if the maximum allowable tension in cable *EADC* is 8000 lb, (*b*) the corresponding values of the horizontal and vertical components of the reaction at *B*.

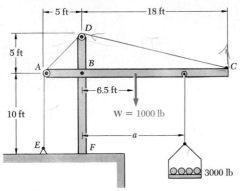

Fig. P4.32

✱4.34 Two wheels *A* and *B*, of weight 2*W* and *W*, respectively, are connected by a rod of negligible weight and are free to roll on the surface shown. If $\theta = 45°$, determine the angle that the rod forms with the horizontal when the system is in equilibrium.

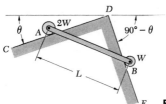

Fig. P4.34 and P4.35

✱4.35 Two wheels *A* and *B*, of weight 2*W* and *W*, respectively, are connected by a rod of negligible weight and are free to roll on the surface shown. Determine the angle θ for which the rod connecting the wheels is horizontal when the system is in equilibrium.

4.36 A slender rod *AB*, of weight *W*, is attached to blocks *A* and *B* which move freely in the guides shown. The blocks are connected by an elastic cord which passes over a pulley at *C*. (*a*) Express the tension in the cord in terms of *W* and θ. (*b*) Determine the value of θ for which the tension in the cord is equal to 2*W*.

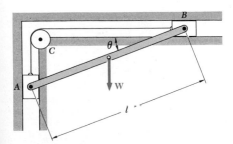

Fig. P4.36

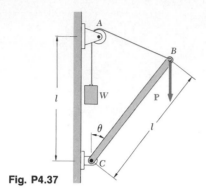

Fig. P4.37

4.37 and 4.38 (a) Derive an expression for the angle θ corresponding to equilibrium. (b) Determine the value of θ corresponding to the equilibrium position if $P = 2W$.

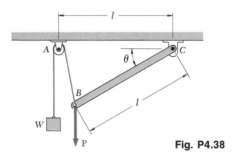

Fig. P4.38

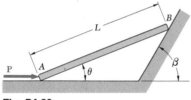

Fig. P4.39

∗4.39 A uniform, slender rod of length L and weight W is held in the position shown by the horizontal force **P**. Neglecting the effect of friction at A and B, determine the angle θ corresponding to equilibrium (a) in terms of P, W, L, and β, (b) if $P = 10$ lb, $W = 20$ lb, $L = 30$ in., and $\beta = 60°$.

4.40 The bracket ABC is supported in eight different ways as shown. All connections consist of frictionless pins, rollers, and short links. In each case, determine whether (a) the bracket is completely, partially, or improperly constrained, (b) the reactions are statically determinate or indeterminate, (c) the equilibrium of the bracket is maintained in the position shown. Also, wherever possible, compute the reactions, assuming that the magnitude of the force **P** is 400 N.

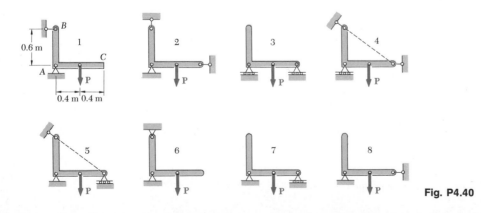

Fig. P4.40

4.41 Nine identical rectangular plates, 2 by 3 ft, weighing 100 lb each are held in a vertical plane as shown. All connections consist of frictionless pins, rollers, or short links. For each case, answer the questions listed in Prob. 4.40, and, wherever possible, compute the reactions.

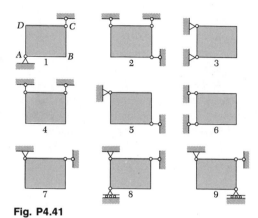

Fig. P4.41

4.6. Equilibrium of a Two-Force Body.

A particular case of equilibrium which is of considerable interest is that of a rigid body subjected to two forces. Such a body is commonly called a *two-force body.* We shall show that, *if a two-force body is in equilibrium, the two forces must have the same magnitude, same line of action, and opposite sense.*

Consider a corner plate subjected to two forces $\mathbf{F}_1$ and $\mathbf{F}_2$ acting at A and B, respectively (Fig. 4.8a). If the plate is to be in equilibrium, the sum of the moments of $\mathbf{F}_1$ and $\mathbf{F}_2$ about any axis must be zero. First, we sum moments about A: Since the moment of $\mathbf{F}_1$ is obviously zero, the moment of $\mathbf{F}_2$ must also be zero and the line of action of $\mathbf{F}_2$ must pass through A (Fig. 4.8b). Summing moments about B, we prove similarly that the line of action of $\mathbf{F}_1$ must pass through B (Fig. 4.8c). Both forces have the same line of action (line AB). From the equation $\Sigma F_x = 0$ or $\Sigma F_y = 0$, it is seen that they must have also the same magnitude but opposite sense.

If several forces act at two points A and B, the forces acting at A may be replaced by their resultant $\mathbf{F}_1$ and those acting at

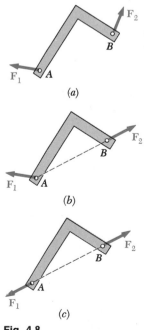

Fig. 4.8

B by their resultant $\mathbf{F}_2$. Thus a two-force body may be more generally defined as *a rigid body subjected to forces acting at only two points*. The resultants $\mathbf{F}_1$ and $\mathbf{F}_2$ then must have the same line of action, same magnitude, and opposite sense.

Although problems dealing with the equilibrium of two-force bodies may be solved by the general methods studied in the preceding sections, it is sometimes desirable to make use of the property we have just established to simplify certain problems so that simple trigonometric or geometric relations can be used.

4.7. Equilibrium of a Three-Force Body. Another case of equilibrium that is of great interest is that of a *three-force body*, i.e., a rigid body subjected to three forces or, more generally, *a rigid body subjected to forces acting at only three points*. Consider a rigid body subjected to a system of forces which may be reduced to three forces $\mathbf{F}_1$, $\mathbf{F}_2$, and $\mathbf{F}_3$ acting at A, B, and C, respectively (Fig. 4.9a). We shall show that, if the body is in equilibrium, *the lines of action of the three forces must be either concurrent or parallel*.

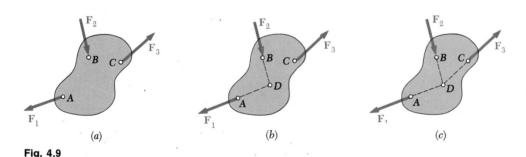

(a)   (b)   (c)

Fig. 4.9

Since the rigid body is in equilibrium, the sum of the moments of $\mathbf{F}_1$, $\mathbf{F}_2$, and $\mathbf{F}_3$ about any axis must be zero. Assuming that the lines of action of $\mathbf{F}_1$ and $\mathbf{F}_2$ intersect, and denoting their point of intersection by D, we sum moments about D (Fig. 4.9b); since the moments of $\mathbf{F}_1$ and $\mathbf{F}_2$ about D are zero, the moment of $\mathbf{F}_3$ about D must also be zero and the line of action of $\mathbf{F}_3$ must pass through D (Fig. 4.9c). The three lines of action are concurrent. The only exception occurs when none of the lines intersect; the lines of action are then parallel.

Although problems concerning three-force bodies may be solved by the general methods of Secs. 4.3 to 4.5, the property just established may be used to solve them either graphically or from simple trigonometric or geometric relations.

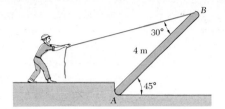

SAMPLE PROBLEM 4.6

A man raises a 10-kg joist, of length 4 m, by pulling on a rope. Find the tension T in the rope and the reaction at A.

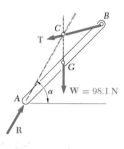

Solution. The joist is a three-force body since it is acted upon by three forces: its weight **W**, the force **T** exerted by the rope, and the reaction **R** of the ground at A. We note that

$$W = mg = (10 \text{ kg})(9.81 \text{ m/s}^2) = 98.1 \text{ N}$$

Three-Force Body. Since the joist is a three-force body, the forces acting on it must be concurrent. The reaction **R**, therefore, will pass through the point of intersection C of the lines of action of the weight **W** and of the tension force **T**. This fact will be used to determine the angle α that **R** forms with the horizontal.

Drawing the vertical BF through B and the horizontal CD through C, we note that

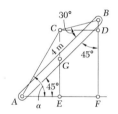

$$AF = BF = (AB) \cos 45° = (4 \text{ m}) \cos 45° = 2.83 \text{ m}$$
$$CD = EF = AE = \tfrac{1}{2}(AF) = 1.415 \text{ m}$$
$$BD = (CD) \cot (45° + 30°) = (1.415 \text{ m}) \tan 15° = 0.379 \text{ m}$$
$$CE = DF = BF - BD = 2.83 \text{ m} - 0.379 \text{ m} = 2.45 \text{ m}$$

We write

$$\tan \alpha = \frac{CE}{AE} = \frac{2.45 \text{ m}}{1.415 \text{ m}} = 1.732$$

$$\alpha = 60.0° \quad \blacktriangleleft$$

We now know the direction of all the forces acting on the joist.

Force Triangle. A force triangle is drawn as shown, and its interior angles are computed from the known directions of the forces. Using the law of sines, we write

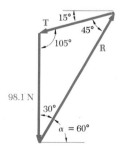

$$\frac{T}{\sin 30°} = \frac{R}{\sin 105°} = \frac{98.1 \text{ N}}{\sin 45°}$$

$$T = 69.4 \text{ N} \quad \blacktriangleleft$$
$$R = 134.0 \text{ N} \; \measuredangle 60.0° \quad \blacktriangleleft$$

PROBLEMS

4.42 Using the method of Sec. 4.7, solve Prob. 4.10.

4.43 Using the method of Sec. 4.7, solve Prob. 4.3.

4.44 In Prob. 4.13, determine (*a*) the value of β for which rollers 1, 2, and 3 may safely be removed, (*b*) the corresponding reactions at *C* and *D*.

4.45 In Prob. 4.13, determine (*a*) the value of β for which rollers 1, 3, and 4 may safely be removed, (*b*) the corresponding reactions at *B* and *D*.

4.46 Determine the reactions at *A* and *E* when $\alpha = 0$.

4.47 Determine (*a*) the value of α for which the reaction at *A* is vertical, (*b*) the corresponding reactions at *A* and *E*.

4.48 A 12-ft ladder, weighing 40 lb, leans against a frictionless vertical wall. The lower end of the ladder rests on rough ground, 4 ft away from the wall. Determine the reactions at both ends.

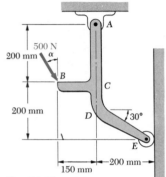

Fig. P4.46 and P4.47

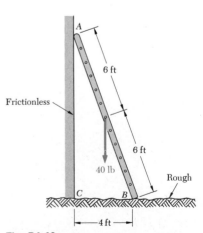

Fig. P4.48

4.49 A 50-lb sign is supported by a pin and bracket at A and by a cable BC. Determine the reaction at A and the tension in the cable.

4.50 A 100-kg roller, of diameter 500 mm, is used on a lawn. Determine the force $\mathbf{F}$ required to make it roll over a 50-mm obstruction (a) if the roller is pushed as shown, (b) if the roller is pulled as shown.

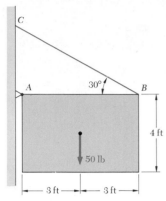

Fig. P4.49

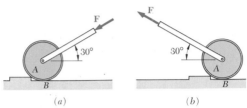

(a)   (b)

Fig. P4.50

4.51 Using the method of Sec. 4.7, solve Prob. 4.7.

4.52 Determine the reactions at A and E.

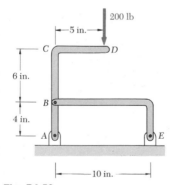

Fig. P4.52

4.53 Solve Prob. 4.52, assuming that the 200-lb force applied at D is directed horizontally to the left.

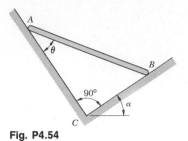

Fig. P4.54

4.54 The uniform rod AB lies in a vertical plane with its ends resting against the frictionless surfaces AC and BC. Determine the angle θ corresponding to equilibrium when (a) $\alpha = 30°$, (b) $\alpha = 40°$, (c) $\alpha = 60°$.

4.55 A slender rod BC of length L and weight W is held by a cable AB and by a pin at C which may slide in a vertical slot. Knowing that $\theta = 30°$, determine (a) the value of β for which the rod is in equilibrium, (b) the corresponding tension in the cable.

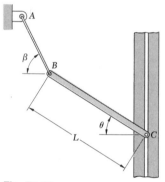

Fig. P4.55

4.56 A slender rod of length L and weight W is lodged between the wall and the peg C. Neglecting friction, determine the angle θ between the rod and the wall corresponding to equilibrium.

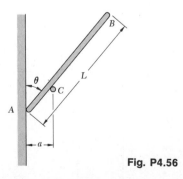

Fig. P4.56

4.57 A 2-m rod, of uniform cross section, is held in equilibrium as shown, with one end against a frictionless vertical wall and the other end attached to a cord. Determine the length of the cord.

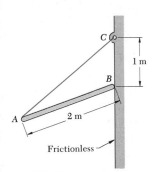

Fig. P4.57

*4.58 A uniform rod AB of length $3R$ rests inside a hemispherical bowl of radius R as shown. Neglecting friction, determine the angle θ corresponding to equilibrium.

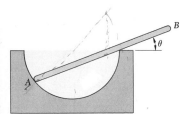

Fig. P4.58

*4.59 A slender rod of length $2r$ and weight W is attached to a collar at B and rests on a circular cylinder of radius r. Knowing that the collar may slide freely along a vertical guide and neglecting friction, determine the value of θ corresponding to equilibrium.

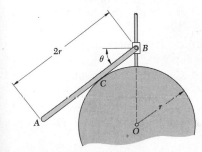

Fig. P4.59

EQUILIBRIUM IN THREE DIMENSIONS

4.8. Reactions at Supports and Connections for a Three-dimensional Structure.
The reactions on a three-dimensional structure range from the single force of known direction exerted by a frictionless surface to the force-couple system exerted by a fixed support. Consequently, the number of unknowns associated with the reaction at a support or connection may vary from one to six in problems involving the equilibrium of a three-dimensional structure. Various types of supports and connections are shown in Fig. 4.10 with the corresponding reactions. A simple way of determining the type of reaction corresponding to a given support or connection and the number of unknowns involved is to find which of the six fundamental motions (translation in x, y, and z directions, rotation about the x, y, and z axes) are allowed and which motions are prevented.

Ball supports, frictionless surfaces, and cables, for example, prevent translation in one direction only and thus exert a single force of known line of action; they each involve one unknown, namely, the magnitude of the reaction. Rollers on rough surfaces and wheels on rails prevent translation in two directions; the corresponding reactions consist of two unknown force components. Rough surfaces in direct contact and ball-and-socket supports prevent translation in three directions; these supports involve three unknown force components.

Some supports and connections may prevent rotation as well as translation; the corresponding reactions include, then, couples as well as forces. The reaction at a fixed support, for example, which prevents any motion (rotation as well as translation), consists of three unknown forces and three unknown couples. A universal joint, which is designed to allow rotation about two axes, will exert a reaction consisting of three unknown force components and one unknown couple.

Other supports and connections are primarily intended to prevent translation; their design, however, is such that they also prevent some rotations. The corresponding reactions consist essentially of force components but *may* also include couples. One group of supports of this type includes hinges and bearings designed to support radial loads only (for example, journal bearings, roller bearings). The corresponding reactions consist of two force components but may also include two couples. Another group includes pin-and-bracket supports, hinges, and bearings designed to support an axial thrust as well as a radial load (for example, ball bearings). The corresponding reactions consist of three force components but may include two couples. However,

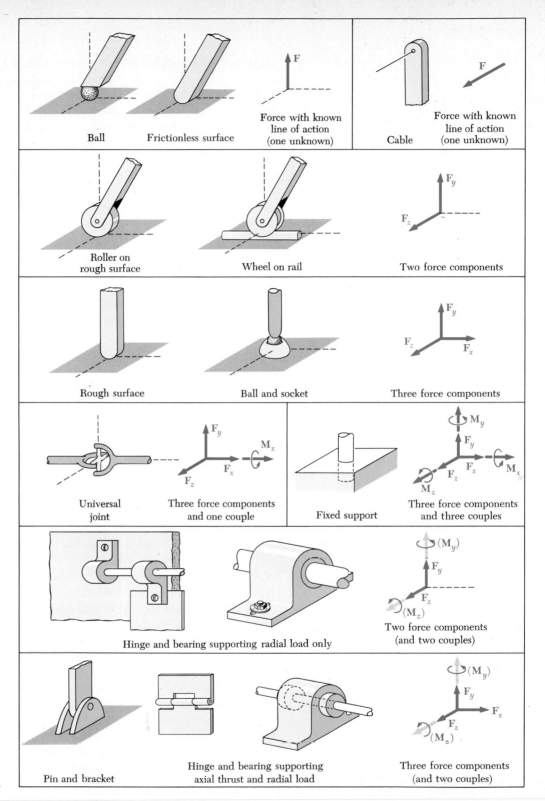

Fig. 4.10 Reactions at supports and connections.

Ball Frictionless surface Force with known line of action (one unknown)

Cable Force with known line of action (one unknown)

Roller on rough surface Wheel on rail Two force components

Rough surface Ball and socket Three force components

Universal joint Three force components and one couple

Fixed support Three force components and three couples

Hinge and bearing supporting radial load only Two force components (and two couples)

Pin and bracket Hinge and bearing supporting axial thrust and radial load Three force components (and two couples)

these supports will not exert any appreciable couples under normal conditions of use. Therefore, *only* force components should be included in their analysis, *unless* it is found that couples are necessary to maintain the equilibrium of the rigid body, or unless the support is known to have been specifically designed to exert a couple (see Probs. 4.82, 4.85, and 4.86).

4.9. Equilibrium of a Rigid Body in Three Dimensions. We saw in Sec. 4.1 that six scalar equations are required to express the conditions for the equilibrium of a rigid body in the general three-dimensional case.

$$\Sigma F_x = 0 \qquad \Sigma F_y = 0 \qquad \Sigma F_z = 0 \qquad (4.2)$$
$$\Sigma M_x = 0 \qquad \Sigma M_y = 0 \qquad \Sigma M_z = 0 \qquad (4.3)$$

These equations may be solved for no more than *six unknowns*, which generally will represent reactions at supports or connections.

In most problems the scalar equations (4.2) and (4.3) will be more conveniently obtained if we first express in vector form the conditions for the equilibrium of the rigid body considered. We write

$$\Sigma \mathbf{F} = 0 \qquad \Sigma \mathbf{M}_O = \Sigma(\mathbf{r} \times \mathbf{F}) = 0 \qquad (4.1)$$

and express the forces **F** and position vectors **r** in terms of scalar components and unit vectors. Next we compute all vector products, either directly or by means of determinants (see Sec. 3.7). Equating to zero the coefficients of the unit vectors in each of the two relations (4.1), we obtain the desired scalar equations.

If the reactions involve more than six unknowns, there are more unknowns than equations, and some of the reactions are *statically indeterminate* (see Probs. 4.84 and 4.105). If the reactions involve less than six unknowns, there are more equations than unknowns, and some of the equations of equilibrium cannot be satisfied under general loading conditions; the rigid body is only *partially constrained*. Under the particular loading conditions corresponding to a given problem, however, the extra equations often reduce to trivial identities such as $0 = 0$ and may be disregarded; although only partially constrained, the rigid body remains in equilibrium (see Sample Probs. 4.7 and 4.8). Even with six or more unknowns, it is possible that some equations of equilibrium will not be satisfied. This may occur when the supports are such that the reactions are forces which must either be parallel or intersect the same line; the rigid body is then *improperly constrained*.

Electronics Designers Casebook Number 4

YOURS FREE
when you subscribe to Electronics.

This **Designers Casebook—** *Number* **4**

1. NAME ☐ Mr. ☐ Mrs. ☐ Ms. _____ TITLE _____

COMPANY _____ DIV. or DEPT. _____

COMPANY ADDRESS _____

CITY _____ STATE _____ ZIP _____

☐ Check here if you wish publication to be sent to home address

STREET _____ CITY _____ STATE _____ ZIP _____

Qualification for above rates is based on answers to all questions listed below. Those not qualifying may pay higher than basic price of $34 one year or $79 for three years.

Signature _____

1a. Please check which best describes your company's business at your location:
☐ Manufacturing ☐ Distribution ☐ Retailing
☐ Other _____

2. ☐ 1 PLANT ☐ 2 DEPARTMENT

Indicate the primary product manufactured or service performed at your plant (Box 1) and in your department (Box 2). Be sure to indicate applicable letter in each of the two boxes even if they are the same letter.

A. Large computers
B. Mini-computers
C. Computer peripheral equipment
D. Data Processing Systems (systems integration)
E. Office and business machines
F. Test and measuring equipment
G. Communications systems and equipment
H. Navigation and guidance or control systems
I. Consumer entertainment electronic equipment
J. Other consumer electronic equip. (appliances, autos, hand tools)

K. Industrial controls, systems and equipment
L. Sub-assemblies
M. Passive electronic components
N. Active electronic components
O. Materials and Hardware
P. Aircraft, Missiles, space and ground support equipment
Q. Oceanography, Geology, Astronomy and support equipment
R. Medical electronics
S. Industrial equipment containing electronic components or products
T. Independent R & D laboratory or consultant

U. Research and development organizations which are a part of an educational institution
V. Government Agency and military
W. Industrial companies using and or incorporating electronic products in their mfg., research or development activities
X. Utilities
Y. Broadcasting, sound and motion pictures and recording studios
Z. Commercial users of electronics equipment (railroads, pipelines, police, airlines, hospitals, banks)
9. 4 yr. College, University
10. Other _____

3. Indicate your principal job function (place applicable number in box. If numbers 9, 10, or 11 are used, fill in name of college or university)

1. General and corporate management
2. Design and development engineering
3. Engineering services (evaluation, quality control, reliability, standards, test)
4. Basic research
5. Manufacturing and production
6. Engineering support (lab, assistant, technician)
7. Purchasing and procurement
8. Marketing and sales
9. Professor at _____
10. Senior student at _____
11. Graduate student at _____

Senior and graduate students are eligible for professional rate for one year subscription only.

4. ☐☐ Indicate your principal job responsibility (place the appropriate number and letter in boxes)

1A. Management 1B. Engineering Management 2A. Engineering

5. ☐☐☐ **Your design function:** *(Insert each letter that applies)*

A. I do electronic design or development engineering work
B. I supervise electronic design or development engineering work
C. I set standards for, or evaluate electronic components, systems and materials

6. Estimated number of employees at this location. (check one)

☐ 1 to 49 ☐ 50 to 249 ☐ 250 to 999 ☐ over 1,000

Subscriptions are normally entered within 2 weeks, but please allow 4 weeks for shipment.

BUSINESS REPLY MAIL
FIRST CLASS   PERMIT NO. 42   HIGHTSTOWN, N.J. 08520

POSTAGE WILL BE PAID BY ADDRESSEE

Electronics

P.O. Box 514
Hightstown, N.J. 08520

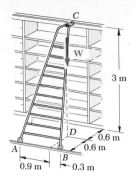

SAMPLE PROBLEM 4.7

A 20-kg ladder used to reach high shelves in a storeroom is supported by two flanged wheels A and B mounted on a rail and by an unflanged wheel C resting against a rail fixed to the wall. An 80-kg man stands on the ladder and leans to the right. The line of action of the combined weight $\mathbf{W}$ of the man and ladder intersects the floor at point D. Determine the components of the reactions at A, B, and C.

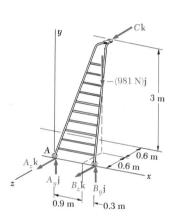

Solution. A free-body diagram of the ladder is drawn. The forces involved are the combined weight of the man and ladder,

$$\mathbf{W} = -mg\mathbf{j} = -(80 \text{ kg} + 20 \text{ kg})(9.81 \text{ m/s}^2)\mathbf{j} = -(981 \text{ N})\mathbf{j}$$

and five unknown reaction components, two at each flanged wheel and one at the unflanged wheel. The ladder is thus only partially constrained; it is free to roll along the rails. It is, however, in equilibrium under the given load since the equation $\Sigma F_x = 0$ is satisfied.

Equilibrium Equations. We express that the forces acting on the ladder form a system equivalent to zero:

$$\Sigma \mathbf{F} = 0: \quad A_y\mathbf{j} + A_z\mathbf{k} + B_y\mathbf{j} + B_z\mathbf{k} - (981 \text{ N})\mathbf{j} + C\mathbf{k} = 0$$
$$(A_y + B_y - 981 \text{ N})\mathbf{j} + (A_z + B_z + C)\mathbf{k} = 0 \tag{1}$$

$$\Sigma \mathbf{M}_A = \Sigma(\mathbf{r} \times \mathbf{F}) = 0:$$
$$1.2\mathbf{i} \times (B_y\mathbf{j} + B_z\mathbf{k}) + (0.9\mathbf{i} - 0.6\mathbf{k}) \times (-981\mathbf{j})$$
$$+ (0.6\mathbf{i} + 3\mathbf{j} - 1.2\mathbf{k}) \times C\mathbf{k} = 0$$

Computing the vector products, we have†

$$1.2B_y\mathbf{k} - 1.2B_z\mathbf{j} - 882.9\mathbf{k} - 588.6\mathbf{i} - 0.6C\mathbf{j} + 3C\mathbf{i} = 0$$
$$(3C - 588.6)\mathbf{i} - (1.2B_z + 0.6C)\mathbf{j} + (1.2B_y - 882.9)\mathbf{k} = 0 \tag{2}$$

Setting the coefficients of $\mathbf{i}$, $\mathbf{j}$, $\mathbf{k}$ equal to zero in Eq. (2), we obtain the following three scalar equations, which express that the sum of the moments about each coordinate axis must be zero:

$$3C - 588.6 = 0 \qquad\qquad C = +196.2 \text{ N} \quad \blacktriangleleft$$
$$1.2B_z + 0.6C = 0 \qquad\qquad B_z = -98.1 \text{ N} \quad \blacktriangleleft$$
$$1.2B_y - 882.9 = 0 \qquad\qquad B_y = +736 \text{ N} \quad \blacktriangleleft$$

Setting the coefficients of $\mathbf{j}$ and $\mathbf{k}$ equal to zero in Eq. (1), we obtain two scalar equations expressing that the sums of the components in the y and z directions are equal to zero. Substituting for B_y, B_z, and C the values obtained above, we write

$$A_y + B_y - 981 = 0 \qquad A_y + 736 - 981 = 0 \quad A_y = +245 \text{ N} \quad \blacktriangleleft$$
$$A_z + B_z + C = 0 \qquad A_z - 98.1 + 196.2 = 0 \quad A_z = -98.1 \text{ N} \quad \blacktriangleleft$$

† The moments in this Sample Problem and in Sample Probs. 4.8 and 4.9 could also be expressed in the form of determinants (see Sample Prob. 3.10).

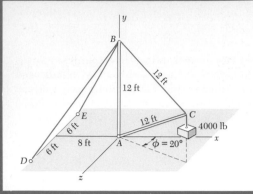

SAMPLE PROBLEM 4.8

The derrick shown supports a 4000-lb load. It is held by a ball and socket at A and by two cables attached at points D and E. In the position shown, the derrick stands in a vertical plane forming an angle $\phi = 20°$ with the xy plane. Determine the tension in each cable and the components of the reaction at A.

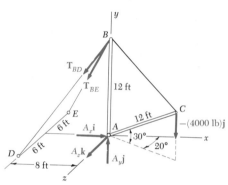

Solution. A free-body diagram of the derrick is drawn. Since the directions of the forces exerted at B by the cables are known, these forces involve one unknown each, namely, the magnitudes T_{BD} and T_{BE}. The reaction at A is a force of unknown direction and is represented by three unknown components. Since there are only five unknowns, the derrick is partially constrained. It may rotate freely about the y axis; it is, however, in equilibrium under the given loading since the equation $\Sigma M_y = 0$ is satisfied.

The components of the forces $\mathbf{T}_{BD}$ and $\mathbf{T}_{BE}$ may be expressed in terms of the unknown magnitudes T_{BD} and T_{BE} by writing

$$\overrightarrow{BD} = -(8 \text{ ft})\mathbf{i} - (12 \text{ ft})\mathbf{j} + (6 \text{ ft})\mathbf{k} \qquad BD = 15.62 \text{ ft}$$
$$\overrightarrow{BE} = -(8 \text{ ft})\mathbf{i} - (12 \text{ ft})\mathbf{j} - (6 \text{ ft})\mathbf{k} \qquad BE = 15.62 \text{ ft}$$
$$\mathbf{T}_{BD} = T_{BD}(\overrightarrow{BD}/BD) = T_{BD}(-0.512\mathbf{i} - 0.768\mathbf{j} + 0.384\mathbf{k})$$
$$\mathbf{T}_{BE} = T_{BE}(\overrightarrow{BE}/BE) = T_{BE}(-0.512\mathbf{i} - 0.768\mathbf{j} - 0.384\mathbf{k})$$

We also resolve the position vector of C into rectangular components:

$$\overrightarrow{AC} = (12 \text{ ft})(\cos 30° \cos 20°\mathbf{i} + \sin 30°\mathbf{j} + \cos 30° \sin 20°\mathbf{k})$$
$$\overrightarrow{AC} = (9.77 \text{ ft})\mathbf{i} + (6 \text{ ft})\mathbf{j} + (3.55 \text{ ft})\mathbf{k}$$

Equilibrium Equations. We express that the forces acting on the derrick form a system equivalent to zero:

$$\Sigma\mathbf{F} = 0: \qquad A_x\mathbf{i} + A_y\mathbf{j} + A_z\mathbf{k} + \mathbf{T}_{BD} + \mathbf{T}_{BE} - (4000 \text{ lb})\mathbf{j} = 0$$
$$(A_x - 0.512T_{BD} - 0.512T_{BE})\mathbf{i}$$
$$+ (A_y - 0.768T_{BD} - 0.768T_{BE} - 4000 \text{ lb})\mathbf{j}$$
$$+ (A_z + 0.384T_{BD} - 0.384T_{BE})\mathbf{k} = 0 \qquad (1)$$
$$\Sigma\mathbf{M}_A = \Sigma(\mathbf{r} \times \mathbf{F}) = 0:$$
$$12\mathbf{j} \times T_{BD}(-0.512\mathbf{i} - 0.768\mathbf{j} + 0.384\mathbf{k})$$
$$+ 12\mathbf{j} \times T_{BE}(-0.512\mathbf{i} - 0.768\mathbf{j} - 0.384\mathbf{k})$$
$$+ (9.77\mathbf{i} + 6\mathbf{j} + 3.55\mathbf{k}) \times (-4000\mathbf{j}) = 0$$
$$(4.61T_{BD} - 4.61T_{BE} + 14,200)\mathbf{i}$$
$$+ (6.14T_{BD} + 6.14T_{BE} - 39,080)\mathbf{k} = 0 \qquad (2)$$

Setting the coefficients of $\mathbf{i}$ and $\mathbf{k}$ equal to zero in Eq. (2), we obtain two scalar equations which may be solved for T_{BD} and T_{BE}:

$$T_{BD} = 1640 \text{ lb} \qquad T_{BE} = 4720 \text{ lb} \quad \blacktriangleleft$$

Setting the coefficients of $\mathbf{i}, \mathbf{j}, \mathbf{k}$ equal to zero in Eq. (1), we obtain three more scalar equations, which yield

$$A_x = +3260 \text{ lb} \qquad A_y = +8880 \text{ lb} \qquad A_z = +1183 \text{ lb} \quad \blacktriangleleft$$

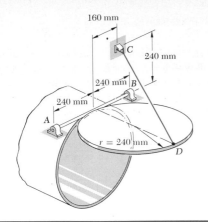

160 mm

C

240 mm

240 mm | B

240 mm

A

$r = 240$ mm

D

SAMPLE PROBLEM 4.9

A uniform pipe cover of radius $r = 240$ mm and mass 30 kg is held in a horizontal position by the cable CD. Assuming that the bearing at B does not exert any axial thrust, determine the tension in the cable and the components of the reactions at A and B.

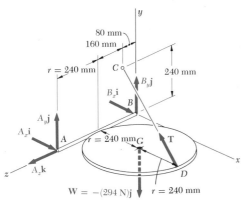

Solution. A free-body diagram is drawn with the coordinate axes shown. The forces acting on the free body are the weight of the cover,

$$\mathbf{W} = -mg\mathbf{j} = -(30 \text{ kg})(9.81 \text{ m/s}^2)\mathbf{j} = -(294 \text{ N})\mathbf{j}$$

and reactions involving six unknowns, namely, the magnitude of the force $\mathbf{T}$ exerted by the cable, three force components at hinge A, and two at hinge B. The components of $\mathbf{T}$ are expressed in terms of the unknown magnitude T by resolving the vector $\overrightarrow{DC}$ into rectangular components and writing

$$\overrightarrow{DC} = -(480 \text{ mm})\mathbf{i} + (240 \text{ mm})\mathbf{j} - (160 \text{ mm})\mathbf{k} \qquad DC = 560 \text{ mm}$$

$$\mathbf{T} = T\frac{\overrightarrow{DC}}{DC} = -\tfrac{6}{7}T\mathbf{i} + \tfrac{3}{7}T\mathbf{j} - \tfrac{2}{7}T\mathbf{k}$$

Equilibrium Equations. We express that the forces acting on the pipe cover form a system equivalent to zero:

$$\Sigma\mathbf{F} = 0: \qquad A_x\mathbf{i} + A_y\mathbf{j} + A_z\mathbf{k} + B_x\mathbf{i} + B_y\mathbf{j} + \mathbf{T} - (294 \text{ N})\mathbf{j} = 0$$

$$(A_x + B_x - \tfrac{6}{7}T)\mathbf{i} + (A_y + B_y + \tfrac{3}{7}T - 294 \text{ N})\mathbf{j}$$
$$+ (A_z - \tfrac{2}{7}T)\mathbf{k} = 0 \qquad (1)$$

$$\Sigma\mathbf{M}_B = \Sigma(\mathbf{r} \times \mathbf{F}) = 0:$$

$$2r\mathbf{k} \times (A_x\mathbf{i} + A_y\mathbf{j} + A_z\mathbf{k})$$
$$+ (2r\mathbf{i} + r\mathbf{k}) \times (-\tfrac{6}{7}T\mathbf{i} + \tfrac{3}{7}T\mathbf{j} - \tfrac{2}{7}T\mathbf{k})$$
$$+ (r\mathbf{i} + r\mathbf{k}) \times (-294 \text{ N})\mathbf{j} = 0$$

$$(-2A_y - \tfrac{3}{7}T + 294 \text{ N})r\mathbf{i} + (2A_x - \tfrac{2}{7}T)r\mathbf{j} + (\tfrac{6}{7}T - 294 \text{ N})r\mathbf{k} = 0 \qquad (2)$$

Setting the coefficients of the unit vectors equal to zero in Eq. (2), we write three scalar equations, which yield

$$T = 343 \text{ N} \qquad A_x = +49.0 \text{ N} \qquad A_y = +73.5 \text{ N} \quad \blacktriangleleft$$

Setting the coefficients of the unit vectors equal to zero in Eq. (1), we obtain three more scalar equations. After substituting the values of T, A_x, and A_y into these equations, we obtain

$$A_z = +98.0 \text{ N} \qquad B_x = +245 \text{ N} \qquad B_y = +73.5 \text{ N} \quad \blacktriangleleft$$

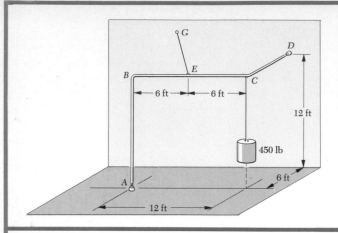

A 450-lb load hangs from the corner C of a rigid piece of pipe $ABCD$ which has been bent as shown. The pipe is supported by the ball-and-socket joints A and D fastened, respectively, to the floor and to a vertical wall, and by a cable attached at the midpoint E of the portion BC of the pipe and at a point G on the wall. Determine (*a*) where G should be located if the tension in the cable is to be minimum, (*b*) the corresponding minimum value of the tension.

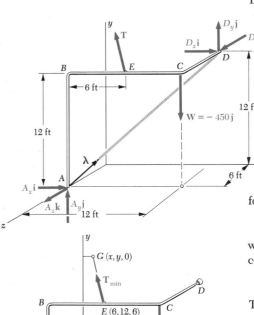

Solution. The free-body diagram of the pipe includes the load $\mathbf{W} = -450\mathbf{j}$, the reactions at A and D, and the force $\mathbf{T}$ exerted by the cable. To eliminate the reactions at A and D from the computations, we express that the sum of the moments of the forces about AD is zero. Denoting by $\boldsymbol{\lambda}$ the unit vector along AD, we write

$$\Sigma M_{AD} = 0: \qquad \boldsymbol{\lambda} \cdot (\overrightarrow{AE} \times \mathbf{T}) + \boldsymbol{\lambda} \cdot (\overrightarrow{AC} \times \mathbf{W}) = 0 \qquad (1)$$

The second term in Eq. (1) may be computed as follows:

$$\overrightarrow{AC} \times \mathbf{W} = (12\mathbf{i} + 12\mathbf{j}) \times (-450\mathbf{j}) = -5400\mathbf{k}$$

$$\boldsymbol{\lambda} = \frac{\overrightarrow{AD}}{AD} = \frac{12\mathbf{i} + 12\mathbf{j} - 6\mathbf{k}}{18} = \tfrac{2}{3}\mathbf{i} + \tfrac{2}{3}\mathbf{j} - \tfrac{1}{3}\mathbf{k}$$

$$\boldsymbol{\lambda} \cdot (\overrightarrow{AC} \times \mathbf{W}) = (\tfrac{2}{3}\mathbf{i} + \tfrac{2}{3}\mathbf{j} - \tfrac{1}{3}\mathbf{k}) \cdot (-5400\mathbf{k}) = +1800$$

Substituting the value obtained into Eq. (1), we write

$$\boldsymbol{\lambda} \cdot (\overrightarrow{AE} \times \mathbf{T}) = -1800 \text{ lb} \cdot \text{ft} \qquad (2)$$

Minimum Value of Tension. Recalling the commutative property for mixed triple products, we rewrite Eq. (2) in the form

$$\mathbf{T} \cdot (\boldsymbol{\lambda} \times \overrightarrow{AE}) = -1800 \text{ lb} \cdot \text{ft} \qquad (3)$$

which shows that the projection of $\mathbf{T}$ on the vector $\boldsymbol{\lambda} \times \overrightarrow{AE}$ is a constant. It follows that $\mathbf{T}$ is minimum when parallel to the vector

$$\boldsymbol{\lambda} \times \overrightarrow{AE} = (\tfrac{2}{3}\mathbf{i} + \tfrac{2}{3}\mathbf{j} - \tfrac{1}{3}\mathbf{k}) \times (6\mathbf{i} + 12\mathbf{j}) = 4\mathbf{i} - 2\mathbf{j} + 4\mathbf{k}$$

The corresponding unit vector being $\tfrac{2}{3}\mathbf{i} - \tfrac{1}{3}\mathbf{j} + \tfrac{2}{3}\mathbf{k}$, we write

$$\mathbf{T}_{\min} = T(\tfrac{2}{3}\mathbf{i} - \tfrac{1}{3}\mathbf{j} + \tfrac{2}{3}\mathbf{k}) \qquad (4)$$

Substituting for $\mathbf{T}$ and $\boldsymbol{\lambda} \times \overrightarrow{AE}$ in Eq. (3), we find $T = -300$. Carrying this value into (4), we obtain

$$\mathbf{T}_{\min} = -200\mathbf{i} + 100\mathbf{j} - 200\mathbf{k} \qquad T_{\min} = 300 \text{ lb} \blacktriangleleft$$

Location of G. Since the vector $\overrightarrow{EG}$ and the force $\mathbf{T}_{\min}$ have the same direction, their components must be proportional. Denoting by $x, y, 0$ the coordinates of G, we write

$$\frac{x-6}{-200} = \frac{y-12}{+100} = \frac{0-6}{-200} \qquad x = 0 \qquad y = 15 \text{ ft} \blacktriangleleft$$

PROBLEMS

4.60 The 10-m pole is acted upon by an 8.4-kN force as shown. It is held by a ball and socket at A and by the two cables BD and BE. Neglecting the weight of the pole, determine the tension in each cable and the reaction at A.

4.61 The boom AB supports a load of 900 N as shown. The boom is held by a ball and socket at A and by the two cables BC and BD. Neglecting the weight of the boom, determine the tension in each cable and the reaction at A.

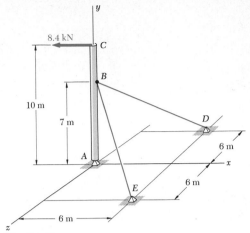

Fig. P4.60

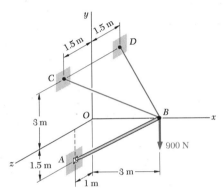

Fig. P4.61

4.62 The boom AB supports a load of weight W = 820 lb. The boom is held by a ball and socket at A and by the two cables BC and BD. Neglecting the weight of the boom, determine the tension in each cable and the reaction at A.

4.63 Solve Prob. 4.62, assuming that the boom weighs 123 lb.

4.64 A 2.5-m boom is held by a ball and socket at A and by two cables EBF and DC; the cable EBF passes around a frictionless pulley at B. Determine the tension in each cable.

4.65 In Sample Prob. 4.8, determine the horizontal angle ϕ for which the tension in BE is maximum (a) if neither cable may become slack, (b) if cables BE and BD are replaced by rods which can withstand either a tension or a compression force.

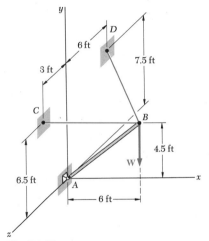

Fig. P4.62

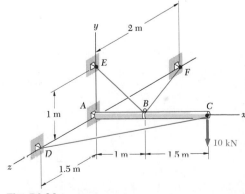

Fig. P4.64

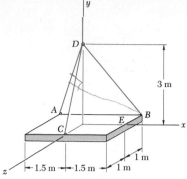

Fig. P4.66

4.66 A 2- by 3-m plate of mass 600 kg is lifted by three cables which are joined at point D directly above the center of the plate. Determine the tension in each cable.

4.67 Solve Prob. 4.66, assuming that cable BD is replaced by a cable connecting points D and E.

4.68 The square plate shown weighs 40 lb and is supported by three wires as shown. Determine the tension in each wire.

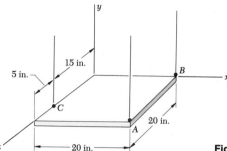

Fig. P4.68, P4.69, and P4.70

4.69 A load W is to be placed on the plate shown. Knowing that the weight of the plate is 40 lb, determine the magnitude of the load W and the point where the load should be placed if the tension is to be 30 lb in each of the three wires.

4.70 Determine the magnitude and location of the smallest load which must be placed on the 40-lb plate if the tensions in the three wires are to be equal.

4.71 The overhead transmission shaft AE is driven at a constant speed by an electric motor connected by a flat belt to pulley B. Pulley C may be used to drive a machine tool located directly below C, while pulley D drives a parallel shaft located at the same height as AE. Knowing that $T_B + T'_B = 36$ lb, $T_C = 40$ lb, $T'_C = 16$ lb, $T_D = 0$, and $T'_D = 0$, determine (a) the tension in each portion of the belt driving pulley B, (b) the reactions at the bearings A and E caused by the tension in the belts.

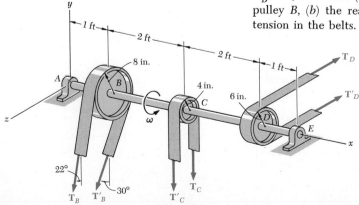

Fig. P4.71

4.72 Solve Prob. 4.71, assuming that $T_B + T'_B = 36$ lb, $T_C = 0$, $T'_C = 0$, $T_D = 36$ lb, and $T'_D = 12$ lb.

4.73 The winch shown is held in equilibrium by a vertical force **P** applied at point E. Knowing that $\theta = 60°$, determine the magnitude of **P** and the reactions at A and B. It is assumed that the bearing at B does not exert any axial thrust.

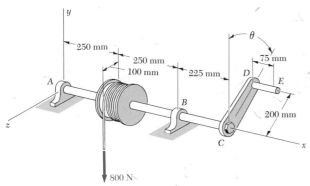

Fig. P4.73

4.74 Solve Prob. 4.73, assuming that the winch is held in equilibrium by a force **P** applied at E in a direction perpendicular to the plane CDE.

4.75 The weight of the uniform rod AB is 56 N and its length 450 mm. It is connected by ball-and-socket joints to collars A and B, which slide freely along the two rods shown. Determine the magnitude of the force **P** required to maintain equilibrium when (*a*) $c = 50$ mm, (*b*) $c = 200$ mm.

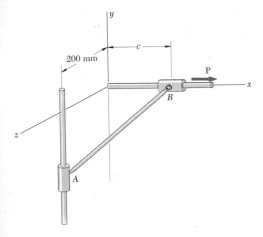

Fig. P4.75

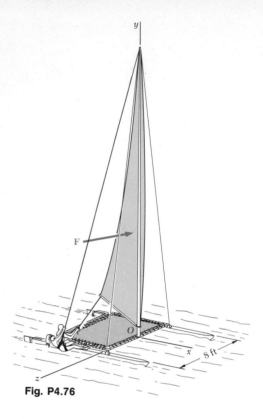

Fig. P4.76

4.76 A 400-lb catamaran is sailing in a straight course and at a constant speed. The effect of the wind on the sail may be represented by the force $\mathbf{F} = (90 \text{ lb})\mathbf{i} - (120 \text{ lb})\mathbf{k}$ applied at the point of coordinates $x = -1.6$ ft, $y = 10$ ft, $z = -1.2$ ft. Due to the efforts of the crew, the mast is maintained in a vertical position and the twin hulls are subjected to equal forces, consisting of buoyant forces $\mathbf{B} = B\mathbf{j}$ along the lines of action $x = -2$ ft, $z = \pm 4$ ft, drag forces $\mathbf{D} = -D\mathbf{i}$ along the lines of action $y = -2$ ft, $z = \pm 4$ ft, and side forces $\mathbf{S} = S\mathbf{k}$ along the common line of action $x = -d$, $y = -2$ ft. Determine (a) the magnitude D of the drag on each of the hulls, (b) the magnitude S of each of the side forces, and the distance d defining their line of action.

4.77 The catamaran of Prob. 4.76 is manned by two sailors, one weighing 160 lb and the other 140 lb. The center of gravity of the boat is located at $x = -3$ ft, $z = 0$, and the 160-lb sailor is sitting at $x = -5$ ft, $z = +4$ ft, while the 140-lb sailor leans overboard to maintain the mast in a vertical position. Determine the x and z coordinates of the center of gravity of the second sailor.

4.78 In order to clean the clogged drainpipe AE, a plumber has disconnected both ends of the pipe and introduced a power snake through the opening at A. The cutting head of the snake is connected by a heavy cable to an electric motor which keeps it rotating at a constant speed while the plumber forces the cable into the pipe. The forces exerted by the plumber and the motor on the end of the cable may be represented by the wrench $\mathbf{F} = -(48 \text{ N})\mathbf{k}$, $\mathbf{M} = -(90 \text{ N} \cdot \text{m})\mathbf{k}$. Determine the additional reactions at B, C, and D caused by the cleaning operation. Assume that the reaction at each support consists of two force components perpendicular to the pipe.

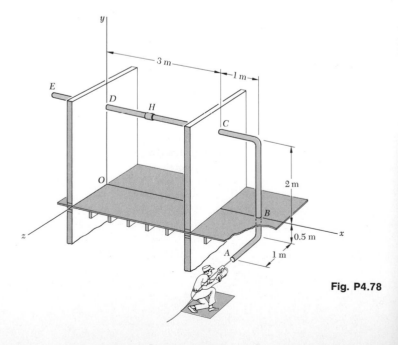

Fig. P4.78

4.79 In Prob. 4.78, the plumber has disconnected the pipe at *H* before introducing the snake through the opening at *A*. Determine the additional reactions caused by the cleaning operation at the remaining supports *B* and *C*, assuming that, in addition to the two force components perpendicular to the pipe, the reaction at *C* now also includes two couple vectors respectively parallel to the *y* and *z* axes.

4.80 A 20-kg door is made self-closing by hanging a 15-kg counterweight from a cable attached at *C*. The door is held open by a force **P** applied at the knob *D*, in a direction perpendicular to the door. Determine the magnitude of **P** and the components of the reactions *A* and *B* when $\theta = 90°$. It is assumed that the hinge at *A* does not exert any axial thrust.

4.81 A 10-kg storm window measuring 900 by 1500 mm is held by hinges at *A* and *B*. In the position shown, it is held away from the side of the house by a 600-mm stick *CD*. Assuming that the hinge at *A* does not exert any axial thrust, determine the magnitude of the force exerted by the stick and the components of the reactions at *A* and *B*.

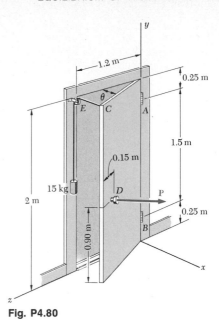
Fig. P4.80

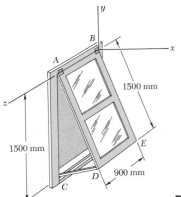
Fig. P4.81

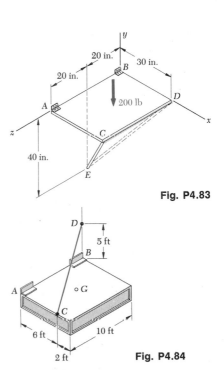
Fig. P4.83

4.82 Solve Prob. 4.81, assuming that the hinge at *A* is removed.

4.83 The horizontal platform *ABCD* weighs 50 lb and supports a 200-lb load at its center. The platform is normally held in position by hinges at *A* and *B* and by braces *CE* and *DE*. If brace *DE* is removed, determine the reactions at the hinges and the force exerted by the remaining brace *CE*. The hinge at *A* does not exert any axial thrust.

4.84 A 500-lb marquee, 8 by 10 ft, is held in a horizontal position by two horizontal hinges at *A* and *B* and by a cable *CD* attached to a point *D* located 5 ft directly above *B*. Determine the tension in the cable and the components of the reactions at the hinges.

4.85 Solve Prob. 4.84, assuming that the hinge at *A* is removed.

Fig. P4.84

4.86 Solve Prob. 4.83, assuming that the brace *DE* and the hinge at *A* are removed.

4.87 The rigid L-shaped member *ABC* is supported by a ball and socket at *A* and by three cables. Determine the tension in each cable and the reaction at *A* caused by the 5-kN load applied at *G*.

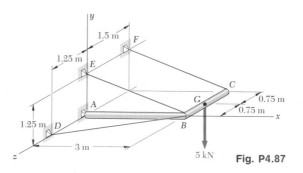

Fig. P4.87

4.88 Solve Prob. 4.87, assuming that cable *BD* is removed and replaced by a cable joining points *E* and *C*.

4.89 Three rods are welded together to form the "corner" shown. The corner is supported by three smooth eyebolts. Determine the reactions at *A*, *B*, and *C* when $P = 250$ lb, $a = 12$ in., $b = 8$ in., and $c = 10$ in.

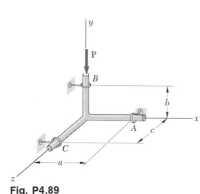

Fig. P4.89

4.90 Solve Prob. 4.89, assuming that the force **P** is removed and is replaced by the couple $\mathbf{M} = +(1500 \text{ lb} \cdot \text{in.})\mathbf{j}$ acting at *B*.

4.91 A uniform circular plate of radius *r* and weight *W* is supported by three vertical wires, each of length *h*, equally spaced around the edge. Determine the horizontal angle θ through which the plate rotates when a couple $\mathbf{M}_0$ is applied in the plane of the plate. Assume that θ is small.

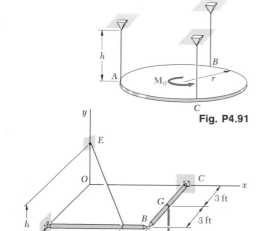

Fig. P4.91

4.92 The 8-ft rod *AB* and the 6-ft rod *BC* are hinged at *B* and supported by the cable *DE* and by ball-and-socket joints at *A* and *C*. Knowing that $h = 3$ ft, determine the tension in the cable for the loading shown.

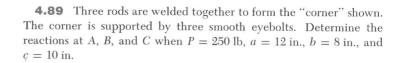

Fig. P4.92

4.93 Solve Prob. 4.92 when $h = 10.5$ ft.

4.94 The 23-kg plate *ABCD* measures 325 by 450 mm; it is held by hinges along edge *AD* and the wire *BE*. Determine the tension in the wire.

4.95 Solve Prob. 4.94, assuming that wire *BE* is replaced by a wire connecting *E* and *C*.

4.96 Two rods are welded together to form a T-shaped lever which leans against a frictionless vertical wall at *D* and is supported by bearings at *A* and *B*. A vertical force **P** of magnitude 400 N is applied at the midpoint of rod *DC*. Determine the reaction at *D*.

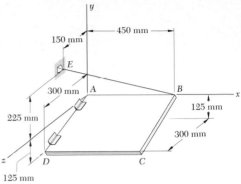

Fig. P4.94

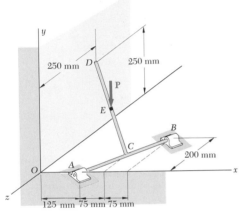

Fig. P4.96

REVIEW PROBLEMS

4.97 Determine the reactions at *A* and *B* for the truss and loading shown.

4.98 Determine the reactions at *A* and *B* for the loading shown.

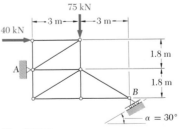

Fig. P4.97

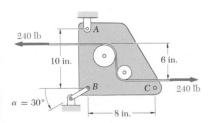

Fig. P4.98

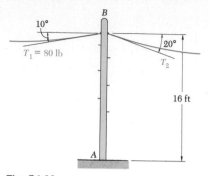

Fig. P4.99

4.99 A 16-ft telephone pole weighing 300 lb is used to support the ends of two wires. The tension in the wire to the left is 80 lb and, at the point of support, the wire forms an angle of 10° with the horizontal. (*a*) If the tension T_2 is zero, determine the reaction at the base *A*. (*b*) Determine the largest and smallest allowable tension T_2, if the magnitude of the couple at *A* may not exceed 600 lb · ft.

4.100 The light bar *AD* is attached to collars *B* and *C* which may move freely on vertical rods. If the surface at *A* is frictionless, determine the reactions at *A*, *B*, and *C* (*a*) if $\alpha = 60°$, (*b*) if $\alpha = 90°$.

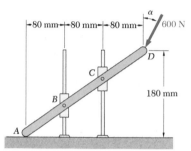

Fig. P4.100

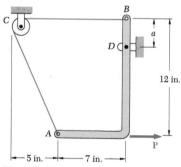

Fig. P4.101

4.101 A force **P** of magnitude 90 lb is applied to the angle *AB* which is supported by a frictionless pin at *D* and by the cable *ACB*. Since the cable passes over a pulley at *C*, the tension may be assumed to be the same in the portions *AC* and *BC* of the cable. Determine the tension in the cable and the reaction at *D* when *a* = 3 in.

*4.102 In the problems listed below, the rigid bodies considered were completely constrained, and the reactions were statically determinate. For each of these rigid bodies it is possible to create an improper set of constraints by changing either a dimension of the body or the direction of a reaction. In each problem determine the value of α or of *a* which results in improper constraints. (*a*) Prob. 4.3, (*b*) Prob. 4.21*d*, (*c*) Prob. 4.97, (*d*) Prob. 4.98, (*e*) Prob. 4.101.

4.103 A 3000-kg uniform plate girder is held by two crane cables as shown. The cable attached at *B* forms an angle of 30° with the vertical. If the girder is to be held in a horizontal position, determine the direction of the cable attached at *A* and the tension in each cable.

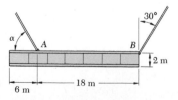

Fig. P4.103

4.104   A 500-lb cylindrical tank, 8 ft in diameter, is to be raised over a 2-ft obstruction. A cable is wrapped around the tank and pulled horizontally as shown. Knowing that the corner of the obstruction at A is rough, find the required tension in the cable and the reaction at A.

4.105   The door of a bank vault weighs 12,000 lb and is supported by two hinges as shown. Determine the components of the reaction at each hinge.

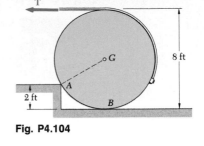

Fig. P4.104

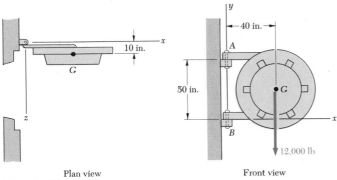

Plan view   Front view
Fig. P4.105

4.106   A 120-kg sign of uniform density measures 1.5 by 2.4 m. It is supported by a ball and socket at A and by two cables. Determine the tension in each cable and the reaction at A.

4.107   The table shown has a radius of 600 mm and a weight of 120 N. It is supported by three legs equally spaced around the edge. A vertical load **P** of magnitude 300 N is applied to the top of the table at D. Determine the maximum value of a if the table is not to tip over. Show, on a sketch, the area of the table over which **P** can act without tipping the table.

4.108   The rod AB is uniform and weighs 25 lb. It is supported by a ball and socket at A and leans against both the rod CD and the vertical wall. Assuming the wall and the rods to be frictionless, determine (a) the force which rod CD exerts on AB, (b) the reactions at A and B. (*Hint*. The force exerted by CD on AB must be perpendicular to both rods.)

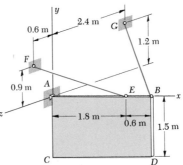

Fig. P4.106

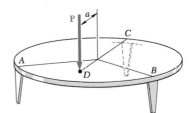

Fig. P4.107

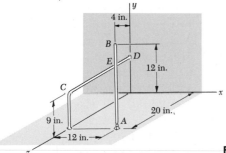

Fig. P4.108

Distributed Forces: Centroids and Centers of Gravity

AREAS AND LINES

5.1. Center of Gravity of a Two-dimensional Body. We have assumed so far that the attraction exerted by the earth on a rigid body could be represented by a single force **W**. This force, called the weight of the body, was to be applied at the *center of gravity* of the body (Sec. 3.1). Actually, the earth exerts a force on each of the particles forming the body. The action of the earth on a rigid body should thus be represented by a large number of small forces distributed over the entire body. We shall see in this chapter, however, that all these small forces may be replaced by a single equivalent force **W**. We shall also learn to determine the center of gravity, i.e., the point of application of the resultant **W**, for various shapes of bodies.

Let us first consider a flat horizontal plate (Fig. 5.1). We may divide the plate into n small elements. The coordinates of the first element are denoted by x_1 and y_1, those of the second element by x_2 and y_2, etc. The forces exerted by the earth on the elements of plate will be denoted, respectively, by $\Delta \mathbf{W}_1$, $\Delta \mathbf{W}_2, \ldots, \Delta \mathbf{W}_n$. These forces or weights are directed toward the center of the earth; however, for all practical purposes they may be assumed parallel. Their resultant is therefore a single force in the same direction. The magnitude W of this force is

166

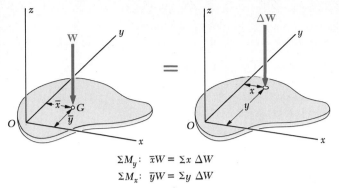

$$\Sigma M_y: \quad \bar{x}W = \Sigma x\,\Delta W$$
$$\Sigma M_x: \quad \bar{y}W = \Sigma y\,\Delta W$$

Fig. 5.1 Center of gravity of a plate.

obtained by adding the magnitudes of the elementary weights,

$$\Sigma F_z: \qquad W = \Delta W_1 + \Delta W_2 + \cdots + \Delta W_n \qquad (5.1)$$

To obtain the coordinates $\bar{x}$ and $\bar{y}$ of the point G where the resultant $\mathbf{W}$ should be applied, we write that the moments of $\mathbf{W}$ about the y and x axes are equal to the sum of the corresponding moments of the elementary weights,

$$\Sigma M_y: \qquad \bar{x}W = x_1\,\Delta W_1 + x_2\,\Delta W_2 + \cdots + x_n\,\Delta W_n$$
$$\Sigma M_x: \qquad \bar{y}W = y_1\,\Delta W_1 + y_2\,\Delta W_2 + \cdots + y_n\,\Delta W_n \qquad (5.2)$$

If we now increase the number of elements into which the plate is divided and simultaneously decrease the size of each element, we obtain at the limit the following expressions:

$$W = \int dW \qquad \bar{x}W = \int x\,dW \qquad \bar{y}W = \int y\,dW \qquad (5.3)$$

These equations define the weight $\mathbf{W}$ and the coordinates $\bar{x}$ and $\bar{y}$ of the center of gravity G of a flat plate. The same equations may be derived for a wire lying in the xy plane (Fig. 5.2). We shall observe, in the latter case, that the center of gravity G will generally not be located on the wire.

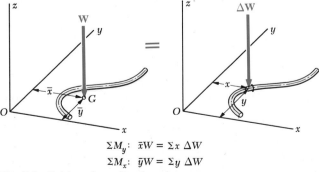

$$\Sigma M_y: \quad \bar{x}W = \Sigma x\,\Delta W$$
$$\Sigma M_x: \quad \bar{y}W = \Sigma y\,\Delta W$$

Fig. 5.2 Center of gravity of a wire.

5.2. Centroids of Areas and Lines. In the case of a homogeneous plate of uniform thickness, the magnitude ΔW of the weight of an element of plate may be expressed as

$$\Delta W = \gamma t\, \Delta A$$

where γ = specific weight (weight per unit volume) of material
t = thickness of plate
ΔA = area of element

Similarly, we may express the magnitude W of the weight of the entire plate in the form

$$W = \gamma t A$$

where A is the total area of the plate.

If U.S. customary units are used, the specific weight γ should be expressed in lb/ft^3, the thickness t in feet, and the areas ΔA and A in square feet. We check that ΔW and W will then be expressed in pounds. If SI units are used, γ should be expressed in N/m^3, t in meters, and the areas ΔA and A in square meters; the weights ΔW and W will then be expressed in newtons.†

Substituting for ΔW and W in the moment equations (5.2) and dividing throughout by γt, we write

$$\begin{aligned}
\Sigma M_y: & \quad \bar{x}A = x_1\, \Delta A_1 + x_2\, \Delta A_2 + \cdots + x_n\, \Delta A_n \\
\Sigma M_x: & \quad \bar{y}A = y_1\, \Delta A_1 + y_2\, \Delta A_2 + \cdots + y_n\, \Delta A_n
\end{aligned} \tag{5.4}$$

If we increase the number of elements into which the area A is divided and simultaneously decrease the size of each element, we obtain at the limit

$$\bar{x}A = \int x\, dA \qquad \bar{y}A = \int y\, dA \tag{5.5}$$

These equations define the coordinates $\bar{x}$ and $\bar{y}$ of the center of gravity of a homogeneous plate. The point of coordinates $\bar{x}$ and $\bar{y}$ is also known as the *centroid C of the area A* of the plate (Fig. 5.3). If the plate is not homogeneous, the equations cannot be used to determine the center of gravity of the plate; they still define, however, the centroid of the area.

---

† It should be noted that in the SI system of units a given material is generally characterized by its density ρ (mass per unit volume) rather than by its specific weight γ. The specific weight of the material can then be obtained by writing

$$\gamma = \rho g$$

where $g = 9.81\ m/s^2$. Since ρ is expressed in kg/m^3, we check that γ will be expressed in $(kg/m^3)(m/s^2)$, that is, in N/m^3.

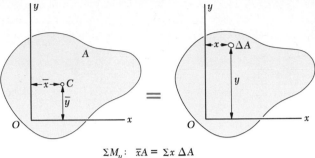

$$\Sigma M_y: \quad \bar{x}A = \Sigma x\,\Delta A$$
$$\Sigma M_x: \quad \bar{y}A = \Sigma y\,\Delta A$$

Fig. 5.3 Centroid of an area.

The integral $\int x\,dA$ is known as the *first moment of the area A with respect to the y axis*. Similarly, the integral $\int y\,dA$ defines the *first moment of A with respect to the x axis*. It is seen from Eqs. (5.5) that, if the centroid of an area is located on a coordinate axis, the first moment of the area with respect to that axis is zero.

In the case of a homogeneous wire of uniform cross section, the magnitude ΔW of the weight of an element of wire may be expressed as

$$\Delta W = \gamma a\,\Delta L$$

where γ = specific weight of material
a = cross-sectional area of wire
ΔL = length of element
The center of gravity of the wire then coincides with the *centroid C of the line L* defining the shape of the wire (Fig. 5.4). The coordinates $\bar{x}$ and $\bar{y}$ of the centroid of the line L are obtained from the equations

$$\bar{x}L = \int x\,dL \qquad \bar{y}L = \int y\,dL \qquad\qquad (5.6)$$

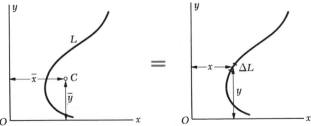

$$\Sigma M_y: \quad \bar{x}L = \Sigma x\,\Delta L$$
$$\Sigma M_x: \quad \bar{y}L = \Sigma y\,\Delta L$$

Fig. 5.4 Centroid of a line.

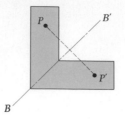

Fig. 5.5

An area A is said to be *symmetrical about an axis BB′* if to every point P of the area corresponds a point P' of the same area such that the line PP' is perpendicular to BB' and is divided into two equal parts by that axis (Fig. 5.5). A line L is said to be symmetrical about BB' if it satisfies similar conditions. When an area A or a line L possesses an axis of symmetry BB', the centroid of the area or line must be located on that axis. Indeed, if the axis of symmetry is chosen as the y axis, the coordinate $\bar{x}$ of the centroid is found to be zero, since to every product $x\,dA$ or $x\,dL$ in the first integral in Eqs. (5.5) or (5.6) will correspond a product of equal magnitude but of opposite sign. It

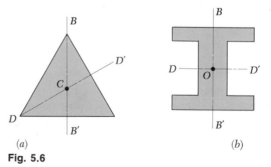

(a)   (b)

Fig. 5.6

Fig. 5.7

follows that, if an area or line possesses two axes of symmetry, the centroid of the area or line is located at the intersection of the two axes of symmetry (Fig. 5.6). This property enables us to determine immediately the centroid of areas such as circles, ellipses, squares, rectangles, equilateral triangles, or other symmetrical figures, as well as the centroid of lines in the shape of the circumference of a circle, the perimeter of a square, etc.

An area is said to be *symmetrical about a center O* if to every point P of the area corresponds a point P' of the same area such that the line PP' is divided into two equal parts by O (Fig. 5.7).

Shape		$\bar{x}$	$\bar{y}$	Area
Triangular area	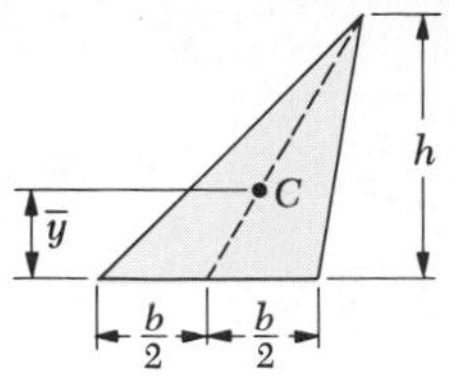		$\dfrac{h}{3}$	$\dfrac{bh}{2}$
Quarter-circular area	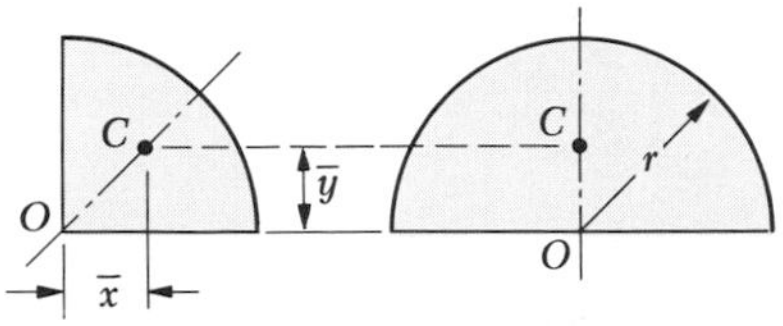	$\dfrac{4r}{3\pi}$	$\dfrac{4r}{3\pi}$	$\dfrac{\pi r^2}{4}$
Semicircular area		0	$\dfrac{4r}{3\pi}$	$\dfrac{\pi r^2}{2}$
Quarter-elliptical area	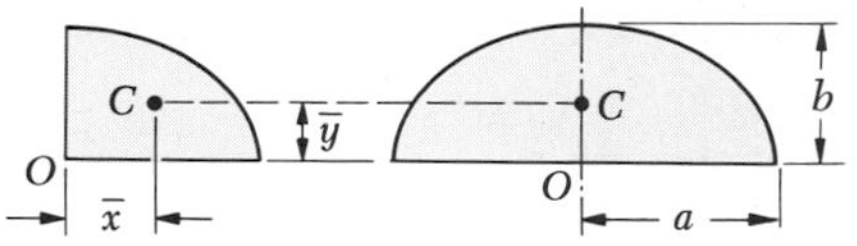	$\dfrac{4a}{3\pi}$	$\dfrac{4b}{3\pi}$	$\dfrac{\pi ab}{4}$
Semielliptical area		0	$\dfrac{4b}{3\pi}$	$\dfrac{\pi ab}{2}$
Semiparabolic area	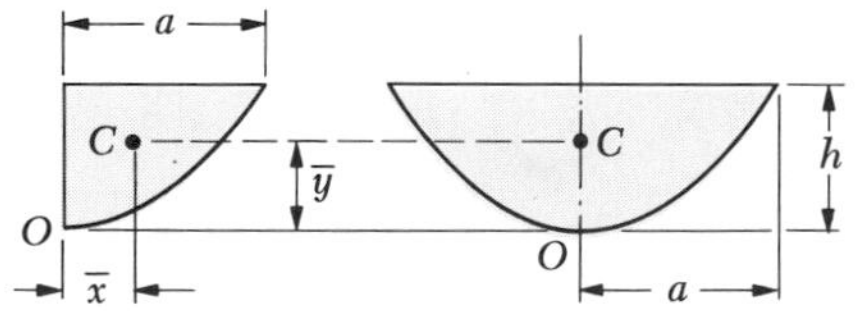	$\dfrac{3a}{8}$	$\dfrac{3h}{5}$	$\dfrac{2ah}{3}$
Parabolic area		0	$\dfrac{3h}{5}$	$\dfrac{4ah}{3}$
Parabolic spandrel	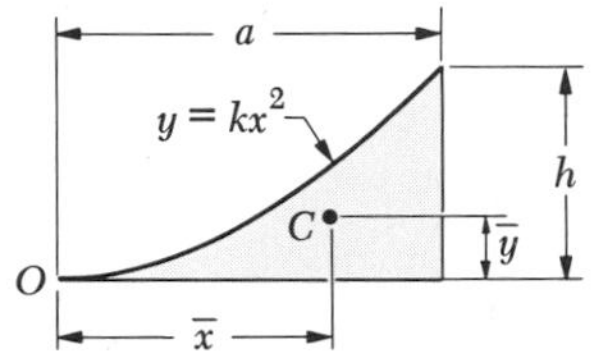	$\dfrac{3a}{4}$	$\dfrac{3h}{10}$	$\dfrac{ah}{3}$
General spandrel	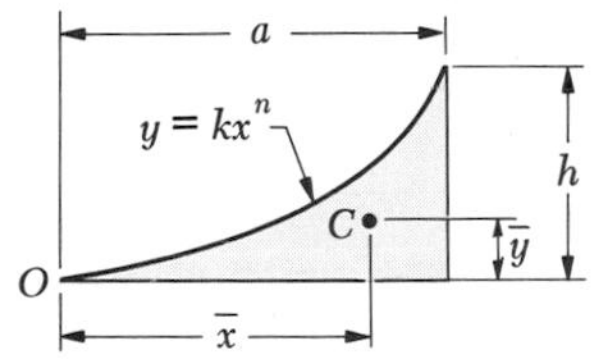	$\dfrac{n+1}{n+2}a$	$\dfrac{n+1}{4n+2}h$	$\dfrac{ah}{n+1}$
Circular sector	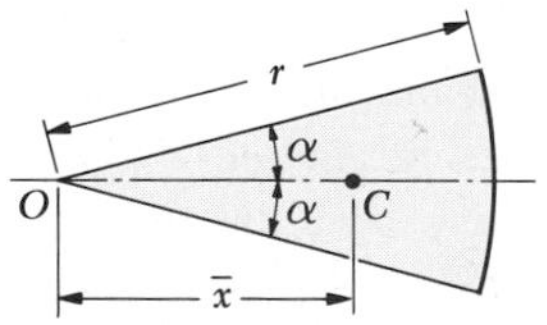	$\dfrac{2r\sin\alpha}{3\alpha}$	0	αr^2

Fig. 5.8A Centroids of common shapes of areas.

Shape		$\bar{x}$	$\bar{y}$	Length
Quarter-circular arc	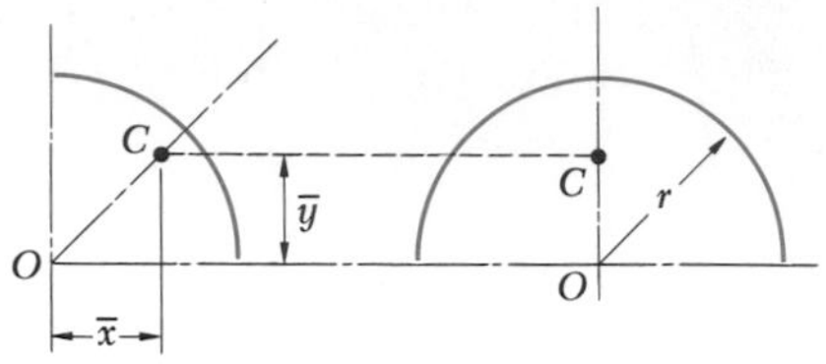	$\dfrac{2r}{\pi}$	$\dfrac{2r}{\pi}$	$\dfrac{\pi r}{2}$
Semicircular arc		0	$\dfrac{2r}{\pi}$	πr
Arc of circle	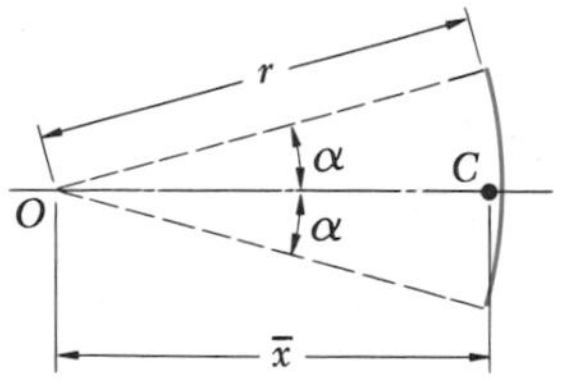	$\dfrac{r \sin \alpha}{\alpha}$	0	$2\alpha r$

Fig. 5.8B Centroids of common shapes of lines.

A line L is said to be symmetrical about O if it satisfies similar conditions. A reasoning similar to that used above would show that, when an area A or line L possesses a center of symmetry O, the point O must be the centroid of the area or line. It should be noted that a figure possessing a center of symmetry does not necessarily possess an axis of symmetry (Fig. 5.7), while a figure possessing two axes of symmetry does not necessarily possess a center of symmetry (Fig. 5.6a). However, if a figure possesses two axes of symmetry at a right angle to each other, the point of intersection of these axes will be a center of symmetry (Fig. 5.6b).

Centroids of unsymmetrical areas and lines and of areas and lines possessing only one axis of symmetry will be determined by the methods of Secs. 5.4 and 5.5. Centroids of common shapes of areas and lines are shown in Fig. 5.8A and B. The formulas defining these centroids will be derived in the Sample Problems and Problems following Secs. 5.4 and 5.5.

5.3. Composite Plates and Wires. In many instances, a flat plate may be divided into rectangles, triangles, or other common shapes shown in Fig. 5.8A. The abscissa $\bar{X}$ of its center of gravity G may be determined from the abscissas $\bar{x}_1, \bar{x}_2, \ldots$ of the centers of gravity of the various parts by expressing that the moment of the weight of the whole plate about the y axis is equal to the sum of the moments of the weights of the various parts about the same axis (Fig. 5.9). The ordinate $\bar{Y}$ of the center

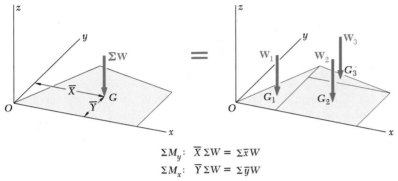

$$\Sigma M_y: \quad \bar{X}\,\Sigma W = \Sigma \bar{x} W$$
$$\Sigma M_x: \quad \bar{Y}\,\Sigma W = \Sigma \bar{y} W$$

Fig. 5.9 Center of gravity of a composite plate.

of gravity of the plate is found in a similar way by equating moments about the x axis.

$$\begin{aligned}
\Sigma M_y: \quad & \bar{X}(W_1 + W_2 + \cdots + W_n) \\
& = \bar{x}_1 W_1 + \bar{x}_2 W_2 + \cdots + \bar{x}_n W_n \\
\Sigma M_x: \quad & \bar{Y}(W_1 + W_2 + \cdots + W_n) \\
& = \bar{y}_1 W_1 + \bar{y}_2 W_2 + \cdots + \bar{y}_n W_n
\end{aligned} \tag{5.7}$$

If the plate is homogeneous and of uniform thickness, the center of gravity coincides with the centroid C of its area. The abscissa $\bar{X}$ of the centroid of the area may then be determined by expressing that the first moment of the composite area with respect to the y axis is equal to the sum of the first moments of the elementary areas with respect to the same axis (Fig. 5.10). The ordinate $\bar{Y}$ of the centroid is found in a similar way by equating first moments of areas with respect to the x axis.

$$\begin{aligned}
\Sigma M_y: \quad & \bar{X}(A_1 + A_2 + \cdots + A_n) \\
& = \bar{x}_1 A_1 + \bar{x}_2 A_2 + \cdots + \bar{x}_n A_n \\
\Sigma M_x: \quad & \bar{Y}(A_1 + A_2 + \cdots + A_n) \\
& = \bar{y}_1 A_1 + \bar{y}_2 A_2 + \cdots + \bar{y}_n A_n
\end{aligned} \tag{5.8}$$

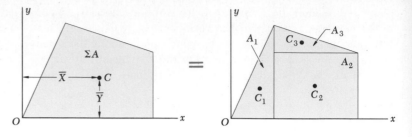

$$\Sigma M_y: \quad \overline{X}\Sigma A = \Sigma \overline{x}A$$
$$\Sigma M_x: \quad \overline{Y}\Sigma A = \Sigma \overline{y}A$$

Fig. 5.10 Centroid of a composite area.

Care should be taken to record the moment of each area with the appropriate sign. First moments of areas, just like moments of forces, may be positive or negative. For example, an area whose centroid is located to the left of the y axis will have a negative first moment with respect to that axis. Also, the area of a hole should be recorded with a negative sign (Fig. 5.11).

Similarly, it is possible in many cases to determine the center of gravity of a composite wire or the centroid of a composite line by dividing the wire or line into simpler elements (Sample Prob. 5.2).

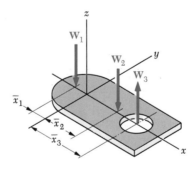

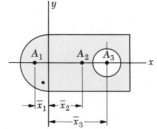

	$\overline{x}$	A	$\overline{x}A$
A_1 Semicircle	−	+	−
A_2 Full rectangle	+	+	+
A_3 Circular hole	+	−	−

Fig. 5.11

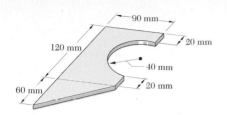

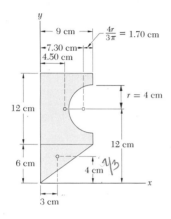

Solution. Since the plate is homogeneous, we may locate its center of gravity by determining the centroid of its area. The area is obtained by adding a rectangle and a triangle and subtracting a semicircle. Coordinate axes are chosen with the origin at the lower left corner of the plate. The area and the coordinates of the centroid of each plate component are determined and entered in the table below. The area of the semicircle is indicated as negative, since it is to be subtracted from the other areas. The moments of the component areas with respect to the coordinate axes are then computed and entered in the table. Note that all dimensions have been converted into *centimeters* (cm) in order to avoid excessively large numerical values in the computation of the areas and moments of areas (1 cm = 10 mm).

Component	A, cm^2	$\bar{x}$, cm	$\bar{y}$, cm	$\bar{x}A$, cm^3	$\bar{y}A$, cm^3
Rectangle	$+108$	4.50	12	$+486$	$+1296$
Triangle	$+27$	3.00	4	$+81$	$+108$
Semicircle	-25.1	7.30	12	-183	-301
	$\Sigma A = 109.9$	...	...	$\Sigma \bar{x}A = 384$	$\Sigma \bar{y}A = 1103$

Substituting the values obtained from the table into the equations defining the centroid of a composite area, we obtain

$\bar{X}\Sigma A = \Sigma \bar{x}A:$ $\bar{X}(109.9 \text{ cm}^2) = 384 \text{ cm}^3$
$\bar{X} = 3.49 \text{ cm}$ $\bar{X} = 34.9 \text{ mm}$ ◄

$\bar{Y}\Sigma A = \Sigma \bar{y}A:$ $\bar{Y}(109.9 \text{ cm}^2) = 1103 \text{ cm}^3$
$\bar{Y} = 10.04 \text{ cm}$ $\bar{Y} = 100.4 \text{ mm}$ ◄

The above values of $\bar{X}$ and $\bar{Y}$ define the centroid of the area and also the center of gravity of the plate. The center of gravity, of course, is actually located halfway between the upper and lower faces of the plate.

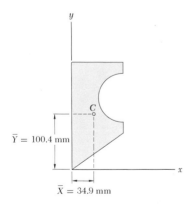

$\bar{Y} = 100.4$ mm

$\bar{X} = 34.9$ mm

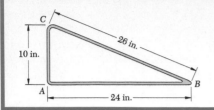

The figure shown is made of a thin homogeneous wire. Determine its center of gravity.

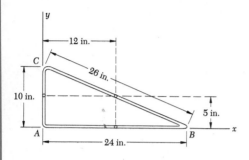

Solution. Since the figure is formed of homogeneous wire, its center of gravity may be located by determining the centroid of the corresponding line. Choosing the coordinate axes shown, with origin at A, we determine the coordinates of the centroid of each segment of line and compute its moments with respect to the coordinate axes.

Segment	L, in.	$\bar{x}$, in.	$\bar{y}$, in.	$\bar{x}L$, in²	$\bar{y}L$, in²
AB	24	12	0	288	0
BC	26	12	5	312	130
CA	10	0	5	0	50
	$\Sigma L = 60$	...	...	$\Sigma \bar{x}L = 600$	$\Sigma \bar{y}L = 180$

Substituting the values obtained from the table into the equations defining the centroid of a composite line, we obtain

$$\bar{X}\Sigma L = \Sigma \bar{x}L: \qquad \bar{X}(60 \text{ in.}) = 600 \text{ in}^2 \qquad \bar{X} = 10 \text{ in.} \blacktriangleleft$$
$$\bar{Y}\Sigma L = \Sigma \bar{y}L: \qquad \bar{Y}(60 \text{ in.}) = 180 \text{ in}^2 \qquad \bar{Y} = 3 \text{ in.} \blacktriangleleft$$

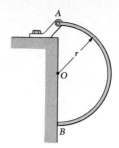

SAMPLE PROBLEM 5.3

A uniform semicircular rod of weight W and radius r is attached to a pin at A and bears against a frictionless surface at B. Determine the reactions at A and B.

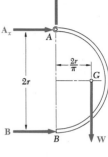

Solution. A free-body diagram of the rod is drawn. The forces acting on the rod consist of its weight $\mathbf{W}$ applied at the center of gravity G, whose position is obtained from Fig. 5.8B, of a reaction at A represented by its components $\mathbf{A}_x$ and $\mathbf{A}_y$, and of a horizontal reaction at B.

$$+\curvearrowleft\Sigma M_A = 0: \qquad B(2r) - W\left(\frac{2r}{\pi}\right) = 0$$

$$B = +\frac{W}{\pi} \qquad\qquad B = \frac{W}{\pi} \rightarrow \quad \blacktriangleleft$$

$$\xrightarrow{+}\Sigma F_x = 0: \qquad A_x + B = 0$$

$$A_x = -B = -\frac{W}{\pi} \qquad A_x = \frac{W}{\pi} \leftarrow$$

$$+\uparrow\Sigma F_y = 0: \qquad A_y - W = 0 \qquad A_y = W \uparrow$$

Adding the two components of the reaction at A:

$$A = \left[W^2 + \left(\frac{W}{\pi}\right)^2\right]^{1/2} \qquad\qquad A = W\left(1 + \frac{1}{\pi^2}\right)^{1/2} \quad \blacktriangleleft$$

$$\tan \alpha = \frac{W}{W/\pi} = \pi \qquad\qquad \alpha = \tan^{-1}\pi \quad \blacktriangleleft$$

The answers may also be expressed as follows:

$$\mathbf{A} = 1.049W \measuredangle 72.3° \qquad \mathbf{B} = 0.318W \rightarrow \quad \blacktriangleleft$$

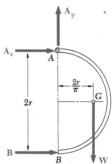

PROBLEMS

5.1 through 5.12   Locate the centroid of the plane area shown.

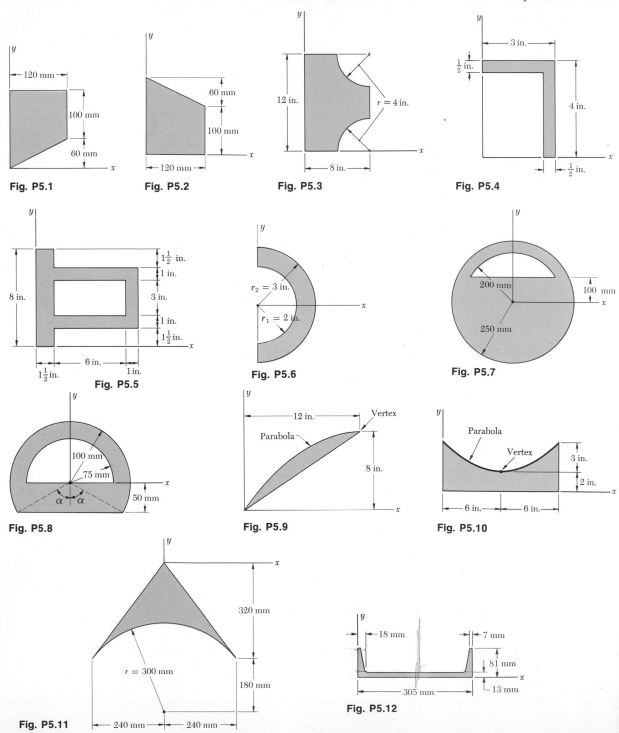

Fig. P5.1

Fig. P5.2

Fig. P5.3

Fig. P5.4

Fig. P5.5

Fig. P5.6

Fig. P5.7

Fig. P5.8

Fig. P5.9

Fig. P5.10

Fig. P5.11

Fig. P5.12

5.13 Determine the abscissa of the centroid of the shaded area in terms of a, b, and h.

5.14 Determine the abscissa of the centroid of the circular segment in terms of r and α.

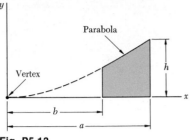

Fig. P5.13

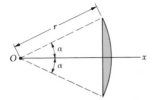

Fig. P5.14

5.15 through 5.18 A thin homogeneous wire is bent to form the *perimeter* of the figure indicated. Locate the center of gravity of the wire figure thus formed.

5.15 Fig. P5.1.
5.16 Fig. P5.2.
5.17 Fig. P5.3.
5.18 Fig. P5.6.

5.19 For the semiannular area of Prob. 5.6, determine the ratio of r_1 to r_2 for which the centroid of the area is located at $x = \frac{1}{2}r_2$ and $y = 0$.

*5.20 Locate the centroid C in Prob. 5.6 in terms of r_1 and r_2 and show that, as r_1 approaches r_2, the location of C approaches that for a semicircular arc of radius $\frac{1}{2}(r_1 + r_2)$.

5.21 For the semiannular area of Prob. 5.6, determine the ratio of r_1 to r_2 for which the centroid of the area is located at the point of intersection of the inner circle and the x axis.

5.22 Knowing that the figure shown is formed of a thin homogeneous wire, determine the angle α for which the center of gravity of the figure is located at the origin O.

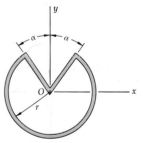

Fig. P5.22

5.23 The homogeneous wire ABC is bent as shown and is attached to a hinge at C. Determine the length L for which portion AB of the wire is horizontal.

5.24 The homogeneous wire ABC is bent as shown and is attached to a hinge at C. Determine the length L for which portion BC of the wire is horizontal.

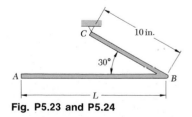

Fig. P5.23 and P5.24

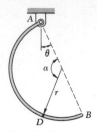

Fig. P5.25

5.25 A semicircular rod of mass m is suspended from a hinge at A. A mass m of negligible dimensions is attached to the rod at D. Determine the value of θ when (*a*) $\alpha = 180°$, (*b*) $\alpha = 90°$.

5.26 In Prob. 5.25 determine the angle α for which the angle θ is maximum. Also determine the corresponding maximum value of θ.

5.27 The plan view of a cam of uniform thickness is shown. Locate by approximate means the center of gravity of the cam.

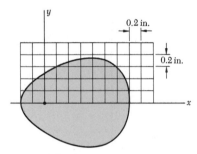

Fig. P5.27

5.28 A plate of uniform thickness is cut as shown. Locate by approximate means the center of gravity of the plate.

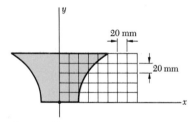

Fig. P5.28

5.29 Divide the parabolic spandrel shown into five vertical sections and determine by approximate means the x coordinate of its centroid; approximate the spandrel by rectangles of the form $bcc'b'$. What is the percentage error in the answer obtained? (See Fig. 5.8A for exact answer.)

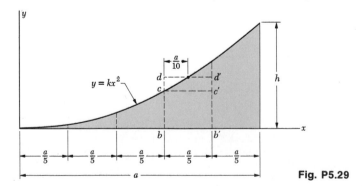

Fig. P5.29

5.30 Solve Prob. 5.29, using rectangles of the form $bdd'b'$.

5.31   Determine by approximate means the x coordinate of the centroid of the area shown.

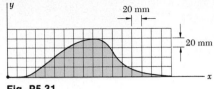

Fig. P5.31

5.4. Determination of Centroids by Integration.

The centroid of an area bounded by analytical curves (i.e., curves defined by algebraic equations) is usually determined by computing the integrals in Eqs. (5.5) of Sec. 5.2.

$$\bar{x}A = \int x\, dA \qquad \bar{y}A = \int y\, dA \qquad (5.5)$$

If the element of area dA is chosen equal to a small square of sides dx and dy, the determination of each of these integrals requires a *double integration* in x and y. A double integration is also necessary if polar coordinates are used and if dA is chosen equal to a small square of sides dr and $r\, d\theta$.

In most cases, however, it is possible to determine the coordinates of the centroid of an area by performing a single integration. This is achieved by choosing for dA a thin rectangle or strip, or a thin sector or pie-shaped element (Fig. 5.12); the centroid of the thin rectangle is located at its center, and the centroid of the thin sector at a distance $\frac{2}{3}r$ from its vertex (as for a triangle). The coordinates of the centroid of the area under consideration are then obtained by expressing that the first moment of the entire area with respect to each of the coordinate axes is equal to the sum (or integral) of the corresponding

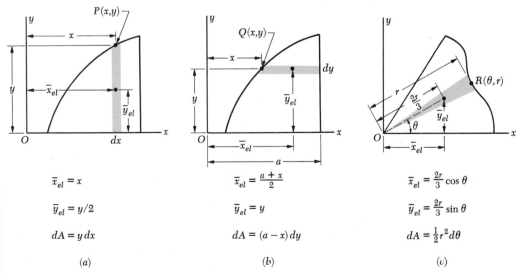

$\bar{x}_{el} = x$

$\bar{y}_{el} = y/2$

$dA = y\, dx$

(a)

$\bar{x}_{el} = \dfrac{a + x}{2}$

$\bar{y}_{el} = y$

$dA = (a - x)\, dy$

(b)

$\bar{x}_{el} = \dfrac{2r}{3}\cos\theta$

$\bar{y}_{el} = \dfrac{2r}{3}\sin\theta$

$dA = \dfrac{1}{2}r^2 d\theta$

(c)

Fig. 5.12   Centroids and areas of differential elements.

moments of the elements of area. Denoting by $\bar{x}_{el}$ and $\bar{y}_{el}$ the coordinates of the centroid of the element dA, we write

$$\Sigma M_y: \qquad \bar{x}A = \int \bar{x}_{el}\, dA$$
$$\Sigma M_x: \qquad \bar{y}A = \int \bar{y}_{el}\, dA \qquad (5.9)$$

If the area itself is not already known, it may also be computed from these elements.

The coordinates $\bar{x}_{el}$ and $\bar{y}_{el}$ of the centroid of the element of area should be expressed in terms of the coordinates of a point located on the curve bounding the area under consideration. Also, the element of area dA should be expressed in terms of the coordinates of the point and their differentials. This has been done in Fig. 5.12 for three common types of elements; the pie-shaped element of part c should be used when the equation of the curve bounding the area is given in polar coordinates. The appropriate expressions should be substituted in formulas (5.9), and the equation of the curve should be used to express one of the coordinates in terms of the other. The integration is thus reduced to a single integration which may be performed according to the usual rules of calculus.

The centroid of a line defined by an algebraic equation may be determined by computing the integrals in Eqs. (5.6) of Sec. 5.2.

$$\bar{x}L = \int x\, dL \qquad \bar{y}L = \int y\, dL \qquad (5.6)$$

The element dL should be replaced by one of the following expressions, depending upon the type of equation used to define the line (these expressions may be derived by using the Pythagorean theorem).

$$dL = \sqrt{1 + \left(\frac{dy}{dx}\right)^2}\, dx$$

$$dL = \sqrt{1 + \left(\frac{dx}{dy}\right)^2}\, dy$$

$$dL = \sqrt{r^2 + \left(\frac{dr}{d\theta}\right)^2}\, d\theta$$

The equation of the line is then used to express one of the coordinates in terms of the other, and the integration may be performed by the methods of calculus.

5.5. Theorems of Pappus-Guldinus. These theorems, which were first formulated by the Greek geometer Pappus during the third century A.D. and later restated by the Swiss mathematician Guldinus, or Guldin (1577–1643), deal with surfaces and bodies of revolution.

A *surface of revolution* is a surface which may be generated by rotating a plane curve about a fixed axis. For example (Fig. 5.13), the surface of a sphere may be obtained by rotating a

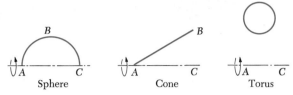

Fig. 5.13 Generating a surface of revolution.

semicircular arc ABC about the diameter AC; the surface of a cone by rotating a straight line AB about an axis AC; the surface of a torus or ring by rotating the circumference of a circle about a nonintersecting axis. A *body of revolution* is a body which may be generated by rotating a plane area about a fixed axis. A solid sphere may be obtained by rotating a semicircular area, a cone by rotating a triangular area, and a solid torus by rotating a full circular area (Fig. 5.14).

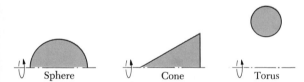

Fig. 5.14 Generating a body of revolution.

THEOREM I. *The area of a surface of revolution is equal to the length of the generating curve times the distance traveled by the centroid of the curve while the surface is being generated.*

Proof. Consider an element dL of the line L (Fig. 5.15) which is revolved about the x axis. The area dA generated by the element dL is equal to $2\pi y\, dL$. Thus, the entire area generated

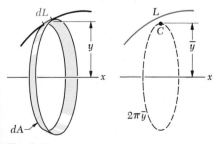

Fig. 5.15

by L is $A = \int 2\pi y\, dL$. But we saw in Sec. 5.2 that the integral $\int y\, dL$ is equal to $\bar{y}L$. We have therefore

$$A = 2\pi\bar{y}L \tag{5.10}$$

where $2\pi\bar{y}$ is the distance traveled by the centroid of L. It should be noted that the generating curve should not cross the axis about which it is rotated; if it did, the two sections on either side of the axis would generate areas of opposite signs and the theorem would not apply.

THEOREM II. *The volume of a body of revolution is equal to the generating area times the distance traveled by the centroid of the area while the body is being generated.*

Proof. Consider an element dA of the area A which is revolved about the x axis (Fig. 5.16). The volume dV generated by the element dA is equal to $2\pi y\, dA$. Thus, the entire volume

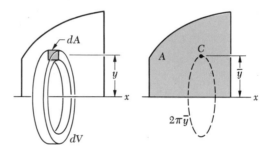

Fig. 5.16

generated by A is $V = \int 2\pi y\, dA$. But since the integral $\int y\, dA$ is equal to $\bar{y}A$ (Sec. 5.2), we have

$$V = 2\pi\bar{y}A \tag{5.11}$$

where $2\pi\bar{y}$ is the distance traveled by the centroid of A. Again, it should be noted that the theorem does not apply if the axis of rotation intersects the generating area.

The theorems of Pappus-Guldinus offer a simple way for computing the area of surfaces of revolution and the volume of bodies of revolution. They may also be used conversely to determine the centroid of a plane curve when the area of the surface generated by the curve is known or to determine the centroid of a plane area when the volume of the body generated by the area is known (see Sample Prob. 5.8).

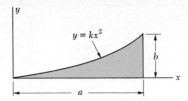

Determine by direct integration the centroid of a parabolic spandrel.

Solution. The value of k is determined by substituting $x = a$ and $y = b$ in the given equation. We have $b = ka^2$ and $k = b/a^2$. The equation of the curve is thus

$$y = \frac{b}{a^2}x^2 \qquad \text{or} \qquad x = \frac{a}{b^{1/2}}y^{1/2}$$

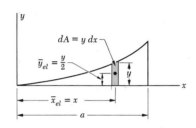

Vertical Differential Element. We choose the differential element shown and find the total area of the figure.

$$A = \int dA = \int y\, dx = \int_0^a \frac{b}{a^2}x^2\, dx = \left[\frac{b}{a^2}\frac{x^3}{3}\right]_0^a = \frac{ab}{3}$$

The moment of the differential element with respect to the y axis is $\bar{x}_{el}\, dA$; hence, the moment of the entire area with respect to this axis is

$$\int \bar{x}_{el}\, dA = \int xy\, dx = \int_0^a x\left(\frac{b}{a^2}x^2\right) dx = \left[\frac{b}{a^2}\frac{x^4}{4}\right]_0^a = \frac{a^2 b}{4}$$

Thus,

$$\bar{x}A = \int \bar{x}_{el}\, dA \qquad \bar{x}\frac{ab}{3} = \frac{a^2 b}{4} \qquad \bar{x} = \tfrac{3}{4}a \quad \blacktriangleleft$$

Likewise, the moment of the differential element with respect to the x axis is $\bar{y}_{el}\, dA$, and the moment of the entire area is

$$\int \bar{y}_{el}\, dA = \int \frac{y}{2}y\, dx = \int_0^a \frac{1}{2}\left(\frac{b}{a^2}x^2\right)^2 dx = \left[\frac{b^2}{2a^4}\frac{x^5}{5}\right]_0^a = \frac{ab^2}{10}$$

Thus,

$$\bar{y}A = \int \bar{y}_{el}\, dA \qquad \bar{y}\frac{ab}{3} = \frac{ab^2}{10} \qquad \bar{y} = \tfrac{3}{10}b \quad \blacktriangleleft$$

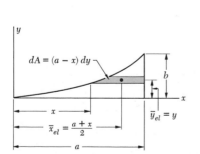

Horizontal Differential Element. The same result may be obtained by considering a horizontal element. The moments of the area are

$$\int \bar{x}_{el}\, dA = \int \frac{a + x}{2}(a - x)\, dy = \int_0^b \frac{a^2 - x^2}{2}\, dy$$

$$= \frac{1}{2}\int_0^b \left(a^2 - \frac{a^2}{b}y\right) dy = \frac{a^2 b}{4}$$

$$\int \bar{y}_{el}\, dA = \int y(a - x)\, dy = \int y\left(a - \frac{a}{b^{1/2}}y^{1/2}\right) dy$$

$$= \int_0^b \left(ay - \frac{a}{b^{1/2}}y^{3/2}\right) dy = \frac{ab^2}{10}$$

These moments are again substituted in the equations defining the centroid of the area to obtain $\bar{x}$ and $\bar{y}$.

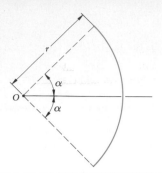

SAMPLE PROBLEM 5.5

Determine the centroid of the arc of circle shown.

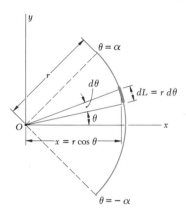

Solution. Since the arc is symmetrical with respect to the x axis, we have $\bar{y} = 0$. A differential element is chosen as shown, and the length of the arc is determined by integration.

$$L = \int dL = \int_{-\alpha}^{\alpha} r\, d\theta = r \int_{-\alpha}^{\alpha} d\theta = 2r\alpha$$

The moment of the arc with respect to the y axis is

$$\int x\, dL = \int_{-\alpha}^{\alpha} (r\cos\theta)(r\, d\theta) = r^2 \int_{-\alpha}^{\alpha} \cos\theta\, d\theta$$

$$= r^2 \left[\sin\theta \right]_{-\alpha}^{\alpha} = 2r^2 \sin\alpha$$

Thus, $\quad \bar{x}L = \int x\, dL \quad\quad \bar{x}(2r\alpha) = 2r^2\sin\alpha \quad\quad \bar{x} = \dfrac{r\sin\alpha}{\alpha}$ ◄

SAMPLE PROBLEM 5.6

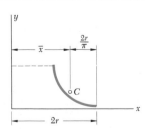

Determine the area of the surface of revolution shown, obtained by rotating a quarter-circular arc about a vertical axis.

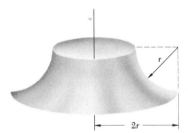

Solution. According to Theorem I of Pappus-Guldinus, the area generated is equal to the product of the length of the arc and the distance traveled by its centroid. Referring to Fig. 5.8B, we have

$$\bar{x} = 2r - \frac{2r}{\pi} = 2r\left(1 - \frac{1}{\pi}\right)$$

$$A = 2\pi\bar{x}L = 2\pi 2r\left(1 - \frac{1}{\pi}\right)\frac{\pi r}{2}$$

$$A = 2\pi r^2(\pi - 1)$$ ◄

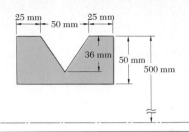

SAMPLE PROBLEM 5.7

The outside diameter of a V-belt pulley is 1 m, and the cross section of its rim is as shown. Determine the mass and weight of the rim, knowing that the pulley is made of steel and that the density of steel is $\rho = 7.85 \times 10^3 \, \text{kg/m}^3$.

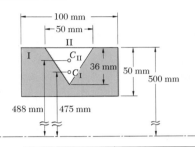

Solution. The volume of the rim may be found by applying Theorem II of Pappus-Guldinus, which states that the volume equals the product of the given cross-sectional area and of the distance traveled by its centroid in one complete revolution. However, the volume will be more easily obtained if we observe that the cross section consists of a rectangle (I) with a positive area and a triangle (II) with a negative area.

Component	Area, mm²	$\bar{y}$, mm	Distance Traveled by C, mm	Volume, mm³
I	+5000	475	$2\pi(475) = 2985$	$(5000)(2985) = 14.92 \times 10^6$
II	−900	488	$2\pi(488) = 3066$	$(-900)(3066) = -2.76 \times 10^6$
				Volume of rim = 12.16×10^6

Since $1 \, \text{mm} = 10^{-3} \, \text{m}$, we have $1 \, \text{mm}^3 = (10^{-3} \, \text{m})^3 = 10^{-9} \, \text{m}^3$, and
$V = 12.16 \times 10^6 \, \text{mm}^3 = (12.16 \times 10^6)(10^{-9} \, \text{m}^3) = 12.16 \times 10^{-3} \, \text{m}^3$

$m = \rho V = (7.85 \times 10^3 \, \text{kg/m}^3)(12.16 \times 10^{-3} \, \text{m}^3) \qquad m = 95.4 \, \text{kg}$ ◄

$W = mg = (95.4 \, \text{kg})(9.81 \, \text{m/s}^2) = 936 \, \text{kg} \cdot \text{m/s}^2 \qquad W = 936 \, \text{N}$ ◄

SAMPLE PROBLEM 5.8

Using the theorems of Pappus-Guldinus, determine (a) the centroid of a semicircular area, (b) the centroid of a semicircular arc. We recall that the volume of a sphere is $\frac{4}{3}\pi r^3$ and that its surface area is $4\pi r^2$.

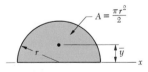

Solution. The volume of a sphere is equal to the product of the area of a semicircle and of the distance traveled by the centroid of the semicircle in one revolution about the x axis.

$$V = 2\pi\bar{y}A \qquad \tfrac{4}{3}\pi r^3 = 2\pi\bar{y}(\tfrac{1}{2}\pi r^2) \qquad \bar{y} = \frac{4r}{3\pi}$$ ◄

Likewise, the area of a sphere is equal to the product of the length of the generating semicircle and of the distance traveled by its centroid in one revolution.

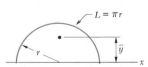

$$A = 2\pi\bar{y}L \qquad 4\pi r^2 = 2\pi\bar{y}(\pi r) \qquad \bar{y} = \frac{2r}{\pi}$$ ◄

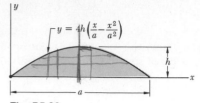

Fig. P5.32

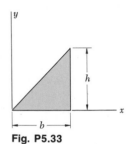

Fig. P5.33

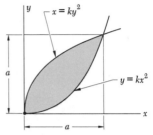

Fig. P5.34

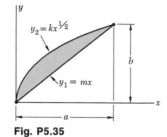

Fig. P5.35

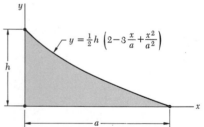

Fig. P5.42 and P5.43

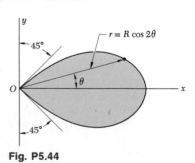

Fig. P5.44

PROBLEMS

5.32 through 5.35 Determine by direct integration the centroid of the area shown.

5.36 through 5.41 Derive by direct integration the expressions for $\bar{x}$ and $\bar{y}$ given in Fig. 5.8 for:

5.36 A general spandrel ($y = kx^n$).
5.37 A quarter-elliptical area.
5.38 A semicircular area.
5.39 A semiparabolic area.
5.40 A circular sector.
5.41 A quarter-circular arc.

5.42 Determine by direct integration the x coordinate of the centroid of the area shown.

5.43 Determine by direct integration the y coordinate of the centroid of the area shown.

*5.44 Determine by direct integration the centroid of the area shown.

*5.45 Determine by direct integration the centroid of the area shown.

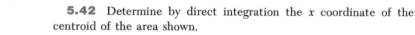

Fig. P5.45

5.46 Determine by direct integration the centroid of the area located in the first quadrant and bounded by the curves $y = x^n$ and $x = y^n$, for $n > 1$.

5.47 Determine the centroid of the area shown.

5.48 Knowing that $b = 3a$, determine the centroid of the area shown.

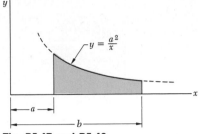

Fig. P5.47 and P5.48

5.49 Determine the volume of the solid obtained by rotating the semiparabolic area shown about (*a*) the y axis, (*b*) the x axis.

5.50 Determine the volume of the solid obtained by rotating the trapezoid of Prob. 5.2 about (*a*) the x axis, (*b*) the y axis.

5.51 Determine the surface area and the volume of the half-torus shown.

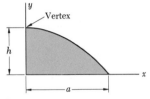

Fig. P5.49

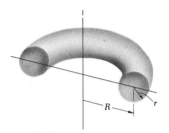

Fig. P5.51

5.52 The spherical cap shown is formed by passing a horizontal plane through a hollow sphere of radius R. Determine the area of the outside surface of the cap in terms of R and ϕ.

5.53 Determine the total surface area of the solid obtained by rotating the area of Prob. 5.3 about the y axis.

5.54 The inside diameter of a spherical tank is 2 m. What volume of liquid is required to fill the tank to a depth of 0.5 m?

5.55 The geometry of the end portion of a proposed embankment is shown. For the portion of the embankment in front of the vertical plane *ABCD*, determine the additional fill material required if the final dimensions are to be $a = 19$ ft, $b = 37$ ft, and $h = 8$ ft.

5.56 Solve Prob. 5.55, assuming that the final dimensions are (*a*) $a = 19$ ft, $b = 34$ ft, $h = 8$ ft, (*b*) $a = 16$ ft, $b = 37$ ft, $h = 8$ ft.

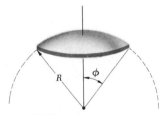

Fig. P5.52

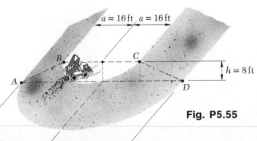

Fig. P5.55

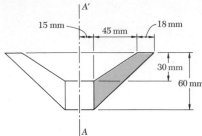

15 mm
45 mm
18 mm
30 mm
60 mm

A'

A

Fig. P5.57

5.57 Determine the volume of the steel collar obtained by rotating the shaded area shown about the vertical axis AA'.

5.58 Determine the volume and the total surface area of the body shown.

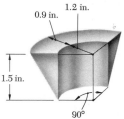

1.2 in.
0.9 in.
1.5 in.
90°

Fig. P5.58

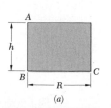

$\bar{y}$

A — — — A'

Fig. P5.59

5.59 The rim of a steel V-belt pulley has a mass of 3.9 kg. Knowing that the area of the cross section of the rim is 522 mm², determine the distance $\bar{y}$ from the axle AA' to the centroid of the cross-sectional area of the rim. (Density of steel = 7850 kg/m³.)

5.60 Determine the volume and total surface area of the portion of ring shown. The cross section of the ring is a semicircle.

25 mm
150 mm
90°

Fig. P5.60

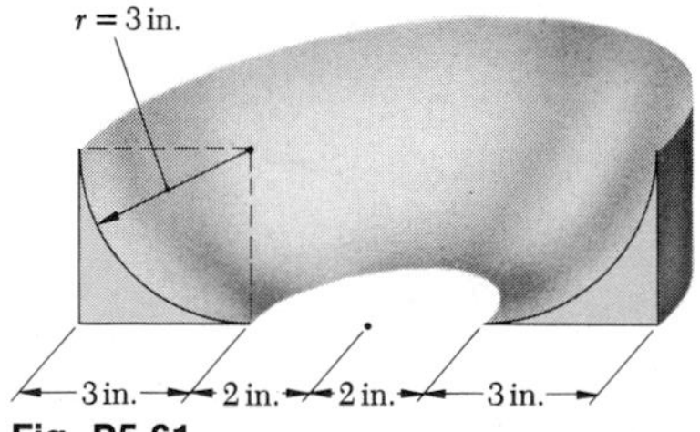

5.61 Determine (*a*) the volume of the body shown, (*b*) the area of its inside curved surface.

5.62 Determine the volume of the solid of revolution formed by revolving each of the plane areas shown about its vertical edge AB. Show that the volumes of the solids formed are in the ratio 6:4:3:2:1.

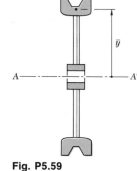

$r = 3$ in.

3 in. — 2 in. — 2 in. — 3 in.

Fig. P5.61

5.63 Determine the volume of the solid formed by revolving each of the plane areas shown about its base BC. Show that the volumes of the solids formed are in the ratio 15:10:8:5:3.

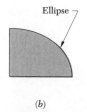

A

h

B

R

C

(*a*)

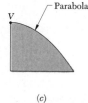

Ellipse

(*b*)

Parabola

V

(*c*)

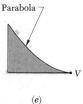

(*d*)

Parabola

V

(*e*)

Fig. P5.62 and P5.63

*5.64   A cylindrical hole is drilled through the center of a steel ball bearing shown here in cross section. The length of the *hole* is L; show that the volume of the steel remaining is $\frac{1}{6}\pi L^3$.

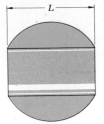

5.65   An experimental high-altitude balloon at a given time has the shape shown. Determine by approximate means (*a*) the volume of gas inside the balloon, (*b*) the surface area of the balloon.

Fig. P5.64

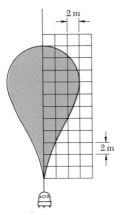

Fig. P5.65

*5.6. Distributed Loads on Beams.   The concept of centroid of an area may be used to solve other problems besides those dealing with the weight of flat plates. Consider, for example, a beam supporting a *distributed load;* this load may consist of the weight of materials supported directly or indirectly by the beam, or it may be caused by wind or hydrostatic pressure. The distributed load may be represented by plotting the load w supported per unit length (Fig. 5.17); this load will be expressed in N/m or in lb/ft. The magnitude of the force exerted on an element of beam of length dx is $dW = w\,dx$, and the total load supported by the beam is

$$W = \int_0^L w\,dx$$

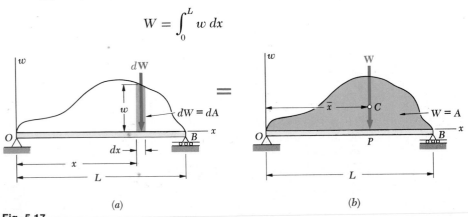

(*a*)   (*b*)

Fig. 5.17

But the product $w\,dx$ is equal in magnitude to the element of area dA shown in Fig. 5.17a, and W is thus equal in magnitude to the total area A under the load curve,

$$W = \int dA = A$$

We shall now determine where a *single concentrated load* **W**, of the same magnitude W as the total distributed load, should be applied on the beam if it is to produce the same reactions at the supports (Fig. 5.17b). This concentrated load **W**, which represents the resultant of the given distributed loading, should be equivalent to this loading as far as the free-body diagram of the entire beam is concerned. The point of application P of the equivalent concentrated load **W** will therefore be obtained by expressing that the moment of **W** about point O is equal to the sum of the moments of the elementary loads $d\mathbf{W}$ about O:

$$(OP)W = \int x\,dW$$

or, since $dW = w\,dx = dA$ and $W = A$,

$$(OP)A = \int_0^L x\,dA \tag{5.12}$$

Since the integral represents the first moment with respect to the w axis of the area under the load curve, it may be replaced by the product $\bar{x}A$. We have therefore $OP = \bar{x}$, where $\bar{x}$ is the distance from the w axis to the centroid C of the area A (this is *not* the centroid of the beam).

A distributed load on a beam may thus be replaced by a concentrated load; the magnitude of this single load is equal to the area under the load curve, and its line of action passes through the centroid of that area. It should be noted, however, that the concentrated load is equivalent to the given loading only as far as external forces are concerned. It may be used to determine reactions but should not be used to compute internal forces and deflections.

∗5.7. Forces on Submerged Surfaces. Another example of the use of first moments and centroids of areas is obtained by considering the forces exerted on a *rectangular surface* submerged in a liquid. Consider the rectangular plate shown in Fig. 5.18; it has a length L, and its width, perpendicular to the plane of the figure, is assumed equal to unity. Since the gage pressure in a liquid is $p = \gamma h$, where γ is the specific weight of the liquid and h the vertical distance from the free surface, the pressure on the plate varies linearly with the distance x. The width of the plate being taken equal to unity, the pressure

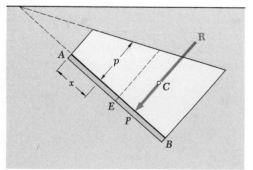

Fig. 5.18

p is equal in magnitude to the load w per unit length used in Sec. 5.6.† The results obtained in that section may thus be used here, and we find that the magnitude of the resultant **R** of the forces exerted on one face of the plate is equal to the area under the pressure curve; we also find that the line of action of **R** passes through the centroid C of that area.

Noting that the area under the pressure curve is equal to $p_E L$, where p_E is the pressure at the center E of the plate and L the length (or area) of the plate, we find that the magnitude R of the resultant may be obtained by multiplying the area of the plate by the pressure at the center E of the plate. The resultant **R**, however, *should not* be applied at E; as indicated above, its line of action passes through the centroid C of the area under the pressure curve. The point of application P of the resultant **R** is known as the *center of pressure*.

We shall consider next the forces exerted by a liquid on a curved surface of constant width (Fig. 5.19*a*). Since the determination of the resultant **R** of these forces by direct integration would not be easy, we shall consider the free body obtained by detaching the volume of liquid ABD bounded by the curved surface AB and by the two plane surfaces AD and DB shown in Fig. 5.19*b*. The forces acting on the free body ABD consist of the weight **W** of the volume of liquid detached, the resultant **R**$_1$ of the forces exerted on AD, the resultant **R**$_2$ of the forces exerted on BD, and the resultant of the forces exerted *by the curved surface on the liquid*. This last resultant is equal and opposite to, and has the same line of action as, the resultant **R** of the forces exerted *by the liquid on the curved surface*; we shall, therefore, denote it by $-$**R**. The forces **W**, **R**$_1$, and **R**$_2$ may be determined by standard methods; after their values have been found, the force $-$**R** will be obtained by solving the equations of equilibrium for the free body of Fig. 5.19*b*. The resultant **R** of the hydrostatic forces exerted on the curved surface will then be obtained by reversing the sense of $-$**R**.

The methods outlined in this section may be used to determine the resultant of the hydrostatic forces exerted on the surface of dams or on rectangular gates and vanes. Resultants of forces on submerged surfaces of variable width should be determined by the methods of Chap. 9.

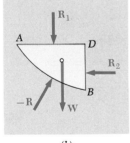

(*a*)

(*b*)

Fig. 5.19

---

† We should note that, while the load w per unit length is expressed in N/m or in lb/ft, the pressure p, which represents a load per unit area, is expressed in N/m² or in lb/ft². The derived SI unit N/m² is called a *pascal* (Pa).

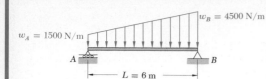

$w_A = 1500 \text{ N/m}$

$w_B = 4500 \text{ N/m}$

A B

$L = 6 \text{ m}$

SAMPLE PROBLEM 5.9

A beam supports a distributed load as shown. (*a*) Determine the equivalent concentrated load. (*b*) Determine the reactions at the supports.

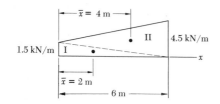

$\bar{x} = 4 \text{ m}$

II 4.5 kN/m

1.5 kN/m I

x

$\bar{x} = 2 \text{ m}$

6 m

a. **Equivalent Concentrated Load.** The magnitude of the resultant of the load is equal to the area under the load curve, and the line of action of the resultant passes through the centroid of the same area. We divide the area under the load curve into two triangles and construct the table below. To simplify the computations and tabulation, the given loads per unit length have been converted into kN/m.

Component	A, kN	$\bar{x}$, m	$\bar{x}A$, kN $\cdot$ m
Triangle I	4.5	2	9
Triangle II	13.5	4	54
	$\Sigma A = 18.0$	. . .	$\Sigma \bar{x}A = 63$

Thus, $\bar{X}\Sigma A = \Sigma \bar{x}A$: $\bar{X}(18 \text{ kN}) = 63 \text{ kN} \cdot \text{m}$ $\bar{X} = 3.5 \text{ m}$

The equivalent concentrated load is

$$W = 18 \text{ kN} \downarrow \quad \blacktriangleleft$$

and its line of action is located at a distance

$$\bar{X} = 3.5 \text{ m to the right of } A \quad \blacktriangleleft$$

18 kN

$\bar{X} = 3.5 \text{ m}$

A B

b. **Reactions.** The reaction at A is vertical and is denoted by **A**; the reaction at B is represented by its components $\mathbf{B}_x$ and $\mathbf{B}_y$. The given load may be considered as the sum of two triangular loads as shown. The resultant of each triangular load is equal to the area of the triangle and acts at its centroid. We write the following equilibrium equations for the free body shown:

$\xrightarrow{+} \Sigma F_x = 0$: $B_x = 0 \quad \blacktriangleleft$

$+\text{\Large\curvearrowright} \Sigma M_A = 0$: $-(4.5 \text{ kN})(2 \text{ m}) - (13.5 \text{ kN})(4 \text{ m}) + B_y(6 \text{ m}) = 0$

$B_y = 10.5 \text{ kN} \uparrow \quad \blacktriangleleft$

$+\text{\Large\curvearrowright} \Sigma M_B = 0$: $+(4.5 \text{ kN})(4 \text{ m}) + (13.5 \text{ kN})(2 \text{ m}) - A(6 \text{ m}) = 0$

$A = 7.5 \text{ kN} \uparrow \quad \blacktriangleleft$

4.5 kN 13.5 kN

B_x

A

2 m B_y

4 m

6 m

Alternate Solution. The given distributed load may be replaced by its resultant, which was found in part *a*. The reactions may be determined by writing the equilibrium equations $\Sigma F_x = 0$, $\Sigma M_A = 0$, and $\Sigma M_B = 0$. We again obtain

$$B_x = 0 \qquad B_y = 10.5 \text{ kN} \uparrow \qquad A = 7.5 \text{ kN} \uparrow \quad \blacktriangleleft$$

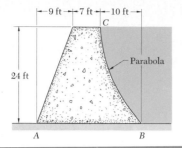

SAMPLE PROBLEM 5.10

The cross section of a concrete dam is as shown. Consider a section of the dam 1 ft thick, and determine (a) the resultant of the reaction forces exerted by the ground on the base of the dam AB, (b) the resultant of the pressure forces exerted by the water on the face BC of the dam. Specific weight of concrete = 150 lb/ft³; of water = 62.4 lb/ft³.

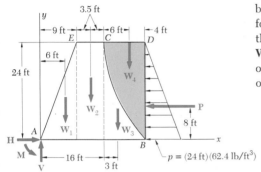

a. Ground Reaction. As a free body we choose a section ABDEA, 1 ft thick, of the dam and water as shown. The reaction forces exerted by the ground on the base AB are represented by an equivalent force-couple system at A. Other forces acting on the free body are the weight of the dam, represented by the weights of its components W_1, W_2, and W_3, the weight of the water W_4, and the resultant **P** of the pressure forces exerted on section BD by the water to the right of section BD. We have

$$W_1 = \tfrac{1}{2}(9\text{ ft})(24\text{ ft})(1\text{ ft})(150\text{ lb/ft}^3) = 16,200\text{ lb}$$
$$W_2 = (7\text{ ft})(24\text{ ft})(1\text{ ft})(150\text{ lb/ft}^3) = 25,200\text{ lb}$$
$$W_3 = \tfrac{1}{3}(10\text{ ft})(24\text{ ft})(1\text{ ft})(150\text{ lb/ft}^3) = 12,000\text{ lb}$$
$$W_4 = \tfrac{2}{3}(10\text{ ft})(24\text{ ft})(1\text{ ft})(62.4\text{ lb/ft}^3) = 9,980\text{ lb}$$
$$P = \tfrac{1}{2}(24\text{ ft})(1\text{ ft})(24\text{ ft})(62.4\text{ lb/ft}^3) = 17,970\text{ lb}$$

Equilibrium Equations

$\Sigma F_x = 0$:　　$H - 17{,}970\text{ lb} = 0$　　　　　　$\mathbf{H} = 17{,}970\text{ lb} \rightarrow$　◄

$\Sigma F_y = 0$:　　$V - 16{,}200\text{ lb} - 25{,}200\text{ lb} - 12{,}000\text{ lb} - 9{,}980\text{ lb} = 0$
　　　　　　　　　　　　　　　　　　　　　　$\mathbf{V} = 63{,}400\text{ lb} \uparrow$　◄

$+ \!\gamma \Sigma M_A = 0$:　　　　$-(16{,}200\text{ lb})(6\text{ ft}) - (25{,}200\text{ lb})(12.5\text{ ft})$
　　$-(12{,}000\text{ lb})(19\text{ ft}) - (9{,}980\text{ lb})(22\text{ ft}) + (17{,}970\text{ lb})(8\text{ ft}) + M = 0$
　　　　　　　　　　　　　　　　$\mathbf{M} = 716{,}000\text{ lb} \cdot \text{ft} \,\gamma$　◄

We may replace the force-couple system obtained by a single force acting at a distance d to the right of A, where

$$d = \frac{716{,}000\text{ lb} \cdot \text{ft}}{63{,}400\text{ lb}} \qquad d = 11.29\text{ ft} \quad ◄$$

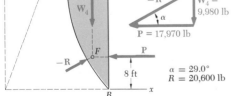

b. Resultant R of Water Forces. The parabolic section of water BCD is chosen as a free body. The forces involved are the resultant −**R** of the forces exerted by the dam on the water, the weight W_4, and the force **P**. Since these forces must be concurrent, −**R** passes through the point of intersection F of W_4 and **P**. A force triangle is drawn from which the magnitude and direction of −**R** are determined. The resultant **R** of the forces exerted by the water on the face BC is equal and opposite:

$$\mathbf{R} - 20{,}600\text{ lb} \,\nearrow\, 29.0° \quad ◄$$

PROBLEMS

5.66 and 5.67 Determine the magnitude and location of the resultant of the distributed load shown. Also calculate the reactions at A and B.

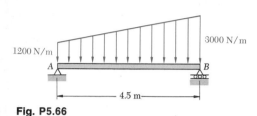

Fig. P5.66

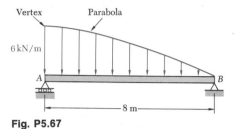

Fig. P5.67

5.68 through 5.71 Determine the reactions at the beam supports for the given loading condition.

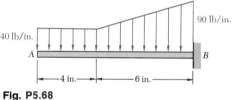

Fig. P5.68

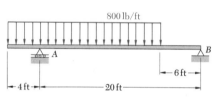

Fig. P5.69

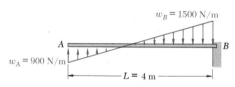

Fig. P5.70

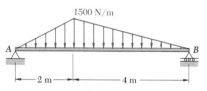

Fig. P5.71

5.72 Solve Sample Prob. 5.9 in terms of the letter quantities w_A, w_B, and L.

5.73 Determine the ratio of w_A to w_B for which the reaction at A is equal to (a) a couple and no force, (b) a force and no couple. In each case express the reaction in terms of w_A and L.

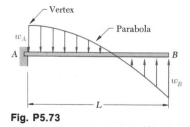

Fig. P5.73

5.74 In Prob. 5.70, determine the ratio of w_A to w_B for which the reaction at B is equal to (a) a force and no couple, (b) a couple and no force. In each case express the reaction in terms of w_B and L.

5.75 A beam supports a uniformly distributed load w_1 and rests on soil which exerts a uniformly varying upward load as shown. (a) Determine w_2 and w_3, corresponding to equilibrium. (b) Knowing that at any point the soil can exert only an upward loading on the beam, state for what range of values of a/L the results obtained are valid.

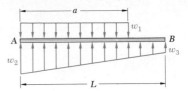

Fig. P5.75

In the following problems, use $\gamma = 62.4 \text{ lb/ft}^3$ for the specific weight of fresh water and $\gamma_c = 150 \text{ lb/ft}^3$ for the specific weight of concrete if U.S. customary units are used. With SI units, use $\rho = 10^3 \text{ kg/m}^3$ for the density of fresh water and $\rho_c = 2.40 \times 10^3 \text{ kg/m}^3$ for the density of concrete. (See footnote page 168 for the determination of the specific weight of a material from its density.)

5.76 The cross section of a concrete dam is as shown. Consider a section of the dam 1 ft thick, and determine (a) the resultant of the reaction forces exerted by the ground on the base of the dam AB, (b) the resultant of the pressure forces exerted by the water on the face BC of the dam.

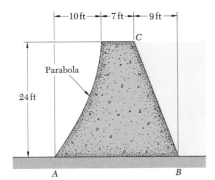

Fig. P5.76

5.77 A 3- by 3-ft gate is placed in a wall below water level as shown. (a) Determine the magnitude and location of the resultant of the forces exerted by the water on the gate. (b) If the gate is hinged at A, determine the force exerted by the sill on the gate at B.

5.78 In Prob. 5.77, determine the depth of water d for which the force exerted by the sill on the gate at B is 2000 lb.

5.79 An automatic valve consists of a square plate, 225 by 225 mm, which is pivoted about a horizontal axis through A located at a distance $h = 100$ mm above the lower edge. Determine the depth of water d for which the valve will open.

5.80 If the valve shown is to open when the depth of water is $d = 300$ mm, determine the distance h from the bottom of the valve to the pivot A.

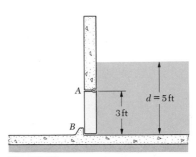

Fig. P5.77

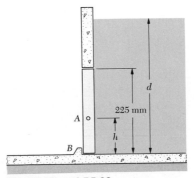

Fig. P5.79 and P5.80

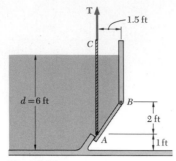

Fig. P5.81

5.81 The quick-acting gate AB is 1.75 ft wide and is held in its closed position by a vertical cable and by hinges located along its top edge B. For a depth of water $d = 6$ ft, determine the minimum tension required in cable AC to prevent the gate from opening.

5.82 For the gate of Prob. 5.81 determine the depth d for which the gate will open when $T = 800$ lb.

5.83 The bent plate $ABCD$ is 2 m wide and is hinged at A. Determine the reactions at A and D for the water level shown.

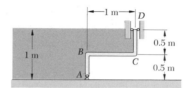

Fig. P5.83

5.84 Solve Prob. 5.83, assuming that the water shown is removed and is replaced by a 1-m depth of water to the right of the plate.

5.85 A uniform rectangular gate of weight W, height r, and length b is hinged at A. Denoting the specific weight of the fluid by γ, determine the required angle θ if the gate is to permit flow when $d = r$.

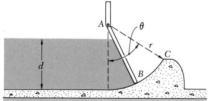

Fig. P5.85

5.86 Determine the minimum allowable value of the width a of the rectangular concrete dam if the dam is not to overturn about point A when $d = h = 4$ m.

5.87 Knowing that the width of the rectangular concrete dam is $a = 1.25$ m and that its height is $h = 4$ m, determine the maximum allowable value of the depth d of water if the dam is not to overturn about A.

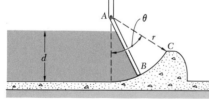

Fig. P5.86, P5.87, and P5.88

5.88 Concrete is a material which is weak in tension. In order to eliminate tension, the line of action of the resultant of the hydrostatic forces and of the weight of the dam must pass through the middle third of the base. Determine the minimum width a for which no tension will occur in the rectangular concrete dam shown when $d = h = 4$ m.

5.89 A block of wood (specific weight $\gamma_1 = 40\ \text{lb/ft}^3$) is placed in a small channel to stop the flow of water. Assuming that $d = h$ and that no water leaks between the block and the floor of the channel, determine the maximum value of the ratio h/a for which the block will not overturn about point B.

5.90 Solve Prob. 5.89 assuming that leakage occurs under the block, causing an upward pressure on the base which varies linearly from zero at B to the full hydrostatic pressure at A.

5.91 The end of a freshwater channel consists of a plate ABC which is hinged at B and is 3 ft wide. Knowing that $b = 2$ ft and $h = 1.5$ ft, determine the reactions at A and B.

5.92 The end of a freshwater channel consists of a plate ABC which is hinged at B and is 3 ft wide. Determine the ratio h/b for which the reaction at A is zero.

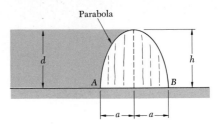
Fig. P5.89

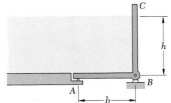
Fig. P5.91 and P5.92

VOLUMES

5.8. Center of Gravity of a Three-Dimensional Body. Centroid of a Volume.

The *center of gravity* G of a three-dimensional body is obtained by dividing the body into small elements and expressing that the weight $\mathbf{W}$ of the body attached at G is equivalent to the system of distributed forces $\Delta\mathbf{W}$ representing the weights of the small elements. Choosing the y axis vertical with positive sense upward (Fig. 5.20), and denoting by $\bar{\mathbf{r}}$ the position vector of G, we write that $\mathbf{W}$ is equal to the sum of the elementary weights $\Delta\mathbf{W}$ and that its moment about O is equal to the sum of the moments about O of the elementary weights:

$$\Sigma\mathbf{F}: \qquad -W\mathbf{j} = \Sigma(-\Delta W\mathbf{j}) \qquad (5.13)$$
$$\Sigma\mathbf{M}_O: \qquad \bar{\mathbf{r}} \times (-W\mathbf{j}) = \Sigma[\mathbf{r} \times (-\Delta W\mathbf{j})]$$

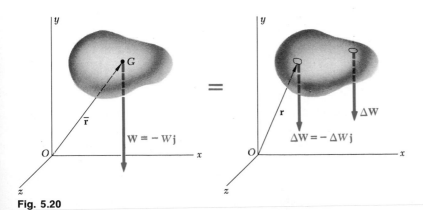
Fig. 5.20

Rewriting the last equation in the form

$$\bar{\mathbf{r}}W \times (-\mathbf{j}) = (\Sigma \mathbf{r} \Delta W) \times (-\mathbf{j}) \qquad (5.14)$$

we observe that the weight $\mathbf{W}$ of the body will be equivalent to the system of the elementary weights $\Delta \mathbf{W}$ if the following conditions are satisfied:

$$W = \Sigma \Delta W \qquad \bar{\mathbf{r}}W = \Sigma \mathbf{r} \Delta W$$

Increasing the number of elements and simultaneously decreasing the size of each element, we obtain at the limit

$$W = \int dW \qquad \bar{\mathbf{r}}W = \int \mathbf{r}\, dW \qquad (5.15)$$

We note that the relations obtained are independent of the orientation of the body. For example, if the body and the axes of coordinates were rotated so that the z axis pointed upward, the unit vector $-\mathbf{j}$ would be replaced by $-\mathbf{k}$ in Eqs. (5.13) and (5.14), but the relations (5.15) would remain unchanged. Resolving the vectors $\bar{\mathbf{r}}$ and $\mathbf{r}$ into rectangular components, we verify that the second of the relations (5.15) is equivalent to the three scalar equations

$$\bar{x}W = \int x\, dW \qquad \bar{y}W = \int y\, dW \qquad \bar{z}W = \int z\, dW \quad (5.16)$$

If the body is made of a homogeneous material of specific weight γ, the magnitude dW of the weight of an infinitesimal element may be expressed in terms of the volume dV of the element, and the magnitude W of the total weight in terms of the total volume V. We write

$$dW = \gamma\, dV \qquad W = \gamma V$$

Substituting for dW and W in the second of the relations (5.15), we write

$$\bar{\mathbf{r}}V = \int \mathbf{r}\, dV \qquad (5.17)$$

or, in scalar form,

$$\bar{x}V = \int x\, dV \qquad \bar{y}V = \int y\, dV \qquad \bar{z}V = \int z\, dV \quad (5.18)$$

The point of coordinates $\bar{x}$, $\bar{y}$, $\bar{z}$ is also known as the *centroid C of the volume V of the body*. If the body is not homogeneous, Eqs. (5.18) cannot be used to determine the center of gravity of the body; they still define, however, the centroid of the volume.

The integral $\int x\, dV$ is known as the *first moment of the volume with respect to the yz plane*. Similarly, the integrals $\int y\, dV$ and $\int z\, dV$ define the first moments of the volume with respect to the zx plane and the xy plane, respectively. It is seen from Eqs. (5.18) that, if the centroid of a volume is located in a coordinate plane, the first moment of the volume with respect to that plane is zero.

A volume is said to be symmetrical with respect to a given plane if to every point P of the volume corresponds a point P' of the same volume, such that the line PP' is perpendicular to the given plane and divided into two equal parts by that plane. The plane is said to be a *plane of symmetry* for the given volume. When a volume V possesses a plane of symmetry, the centroid of the volume must be located in that plane. When a volume possesses two planes of symmetry, the centroid of the volume must be located on the line of intersection of the two planes. Finally, when a volume possesses three planes of symmetry which intersect in a well-defined point (i.e., not along a common line), the point of intersection of the three planes must coincide with the centroid of the volume. This property enables us to determine immediately the centroid of the volume of spheres, ellipsoids, cubes, rectangular parallelepipeds, etc.

Centroids of unsymmetrical volumes or of volumes possessing only one or two planes of symmetry should be determined by integration (Sec. 5.10). Centroids of common shapes of volumes are shown in Fig. 5.21. It should be observed that the centroid of a volume of revolution in general *does not coincide* with the centroid of its cross section. Thus, the centroid of a hemisphere is different from that of a semicircular area, and the centroid of a cone is different from that of a triangle.

5.9. Composite Bodies. If a body can be divided into several of the common shapes shown in Fig. 5.21, its center of gravity G may be determined by expressing that the moment about O of its total weight is equal to the sum of the moments about O of the weights of the various component parts. Proceeding as in Sec. 5.8, we obtain the following equations defining the coordinates $\bar{X}$, $\bar{Y}$, $\bar{Z}$ of the center of gravity G:

$$\bar{X}\Sigma W = \Sigma\bar{x}W \qquad \bar{Y}\Sigma W = \Sigma\bar{y}W \qquad \bar{Z}\Sigma W = \Sigma\bar{z}W \quad (5.19)$$

If the body is made of a homogeneous material, its center of gravity coincides with the centroid of its volume and the following equations may be used:

$$\bar{X}\Sigma V = \Sigma\bar{x}V \qquad \bar{Y}\Sigma V = \Sigma\bar{y}V \qquad \bar{Z}\Sigma V = \Sigma\bar{z}V \quad (5.20)$$

Shape		$\bar{x}$	Volume
Hemisphere	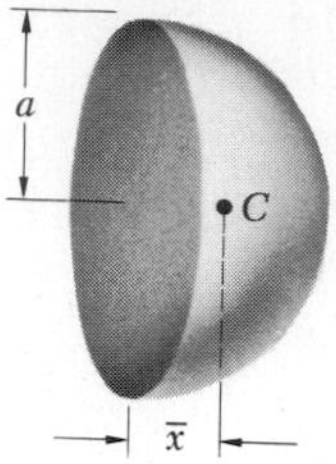	$\dfrac{3a}{8}$	$\frac{2}{3}\pi a^3$
Semiellipsoid of revolution	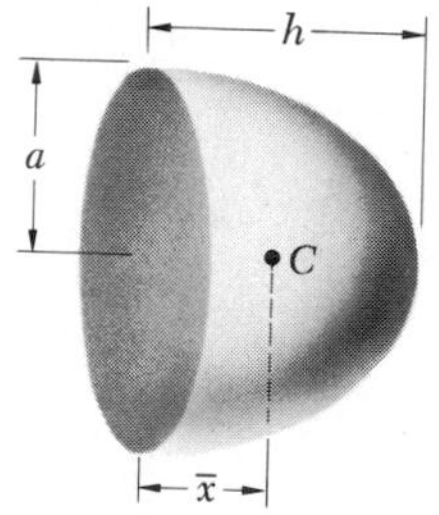	$\dfrac{3h}{8}$	$\frac{2}{3}\pi a^2 h$
Paraboloid of revolution	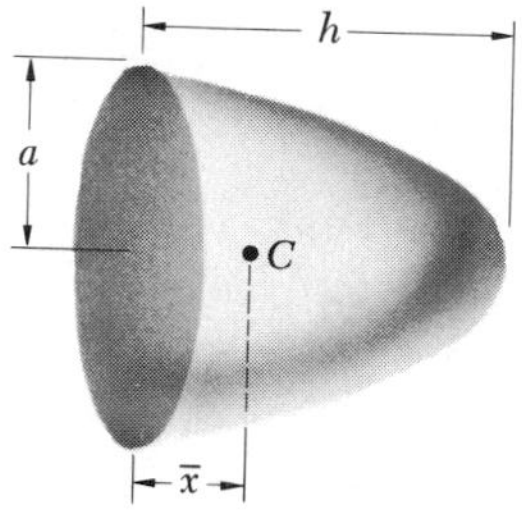	$\dfrac{h}{3}$	$\frac{1}{2}\pi a^2 h$
Cone	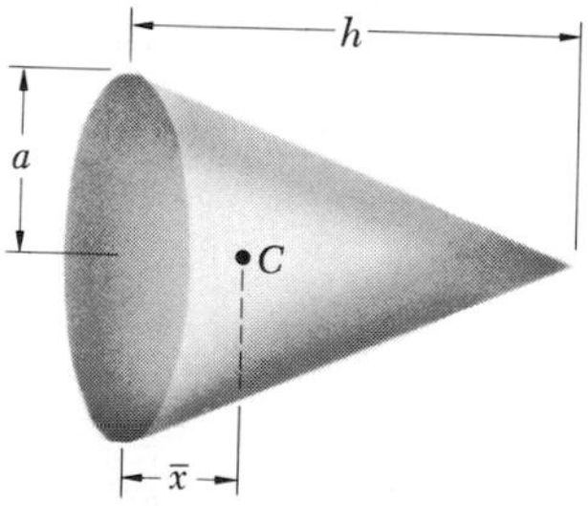	$\dfrac{h}{4}$	$\frac{1}{3}\pi a^2 h$
Pyramid	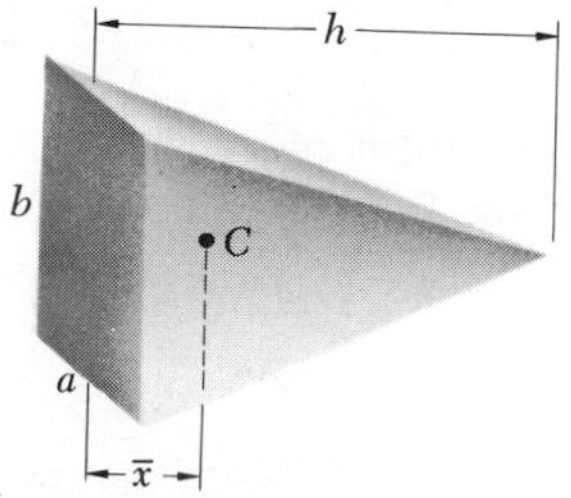	$\dfrac{h}{4}$	$\frac{1}{3}abh$

Fig. 5.21 Centroids of common shapes of volumes.

5.10. Determination of Centroids of Volumes by Integration.

The centroid of a volume bounded by analytical surfaces may be determined by computing the integrals given in Sec. 5.8.

$$\bar{x}V = \int x\, dV \qquad \bar{y}V = \int y\, dV \qquad \bar{z}V = \int z\, dV \quad (5.21)$$

If the element of volume dV is chosen equal to a small cube of sides dx, dy, and dz, the determination of each of these integrals requires a *triple integration* in x, y, and z. However, it is possible to determine the coordinates of the centroid of most volumes by *double integration* if dV is chosen equal to the volume of a thin filament as shown in Fig. 5.22. The coordinates of the centroid of the volume are then obtained by writing

$$\bar{x}V = \int \bar{x}_{el}\, dV \qquad \bar{y}V = \int \bar{y}_{el}\, dV \qquad \bar{z}V = \int \bar{z}_{el}\, dV \quad (5.22)$$

and substituting for the volume dV and the coordinates $\bar{x}_{el}$, $\bar{y}_{el}$, $\bar{z}_{el}$ the expressions given in Fig. 5.22. Using the equation of the surface to express z in terms of x and y, the integration is reduced to a double integration in x and y.

If the volume under consideration possesses *two planes of symmetry*, its centroid must be located on their line of intersection. Choosing the x axis along this line, we have

$$\bar{y} = \bar{z} = 0$$

and the only coordinate to determine is $\bar{x}$. This will be done most conveniently by dividing the given volume into thin slabs parallel to the yz plane. In the particular case of a body of revolution, these slabs are circular; their volume dV is given in Fig. 5.23. Substituting for $\bar{x}_{el}$ and dV into the equation

$$\bar{x}V = \int \bar{x}_{el}\, dV \quad (5.23)$$

and expressing the radius r of the slab in terms of x, we may determine $\bar{x}$ by a single integration.

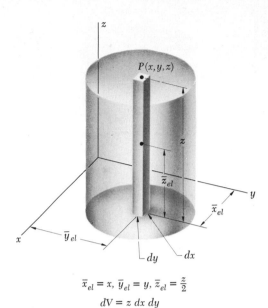

$$\bar{x}_{el} = x, \; \bar{y}_{el} = y, \; \bar{z}_{el} = \frac{z}{2}$$
$$dV = z\, dx\, dy$$

Fig. 5.22 Determination of the centroid of a volume by double integration.

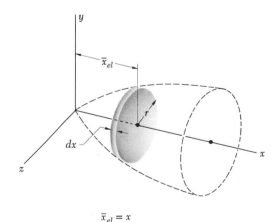

$$\bar{x}_{el} = x$$
$$dV = \pi r^2\, dx$$

Fig. 5.23 Determination of the centroid of a body of revolution.

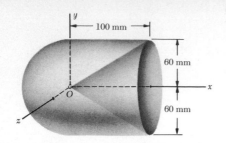

100 mm

60 mm

60 mm

SAMPLE PROBLEM 5.11

Determine the center of gravity of the homogeneous body of revolution shown.

Solution. Because of symmetry, the center of gravity lies on the x axis. The body is seen to consist of a hemisphere, plus a cylinder, minus a cone, as shown. The volume and the abscissa of the centroid of each of these components are obtained from Fig. 5.21 and entered in the table below. All dimensions have been converted into centimeters to avoid excessively large numerical values in the computation of volumes and moments of volumes.

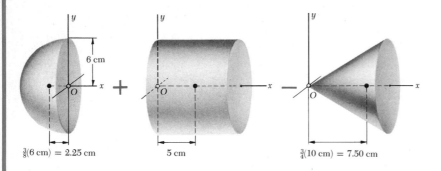

6 cm

$\frac{3}{8}(6 \text{ cm}) = 2.25 \text{ cm}$ 5 cm $\frac{3}{4}(10 \text{ cm}) = 7.50 \text{ cm}$

Component	Volume, cm³	$\bar{x}$, cm	$\bar{x}V$, cm⁴
Hemisphere	$\frac{1}{2}\frac{4\pi}{3}(6)^3 = 452$	-2.25	-1017
Cylinder	$\pi(6)^2(10) = 1131$	$+5.00$	$+5655$
Cone	$-\frac{\pi}{3}(6)^2(10) = -377$	$+7.50$	-2828
	$\Sigma V = 1206$	$\ldots$	$\Sigma \bar{x}V = +1810$

Thus,

$$\bar{X}\Sigma V = \Sigma \bar{x}V: \qquad \bar{X}(1206 \text{ cm}^3) = 1810 \text{ cm}^4$$
$$\bar{X} = 1.5 \text{ cm} \qquad\qquad \bar{X} = 15 \text{ mm} \blacktriangleleft$$

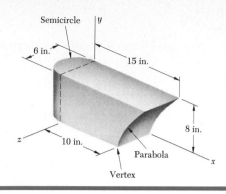

Semicircle
6 in.
15 in.
8 in.
10 in.
Parabola
Vertex

SAMPLE PROBLEM 5.12

Determine the center of gravity of the homogeneous body shown.

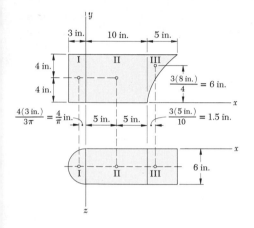

3 in.
10 in.
5 in.
4 in.
I   II   III
4 in.
$\frac{3(8 \text{ in.})}{4} = 6 \text{ in.}$
$\frac{4(3 \text{ in.})}{3\pi} = \frac{4}{\pi} \text{ in.}$
5 in.   5 in.
$\frac{3(5 \text{ in.})}{10} = 1.5 \text{ in.}$
I   II   III
6 in.

Solution. Because of symmetry, the center of gravity lies in a plane parallel to the xy plane, with $\bar{Z} = 3$ in. The body is seen to consist of a half cylinder (I), a rectangular parallelepiped (II), and a cylindrical shape having for a cross section a parabolic spandrel (III). Using the information given in Fig. 5.8A for a semicircle and a parabolic spandrel, we determine the coordinates of the centroids of the various components of the body. These coordinates are shown in the adjoining sketch. The total volume and the moments of the volume with respect to the zx and yz planes are determined in the table below.

Component	V, in^3	$\bar{x}$, in.	$\bar{y}$, in.	$\bar{x}V$, in^4	$\bar{y}V$, in^4
I	$(8)\dfrac{\pi(3^2)}{2} = 113$	$-\dfrac{4}{\pi}$	4	-144	452
II	$(10)(8)(6) = 480$	5	4	2400	1920
III	$\frac{1}{3}(5)(8)(6) = 80$	11.5	6	920	480
	$\Sigma V = 673$	...	...	$\Sigma \bar{x}V = 3176$	$\Sigma \bar{y}V = 2852$

Thus,

$\bar{X}\Sigma V = \Sigma \bar{x}V$: $\bar{X}(673 \text{ in}^3) = 3176 \text{ in}^4$ $\bar{X} = 4.72 \text{ in.}$ ◀

$\bar{Y}\Sigma V = \Sigma \bar{y}V$: $\bar{Y}(673 \text{ in}^3) = 2852 \text{ in}^4$ $\bar{Y} = 4.24 \text{ in.}$ ◀

$\bar{Z} = 3.00 \text{ in.}$ ◀

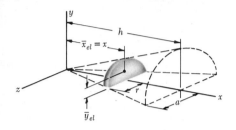

Determine the location of the centroid of the half right circular cone shown.

Solution.   Since the xy plane is a plane of symmetry, the centroid lies in this plane and $\bar{z} = 0$. A slab of thickness dx is chosen as a differential element. The volume of this element is

$$dV = \tfrac{1}{2}\pi r^2 \, dx$$

The coordinates $\bar{x}_{el}$ and $\bar{y}_{el}$ of the centroid of the element are obtained from Fig. 5.8 (semicircular area).

$$\bar{x}_{el} = x \qquad \bar{y}_{el} = \frac{4r}{3\pi}$$

We observe that r is proportional to x and write

$$\frac{r}{x} = \frac{a}{h} \qquad r = \frac{a}{h}x$$

The volume of the body is

$$V = \int dV = \int_0^h \tfrac{1}{2}\pi r^2 \, dx = \int_0^h \tfrac{1}{2}\pi \left(\frac{a}{h}x\right)^2 dx = \frac{\pi a^2 h}{6}$$

The moment of the differential element with respect to the yz plane is $\bar{x}_{el}\, dV$; and the total moment of the body with respect to this plane is

$$\int \bar{x}_{el}\, dV = \int_0^h x(\tfrac{1}{2}\pi r^2)\, dx = \int_0^h x(\tfrac{1}{2}\pi)\left(\frac{a}{h}x\right)^2 dx = \frac{\pi a^2 h^2}{8}$$

Thus, $\qquad \bar{x}V = \int \bar{x}_{el}\, dV \qquad \bar{x}\frac{\pi a^2 h}{6} = \frac{\pi a^2 h^2}{8} \qquad \bar{x} = \tfrac{3}{4}h \quad \blacktriangleleft$

Likewise, the moment of the differential element with respect to the zx plane is $\bar{y}_{el}\, dV$; and the total moment is

$$\int \bar{y}_{el}\, dV = \int_0^h \frac{4r}{3\pi}(\tfrac{1}{2}\pi r^2)\, dx = \frac{2}{3}\int_0^h \left(\frac{a}{h}x\right)^3 dx = \frac{a^3 h}{6}$$

Thus, $\qquad \bar{y}V = \int \bar{y}_{el}\, dV \qquad \bar{y}\frac{\pi a^2 h}{6} = \frac{a^3 h}{6} \qquad \bar{y} = \frac{a}{\pi} \quad \blacktriangleleft$

PROBLEMS

5.93   A cone and a cylinder of the same radius a and height h are attached as shown. Determine the location of the centroid of the composite body.

5.94   A hemisphere and a cylinder are placed together as shown. Determine the ratio h/r for which the centroid of the composite body is located in the plane between the hemisphere and the cylinder.

Fig. P5.93

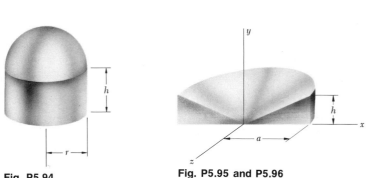

Fig. P5.94

Fig. P5.95 and P5.96

5.95   Determine the y coordinate of the centroid of the solid shown.

5.96   Determine the z coordinate of the solid shown. (*Hint.* Use the result of Sample Prob. 5.13.)

5.97   Locate the centroid of the frustum of a right circular cone when $r_1 = 100$ mm, $r_2 = 125$ mm, and $h = 150$ mm.

Fig. P5.97

5.98 and 5.99   Locate the center of gravity of the machine element shown.

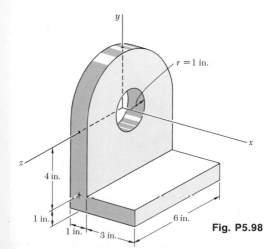

Fig. P5.98

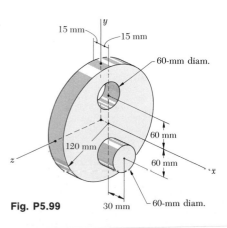

Fig. P5.99

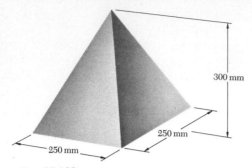

Fig. P5.100

5.100 A regular pyramid 300 mm high, with a square base of side 250 mm, is made of wood. Its four triangular faces are covered with steel sheets 1 mm thick. Locate the center of gravity of the composite body. (Densities: steel = 7850 kg/m³; wood = 550 kg/m³.)

5.101 The brass sleeve is to be mounted on the pin of a machine part made of aluminum. Locate the center of gravity of the assembly. (Specific weights: brass = 530 lb/ft³; aluminum = 170 lb/ft³.)

5.102 and 5.103 Locate the center of gravity of the sheet-metal form shown.

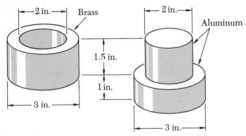

Fig. P5.101

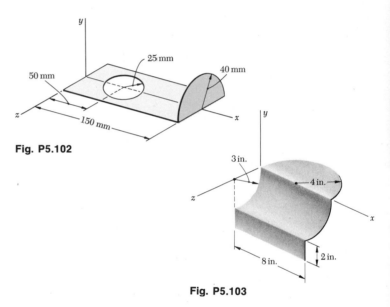

Fig. P5.102

Fig. P5.103

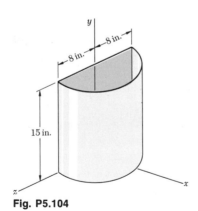

Fig. P5.104

5.104 A wastebasket, designed to fit along a wall, is 15 in. high and has a base in the shape of a semicircle of radius 8 in. Locate the center of gravity of the wastebasket, knowing that it is made of sheet metal of uniform thickness.

5.105 Locate the centroid of the frustum of the right circular cone of Prob. 5.97, expressing the result in terms of r_1, r_2, and h.

5.106 Locate the center of gravity of a thin hemispherical shell of radius r and thickness t. (*Hint.* Consider the shell as formed by removing a hemisphere of radius r from a hemisphere of radius $r + t$; then neglect the terms containing t^2 and t^3, and keep those terms containing t.)

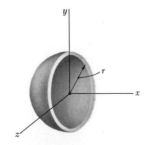

Fig. P5.106

5.107 Sheet metal of thickness t is used to form the half-conical shell shown. Locate the center of gravity of the shell. (*Hint.* Use the differential element shown.)

5.108 Derive by direct integration the expression given for $\bar{x}$ in Fig. 5.21 for a semiellipsoid of revolution.

5.109 Locate the centroid of the volume obtained by rotating the area shown about the x axis. The expression obtained may be used to confirm the values given in Fig. 5.21 for the paraboloid (with $n = \frac{1}{2}$) and the cone (with $n = 1$).

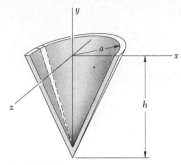

Fig. P5.107

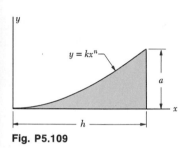

Fig. P5.109

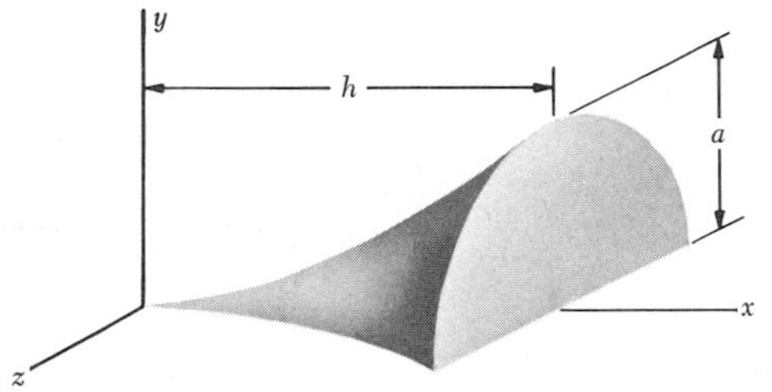

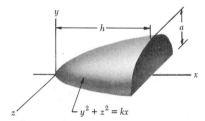

Fig. P5.110

5.110 The volume shown was obtained by rotating the area of Prob. 5.109, with $n = 2$, through $180°$ about the x axis. Locate its centroid by direct integration.

5.111 Locate the centroid of the semiparaboloid of revolution shown.

✱**5.112** A thin spherical cup is of radius a and constant thickness t. Show by direct integration that the center of gravity of the cup is located at a distance $\frac{1}{2}h$ above the base of the cup.

Fig. P5.111

Fig. P5.112

✱**5.113** A hemispherical tank of radius a is filled with water to a depth h. Determine by direct integration the center of gravity of the water in the tank.

Fig. P5.113

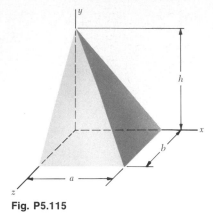

Fig. P5.115

*5.114 A spherical tank is 2 m in diameter and is filled with water to a depth of 1.5 m. Determine by direct integration the center of gravity of the water in the tank.

5.115 Locate the centroid of the volume of the irregular pyramid shown.

5.116 Determine by direct integration the location of the centroid of the volume between the xy plane and the portion of the hyperbolic paraboloid shown.

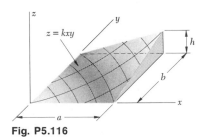

Fig. P5.116

5.117 Locate the centroid of the section shown, which was cut from a circular cylinder by an oblique plane.

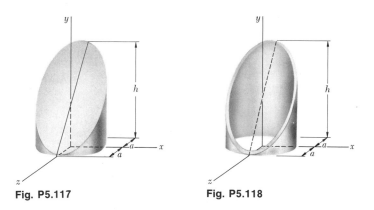

Fig. P5.117   **Fig. P5.118**

5.118 Locate the centroid of the section shown, which was cut from a thin circular pipe by an oblique plane.

Fig. P5.119

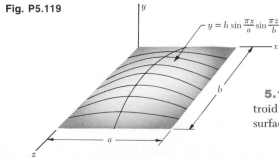

$y = h \sin \dfrac{\pi x}{a} \sin \dfrac{\pi z}{b}$

5.119 Determine by direct integration the location of the centroid of the volume between the xz plane and the portion shown of the surface $y = h \sin (\pi x/a) \sin (\pi z/b)$.

REVIEW PROBLEMS

5.120 A brass plug is made from a 50-mm-diameter cylinder by machining it as shown. Determine the volume of the material removed in forming (a) the semicircular groove, (b) the quarter-circular rounding on the left end of the plug.

5.121 Locate the center of gravity of the brass plug shown.

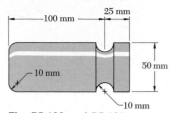

Fig. P5.120 and P5.121

5.122 Locate the center of gravity of the sheet-metal form shown.

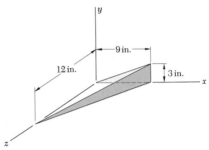

Fig. P5.122

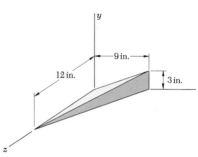

Fig. P5.123

5.123 Locate the centroid of the irregular pyramid shown. (*Hint.* See Fig. 5.21.)

5.124 The gate shown is supported by hinges along its bottom edge at A and rests against a wall at B. Knowing that the gate is 2 m wide, determine the reactions at A and B due to the hydrostatic forces acting on the gate.

5.125 An oil-storage tank in the form of a shell of constant strength has the cross section shown. (This is also the cross section of a drop of water on an unwetted surface.) Determine by approximate means (a) the volume of the tank, (b) the surface area of the tank.

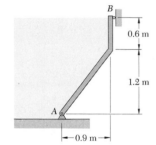

Fig. P5.124

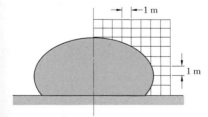

Fig. P5.125

5.126 Determine the reactions at A and B.

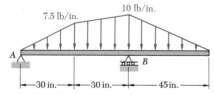

Fig. P5.126

5.127 The production-line balancing of an automobile speedometer cup is done as follows. The unbalanced cup is placed on a frictionless shaft at O and is allowed to come to rest. A hole is then punched at A; the cup rotates through an angle θ and again comes to rest. The balancing is completed by punching additional holes at B and C. Knowing that the three holes are equal in size and are at the same distance from the shaft O, determine the required angle α in terms of the angle θ.

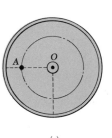

(a)   (b)

Fig. P5.127

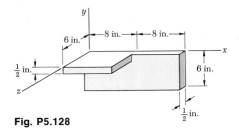

Fig. P5.128

5.128 A portion of one leg of a 6- by 6- by $\frac{1}{2}$-in. angle is cut off. Determine the coordinates of the center of gravity of the remaining portion of the angle.

5.129 Locate the center of gravity of the sheet-metal form shown when $a = 0.2$ m.

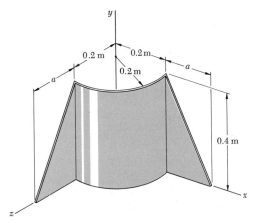

Fig. P5.129 and P5.130

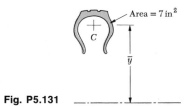

Fig. P5.131

5.130 Determine the distance a so that the center of gravity of the sheet-metal form is located 0.2 m from the y axis.

5.131 An automobile tire weighs 22 lb and has a cross-sectional area of 7 in². The specific weight of the rubber used is 80 lb/ft³; determine the location of the centroid of the cross-sectional area.

Analysis of Structures

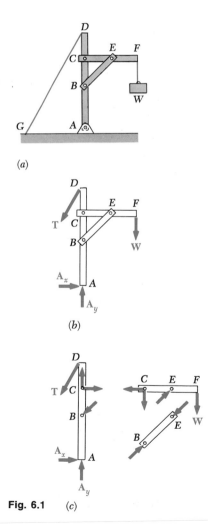

(a)

(b)

Fig. 6.1   (c)

6.1. Internal Forces. Newton's Third Law. The problems considered in the preceding chapters concerned the equilibrium of a single rigid body, and all forces involved were external to the rigid body. We shall now consider problems dealing with the equilibrium of structures made of several connected parts. These problems call not only for the determination of the external forces acting on the structure but also for the determination of the forces which hold together the various parts of the structure. From the point of view of the structure as a whole, these forces are *internal forces*.

Consider, for example, the crane shown in Fig. 6.1a, which carries a load W. The crane consists of three beams AD, CF, and BE connected by frictionless pins; it is supported by a pin at A and by a cable DG. The free-body diagram of the crane has been drawn in Fig. 6.1b. The external forces are shown in the diagram and include the weight **W**, the two components **A**$_x$ and **A**$_y$ of the reaction at A, and the force **T** exerted by the cable at D. The internal forces holding the various parts of the crane together do not appear in the diagram. If, however, the crane is dismembered and if a free-body diagram is drawn for each of its component parts, the forces holding the three beams together must also be represented, since these forces are external forces from the point of view of each component part (Fig. 6.1c).

It will be noted that the force exerted at B by member BE on member AD has been represented as equal and opposite to the force exerted at the same point by member AD on member

213

BE; similarly, the force exerted at *E* by *BE* on *CF* is shown equal and opposite to the force exerted by *CF* on *BE*; and the components of the force exerted at *C* by *CF* on *AD* are shown equal and opposite to the components of the force exerted by *AD* on *CF*. This is in conformity with Newton's third law, which states that *the forces of action and reaction between bodies in contact have the same magnitude, same line of action, and opposite sense*. As pointed out in Chap. 1, this law is one of the six fundamental principles of elementary mechanics and is based on experimental evidence. Its application is essential to the solution of problems involving connected bodies.

TRUSSES

6.2. Definition of a Truss. The truss is one of the major types of engineering structures. It provides both a practical and an economical solution to many engineering situations, especially in the design of bridges and buildings. A truss consists of straight members connected at joints; a typical truss is shown in Fig. 6.2*a*. Truss members are connected at their extremities only; thus no member is continuous through a joint. In Fig. 6.2*a*, for example, there is no member *AB*; there are instead two distinct members *AD* and *DB*. Actual structures are made of several trusses joined together to form a space framework. Each truss is designed to carry those loads which act in its plane and thus may be treated as a two-dimensional structure.

In general, the members of a truss are slender and can support little lateral load; all loads, therefore, must be applied to the various joints, and not to the members themselves. When a concentrated load is to be applied between two joints, or when a distributed load is to be supported by the truss, as in the case of a bridge truss, a floor system must be provided which, through the use of stringers and floor beams, transmits the load to the joints (Fig. 6.3).

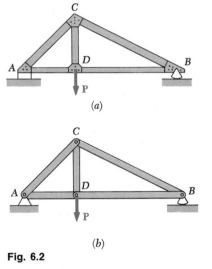

Fig. 6.2

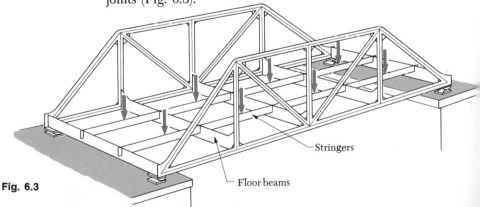

Fig. 6.3

The weights of the members of the truss are also assumed to be applied to the joints, half of the weight of each member being applied to each of the two joints the member connects. Although the members are actually joined together by means of riveted and welded connections, it is customary to assume that the members are pinned together; therefore, the forces acting at each end of a member reduce to a single force and no couple. Thus, the only forces assumed to be applied to a truss member are a single force at each end of the member. Each member may then be treated as a two-force member, and the entire truss may be considered as a group of pins and two-force members (Fig. 6.2b). An individual member may be acted upon as shown in either of the two sketches of Fig. 6.4. In the first sketch, the forces tend to pull the member apart, and the member is in tension, while, in the second sketch, the forces tend to compress the member, and the member is in compression. Several typical trusses are shown in Fig. 6.5.

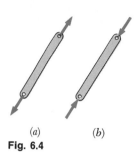

(a)   (b)

Fig. 6.4

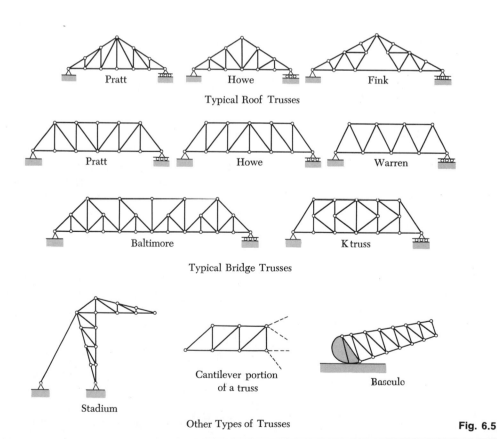

Pratt   Howe   Fink

Typical Roof Trusses

Pratt   Howe   Warren

Baltimore   K truss

Typical Bridge Trusses

Stadium

Cantilever portion
of a truss

Bascule

Other Types of Trusses

Fig. 6.5

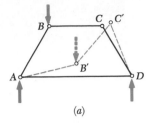

(a)

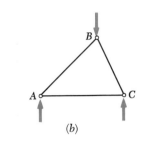

(b)

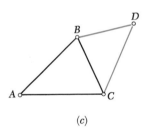

(c)

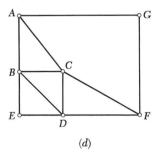

(d)

Fig. 6.6

6.3. Simple Trusses. Consider the truss of Fig. 6.6*a*, which is made of four members connected by pins at *A*, *B*, *C*, and *D*. If a load is applied at *B*, the truss will greatly deform and lose completely its original shape. On the other hand, the truss of Fig. 6.6*b*, which is made of three members connected by pins at *A*, *B*, and *C*, will deform only slightly under a load applied at *B*. The only possible deformation for this truss is one involving small changes in the length of its members. The truss of Fig. 6.6*b* is said to be a *rigid truss*, the term rigid being used here to indicate that the truss *will not collapse*.

As shown in Fig. 6.6*c*, a larger rigid truss may be obtained by adding two members *BD* and *CD* to the basic triangular truss of Fig. 6.6*b*. This procedure may be repeated as many times as desired, and the resulting truss will be rigid if, each time we add two new members, we attach them to separate existing joints and connect them at a new joint.† A truss which may be constructed in this manner is called a *simple truss*.

It should be noted that a simple truss is not necessarily made only of triangles. The truss of Fig. 6.6*d*, for example, is a simple truss which was constructed from triangle *ABC* by adding successively the joints *D*, *E*, *F*, and *G*. On the other hand, rigid trusses are not always simple trusses, even when they appear to be made of triangles. The Fink and Baltimore trusses shown in Fig. 6.5, for instance, are not simple trusses, since they cannot be constructed from a single triangle in the manner described above. All the other trusses shown in Fig. 6.5 are simple trusses, as may be easily checked. (For the K truss, start with one of the central triangles.)

Returning to the basic triangular truss of Fig. 6.6*b*, we note that this truss has three members and three joints. The truss of Fig. 6.6*c* has two more members and one more joint, i.e., altogether five members and four joints. Observing that every time two new members are added, the number of joints is increased by one, we find that in a simple truss the total number of members is $m = 2n - 3$, where n is the total number of joints.

6.4. Analysis of Trusses by the Method of Joints. We saw in Sec. 6.2 that a truss may be considered as a group of pins and two-force members. The truss of Fig. 6.2, whose free-body diagram is shown in Fig. 6.7*a*, may thus be dismembered, and a free-body diagram can be drawn for each pin and each member (Fig. 6.7*b*). Each member is acted upon by two forces, one at each end; these forces have the same magnitude, same line of action, and opposite sense (Sec. 4.6). Besides, Newton's third law indicates that the forces of action and reac-

†The three joints must not be in a straight line.

tion between a member and a pin are equal and opposite. Therefore, the forces exerted by a member on the two pins it connects must be directed along that member and be equal and opposite. The common magnitude of the forces exerted by a member on the two pins it connects is commonly referred to as the *force in the member* considered, even though this quantity is actually a scalar. Since the lines of action of all the internal forces in a truss are known, the analysis of a truss reduces to the computation of the forces in its various members and to the determination of whether each of its members is in tension or in compression.

Since the entire truss is in equilibrium, each pin must be in equilibrium. The fact that a pin is in equilibrium may be expressed by drawing its free-body diagram and writing two equilibrium equations (Sec. 2.8). If the truss contains n pins, there will be therefore $2n$ equations available, which may be solved for $2n$ unknowns. In the case of a simple truss, we have $m = 2n - 3$, that is, $2n = m + 3$, and the number of unknowns which may be determined from the free-body diagrams of the pins is thus $m + 3$. This means that the forces in all the members, as well as the two components of the reaction $\mathbf{R}_A$, and the reaction $\mathbf{R}_B$ may be found by considering the free-body diagrams of the pins.

The fact that the entire truss is a rigid body in equilibrium may be used to write three more equations involving the forces shown in the free-body diagram of Fig. 6.7a. Since they do not contain any new information, these equations are not independent from the equations associated with the free-body diagrams of the pins. Nevertheless, they may be used to determine immediately the components of the reactions at the supports. The arrangement of pins and members in a simple truss is such that it will then always be possible to find a joint involving only two unknown forces. These forces may be determined by the methods of Sec. 2.10 and their values transferred to the adjacent joints and treated as known quantities at these joints. This procedure may be repeated until all unknown forces have been determined.

As an example, we shall analyze the truss of Fig. 6.7 by considering successively the equilibrium of each pin, starting with a joint at which only two forces are unknown. In the truss considered, all pins are subjected to at least three unknown forces. Therefore, the reactions at the supports must first be determined by considering the entire truss as a free body and using the equations of equilibrium of a rigid body. We find in this way that $\mathbf{R}_A$ is vertical and determine the magnitudes of $\mathbf{R}_A$ and $\mathbf{R}_B$.

The number of unknown forces at joint A is thus reduced to

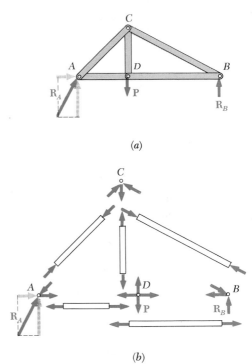

(a)

(b)

Fig. 6.7

two, and these forces may be determined by considering the equilibrium of pin A. The reaction $\mathbf{R}_A$ and the forces $\mathbf{F}_{AC}$ and $\mathbf{F}_{AD}$ exerted on pin A by members AC and AD, respectively, must form a force triangle. First we draw $\mathbf{R}_A$ (Fig. 6.8); noting that $\mathbf{F}_{AC}$ and $\mathbf{F}_{AD}$ are directed along AC and AD, respectively, we complete the triangle and determine the magnitude and sense of $\mathbf{F}_{AC}$ and $\mathbf{F}_{AD}$. The magnitudes F_{AC} and F_{AD} represent the forces in members AC and AD. Since $\mathbf{F}_{AC}$ is directed down and to the left, that is, toward joint A, member AC pushes on pin A and is in compression. On the other hand, since $\mathbf{F}_{AD}$ is directed away from the joint, member AD pulls on pin A and is in tension.

We may now proceed to joint D, where only two forces, $\mathbf{F}_{DC}$ and $\mathbf{F}_{DB}$, are still unknown. The other forces are the load P,

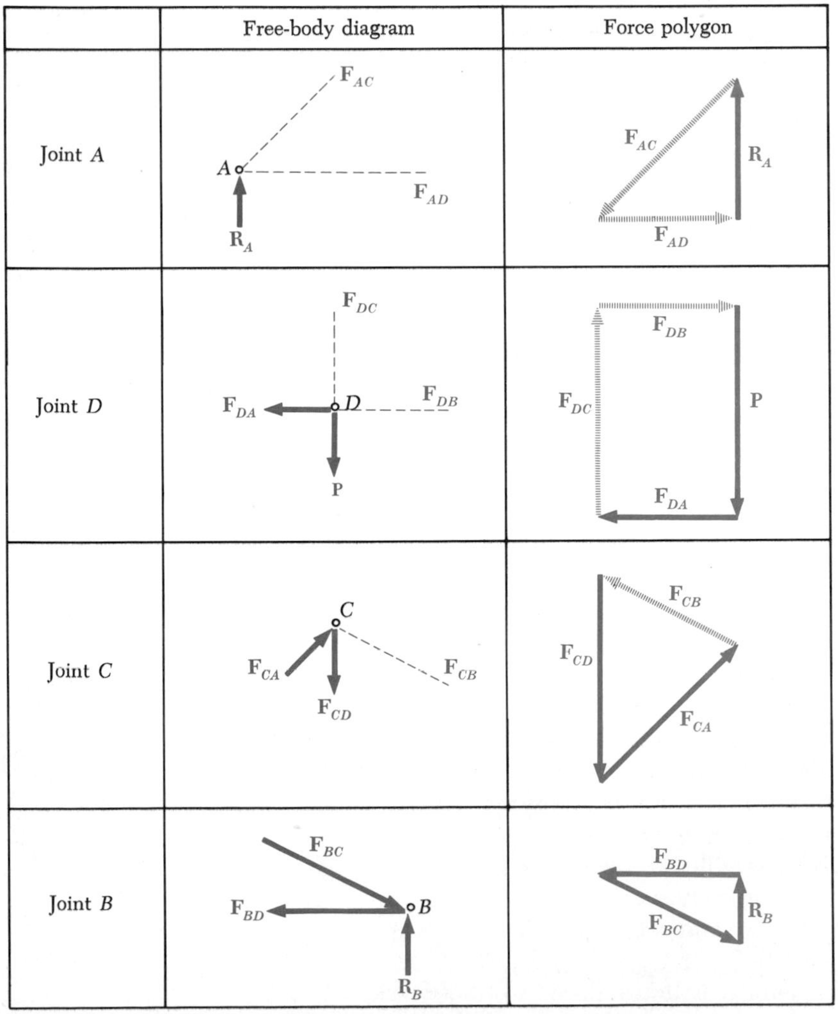

	Free-body diagram	Force polygon
Joint A		
Joint D		
Joint C		
Joint B		

Fig. 6.8

which is given, and the force $\mathbf{F}_{DA}$ exerted on the pin by member
AD. As indicated above, this force is equal and opposite to the
force $\mathbf{F}_{AD}$ exerted by the same member on pin *A*. We may draw
the force polygon corresponding to joint *D*, as shown in Fig. 6.8,
and determine the forces $\mathbf{F}_{DC}$ and $\mathbf{F}_{DB}$ from that polygon.
However, when more than three forces are involved, it is usually
more convenient to write the equations of equilibrium $\Sigma F_x = 0$,
$\Sigma F_y = 0$ and solve these equations for the two unknown forces.
Since both of these forces are found to be directed away from
joint *D*, members *DC* and *DB* pull on the pin and are in tension.

Next, joint *C* is considered; its free-body diagram is shown in
Fig. 6.8. It is noted that both $\mathbf{F}_{CD}$ and $\mathbf{F}_{CA}$ are known from the
analysis of the preceding joints and that only $\mathbf{F}_{CB}$ is unknown.
Since the equilibrium of each pin provides sufficient information
to determine two unknowns, a check of our analysis is obtained
at this joint. The force triangle is drawn, and the magnitude and
sense of $\mathbf{F}_{CB}$ are determined. Since $\mathbf{F}_{CB}$ is directed toward joint
C, member *CB* pushes on pin *C* and is in compression. The check
is obtained by verifying that the force $\mathbf{F}_{CB}$ and member *CB* are
parallel.

At joint *B*, all the forces are known. Since the corresponding
pin is in equilibrium, the force triangle must close and an additional check of the analysis is obtained.

It should be noted that the force polygons shown in Fig. 6.8
are not unique. Each of them could be replaced by an alternate
configuration. For example, the force triangle corresponding to

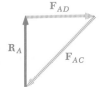

Fig. 6.9

joint *A* could be drawn as shown in Fig. 6.9. The triangle actually shown in Fig. 6.8 was obtained by drawing the three forces
$\mathbf{R}_A$, $\mathbf{F}_{AC}$, and $\mathbf{F}_{AD}$ in tip-to-tail fashion in the order in which their
lines of action are encountered when moving clockwise around
joint *A*. The other force polygons in Fig. 6.8, having been drawn
in the same way, can be made to fit into a single diagram, as
shown in Fig. 6.10. Such a diagram, known as *Maxwell's diagram*, greatly facilitates the *graphical analysis* of truss problems.†

† For a complete discussion of Maxwell's diagrams, see F. P. Beer and E. R.
Johnston, "Mechanics for Engineers," sec. 6.7, McGraw-Hill Book Company,
1976.

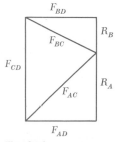

Fig. 6.10

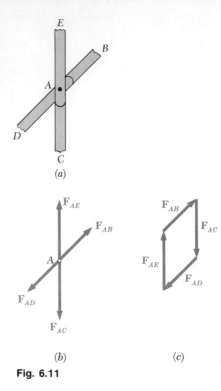

(a)

(b)   (c)

Fig. 6.11

*6.5. Joints under Special Loading Conditions.
Consider the joint shown in Fig. 6.11a, which connects four
members lying in two intersecting straight lines. The free-body
diagram of Fig. 6.11b shows that pin A is subjected to two pairs
of directly opposite forces. The corresponding force polygon,
therefore, must be a parallelogram (Fig. 6.11c), and *the forces in
opposite members must be equal.*

Consider next the joint shown in Fig. 6.12a, which connects
three members and supports a load **P.** Two of the members lie in
the same line, and the load **P** acts along the third member. The
free-body diagram of pin A and the corresponding force polygon
will be as shown in Fig. 6.11b and c with $\mathbf{F}_{AE}$ replaced by the
load **P.** Thus, *the forces in the two opposite members must be
equal, and the force in the other member must equal P.* A particular case of special interest is shown in Fig. 6.12b. Since, in this
case, no external load is applied to the joint, we have $P = 0$, and
the force in member AC is zero. Member AC is said to be a
zero-force member.

Consider now a joint connecting two members only. From
Sec. 2.8, we know that a particle which is acted upon by two
forces will be in equilibrium if the two forces have the same
magnitude, same line of action, and opposite sense. In the case
of the joint of Fig. 6.13a, which connects two members AB and AD
lying in the same line, the equilibrium of pin A requires therefore

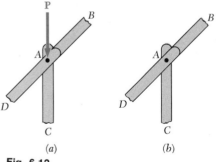

(a)   (b)

Fig. 6.12

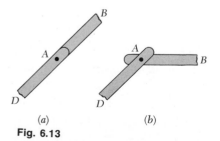

(a)   (b)

Fig. 6.13

that *the forces in the two members be equal.* In the case of the joint of Fig. 6.13*b*, the equilibrium of pin *A* is impossible unless the forces in both members are zero. Members connected as shown in Fig. 6.13*b*, therefore, must be *zero-force members.*

Spotting the joints which are under the special loading conditions listed above will expedite the analysis of a truss. Consider, for example, a Howe truss loaded as shown in Fig. 6.14. All the members represented by colored lines will be recognized as zero-force members. Joint *C* connects three members, two of which lie in the same line, and is not subjected to any external load; member *BC* is thus a zero-force member. Applying the same reasoning to joint *K*, we find that member *JK* is also a zero-force member. But joint *J* is now in the same situation as joints *C* and *K*, and member *IJ* must be a zero-force member. The examination of joints *C*, *J*, and *K* also shows that the forces in members *AC* and *CE* are equal, that the forces in members *HJ* and *JL* are equal, and that the forces in members *IK* and *KL* are equal. Furthermore, now turning our attention to joint *I*, where the 20-kN load and member *HI* are collinear, we note that the force in member *HI* is 20 kN (tension) and that the forces in members *GI* and *IK* are equal. Hence, the forces in members *GI*, *IK*, and *KL* are equal.

Students, however, should be warned against misusing the rules established in this section. For example, it would be wrong to assume that the force in member *DE* is 25 kN or that the forces in members *AB* and *BD* are equal. The conditions discussed above do not apply to joints *B* and *D*. The forces in these members and in all remaining members should be found by carrying out the analysis of joints *A*, *B*, *D*, *E*, *F*, *G*, *H*, and *L* in the

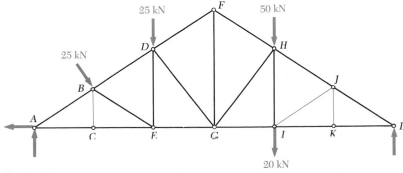

Fig. 6.14

usual manner. Until they have become thoroughly familiar with the conditions of application of the rules established in this section, students would be well advised to draw the free-body diagrams of all pins and to write the corresponding equilibrium equations (or draw the corresponding force polygons), whether or not the joints considered fall into the categories listed above.

A final remark concerning zero-force members: These members are not useless. While they do not carry any load under the particular loading conditions shown, the zero-force members of Fig. 6.14 will probably carry loads if the loading conditions are changed. Besides, even in the case considered, these members are needed to support the weight of the truss and to maintain the truss in the desired shape.

∗6.6. Space Trusses. When several straight members are joined together at their extremities to form a three-dimensional configuration, the structure obtained is called a *space truss*.

We recall from Sec. 6.3 that the most elementary two-dimensional rigid truss consisted of three members joined at their extremities to form the sides of a triangle; by adding two members at a time to this basic configuration, and connecting them at a new joint, it was possible to obtain a larger rigid structure which was defined as a simple truss. Similarly, the most elementary rigid space truss consists of six members joined at their extremities to form the edges of a tetrahedron *ABCD* (Fig. 6.15*a*). By adding three members at a time to this basic configuration, such as *AE*, *BE*, and *CE*, attaching them at separate existing joints, and connecting them at a new joint, we can obtain a larger rigid structure which is defined as a *simple space truss* (Fig. 6.15*b*).† Observing that the basic tetrahedron has six members and four joints, and that, every time three members are added, the number of joints is increased by one, we conclude that in a simple space truss the total number of members is $m = 3n - 6$, where n is the total number of joints.

If a space truss is to be completely constrained and if the reactions at its supports are to be statically determinate, the supports should consist of a combination of balls, rollers, and balls and sockets which provides six unknown reactions (see Sec. 4.8). These unknown reactions may be readily determined by solving the six equations expressing that the three-dimensional truss is in equilibrium.

Although the members of a space truss are actually joined together by means of riveted or welded connections, it is assumed that each joint consists of a ball-and-socket connection.

† The four joints must not lie in a plane.

Thus, no couple will be applied to the members of the truss, and each member may be treated as a two-force member. The conditions of equilibrium for each joint will be expressed by the three equations $\Sigma F_x = 0$, $\Sigma F_y = 0$, and $\Sigma F_z = 0$. In the case of a simple space truss containing n joints, writing the conditions of equilibrium for each joint will thus yield $3n$ equations. Since $m = 3n - 6$, these equations suffice to determine all unknown forces (forces in m members and six reactions at the supports). However, to avoid solving many simultaneous equations, care should be taken to select joints in such an order that no selected joint will involve more than three unknown forces.

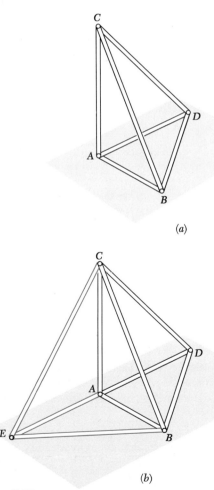

(a)

(b)

Fig. 6.15

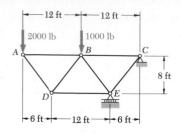

SAMPLE PROBLEM 6.1

Using the method of joints, determine the force in each member of the truss shown.

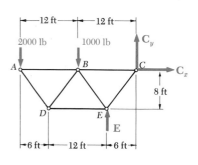

Solution. A free-body diagram of the entire truss is drawn; external forces acting on this free body consist of the applied loads and the reactions at C and E.

Equilibrium of Entire Truss.

$+\sum M_C = 0$: $(2000 \text{ lb})(24 \text{ ft}) + (1000 \text{ lb})(12 \text{ ft}) - E(6 \text{ ft}) = 0$

$\qquad\qquad\qquad\qquad E = +10,000 \text{ lb}$ $E = 10,000 \text{ lb} \uparrow$

$\xrightarrow{+} \sum F_x = 0$: $\qquad\qquad\qquad\qquad\qquad\qquad C_x = 0$

$+\uparrow \sum F_y = 0$: $-2000 \text{ lb} - 1000 \text{ lb} + 10,000 \text{ lb} + C_y = 0$

$\qquad\qquad\qquad C_y = -7000 \text{ lb}$ $\mathbf{C}_y = 7000 \text{ lb} \downarrow$

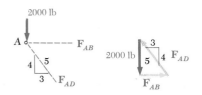

Joint A. This joint is subjected to only two unknown forces, namely, the forces exerted by members AB and AD. A force triangle is used to determine $\mathbf{F}_{AB}$ and $\mathbf{F}_{AD}$. We note that member AB pulls on the joint and thus is in tension and that member AD pushes on the joint and thus is in compression. The magnitudes of the two forces are obtained from the proportion

$$\frac{2000 \text{ lb}}{4} = \frac{F_{AB}}{3} = \frac{F_{AD}}{5}$$

$$F_{AB} = 1500 \text{ lb } T \blacktriangleleft$$
$$F_{AD} = 2500 \text{ lb } C \blacktriangleleft$$

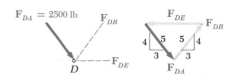

Joint D. Since the force exerted by member AD has been determined, only two unknown forces are now involved at this joint. Again, a force triangle is used to determine the unknown forces in members DB and DE.

$$F_{DB} = F_{DA} \qquad\qquad F_{DB} = 2500 \text{ lb } T \blacktriangleleft$$
$$F_{DE} = 2(\tfrac{3}{5})F_{DA} \qquad F_{DE} = 3000 \text{ lb } C \blacktriangleleft$$

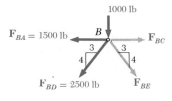

Joint B. Since more than three forces act at this joint, we determine the two unknown forces $\mathbf{F}_{BC}$ and $\mathbf{F}_{BE}$ by solving the equilibrium equations $\Sigma F_x = 0$ and $\Sigma F_y = 0$. We arbitrarily assume that both unknown forces act away from the joint, i.e., that the members are in tension. The positive value obtained for F_{BC} indicates that our assumption was correct; member BC is in tension. The negative value of F_{BE} indicates that our assumption was wrong; member BE is in compression.

$$+\uparrow\Sigma F_y = 0: \qquad -1000 - \tfrac{4}{5}(2500) - \tfrac{4}{5}F_{BE} = 0$$
$$F_{BE} = -3750 \text{ lb} \qquad F_{BE} = 3750 \text{ lb } C \quad \blacktriangleleft$$

$$\overset{+}{\rightarrow}\Sigma F_x = 0: \quad F_{BC} - 1500 - \tfrac{3}{5}(2500) - \tfrac{3}{5}(3750) = 0$$
$$F_{BC} = +5250 \text{ lb} \qquad F_{BC} = 5250 \text{ lb } T \quad \blacktriangleleft$$

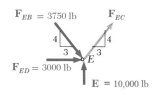

Joint E. The unknown force $\mathbf{F}_{EC}$ is assumed to act away from the joint. Summing x components, we write

$$\overset{+}{\rightarrow}\Sigma F_x = 0: \qquad \tfrac{3}{5}F_{EC} + 3000 + \tfrac{3}{5}(3750) = 0$$
$$F_{EC} = -8750 \text{ lb} \qquad F_{EC} = 8750 \text{ lb } C \quad \blacktriangleleft$$

Summing y components, we obtain a check of our computations:

$$+\uparrow\Sigma F_y = 10{,}000 - \tfrac{4}{5}(3750) - \tfrac{4}{5}(8750)$$
$$= 10{,}000 - 3000 - 7000 = 0 \qquad \text{(checks)}$$

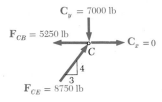

Joint C. Using the computed values of $\mathbf{F}_{CB}$ and $\mathbf{F}_{CE}$, we may determine the reactions $\mathbf{C}_x$ and $\mathbf{C}_y$ by considering the equilibrium of this joint. Since these reactions have already been determined from the equilibrium of the entire truss, we will obtain two checks of our computations. We may also merely use the computed values of all forces acting on the joint (forces in members and reactions) and check that the joint is in equilibrium:

$$\overset{+}{\rightarrow}\Sigma F_x = -5250 + \tfrac{3}{5}(8750) = -5250 + 5250 = 0 \qquad \text{(checks)}$$
$$+\uparrow\Sigma F_y = -7000 + \tfrac{4}{5}(8750) = -7000 + 7000 = 0 \qquad \text{(checks)}$$

PROBLEMS

6.1 through 6.12 Using the method of joints, determine the force in each member of the truss shown. State whether each member is in tension or compression.

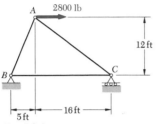

Fig. P6.1

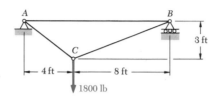

Fig. P6.2

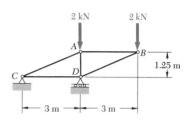

Fig. P6.3

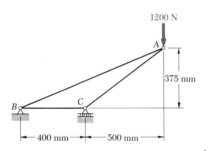

Fig. P6.4

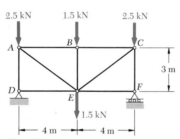

Fig. P6.5

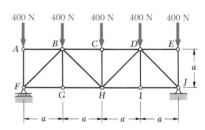

Fig. P6.6

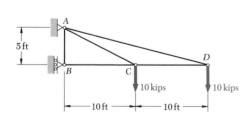

Fig. P6.7

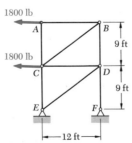

Fig. P6.8

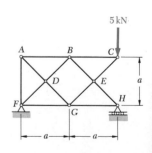

Fig. P6.9

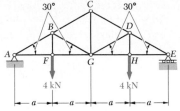

4 kN   4 kN

← a → a → a → a →

Fig. P6.10

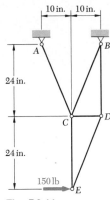

10 in. 10 in.

24 in.

24 in.

150 lb

Fig. P6.11

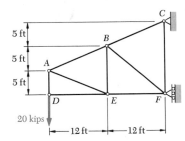

5 ft

5 ft

5 ft

20 kips

←12 ft→←12 ft→

Fig. P6.12

6.13 Determine whether the trusses given in Probs. 6.7, 6.9, 6.14, 6.15, and 6.16 are simple trusses.

6.14 through 6.16 Determine the zero-force members in the truss shown for the given loading.

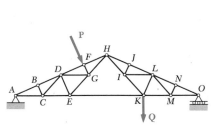

Fig. P6.14

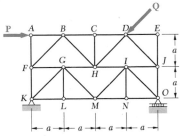

←a→a→a→a→

Fig. P6.15

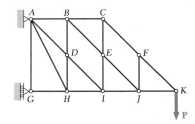

Fig. P6.16

∗6.17 Twelve members, each of length L, are connected to form a regular octahedron. Determine the force in each member if two vertical loads are applied as shown.

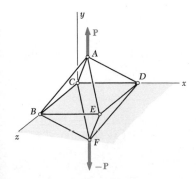

Fig. P6.17

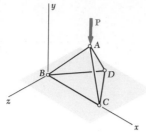

Fig. P6.18

*6.18 Six bars, each of length L, are connected to form a regular tetrahedron which rests on a frictionless horizontal surface. Determine the force in each of the six members when a vertical force **P** is applied at A.

*6.19 The three-dimensional truss is supported by the six reactions shown. If a 240-lb load is applied at E in a direction parallel to the x axis, determine (a) the reactions, (b) the force in each member.

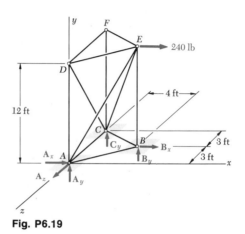

Fig. P6.19

*6.20 Solve Prob. 6.19 assuming that the horizontal 240-lb load is applied at E in a direction parallel to member FE and is directed away from the truss.

6.7. Analysis of Trusses by the Method of Sections.

The method of joints is most effective when the forces in all the members of a truss are to be determined. If, however, the force in only one member or the forces in a very few members are desired, another method, the method of sections, will prove more efficient.

Assume, for example, that we want to determine the force in member BD of the truss shown in Fig. 6.16a. To do this, we must determine the force with which member BD acts on either joint B or joint D. If we were to use the method of joints, we would choose either joint B or joint D as a free body. However, we may also choose as a free body a larger portion of the truss, composed of several joints and members, provided that the desired force is one of the external forces acting on that portion. If, in addition, the portion of the truss is chosen so that there is a total of only three unknown forces acting upon it, the desired force may be obtained by solving the equations of equilibrium for this portion of the truss. In practice, the portion of the truss to be utilized is obtained by *passing a section* through three members of the truss, one of which is the desired member, i.e., by drawing a line which divides the truss into two completely separate parts but does not intersect more than three members. Either of the two portions of the truss obtained after the intersected members have been removed may then be used as a free body.†

In Fig. 6.16a, the section nn has been passed through members BD, BE, and CE, and the portion ABC of the truss is chosen as the free body (Fig. 6.16b). The forces acting on the free body are the loads P_1 and P_2 at points A and B and the three unknown forces $\mathbf{F}_{BD}$, $\mathbf{F}_{BE}$, and $\mathbf{F}_{CE}$. Since it is not known whether the members removed were in tension or compression, the three forces have been arbitrarily drawn away from the free body as if the members were in tension.

The fact that the rigid body ABC is in equilibrium can be expressed by writing three equations which may be solved for the three unknown forces. If only the force $\mathbf{F}_{BD}$ is desired, we need write only one equation, provided that the equation does not contain the other unknowns. Thus the equation $\Sigma M_E = 0$ yields the value of the magnitude F_{BD} of the force $\mathbf{F}_{BD}$. A positive sign in the answer will indicate that our original assumption regarding the sense of $\mathbf{F}_{BD}$ was correct and that member BD is in tension; a negative sign will indicate that our assumption was incorrect and that BD is in compression.

On the other hand, if only the force $\mathbf{F}_{CE}$ is desired, an equation which does not involve $\mathbf{F}_{BD}$ or $\mathbf{F}_{BE}$ should be written; the appropriate equation is $\Sigma M_B = 0$. Again a positive sign for the magnitude F_{CE} of the desired force indicates a correct assumption, hence tension; and a negative sign indicates an incorrect assumption, hence compression.

If only the force $\mathbf{F}_{BE}$ is desired, the appropriate equation is $\Sigma F_y = 0$. Whether the member is in tension or compression is again determined from the sign of the answer.

When the force in only one member is determined, no independent check of the computation is available. However, when

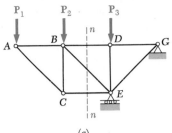

(a)

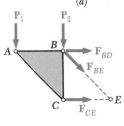

(b)

Fig. 6.16

---

† In the analysis of certain trusses, sections are passed which intersect more than three members; the forces in one, or possibly two, of the intersected members may be obtained if equilibrium equations can be found, each of which involves only one unknown (see Probs. 6.34 to 6.36).

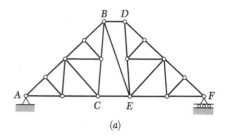

(a)

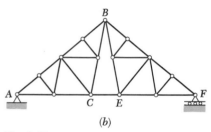

(b)

Fig. 6.17

all the unknown forces acting on the free body are determined, the computations can be checked by writing an additional equation. For instance, if $\mathbf{F}_{BD}$, $\mathbf{F}_{BE}$, and $\mathbf{F}_{CE}$ are determined as indicated above, the computation can be checked by verifying that $\Sigma F_x = 0$.

*6.8. Trusses Made of Several Simple Trusses.

Consider two simple trusses *ABC* and *DEF*. If they are connected by three bars *BD*, *BE*, and *CE* as shown in Fig. 6.17*a*, they will form together a rigid truss *ABDF*. The trusses *ABC* and *DEF* can also be combined into a single rigid truss by joining joints *B* and *D* into a single joint *B* and by connecting joints *C* and *E* by a bar *CE* (Fig. 6.17*b*). The truss thus obtained is known as a Fink truss. It should be noted that the trusses of Fig. 6.17*a* and *b* are *not* simple trusses; they cannot be constructed from a triangular truss by adding successive pairs of members as prescribed in Sec. 6.3. They are rigid trusses, however, as we may check by comparing the systems of connections used to hold the simple trusses *ABC* and *DEF* together (three bars in Fig. 6.17*a*, one pin and one bar in Fig. 6.17*b*) with the systems of supports discussed in Secs. 4.4 and 4.5. Trusses made of several simple trusses rigidly connected are known as *compound trusses*.

It may be checked that in a compound truss the number of members *m* and the number of joints *n* are still related by the formula $m = 2n - 3$. If a compound truss is supported by a frictionless pin and a roller (involving three unknown reactions), the total number of unknowns is $m + 3$ and this number is therefore equal to the number $2n$ of equations obtained by expressing that the *n* pins are in equilibrium. Compound trusses supported by a pin and a roller, or by an equivalent system of supports, are *statically determinate, rigid,* and *completely constrained.* This means that all unknown reactions and forces in members can be determined by the methods of statics and that, all equilibrium equations being satisfied, the truss will neither collapse nor move. The forces in the members, however, cannot all be determined by the method of joints, except by solving a large number of simultaneous equations. In the case of the compound truss of Fig. 6.17*a*, for example, it will be found more expeditious to pass a section through members *BD*, *BE*, and *CE* to determine their forces.

Suppose, now, that the simple trusses *ABC* and *DEF* are connected by *four* bars *BD*, *BE*, *CD*, and *CE* (Fig. 6.18). The number of members *m* is now larger than $2n - 3$; the truss obtained is *overrigid*, and one of the four members *BD*, *BE*, *CD*, or *CE* is said to be *redundant*. If the truss is supported by a pin at *A* and a roller at *F*, the total number of unknowns is $m + 3$.

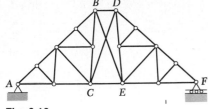

Fig. 6.18

This number is now larger than the number $2n$ of available independent equations; the truss is *statically indeterminate*.

Finally, we shall assume that the two simple trusses ABC and DEF are joined by a pin as shown in Fig. 6.19a. The number of members m is smaller than $2n - 3$. If the truss is supported by a pin at A and a roller at F, the total number of unknowns is $m + 3$. This number is now smaller than the number $2n$ of equilibrium equations which should be satisfied; the truss is *nonrigid* and will collapse under its own weight. However, if two pins are used to support it, the truss becomes *rigid* and will not collapse (Fig. 6.19b). We note that the total number of unknowns is now $m + 4$ and is thus equal to the number of equations. While necessary, this condition, however, is not sufficient for the equilibrium of a structure which ceases to be rigid when detached from its supports (see Sec. 6.11).

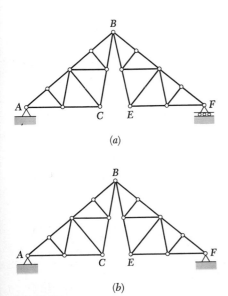

Fig. 6.19

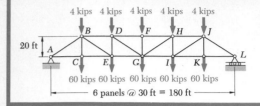

4 kips 4 kips 4 kips 4 kips 4 kips

20 ft

60 kips 60 kips 60 kips 60 kips 60 kips

6 panels @ 30 ft = 180 ft

SAMPLE PROBLEM 6.2

Determine the forces in members *DE* and *HJ* of the truss shown.

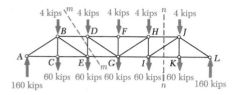

4 kips *m* 4 kips 4 kips 4 kips *n* 4 kips

60 kips 60 kips 60 kips 60 kips 60 kips

160 kips   160 kips

Solution. Considering the entire truss as a free body, we determine the reactions at *A* and *L*.

$$\mathbf{A} = 160 \text{ kips} \uparrow$$
$$\mathbf{L} = 160 \text{ kips} \uparrow$$

Force in Member HJ. Section *nn* is passed through the truss so that it intersects member *HJ* and only two additional members. After the intersected members have been removed, we choose the right-hand portion of the truss as a free body. Three unknown forces are involved; to eliminate the two forces passing through point *I*, we write

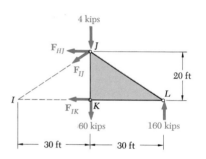

4 kips

F_{HJ}

F_{IJ}

20 ft

F_{IK} *K*   *L*

60 kips   160 kips

30 ft   30 ft

$+\circlearrowleft \Sigma M_I = 0$:

$(160 \text{ kips})(60 \text{ ft}) - (60 \text{ kips})(30 \text{ ft}) - (4 \text{ kips})(30 \text{ ft}) + F_{HJ}(20 \text{ ft}) = 0$
$$F_{HJ} = -384 \text{ kips}$$

The sense of $\mathbf{F}_{HJ}$ was chosen assuming member *HJ* to be in tension; the negative sign obtained indicates that the member is in compression.

$$F_{HJ} = 384 \text{ kips } C \quad \blacktriangleleft$$

Force in Member DE. Section *mm* is passed through the truss so that it intersects member *DE* and only two additional members. After the intersected members have been removed, the left-hand portion of the truss is chosen as a free body. Three unknown forces are again involved; since the equation $\Sigma F_y = 0$ involves only F_{DE} as an unknown, we write

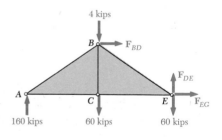

4 kips

B F_{BD}

F_{DE}

A
C   *E* F_{EG}

160 kips   60 kips   60 kips

$+\uparrow \Sigma F_y = 0$:

$+160 \text{ kips} - 60 \text{ kips} - 4 \text{ kips} - 60 \text{ kips} + F_{DE} = 0$
$$F_{DE} = -36 \text{ kips} \qquad F_{DE} = 36 \text{ kips } C \quad \blacktriangleleft$$

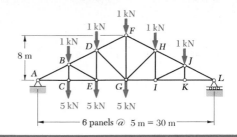

SAMPLE PROBLEM 6.3

Determine the forces in members *FH*, *GH*, and *GI* of the roof truss shown.

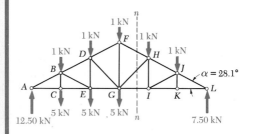

Solution. Section *nn* is passed through the truss as shown. The right-hand portion of the truss will be taken as a free body. Since the reaction at *L* acts on this free body, the value of **L** must be calculated separately, using the entire truss as a free body; the equation $\Sigma M_A = 0$ yields **L** = 7.50 kN ↑.

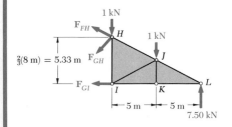

Force in Member GI. Using the portion *HLI* of the truss as a free body, the value of F_{GI} is obtained by writing

$+\!\uparrow\!\Sigma M_H = 0:$
$$(7.50 \text{ kN})(10 \text{ m}) - (1 \text{ kN})(5 \text{ m}) - F_{GI}(5.33 \text{ m}) = 0$$
$$F_{GI} = +13.13 \text{ kN} \qquad F_{GI} = 13.13 \text{ kN } T \quad \blacktriangleleft$$

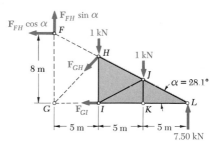

Force in Member FH. The value of F_{FH} is obtained from the equation $\Sigma M_G = 0$. We move $\mathbf{F}_{FH}$ along its line of action until it acts at point *F*, where it is resolved into its *x* and *y* components. The moment of $\mathbf{F}_{FH}$ with respect to point *G* is now equal to $(F_{FH} \cos \alpha)(8 \text{ m})$.

$+\!\uparrow\!\Sigma M_G = 0:$
$$(7.50 \text{ kN})(15 \text{ m}) - (1 \text{ kN})(10 \text{ m}) - (1 \text{ kN})(5 \text{ m}) + (F_{FH} \cos \alpha)(8 \text{ m}) = 0$$
$$F_{FH} = -13.82 \text{ kN} \qquad F_{FH} = 13.82 \text{ kN } C \quad \blacktriangleleft$$

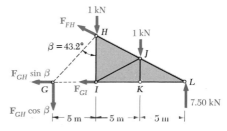

Force in Member GH. The value of F_{GH} is determined by first resolving the force $\mathbf{F}_{GH}$ into *x* and *y* components at point *G* and then solving the equation $\Sigma M_L = 0$.

$+\!\uparrow\!\Sigma M_L = 0: \qquad (1 \text{ kN})(10 \text{ m}) + (1 \text{ kN})(5 \text{ m}) + (F_{GH} \cos \beta)(15 \text{ m}) = 0$
$$F_{GH} = -1.372 \text{ kN} \qquad F_{GH} = 1.372 \text{ kN } C \quad \blacktriangleleft$$

PROBLEMS

6.21 Determine the force in members *BD* and *CD* of the truss shown.

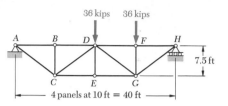

Fig. P6.21 and P6.22

6.22 Determine the force in members *DF* and *DG* of the truss shown.

6.23 Determine the force in members *DF* and *DE* of the truss shown.

6.24 Determine the force in members *CD* and *CE* of the truss shown.

6.25 Determine the force in members *EF* and *DF* of the truss shown.

6.26 Determine the force in member *CD* of the Fink roof truss shown.

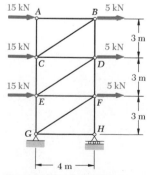

Fig. P6.23, P6.24, and P6.25

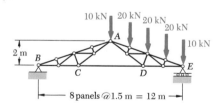

Fig. P6.26

6.27 Determine the force in members *CE*, *CD*, and *BD* of the truss shown.

6.28 Determine the force in members *EG*, *EF*, and *DF* of the truss shown.

Fig. P6.27 and P6.28

6.29 Determine the force in members *FH*, *GH*, and *GI* of the stadium truss shown.

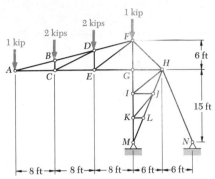

Fig. P6.29 and P6.30

6.30 Determine the force in members *DF*, *DE*, and *CE* of the stadium truss shown.

6.31 Determine the force in members *BD* and *DE* of the truss shown.

6.32 Determine the force in members *FH* and *DH* of the truss shown.

6.33 Determine the force in members *DF* and *EG* of the truss shown.

6.34 Determine the force in members *AB* and *KL* of the truss shown. (*Hint.* Use section *a-a*.)

6.35 Solve Prob. 6.34, assuming that **P** = 0 and that a vertical load **Q** is applied at joint *B*.

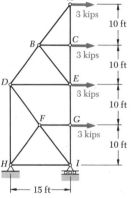

Fig. P6.31, P6.32, and P6.33

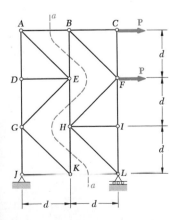

Fig. P6.34

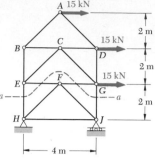

Fig. P6.36

6.36 Determine the force in member *GJ* of the truss shown. (*Hint.* Use section *a-a*.)

6.37 Determine the force in members *AB* and *EJ* of the truss shown, if α = 0°. (*Hint.* Use portion *IBE* of the truss as a free body and apply to joints *C*, *F*, *G*, and *H* the results obtained in Sec. 6.5.)

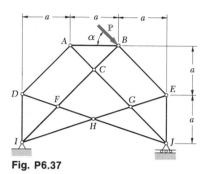

Fig. P6.37

6.38 Solve Prob. 6.37, assuming that α = 90°.

6.39 The diagonal members in the center panel of the truss shown are very slender and can act only in tension; such members are known as *counters*. Determine the force in members *BD* and *CE* and in the counter which is acting under the given loading.

6.40 Solve Prob. 6.39, assuming that the 60-kN load has been removed.

6.41 Determine the force in members *BE* and *CG* and in the counter which is acting under the given loading. (See Prob. 6.39 for the definition of a counter.)

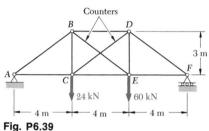

Fig. P6.39

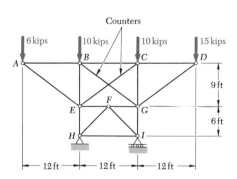

Fig. P6.41

6.42 Solve Prob. 6.41, assuming that the 15-kip load has been removed.

∗6.43 through 6.46 Classify each of the given structures as completely, partially, or improperly constrained; if completely constrained, further classify as determinate or indeterminate. (All members can act both in tension and in compression.)

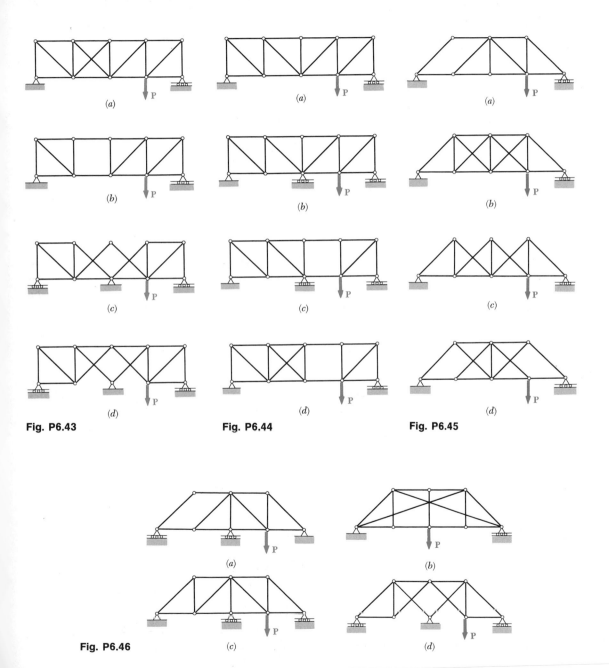

Fig. P6.43

Fig. P6.44

Fig. P6.45

Fig. P6.46

FRAMES AND MACHINES

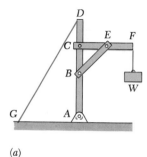

(a)

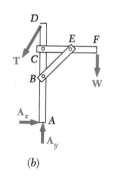

(b)

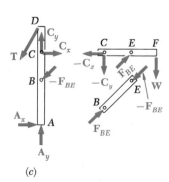

(c)

Fig. 6.20

6.9. Structures Containing Multiforce Members. Under Trusses, we have considered structures consisting entirely of pins and of straight two-force members. The forces acting on the two-force members were known to be directed along the members themselves. We shall now consider structures in which at least one of the members is a *multiforce* member, i.e., a member acted upon by three or more forces. These forces will generally not be directed along the members on which they act; their direction is unknown, and they should be represented therefore by two unknown components.

Frames and machines are structures containing multiforce members. *Frames* are designed to support loads and are usually stationary, fully constrained structures. *Machines* are designed to transmit and modify forces; they may or may not be stationary and will always contain moving parts.

6.10. Analysis of a Frame. As a first example of analysis of a frame, we shall consider again the crane described in Sec. 6.1, which carries a given load W (Fig. 6.20a). The free-body diagram of the entire frame is shown in Fig. 6.20b. This diagram may be used to determine the external forces acting on the frame. Summing moments about A, we first determine the force $\mathbf{T}$ exerted by the cable; summing x and y components, we then determine the components $\mathbf{A}_x$ and $\mathbf{A}_y$ of the reaction at the pin A.

In order to determine the internal forces holding the various parts of a frame together, we must dismember the frame and draw a free-body diagram for each of its component parts (Fig. 6.20c). First, the two-force members should be considered. In this frame, member BE is the only two-force member. The forces acting at each end of this member must have the same magnitude, same line of action, and opposite sense (Sec. 4.6). They are therefore directed along BE and will be denoted, respectively, by $\mathbf{F}_{BE}$ and $-\mathbf{F}_{BE}$. Their sense will be arbitrarily assumed as shown in Fig. 6.20c, and the correctness of this assumption will be checked later by the sign obtained for the common magnitude F_{BE} of the two forces.

Next, we consider the multiforce members, i.e., the members which are acted upon by three or more forces. According to Newton's third law, the force exerted at B by member BE on member AD must be equal and opposite to the force $\mathbf{F}_{BE}$ exerted by AD on BE. Similarly, the force exerted at E by member BE on member CF must be equal and opposite to the force $-\mathbf{F}_{BE}$

exerted by *CF* on *BE*. The forces that the two-force member *BE* exerts on *AD* and *CF* are therefore respectively equal to $-\mathbf{F}_{BE}$ and $\mathbf{F}_{BE}$; they have the same magnitude F_{BE} and opposite sense, and should be directed as shown in Fig. 6.20c.

At *C* two multiforce members are connected. Since neither the direction nor the magnitude of the forces acting at *C* is known, these forces will be represented by their *x* and *y* components. The components $\mathbf{C}_x$ and $\mathbf{C}_y$ of the force acting on member *AD* will be arbitrarily directed to the right and upward. Since, according to Newton's third law, the forces exerted by member *CF* on *AD* and by member *AD* on *CF* are equal and opposite, the components of the force acting on member *CF must* be directed to the left and downward; they will be denoted, respectively, by $-\mathbf{C}_x$ and $-\mathbf{C}_y$. Whether the force $\mathbf{C}_x$ is actually directed to the right and the force $-\mathbf{C}_x$ is actually directed to the left will be determined later from the sign of their common magnitude C_x, a plus sign indicating that the assumption made was correct, and a minus sign that it was wrong. The free-body diagrams of the multiforce members are completed by showing the external forces acting at *A*, *D*, and *F*.†

The internal forces may now be determined by considering the free-body diagram of either of the two multiforce members. Choosing the free-body diagram of *CF*, for example, we write the equations $\Sigma M_C = 0$, $\Sigma M_E = 0$, and $\Sigma F_x = 0$, which yield the values of the magnitudes F_{BE}, C_y, and C_x, respectively. These values may be checked by verifying that member *AD* is also in equilibrium.

It should be noted that the free-body diagrams of the pins were not shown in Fig. 6.20c. This was because the pins were assumed to form an integral part of one of the two members they connected. This assumption can always be used to simplify the analysis of frames and machines. When a pin connects three or more members, however, or when a pin connects a support and two or more members, or when a load is applied to a pin, a clear decision must be made in choosing the member to which the pin will be assumed to belong. (If multiforce members are involved, the pin should be attached to one of these members.) The various forces exerted on the pin should then be clearly identified. This is illustrated in Sample Prob. 6.6.

---

† The use of a minus sign to distinguish the force exerted by one member on another from the equal and opposite force exerted by the second member on the first is not strictly necessary, since the two forces belong to different free-body diagrams and thus cannot easily be confused. In the Sample Problems, we shall represent by the same symbol equal and opposite forces which are applied to different free bodies.

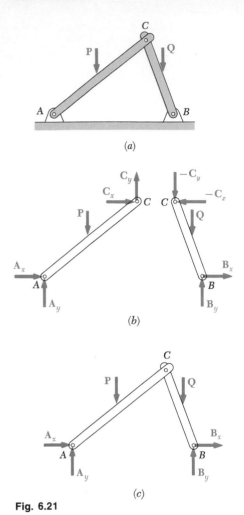

(a)

(b)

(c)

Fig. 6.21

6.11. Frames Which Cease to Be Rigid When Detached from Their Supports.

The crane analyzed in Sec. 6.10 was so constructed that it could keep the same shape without the help of its supports; it was therefore considered as a rigid body. Many frames, however, will collapse if detached from their supports; such frames cannot be considered as rigid bodies. Consider, for example, the frame shown in Fig. 6.21a, which consists of two members AC and CB carrying loads P and Q at their midpoints; the members are supported by pins at A and B and are connected by a pin at C. If detached from its supports, this frame will not maintain its shape; it should therefore be considered as made of *two distinct rigid parts AC and CB*.

The equations $\Sigma F_x = 0$, $\Sigma F_y = 0$, $\Sigma M = 0$ (about any given point) express the conditions for the *equilibrium of a rigid body* (Chap. 4); we should use them, therefore, in connection with the free-body diagrams of rigid bodies, namely, the free-body diagrams of members AC and CB (Fig. 6.21b). Since these members are multiforce members, and since pins are used at the supports and at the connection, the reactions at A and B and the forces at C will each be represented by two components. In accordance with Newton's third law, the components of the force exerted by CB on AC and the components of the force exerted by AC on CB will be represented by vectors of the same magnitude and opposite sense; thus, if the first pair of components consists of C_x and C_y, the second pair will be represented by $-C_x$ and $-C_y$. We note that four unknown force components act on free body AC, while only three independent equations may be used to express that the body is in equilibrium; similarly, four unknowns, but only three equations, are associated with CB. However, only six different unknowns are involved in the analysis of the two members, and altogether six equations are available to express that the members are in equilibrium. Writing $\Sigma M_A = 0$ for free body AC and $\Sigma M_B = 0$ for CB, we obtain two simultaneous equations which may be solved for the common magnitude C_x of the components C_x and $-C_x$, and for the common magnitude C_y of the components C_y and $-C_y$. Writing, then, $\Sigma F_x = 0$ and $\Sigma F_y = 0$ for each of the two free bodies, we obtain successively the magnitudes A_x, A_y, B_x, and B_y.

We shall observe now that, since the equations of equilibrium $\Sigma F_x = 0$, $\Sigma F_y = 0$, $\Sigma M = 0$ (about any given point) are satisfied by the forces acting on free body AC, and since they are also satisfied by the forces acting on free body CB, they must be satisfied when the forces acting on the two free bodies are considered simultaneously. Since the internal forces at C cancel each other, we find that the equations of equilibrium must be satisfied by the external forces shown on the free-body diagram of the frame ACB itself (Fig. 6.21c), although the frame is not a rigid body. These equations may be used to determine some of the components of the reactions at A and B. We shall note, however, that *the reactions cannot be completely determined from the free-body diagram of the whole frame.* It is thus necessary to dismember the frame and to consider the free-body diagrams of its component parts (Fig. 6.21b), even when we are interested only in finding external reactions. This may be explained by the fact that the equilibrium equations obtained for free body ACB are *necessary conditions* for the equilibrium of a nonrigid structure, *but not sufficient conditions.*

The method of solution outlined in the second paragraph of this section involved simultaneous equations. We shall now discuss a more expeditious method, which utilizes the free body ACB as well as the free bodies AC and CB. Writing $\Sigma M_A = 0$ and $\Sigma M_B = 0$ for free body ACB, we obtain B_y and A_y. Writing $\Sigma M_C = 0$, $\Sigma F_x = 0$, and $\Sigma F_y = 0$ for free body AC, we obtain successively A_x, C_x, and C_y. Finally, writing $\Sigma F_x = 0$ for ACB, we obtain B_x.

We noted above that the analysis of the frame of Fig. 6.21 involves six unknown force components and six independent equilibrium equations (the equilibrium equations for the whole frame were obtained from the original six equations and, therefore, are not independent). Moreover, we checked that all unknowns could be actually determined and that all equations could be satisfied. The frame considered is *statically determinate and rigid.*† In general, to determine whether a structure is statically determinate and rigid, we should draw a free-body diagram for each of its component parts and count the reactions and internal forces involved. We should also determine the number of independent equilibrium equations (excluding equations expressing the equilibrium of the whole structure or of groups of component parts already analyzed). If there are more unknowns than equations, the structure is *statically indeterminate.* If there are fewer unknowns than equations, the structure is *nonrigid.* If there are as many unknowns as equations, *and if all unknowns may be determined and all equations satisfied* under general loading conditions, the structure is *statically determinate and rigid;* if, however, due to an *improper arrangement* of members and supports, all unknowns cannot be determined and all equations cannot be satisfied, the structure is *statically indeterminate and nonrigid.*

---

† The word "rigid" is used here to indicate that the frame will maintain its shape as long as it remains attached to its supports.

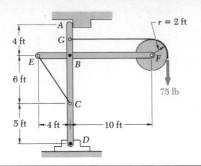

SAMPLE PROBLEM 6.4

In the small frame shown, members *EBF* and *ABCD* are connected by a pin at *B* and by the cable *EC*. A 75-lb load is supported by a second cable which passes over a pulley at *F* and is attached to the vertical member at *G*. Determine the tension in cable *EC* and the components of the pin reaction at *B*.

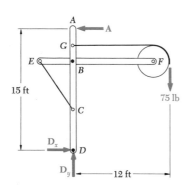

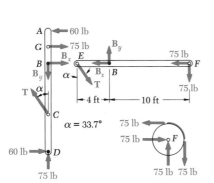

Entire Frame. The external reactions on the frame involve three unknowns; these reactions are determined by taking the entire frame (including the cables) as a free body.

$+\uparrow\Sigma F_y = 0$: $D_y - 75\,\text{lb} = 0$
$$D_y = +75\,\text{lb} \qquad \mathbf{D}_y = 75\,\text{lb} \uparrow$$

$+\curvearrowleft\Sigma M_D = 0$: $-(75\,\text{lb})(12\,\text{ft}) + A(15\,\text{ft}) = 0$
$$A = +60\,\text{lb} \qquad \mathbf{A} = 60\,\text{lb} \leftarrow$$

$\xrightarrow{+}\Sigma F_x = 0$: $-60\,\text{lb} + D_x = 0$
$$D_x = +60\,\text{lb} \qquad \mathbf{D}_x = 60\,\text{lb} \rightarrow$$

Since the values obtained are positive, the forces are directed as assumed in the diagram, that is, $\mathbf{D}_x$ to the right, $\mathbf{D}_y$ up, and $\mathbf{A}$ to the left.

Members. The frame is dismembered; since only two members are connected at *B*, the components of the unknown forces acting on *EBF* and *ABCD* at *B* are, respectively, equal and opposite. The forces exerted at *E* and *C* by the cable *EC* are equal and opposite, and their direction is known. From the free-body diagram of the pulley, it is seen that the force exerted at *F* by the pulley on member *EBF* may be resolved into two 75-lb components as shown. The cable also exerts a 75-lb force on *ABCD* at point *G*.

Member EBF. Using the free body *EBF*, we write

$+\curvearrowleft\Sigma M_E = 0$: $B_y(4\,\text{ft}) - (75\,\text{lb})(14\,\text{ft}) = 0$ $B_y = +263\,\text{lb}$ ◀

$+\curvearrowleft\Sigma M_B = 0$: $(T\cos\alpha)(4\,\text{ft}) - (75\,\text{lb})(10\,\text{ft}) = 0$
$$T = +225\,\text{lb} \quad ◀$$

$\xrightarrow{+}\Sigma F_x = 0$: $+T\sin\alpha - B_x - 75\,\text{lb} = 0$ $B_x = +50.0\,\text{lb}$ ◀

Since the values obtained are positive, the forces are directed as shown on the diagram: the forces $\mathbf{B}_x$ and $\mathbf{B}_y$ acting on member *ABCD* are directed, respectively, to the right and down, while the forces $\mathbf{B}_x$ and $\mathbf{B}_y$ acting on member *EBF* are directed, respectively, to the left and up.

Member ABCD (Check). The computations are checked by considering the free body *ABCD*. For example,

$\xrightarrow{+}\Sigma F_x = -60\,\text{lb} + 75\,\text{lb} + B_x - T\sin\alpha + 60\,\text{lb}$
$$= -60\,\text{lb} + 75\,\text{lb} + 50\,\text{lb} - (225\,\text{lb})\sin 33.7° + 60\,\text{lb} = 0$$
$$\text{(checks)}$$

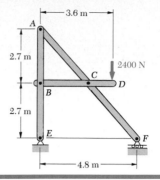

Determine the components of the forces acting on each member of the frame shown.

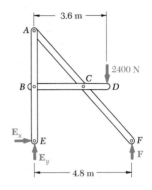

Entire Frame. Since the external reactions involve only three unknowns, we compute the reactions by considering the free-body diagram of the entire frame.

$$+\!\uparrow\Sigma M_E = 0: \qquad -(2400\text{ N})(3.6\text{ m}) + F(4.8\text{ m}) = 0$$
$$F = +1800\text{ N} \qquad\qquad \mathbf{F} = 1800\text{ N}\uparrow \blacktriangleleft$$

$$+\!\uparrow\Sigma F_y = 0: \qquad -2400\text{ N} + 1800\text{ N} + E_y = 0$$
$$E_y = +600\text{ N} \qquad\qquad \mathbf{E}_y = 600\text{ N}\uparrow \blacktriangleleft$$

$$\xrightarrow{+} \Sigma F_x = 0: \qquad\qquad\qquad\qquad\qquad \mathbf{E}_x = 0 \blacktriangleleft$$

The frame is now dismembered; since only two members are connected at each joint, equal and opposite components are shown on each member at each joint.

Member BCD

$$+\!\uparrow\Sigma M_B = 0: \ -(2400\text{ N})(3.6\text{ m}) + C_y(2.4\text{ m}) = 0 \quad C_y = +3600\text{ N} \blacktriangleleft$$

$$+\!\uparrow\Sigma M_C = 0: \ -(2400\text{ N})(1.2\text{ m}) + B_y(2.4\text{ m}) = 0 \quad B_y = +1200\text{ N} \blacktriangleleft$$

$$\xrightarrow{+} \Sigma F_x = 0: \qquad -B_x + C_x = 0$$

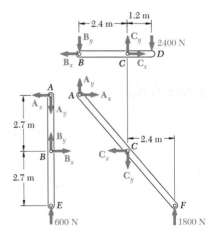

We note that neither B_x nor C_x can be obtained by considering only member *BCD*. The positive values obtained for B_y and C_y indicate that the force components $\mathbf{B}_y$ and $\mathbf{C}_y$ are directed as assumed.

Member ABE

$$+\!\uparrow\Sigma M_A = 0: \qquad B_x(2.7\text{ m}) = 0 \qquad\qquad B_x = 0 \blacktriangleleft$$

$$\xrightarrow{+} \Sigma F_x = 0: \qquad +B_x - A_x = 0 \qquad\qquad A_x = 0 \blacktriangleleft$$

$$+\!\uparrow\Sigma F_y = 0: \qquad -A_y + B_y + 600\text{ N} = 0$$
$$-A_y + 1200\text{ N} + 600\text{ N} = 0 \quad A_y = +1800\text{ N} \blacktriangleleft$$

Member BCD. Returning now to member *BCD*, we write

$$\xrightarrow{+} \Sigma F_x = 0: \qquad -B_x + C_x = 0 \qquad 0 + C_x = 0 \qquad C_x = 0 \blacktriangleleft$$

Member ACF (Check). All unknown components have now been found; to check the results, verify that member *ACF* is in equilibrium.

$$+\!\uparrow\Sigma M_C = (1800\text{ N})(2.4\text{ m}) - A_y(2.4\text{ m}) - A_x(2.7\text{ m})$$
$$= (1800\text{ N})(2.4\text{ m}) - (1800\text{ N})(2.4\text{ m}) - 0 = 0 \qquad \text{(checks)}$$

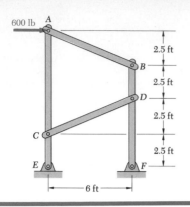

600 lb

A

2.5 ft

B

2.5 ft

D

2.5 ft

C

2.5 ft

E   F

6 ft

SAMPLE PROBLEM 6.6

A 600-lb horizontal force is applied to pin A of the frame shown. Determine the forces acting on the two vertical members of the frame.

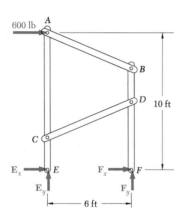

600 lb

A

B

D   10 ft

C

E_x E F_x F

E_y F_y

6 ft

Entire Frame. The entire frame is chosen as a free body; although the reactions involve four unknowns, $\mathbf{E}_y$ and $\mathbf{F}_y$ may be determined by writing

$+\uparrow\Sigma M_E = 0$: $-(600\text{ lb})(10\text{ ft}) + F_y(6\text{ ft}) = 0$
$F_y = +1000\text{ lb}$ $\mathbf{F}_y = 1000\text{ lb} \uparrow$ ◄

$+\uparrow\Sigma F_y = 0$: $E_y + F_y = 0$
$E_y = -1000\text{ lb}$ $\mathbf{E}_y = 1000\text{ lb} \downarrow$ ◄

The equations of equilibrium of the entire frame are not sufficient to determine $\mathbf{E}_x$ and $\mathbf{F}_x$. The equilibrium of the various members must now be considered in order to proceed with the solution. In dismembering the frame we shall assume that pin A is attached to the multiforce member ACE and, thus, that the 600-lb force is applied to that member. We also note that AB and CD are two-force members.

Member ACE

$+\uparrow\Sigma F_y = 0$: $-\tfrac{5}{13}F_{AB} + \tfrac{5}{13}F_{CD} - 1000\text{ lb} = 0$

$+\uparrow\Sigma M_E = 0$: $-(600\text{ lb})(10\text{ ft}) - (\tfrac{12}{13}F_{AB})(10\text{ ft}) - (\tfrac{12}{13}F_{CD})(2.5\text{ ft}) = 0$

Solving these equations simultaneously, we find

$$F_{AB} = -1040\text{ lb} \qquad F_{CD} = +1560\text{ lb}$$ ◄

The signs obtained indicate that the sense assumed for F_{CD} was correct and the sense for F_{AB} incorrect. Summing now x components:

$\overset{+}{\rightarrow}\Sigma F_x = 0$: $600\text{ lb} + \tfrac{12}{13}(-1040\text{ lb}) + \tfrac{12}{13}(+1560\text{ lb}) + E_x = 0$
$E_x = -1080\text{ lb}$ $\mathbf{E}_x = 1080\text{ lb} \leftarrow$ ◄

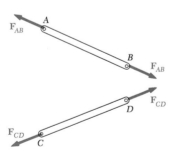

F_{AB}

A

B

F_{AB}

F_{CD}

D

F_{CD}

C

Entire Frame. Since $\mathbf{E}_x$ has been determined, we may return to the free-body diagram of the entire frame and write

$\overset{+}{\rightarrow}\Sigma F_x = 0$: $600\text{ lb} - 1080\text{ lb} + F_x = 0$
$F_x = +480\text{ lb}$ $\mathbf{F}_x = 480\text{ lb} \rightarrow$ ◄

Member BDF (Check). We may check our computations by verifying that the equation $\Sigma M_B = 0$ is satisfied by the forces acting on member BDF.

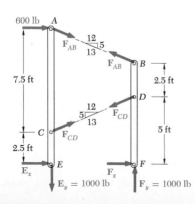

600 lb   A

12 / 5 / 13 F_{AB}

F_{AB} B

7.5 ft

2.5 ft

D

12 / 5 / 13 F_{CD}

5 ft

C F_{CD}

2.5 ft

E_x E F_x F

$E_y = 1000\text{ lb}$ $F_y = 1000\text{ lb}$

$+\uparrow\Sigma M_B = -(\tfrac{12}{13}F_{CD})(2.5\text{ ft}) + (F_x)(7.5\text{ ft})$
$= -\tfrac{12}{13}(1560\text{ lb})(2.5\text{ ft}) + (480\text{ lb})(7.5\text{ ft})$
$= -3600\text{ lb}\cdot\text{ft} + 3600\text{ lb}\cdot\text{ft} = 0$ (checks)

PROBLEMS

6.47 and 6.48 Determine the force in member *BD* and the components of the reaction at *C*.

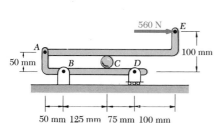

Fig. P6.47

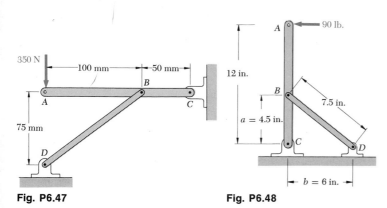

Fig. P6.48

6.49 Determine the components of all forces acting on member *AE* of the frame shown.

6.50 Determine the components of all forces acting on member *ABCD* of the assembly shown.

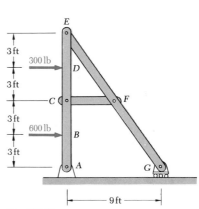

Fig. P6.49

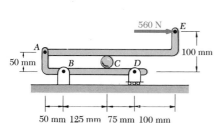

Fig. P6.50

6.51 Solve Prob. 6.50, assuming that the 560-N force applied at *E* is directed vertically downward.

6.52 The low-bed trailer shown is designed so that the rear end of the bed can be lowered to ground level in order to facilitate the loading of equipment or wrecked vehicles. A 2500-lb vehicle has been hauled to the position shown by a winch; the trailer is then returned to a traveling position where $\alpha = 0$ and both *AB* and *BE* are horizontal. Considering only the weight of the disabled automobile, determine the force which must be exerted by the hydraulic cylinder to maintain a position with $\alpha = 0$.

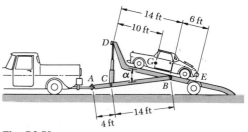

Fig. P6.52

6.53 For the marine crane shown, which is used in offshore drilling operations, determine (*a*) the force in link *CD*, (*b*) the force in the brace *AC*, (*c*) the force exerted at *A* on the boom *AB*.

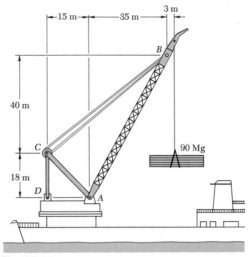

Fig. P6.53

6.54 Determine the components of the force exerted at *B* on member *BE* (*a*) if the 400-N load is applied as shown, (*b*) if the 400-N load is moved along its line of action and is applied at point *F*.

6.55 Determine the forces exerted on member *AB* if the frame is loaded by a clockwise couple of magnitude 6 N · m applied (*a*) at point *D*, (*b*) at point *E*. (*c*) Determine the forces exerted on member *AB* if the frame is loaded by vertical forces applied at *D* and *E* which are equivalent to a 6-N · m clockwise couple.

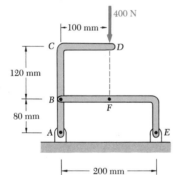

Fig. P6.54

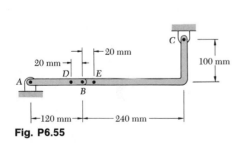

Fig. P6.55

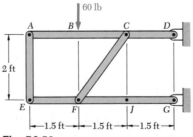

Fig. P6.56

6.56 Determine the components of all forces acting on member *EFG* of the frame shown.

6.57 The hydraulic cylinder *CD*, which partially controls the position of rod *AB*, has been locked in the position shown. Knowing that *P* = 500 N and *θ* = 15°, determine (*a*) the force in link *EF*, (*b*) the force exerted on member *ADF* at point *D*.

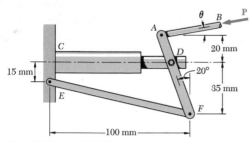

Fig. P6.57 and P6.58

6.58 The hydraulic cylinder has been locked in the position shown. Knowing that *θ* = 10°, determine (*a*) the force **P** for which the tension in link *EF* is 450 N, (*b*) the corresponding force exerted on member *ADF* at point *D*.

6.59 and 6.60 Determine the reactions at the supports of the beam shown.

6.61 A pipe weighs 40 lb/ft and is supported every 30 ft by a small frame; a typical frame is shown. Knowing that *θ* = 30°, determine the components of the reactions and the components of the force exerted at *B* on member *AB*.

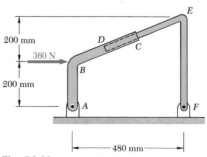

Fig. P6.59

Fig. P6.60

Fig. P6.61

6.62 In Prob. 6.61, determine (*a*) the value of *θ* for which the reaction at *A* is vertical, (*b*) the corresponding forces acting on member *ADB*.

6.63 The bent rod *DEF* fits into the bent pipe *ABC* as shown. Neglecting the effect of friction, determine the reactions at *A* and *F* due to the 360-N force applied at *B*.

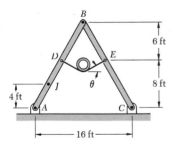

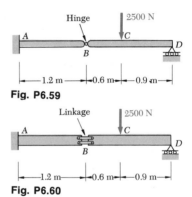

Fig. P6.63

6.64 The tractor and scraper units shown are connected by a vertical pin located 0.6 m behind the tractor wheels. The distance from C to D is 0.75 m. The center of gravity of the 10-Mg tractor unit is located at G_t. The scraper unit and load have a total mass of 50 Mg and a combined center of gravity located at G_s. Knowing that the machine is at rest, with its brakes released, determine (a) the reactions at each of the four wheels, (b) the forces exerted on the tractor unit at C and D.

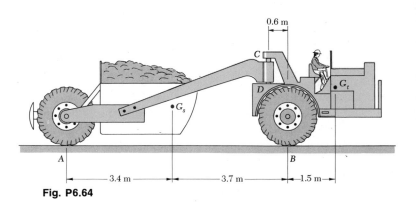

Fig. P6.64

6.65 A trailer weighing 2750 lb is attached to a 3200-lb automobile by a ball-and-socket trailer hitch at D. Determine (a) the reactions at each of the six wheels when the automobile and trailer are at rest, (b) the additional load on each of the automobile wheels due to the trailer.

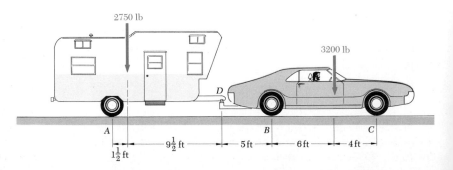

Fig. P6.65

6.66 In order to obtain a better weight distribution over the four wheels of the automobile of Prob. 6.65, a compensating hitch of the type shown is used to attach the trailer to the automobile. This hitch consists of two bar springs (only one is shown in the figure) which fit into bearings inside a support rigidly attached to the automobile. The springs are also connected by chains to the trailer frame, and specially designed hooks make it possible to place both chains under a tension *T*. Solve Prob. 6.65 assuming that such a compensating hitch is used and that the tension *T* in each chain is 440 lb.

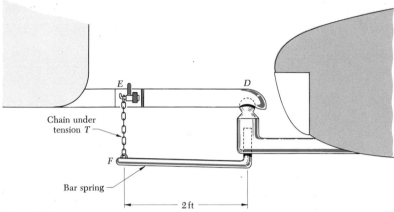

Fig. P6.66

6.67 Determine the components of all forces acting on member *CDEF*.

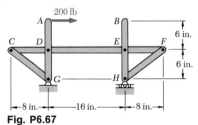

6.68 Solve Prob. 6.67, assuming that the 200-lb force is attached at *B* instead of *A* and is directed horizontally to the right.

Fig. P6.67

6.69 A vertical load **P** of magnitude 600 N is applied to member *AB*. Member *AB* is placed between two frictionless walls and is pin-connected at *C* to a link *CD*. Determine all forces exerted on member *AB*.

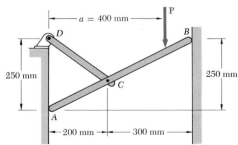

6.70 In Prob. 6.69, determine the range of values of the distance *a* for which the load **P** can be supported.

Fig. P6.69

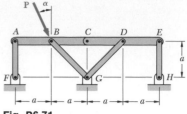

Fig. P6.71

6.71 Members ABC and CDE are pin-connected at C and supported by the four links AF, BG, GD, and EH. Determine the force in each link when $\alpha = 0°$.

6.72 Solve Prob. 6.71 when $\alpha = 90°$.

6.73 Solve Prob. 6.71, assuming that the force $\mathbf{P}$ is replaced by a clockwise couple of moment $\mathbf{M}_0$ applied to member CDE at D.

6.74 The axis of the three-hinged arch ABC is a parabola with vertex at B. Knowing that $P = 20$ kN and $Q = 0$, determine (a) the components of the reaction at C, (b) the components of the force exerted at B on segment AB.

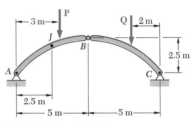

Fig. P6.74 and P6.75

6.75 The axis of the three-hinged arch ABC is a parabola with vertex at B. Knowing that $P = 20$ kN and $Q = 10$ kN, determine (a) the components of the reaction at C, (b) the components of the force exerted at B on segment AB.

6.76 (a) Show that, when a frame supports a pulley at A, an equivalent loading of the frame and of each of its component parts may be obtained by removing the pulley and applying at A two forces equal and parallel to the forces of tension in the cable. (b) Further, show that, if one end of the cable is attached to the frame at a point B, a force of magnitude equal to the tension should also be applied at B.

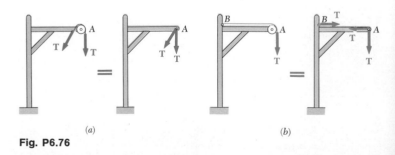

(a)   (b)

Fig. P6.76

6.77 through 6.79 Determine the reactions at the supports for each of the trusses shown. Indicate whether the truss is rigid. The height of each truss is 4 m; the length of each panel is 4 m; and the magnitude of **P** is 40 kN.

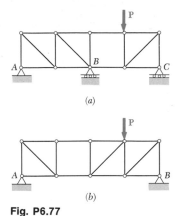

(a)

(b)

Fig. P6.77

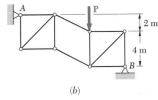

(a)

(b)

Fig. P6.78

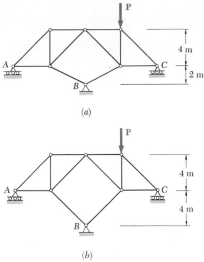

(a)

(b)

Fig. P6.79

6.80 The rigid bar *EFG* is supported by the truss system shown. Determine the force in each of the two-force members.

6.81 Solve Prob. 6.80, assuming that the 1200-lb load is replaced by a clockwise couple of magnitude 4800 lb · ft applied to member *EFG* at *F*.

6.82 The frame shown consists of two members *ABC* and *DEF* connected by the links *AE* and *BF*. Determine the reactions at *C* and *D* and the force in links *AE* and *BF*.

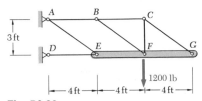

Fig. P6.80

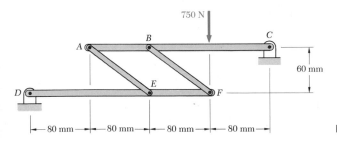

Fig. P6.82

6.83 Solve Prob. 6.82, assuming that the 750-N load is replaced by a counterclockwise couple of magnitude 60 N · m applied to member *ABC*.

6.84 Solve Prob. 6.82, assuming that the 750-N load is applied at *F*.

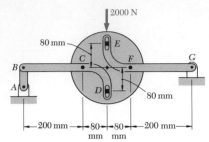

Fig. P6.85

6.85 Two arms BCD and EFG are connected to a 200-mm-diameter disk by four pins which are attached to the disk. Assuming that the pins at D and E may slide in the vertical slots, determine the components of all forces exerted on the disk when a 2000-N load is applied to the disk as shown.

6.86 Solve Prob. 6.85, assuming that the 2000-N load is applied to the top edge of member EFG.

6.87 In Prob. 6.48, knowing that the length of rod BD must be 7.5 in., determine (a) other values of a and b for which the tension in BD has the same value as when $a = 4.5$ in. and $b = 6$ in., (b) the values of a and b for which the tension in BD is as small as possible.

6.88 Determine the vertical reactions at points A, B, C, and D of the runway girders for the overhead traveling crane shown if $P = 12$ kips, $a = 15$ ft, $L = 40$ ft, $b = 10$ ft, and $l = 30$ ft.

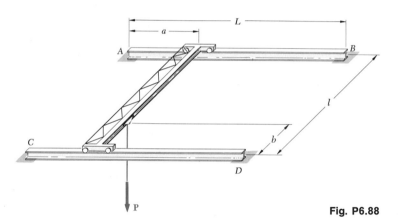

Fig. P6.88

6.89 Determine the vertical reactions at points A and B of the runway girder AB of Prob. 6.88 in terms of P, a, b, l, and L.

6.90 Four beams, each of length $2a$, are nailed together at their midpoints to form the support system shown. Assuming that only vertical forces are exerted at the connections, determine the vertical reactions at A, D, E, and H.

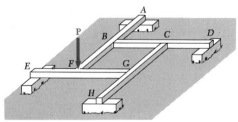

Fig. P6.90

6.91 Three beams, each of length L, are held together by single nails at A, B, and C. Assuming that a single vertical force is exerted at each of the connections, determine the vertical reactions at D, E, and F.

6.92 Solve Prob. 6.91 if $a = 0.6$ m, $b = 1.2$ m, and $P = 600$ N.

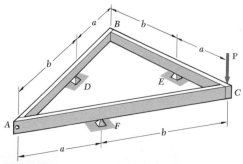

Fig. P6.91

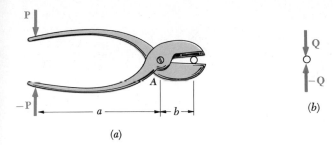

(a)

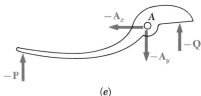

(b)

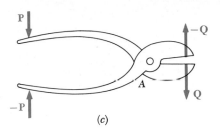

(c)

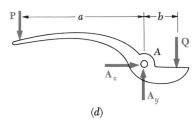

(d)

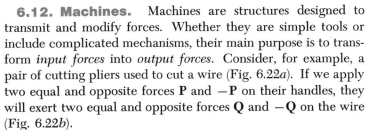

(e)

Fig. 6.22

6.12. Machines.   Machines are structures designed to transmit and modify forces. Whether they are simple tools or include complicated mechanisms, their main purpose is to transform *input forces* into *output forces*. Consider, for example, a pair of cutting pliers used to cut a wire (Fig. 6.22a). If we apply two equal and opposite forces **P** and −**P** on their handles, they will exert two equal and opposite forces **Q** and −**Q** on the wire (Fig. 6.22b).

To determine the magnitude Q of the output forces when the magnitude P of the input forces is known (or, conversely, to determine P when Q is known), we draw a free-body diagram of the pliers *alone*, showing the input forces **P** and −**P** and the *reactions* −**Q** and **Q** that the wire exerts on the pliers (Fig. 6.22c). However, since a pair of pliers forms a nonrigid structure, we must use one of the component parts as a free body in order to determine the unknown forces. Considering Fig. 6.22d, for example, and taking moments about A, we obtain the relation $Pa = Qb$, which defines the magnitude Q in terms of P or P in terms of Q. The same free-body diagram may be used to determine the components of the internal force at A; we find $A_x = 0$ and $A_y = P + Q$.

In the case of more complicated machines, it generally will be necessary to use several free-body diagrams and, possibly, to solve simultaneous equations involving various internal forces. The free bodies should be chosen to include the input forces and the reactions to the output forces, and the total number of unknown force components involved should not exceed the number of available independent equations. While it is advisable to check whether the problem is determinate before attempting to solve it, there is no point in discussing the rigidity of a machine. A machine includes moving parts and thus must be nonrigid.

A hydraulic-lift table is used to raise a 1000-kg crate. It consists of a platform and two identical linkages on which hydraulic cylinders exert equal forces. (Only one linkage and one cylinder are shown.) Members EDB and CG are each of length $2a$, and member AD is pinned to the midpoint of EDB. If the crate is placed on the table, so that half of its weight is supported by the system shown, determine the force exerted by each cylinder in raising the crate for $\theta = 60°$, $a = 0.70$ m, and $L = 3.20$ m. Show that the result obtained is independent of the distance d.

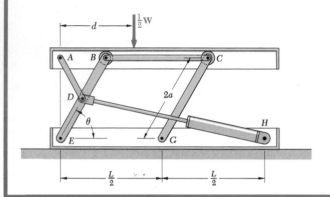

Solution. The machine considered consists of the platform and of the linkage, with an input force $\mathbf{F}_{DH}$ exerted by the cylinder and an output force equal and opposite to $\frac{1}{2}\mathbf{W}$. Since more than three unknowns would be involved, the entire mechanism is not used as a free body. The mechanism is dismembered and free-body diagrams are drawn for platform ABC, for roller C, and for member EDB. Since AD, BC, and CG are two-force members, their free-body diagrams have been omitted, and the forces they exert on the other parts of the mechanism have been drawn parallel to these members.

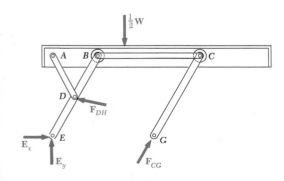

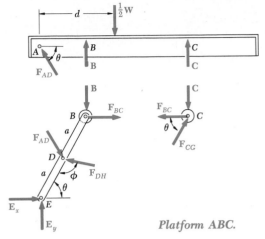

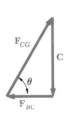

Platform ABC.

$$\xrightarrow{+} \Sigma F_x = 0: \qquad -F_{AD} \cos \theta = 0 \qquad F_{AD} = 0$$
$$+\uparrow \Sigma F_y = 0: \qquad B + C - \tfrac{1}{2}W = 0 \qquad B + C = \tfrac{1}{2}W \qquad (1)$$

Roller C. We draw a force triangle and obtain $F_{BC} = C \cot \theta$.

Member EDB. Recalling that $F_{AD} = 0$,

$$+\,\gamma\Sigma M_E = 0: \quad F_{DH} \cos (\phi - 90°)a - B(2a \cos \theta) - F_{BC}(2a \sin \theta) = 0$$
$$F_{DH}a \sin \phi - B(2a \cos \theta) - (C \cot \theta)(2a \sin \theta) = 0$$
$$F_{DH} \sin \phi - 2(B + C) \cos \theta = 0$$

Recalling Eq. (1), we have

$$F_{DH} = W \frac{\cos \theta}{\sin \phi} \qquad (2)$$

and we observe that the result obtained is independent of d. ◄

Applying first the law of sines to triangle *EDH*, we write

$$\frac{\sin \phi}{EH} = \frac{\sin \theta}{DH} \qquad \sin \phi = \frac{EH}{DH} \sin \theta \qquad (3)$$

Using now the law of cosines, we have

$$(DH)^2 = a^2 + L^2 - 2aL \cos \theta$$
$$= (0.70)^2 + (3.20)^2 - 2(0.70)(3.20) \cos 60°$$
$$(DH)^2 = 8.49 \qquad DH = 2.91 \text{ m}$$

We also note that

$$W = mg = (1000 \text{ kg})(9.81 \text{ m/s}^2) = 9810 \text{ N} = 9.81 \text{ kN}$$

Substituting for $\sin \phi$ from (3) into (2) and using the numerical data, we write

$$F_{DH} = W \frac{DH}{EH} \cot \theta = (9.81 \text{ kN}) \frac{2.91 \text{ m}}{3.20 \text{ m}} \cot 60°$$

$$F_{DH} = 5.15 \text{ kN} \qquad ◄$$

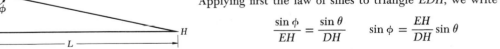

PROBLEMS

6.93 A 360-N force is applied to the toggle vise at C. Determine (a) the horizontal force exerted on the block at D, (b) the force exerted on member ABC at B.

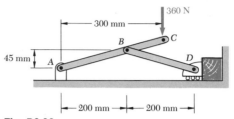

Fig. P6.93

6.94 The control rod CE passes through a horizontal hole in the body of the toggle clamp shown. Determine (a) the force $\mathbf{Q}$ required to hold the clamp in equilibrium, (b) the corresponding force in link BD.

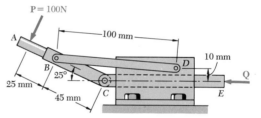

Fig. P6.94

6.95 Water pressure in the supply system exerts a downward force of 25 lb on the vertical plug at A. Determine the tension in the fusible link DE and the force exerted on member BCE at B.

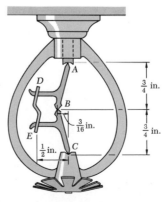

Fig. P6.95

6.96 A cylinder weighs 400 lb and is lifted by a pair of tongs as shown. Determine the forces exerted at D and C on the tong BCD.

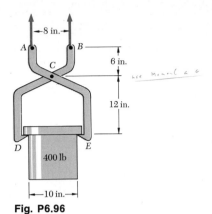

Fig. P6.96

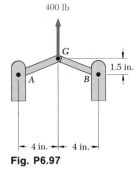

Fig. P6.97

6.97 If the toggle shown is added to the tongs of Prob. 6.96 and the load is lifted by applying a single force at G, determine the forces exerted at D and C on the tong BCD.

6.98 In using the boltcutter shown, a worker applies two 100-lb forces to the handles. Determine the magnitude of the forces exerted by the cutter on the bolt.

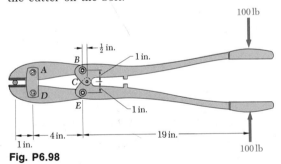

Fig. P6.98

6.99 Two 300-N forces are applied to the handles of the pliers as shown. Determine (*a*) the magnitude of the forces exerted on the rod, (*b*) the force exerted by the pin at A on portion AB of the pliers.

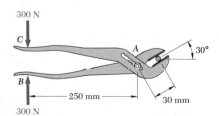

Fig. P6.99

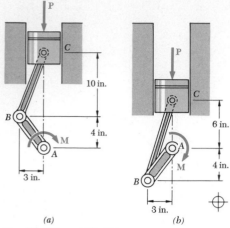

10 in.

4 in.

3 in.

(a)

6 in.

4 in.

3 in.

(b)

Fig. P6.100 and P6.101

6.100 A force **P** of magnitude 400 lb is applied to the piston of the engine system shown. For each of the two positions shown, determine the couple **M** required to hold the system in equilibrium.

6.101 A couple **M** of magnitude 210 lb · ft is applied to the crank of the engine system shown. For each of the two positions shown, determine the force **P** required to hold the system in equilibrium.

6.102 and 6.103 Two rods are connected by a frictionless collar B. Knowing that the magnitude of the couple M_A is 20 N · m, determine (a) the couple M_C required for equilibrium, (b) the corresponding components of the reaction at C.

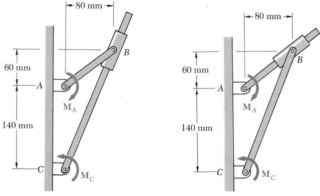

Fig. P6.102   **Fig. P6.103**

6.104 A pipe of diameter 50 mm is gripped by the Stillson wrench shown. Portions AB and DE of the wrench are rigidly attached to each other and portion CF is connected by a pin at D. Assuming that no slipping occurs between the pipe and the wrench, determine the components of the forces exerted on the pipe at A and at C.

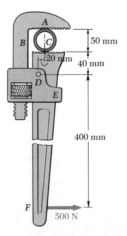

Fig. P6.104

6.105 The specialized plumbing wrench shown is used in confined areas (e.g., under a basin or sink). It consists essentially of a jaw BC pinned at B to a long rod. Knowing that the forces exerted on the nut are equivalent to a clockwise couple (when viewed from above) of magnitude 135 lb · in., determine (a) the magnitude of the force exerted by pin B on jaw BC, (b) the couple $\mathbf{M}_0$ which is applied to the wrench.

6.106 Determine the magnitude of the gripping forces produced when two 60-lb forces are applied as shown.

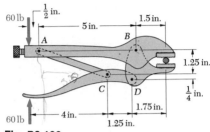

Fig. P6.106

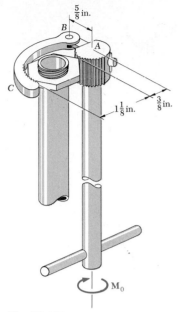

Fig. P6.105

6.107 In the pliers shown, the clamping jaws remain parallel as objects of various sizes are held. If gripping forces of magnitude $Q = 450$ lb are desired, determine the magnitude P of the forces which must be applied. Assume that pins B and E slide freely in the slots cut in the jaws.

6.108 The retractable shelf is maintained in the position shown by two identical linkage-and-spring systems; only one of the systems is shown. A 20-kg machine is placed on the shelf so that half of its weight is supported by the system shown. Determine (a) the force in link AB, (b) the tension in the spring.

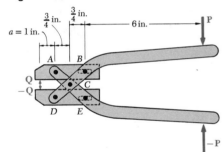

Fig. P6.107

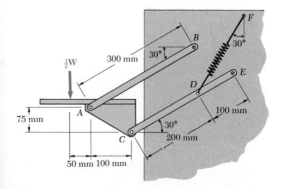

Fig. P6.108

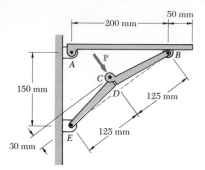

Fig. P6.109

6.109 A 20-kg shelf is held horizontally by a self-locking brace which consists of two parts EDC and CDB hinged at C and bearing against each other at D. Determine the force **P** required to release the brace. (*Hint.* To release the brace, the forces of contact at D must be zero.)

6.110 In the boring rig shown, the center of gravity of the 3000-kg tower is located at point G. For the position shown, determine the force exerted by the hydraulic cylinder AB.

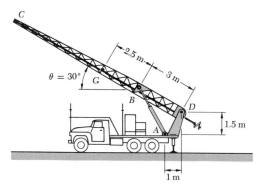

Fig. P6.110

6.111 Solve Prob. 6.110 for the position $\theta = 60°$.

6.112 The action of the backhoe bucket is controlled by the three hydraulic cylinders shown. Determine the force exerted by each cylinder in supporting the 3000-lb load shown.

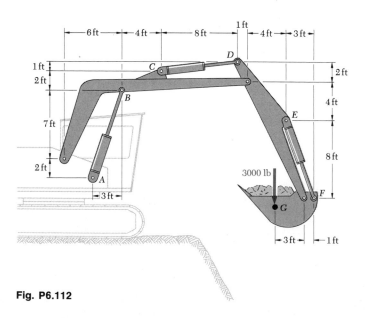

Fig. P6.112

6.113 Two identical linkage-and-hydraulic-cylinder systems control the position of the forks of a fork-lift truck; only one system is shown. Knowing that the load supported by the one system shown is 1500 lb, determine (a) the force exerted by the hydraulic cylinder on point D, (b) the components of the force exerted on member ACE at point C.

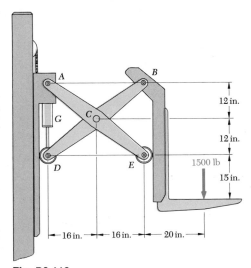

Fig. P6.113

6.114 In order to provide a mobile overhead working platform JK, the system shown is mounted on the bed of a truck. The elevation of the platform is controlled by two identical mechanisms, only one of which is shown. A load of 5 kN is applied to the mechanism shown. Knowing that the pin at C can transmit only a horizontal force, determine (a) the force in link BE, (b) the components of the force exerted by the hydraulic cylinder on pin H.

6.115 Solve Prob. 6.114 when the distance from C to the point of application of the load is a = 1.5 m.

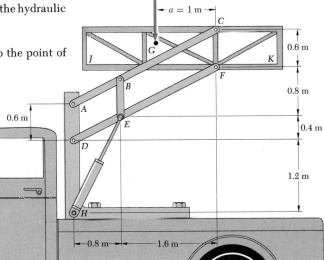

Fig. P6.114

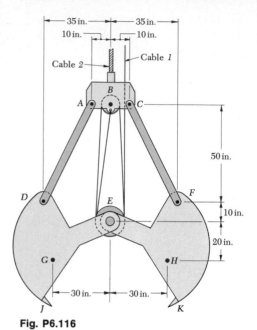

Fig. P6.116

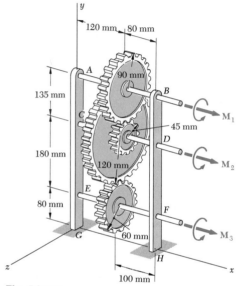

Fig. P6.117

6.116 The total weight of the 1-yd clamshell bucket shown is 4500 lb. The centers of gravity of sections *DEJ* and *EFK*, which weigh 2000 lb each, are located at *G* and *H*, respectively. The double-sheave pulley *E* and a counterweight located at *E* together weigh 400 lb. Determine the tension in cable *1* and cable *2* for the position shown. (Neglect the effect of the horizontal distance between the cables.)

6.117 The four gears are rigidly attached to shafts which are held by frictionless bearings. If $M_1 = 24$ N · m and $M_2 = 0$, determine (a) the couple $\mathbf{M}_3$ which must be applied for equilibrium, (b) the reactions at *G* and *H*.

6.118 Solve Prob. 6.117 if $M_1 = 48$ N · m and $M_2 = -18$ N · m.

6.119 In the planetary-gear system shown, the radius of the central gear *A* is *a*, the radius of each planetary gear is *b*, and the radius of the outer gear *E* is $(a + 2b)$. In a particular gear system where $a = b = 40$ mm, a clockwise couple $\mathbf{M}_A$ is applied to gear *A*. If the system is to be in equilibrium, determine (a) the couple $\mathbf{M}_S$ which must be applied to the spider *BCD*, (b) the couple $\mathbf{M}_E$ which must be applied to the outer gear *E*.

6.120 In the planetary-gear system shown, the radius of the central gear *A* is *a*, the radius of each of the planetary gears is *b*, and the radius of the outer gear *E* is $(a + 2b)$. A clockwise couple of magnitude M_A is applied to the central gear *A* and a counterclockwise couple of magnitude $5M_A$ is applied to the spider *BCD*. If the system is to be in equilibrium, determine (a) the required ratio b/a, (b) the couple $\mathbf{M}_E$ which must be applied to the outer gear *E*.

Fig. P6.119 and P6.120

*6.121 Two shafts AC and CF, which lie in the vertical xy plane, are connected by a universal joint at C. The bearings at B and D do not exert any axial force. A couple of magnitude 50 N · m (clockwise when viewed from the positive x axis) is applied to shaft CF at F. At a time when the arm of the crosspiece attached to shaft CF is horizontal, determine (a) the magnitude of the couple which must be applied to shaft AC at A to maintain equilibrium, (b) the reactions at B, D, and E. (*Hint.* The sum of the couples exerted on the crosspiece must be zero.)

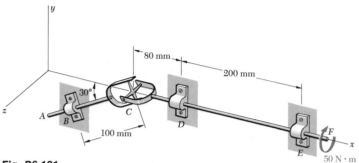

Fig. P6.121

*6.122 Solve Prob. 6.121, assuming that the arm of the crosspiece attached to shaft CF is vertical.

6.123 The weight, in ounces, of letters placed on the postal scale shown is indicated on the moving dial by the stationary pointer P. The scale is shown in its unloaded position. It is known that the dial, the arm AB, and a counterweight together weigh 10 oz and have a combined center of gravity at G. The distance AG is $1\frac{1}{4}$ in., and the length of the arm AB and of the link CD is 1.00 in. At what angle θ should the 2-oz number be painted on the dial? (*Hint.* The weights of the tray, of BC, and of CD are unknown, but their effect must be considered.)

6.124 A letter of unknown weight is placed on the postal scale of Prob. 6.123. Knowing that the dial rotates counterclockwise through 45° before coming to rest, determine the weight of the letter. (See hint of Prob. 6.123.)

REVIEW PROBLEMS

6.125 Determine the reaction at F and the force in members AE and BD.

6.126 Solve Prob. 6.125 if the 180-lb load is applied at D and is directed to the right.

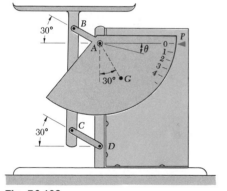

Fig. P6.123

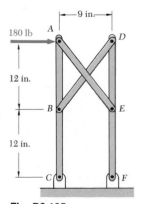

Fig. P6.125

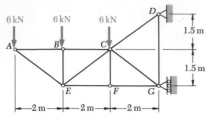

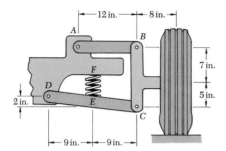

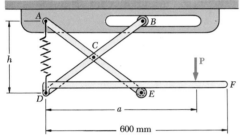

Fig. P6.127 and P6.128

Fig. P6.129

Fig. P6.130

6.127 Determine the force in members *BC*, *CE*, and *EF*.

6.128 Determine the force in members *CD* and *CG*.

6.129 An automobile front-wheel assembly supports 750 lb. Determine the force exerted by the spring and the components of the forces exerted on the frame at points *A* and *D*.

6.130 Members *ACE* and *DCB* are each of length 400 mm and are connected by a pin at their midpoints *C*. A load **P** of magnitude 1600 N is applied to member *DF*. If $h = 240$ mm and $a = 500$ mm, determine (*a*) all forces acting on member *DCB*, (*b*) the tension in the spring *AD*, (*c*) the unstretched length of the spring knowing that the spring constant is 20 kN/m.

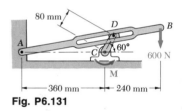

Fig. P6.131

6.131 Determine the couple **M** which must be applied to the crank *CD* to hold the mechanism in equilibrium. The block at *D* is pinned to the crank *CD* and is free to slide in a slot cut in member *AB*.

6.132 A 400-kg block may be supported by a small frame in each of the four ways shown. The diameter of the pulley is 300 mm. For each case, determine the force components and the couple representing the reaction at *A* and also the force exerted at *D* on the vertical member.

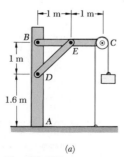

(*a*)

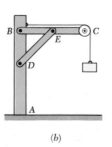

(*b*)

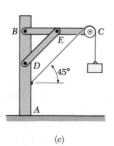

(*c*)

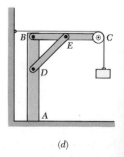

(*d*)

Fig. P6.132

6.133 The truck shown is used to facilitate work on overhead wires by raising a pair of buckets to the required elevation. Two tubular members AB and BC, each of length 12 ft, form the main supporting mechanism; the position of AB is controlled by means of the hydraulic cylinder DE. Two 2-ft-diameter sheaves, one on each side, are rigidly attached to member BC at B. Cables are fastened to the sheaves, pass over pulleys at H, and are fastened to a common movable block at F. The position of block F is controlled by a second hydraulic cylinder EF. Knowing that the workers, buckets, and equipment attached to the buckets together weigh 700 lb and have a combined center of gravity at G, determine the force which must be exerted by each hydraulic cylinder to maintain the position shown. Neglect the weight of the mechanism.

6.134 Solve Prob. 6.133 when $\theta = 60°$.

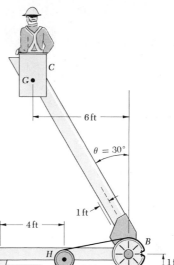

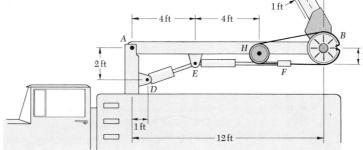

Fig. P6.133

6.135 Knowing that the surfaces at A, B, D, and E are frictionless, determine (*a*) the reactions, (*b*) the components of the force exerted on member ACE at point C.

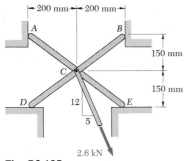

Fig. P6.135

*6.136 In the folding chair shown, members $ABEH$ and CFK are parallel. Determine the components of all forces acting on member $ABEH$ when a 160-lb person sits in the chair. It may be assumed that the floor is frictionless and that half the person's weight is carried by each side of the chair and is applied at point M.

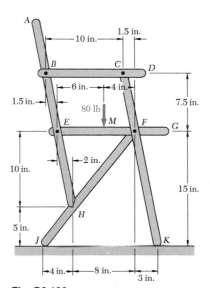

Fig. P6.136

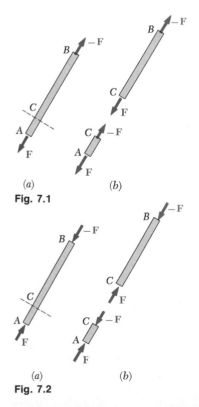

B −F

B −F

C

A

F

C −F

A

F

(a)   (b)

Fig. 7.1

B −F

B −F

C

C

A

F

C −F

A

F

(a)   (b)

Fig. 7.2

CHAPTER

7 Forces in Beams and Cables

*7.1. Introduction. Internal Forces in Members.

In preceding chapters, two basic problems involving structures were considered, (1) the determination of the external forces acting on a structure (Chap. 4) and (2) the determination of the forces which hold together the various members forming a structure (Chap. 6). We shall now consider the problem of determining the internal forces which hold together the various parts of a given member.

We shall first consider a *straight two-force member AB* (Fig. 7.1a). From Sec. 4.6, we know that the forces **F** and −**F** acting at A and B, respectively, must be directed along AB in opposite sense and have the same magnitude F. Now, let us cut the member at C. To maintain the equilibrium of the free bodies AC and CB thus obtained, we must apply to AC a force −**F** equal and opposite to **F**, and to CB a force **F** equal and opposite to −**F** (Fig. 7.1b). These new forces are directed along AB in opposite sense and have the same magnitude F. Since the two parts AC and CB were in equilibrium before the member was cut, *internal forces* equivalent to these new forces must have existed in the member itself. We see that, in the case of a straight two-force member, the internal forces acting on each part of the member are equivalent to an axial force. The magnitude F of this force does not depend upon the location of the section C and is referred to as the *force in member AB*. In the case considered, the member is in tension and will elongate under

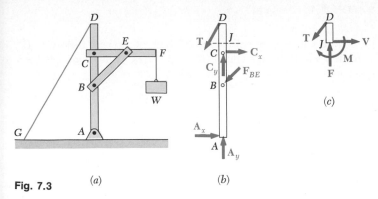

Fig. 7.3   (a)   (b)   (d)   (e)

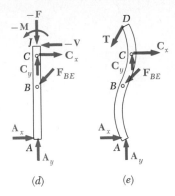

(c)

the action of the internal forces. In the case represented in Fig. 7.2, the member is in compression and will decrease in length under the action of the internal forces.

Next we shall consider a *multiforce member*. Take, for instance, member AD of the crane analyzed in Sec. 6.10. This crane is shown again in Fig. 7.3a, and the free-body diagram of member AD is drawn in Fig. 7.3b. We now cut member AD at J and draw a free-body diagram for each of the portions JD and AJ of the member (Fig. 7.3c and d). Considering the free body JD, we find that its equilibrium will be maintained if we apply at J a force **F** to balance the vertical component of **T**, a force **V** to balance the horizontal component of **T**, and a couple **M** to balance the moment of **T** about J. Again we conclude that internal forces must have existed at J before member AD was cut. The internal forces acting on the portion JD of member AD are equivalent to the force-couple system shown in Fig. 7.3c. According to Newton's third law, the internal forces acting on AJ must be equivalent to an equal and opposite force-couple system, as shown in Fig. 7.3d. It clearly appears that the action of the internal forces in member AD *is not limited to producing tension or compression* as in the case of straight two-force members; the internal forces *also produce shear and bending*. The force **F** is again in this case called an *axial force;* the force **V** is called a *shearing force;* and the moment **M** of the couple is known as the *bending moment* at J. We note that, when determining internal forces in a member, we should clearly indicate on which portion of the member the forces are supposed to act. The deformation which will occur in member AD is sketched in Fig. 7.3e. The actual analysis of such a deformation is part of the study of mechanics of materials.

It should be noted that, in a *two-force member which is not straight*, the internal forces are also equivalent to a force-couple system. This is shown in Fig. 7.4, where the two-force member ABC has been cut at D.

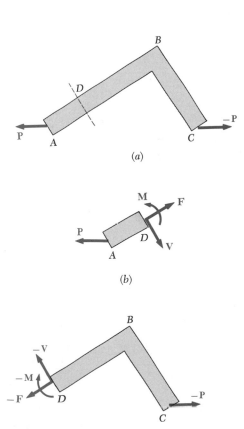

(a)

(b)

(c)

Fig. 7.4

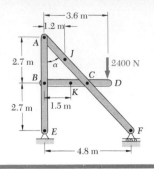

In the frame shown, determine the internal forces (*a*) in member *ACF* at point *J* and (*b*) in member *BCD* at point *K*. This frame has been previously considered in Sample Prob. 6.5.

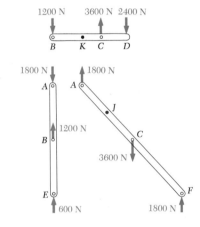

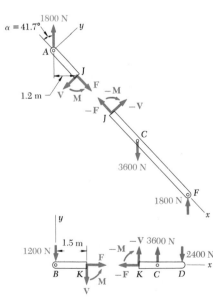

Solution. The reactions and the forces acting on each member of the frame are determined; this has been previously done in Sample Prob. 6.5, and the results are repeated here.

a. Internal Forces at J. Member *ACF* is cut at point *J*, and the two parts shown are obtained. The internal forces at *J* are represented by an equivalent force-couple system and may be determined by considering the equilibrium of either part. Considering the *free body AJ*, we write

$$+\curvearrowleft\Sigma M_J = 0: \qquad -(1800\text{ N})(1.2\text{ m}) + M = 0$$
$$M = +2160\text{ N}\cdot\text{m} \qquad \mathbf{M} = 2160\text{ N}\cdot\text{m}\,\curvearrowleft \;\blacktriangleleft$$

$$+\searrow\Sigma F_x = 0: \qquad F - (1800\text{ N})\cos 41.7° = 0$$
$$F = +1344\text{ N} \qquad \mathbf{F} = 1344\text{ N}\,\searrow \;\blacktriangleleft$$

$$+\nearrow\Sigma F_y = 0: \qquad -V + (1800\text{ N})\sin 41.7° = 0$$
$$V = +1197\text{ N} \qquad \mathbf{V} = 1197\text{ N}\,\nearrow \;\blacktriangleleft$$

The internal forces at *J* are therefore equivalent to a couple **M**, an axial force **F**, and a shearing force **V**. The internal force-couple system acting on part *JCF* is equal and opposite.

b. Internal Forces at K. We cut member *BCD* at *K* and obtain the two parts shown. Considering the *free body BK*, we write

$$+\curvearrowleft\Sigma M_K = 0: \qquad (1200\text{ N})(1.5\text{ m}) + M = 0$$
$$M = -1800\text{ N}\cdot\text{m} \qquad \mathbf{M} = 1800\text{ N}\cdot\text{m}\,\curvearrowright \;\blacktriangleleft$$

$$\xrightarrow{+}\Sigma F_x = 0: \qquad F = 0 \qquad\qquad\qquad \mathbf{F} = 0 \;\blacktriangleleft$$

$$+\uparrow\Sigma F_y = 0: \qquad -1200\text{ N} - V = 0$$
$$V = -1200\text{ N} \qquad \mathbf{V} = 1200\text{ N}\uparrow \;\blacktriangleleft$$

PROBLEMS

7.1 through 7.4 Determine the internal forces (axial force, shearing force, and bending moment) at point J of the structure indicated:

7.1	Frame and loading of Prob. 6.61.	
7.2	Frame and loading of Prob. 6.56.	
7.3	Frame and loading of Prob. 6.75.	
7.4	Frame and loading of Prob. 6.74.	

7.5 The bracket AD is supported by a pin at A and by the cable DE. Determine the internal forces just to the left of the load if $a = 200$ mm.

7.6 The bracket AD is supported by a pin at A and by the cable DE. Determine the distance a for which the bending moment at B is equal to the bending moment at C.

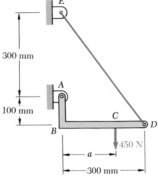

Fig. P7.5 and P7.6

7.7 Determine the internal forces at point J of the adjustable hanger shown.

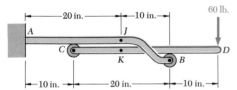

Fig. P7.7 and P7.8

7.8 Determine the internal forces at point K of the adjustable hanger shown.

7.9 Determine the internal forces at point J when $\alpha = 90°$.

7.10 Determine the internal forces at point J when $\alpha = 0$.

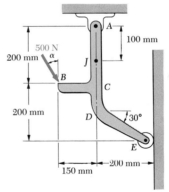

Fig. P7.9 and P7.10

7.11 A uniform rod ABC has been bent and is held by a wire sling as shown. Knowing that the rod weighs 9 lb, determine the internal forces at point B.

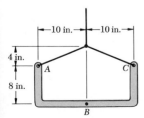

Fig. P7.11

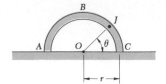

Fig. P7.12 and P7.14

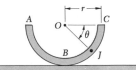

Fig. P7.13 and P7.15

7.12 and 7.13 A half section of pipe rests on a frictionless horizontal surface as shown. If the half section of pipe weighs 30 lb and has a diameter of 20 in., determine the bending moment at point J when $\theta = 90°$.

∗7.14 and 7.15 A half section of pipe, of weight W and of unit length, rests on a frictionless horizontal surface. Determine the internal forces at point J in terms of W, r, and θ.

∗7.16 In Prob. 7.15, determine the magnitude and location of the maximum internal axial force.

BEAMS

∗7.2. Various Types of Loading and Support. A structural member designed to support loads applied at various points along the member is known as a *beam*. In most cases, the loads are perpendicular to the axis of the beam and will cause only shear and bending in the beam. When the loads are not at a right angle to the beam, they will also produce axial forces in the beam. Axial forces, however, may usually be neglected in the design of beams, since the ability of a beam to resist shear and especially bending is more critical than its ability to resist axial forces.

Beams are usually long, straight prismatic bars. Designing a beam consists essentially in selecting the cross section which will provide the most effective resistance to the shear and bending produced by the applied loads. The design of the beam, therefore, includes two distinct parts. In the first part, the shearing forces and bending moments produced by the loads are determined. The second part is concerned with the selection of the cross section best suited to resist the shearing forces and bending moments determined in the first part. This portion of the chapter, Beams, deals with the first part of the problem of beam design, namely, the determination of the shearing forces and bending moments in beams subjected to various loading conditions and supported in various ways. The second part of the problem belongs to the study of mechanics of materials.

A beam may be subjected to *concentrated loads* P_1, P_2, . . . , expressed in newtons, pounds, or their multiples kilonewtons and kips (Fig. 7.5a), to a *distributed load* w, expressed in N/m, kN/m, lb/ft, or kips/ft (Fig. 7.5b), or to a combination of both. When the load w per unit length has a constant value over part of the beam (as between A and B in Fig. 7.5b), the load is said to be *uniformly distributed* over that part of the beam. The determination of the reactions at the supports may be consid-

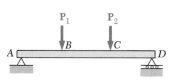

(a) Concentrated loads

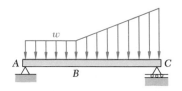

(b) Distributed load

Fig. 7.5 Types of loadings

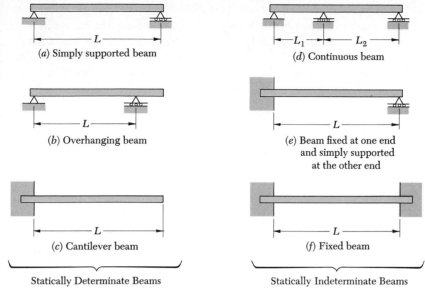

(a) Simply supported beam

(d) Continuous beam

(b) Overhanging beam

(e) Beam fixed at one end
and simply supported
at the other end

(c) Cantilever beam

(f) Fixed beam

Statically Determinate Beams   Statically Indeterminate Beams

Fig. 7.6 Types of beams

erably simplified if distributed loads are replaced by equivalent concentrated loads, as explained in Sec. 5.6. This substitution, however, should not be performed, or at least should be performed with care, when internal forces are being computed (see Sample Prob. 7.3).

Beams are classified according to the way in which they are supported. Several types of beams frequently used are shown in Fig. 7.6. The distance L between supports is called the *span*. It should be noted that the reactions will be determinate if the supports involve only three unknowns. The reactions will be statically indeterminate if more unknowns are involved; the methods of statics are not sufficient then to determine the reactions, and the properties of the beam with regard to its resistance to bending must be taken into consideration. Beams supported by two rollers are not shown here; such beams are only partially constrained and will move under certain loading conditions.

Sometimes two or more beams are connected by hinges to form a single continuous structure. Two examples of beams hinged at a point H are shown in Fig. 7.7. It will be noted that the reactions at the supports involve four unknowns and cannot be determined from the free-body diagram of the two-beam system. They can be determined, however, by considering the free-body diagram of each beam separately; six unknowns are involved (including two force components at the hinge), and six equations are available.

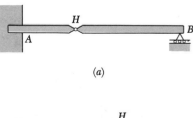

(a)

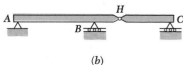

(b)

Fig. 7.7 Combined beams

*7.3. Shear and Bending Moment in a Beam.

Consider a beam AB subjected to various concentrated and distributed loads (Fig. 7.8a). We propose to determine the shearing force and bending moment at any point of the beam. In the example considered here, the beam is simply supported, but the method used could be applied to any type of statically determinate beam.

First we determine the reactions at A and B by choosing the entire beam as a free body (Fig. 7.8b); writing $\Sigma M_A = 0$ and $\Sigma M_B = 0$, we obtain, respectively, $\mathbf{R}_B$ and $\mathbf{R}_A$.

To determine the internal forces at C, we cut the beam at C and draw the free-body diagrams of the portions AC and CB of the beam (Fig. 7.8c). Using the free-body diagram of AC, we may determine the shearing force $\mathbf{V}$ at C by equating to zero the sum of the vertical components of all forces acting on AC. Similarly, the bending moment $\mathbf{M}$ at C may be found by equating to zero the sum of the moments about C of all forces and couples acting on AC. We could have used just as well, however, the free-body diagram of CB† and determined the shearing force $\mathbf{V}'$ and the bending moment $\mathbf{M}'$ by equating to zero the sum of the vertical components and the sum of the moments about C of all forces and couples acting on CB. While this possible choice of alternate free bodies may facilitate the

---

† The force and couple representing the internal forces acting on CB will now be denoted by $\mathbf{V}'$ and $\mathbf{M}'$, rather than by $-\mathbf{V}$ and $-\mathbf{M}$ as done earlier, in order to avoid confusion when applying the sign convention which we are about to introduce.

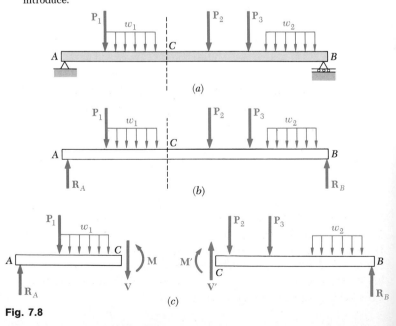

Fig. 7.8

computation of the numerical values of the shearing force and bending moment, it makes it necessary to indicate on which portion of the beam the internal forces considered are acting. If the shearing force and bending moment, however, are to be computed at every point of the beam and efficiently recorded, we should not have to specify every time which portion of the beam is used as a free body. We shall adopt, therefore, the following convention:

To determine the shearing force in a beam, *we shall always assume* that the internal forces V and V' are directed as shown in Fig. 7.8c. A positive value obtained for their common magnitude V will indicate that this assumption was correct and that the shearing forces are actually directed as shown. A negative value obtained for V will indicate that the assumption was wrong and that the shearing forces are directed in the opposite way. Thus, only the magnitude V, together with a plus or minus sign, needs to be recorded to define completely the shearing forces at a given point of the beam. The scalar V is commonly referred to as the *shear* at the given point of the beam.

Similarly, *we shall always assume* that the internal couples M and M' are directed as shown in Fig. 7.8c. A positive value obtained for their magnitude M, commonly referred to as the bending moment, will indicate that this assumption was correct, and a negative value that it was wrong. Summarizing the sign convention we have presented, we state:

The shear V and the bending moment M at a given point of a beam are said to be positive when the internal forces and couples acting on each portion of the beam are directed as shown in Fig. 7.9a.

This convention may be more easily remembered if we note that:

1. *The shear at C is positive when the **external** forces (loads and reactions) acting on the beam tend to shear off the beam at C as indicated in Fig. 7.9b.*

2. *The bending moment at C is positive when the **external** forces acting on the beam tend to bend the beam at C as indicated in Fig. 7.9c.*

It may also help to note that the situation described in Fig. 7.9, and corresponding to positive values of the shear and of the bending moment, is precisely the situation which occurs in the left half of a simply supported beam carrying a single concentrated load at its midpoint. This particular example is fully discussed in the following section.

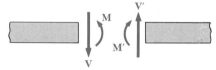

(a) Internal forces at section
(positive shear and positive bending moment)

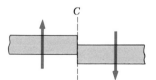

(b) Effect of external forces
(positive shear)

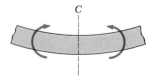

(c) Effect of external forces
(positive bending moment)

Fig. 7.9

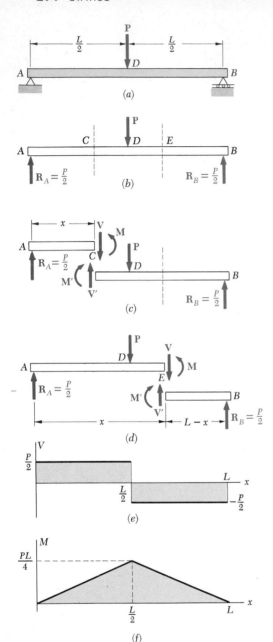

Fig. 7.10

7.4. Shear and Bending-Moment Diagrams. Now that shear and bending moment have been clearly defined in sense as well as in magnitude, we may easily record their values at any point of a beam by plotting these values against the distance x measured from one end of the beam. The graphs obtained in this way are called, respectively, the *shear diagram* and the *bending-moment diagram.* As an example, consider a simply supported beam AB of span L subjected to a single concentrated load $\mathbf{P}$ applied at its midpoint D (Fig. 7.10a). We first determine the reactions at the supports from the free-body diagram of the entire beam (Fig. 7.10b); we find that the magnitude of each reaction is equal to $P/2$.

Next we cut the beam at a point C between A and D and draw the free-body diagrams of AC and CB (Fig. 7.10c). *Assuming that shear and bending moment are positive,* we direct the internal forces $\mathbf{V}$ and $\mathbf{V'}$ and the internal couples $\mathbf{M}$ and $\mathbf{M'}$ as indicated in Fig. 7.9a. Considering the free body AC and writing that the sum of the vertical components and the sum of the moments about C of the forces acting on the free body are zero, we find $V = +P/2$ and $M = +Px/2$. Both the shear and the bending moment are therefore positive; this may be checked by observing that the reaction at A tends to shear off and to bend the beam at C as indicated in Fig. 7.9b and c. We may plot V and M between A and D (Fig. 7.10e and f); the shear has a constant value $V = P/2$, while the bending moment increases linearly from $M = 0$ at $x = 0$ to $M = PL/4$ at $x = L/2$.

Cutting, now, the beam at a point E between D and B and considering the free body EB (Fig. 7.10d), we write that the sum of the vertical components and the sum of the moments about E of the forces acting on the free body are zero. We obtain $V = -P/2$ and $M = P(L - x)/2$. The shear is therefore negative and the bending moment positive; this may be checked by observing that the reaction at B bends the beam at E as indicated in Fig. 7.9c but tends to shear it off in a manner opposite to that shown in Fig. 7.9b. We can complete, now, the shear and bending-moment diagrams of Fig. 7.10e and f; the shear has a constant value $V = -P/2$ between D and B, while the bending moment decreases linearly from $M = PL/4$ at $x = L/2$ to $M = 0$ at $x = L$.

We shall note that, when a beam is subjected only to concentrated loads, the shear is of constant value between loads and the bending moment varies linearly between loads. On the other hand, when a beam is subjected to distributed loads, the shear and bending moment vary quite differently (see Sample Prob. 7.3).

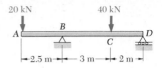

20 kN   40 kN

A   B   C   D

2.5 m — 3 m — 2 m

SAMPLE PROBLEM 7.2

Draw the shear and bending-moment diagram for the beam and loading shown.

Solution. The reactions are determined by considering the entire beam as a free body; they are

$$R_B = 46 \text{ kN} \uparrow \qquad R_D = 14 \text{ kN} \uparrow$$

We first determine the internal forces just to the right of the 20-kN load at A. Considering the stub of beam to the left of section 1 as a free body and assuming V and M to be positive (according to the standard convention), we write

$$+\uparrow\Sigma F_y = 0: \qquad -20 \text{ kN} - V_1 = 0 \qquad\qquad V_1 = -20 \text{ kN}$$
$$+\,\gamma\Sigma M_1 = 0: \qquad (20 \text{ kN})(0 \text{ m}) + M_1 = 0 \qquad\qquad M_1 = 0$$

We next consider as a free body the portion of beam to the left of section 2 and write

$$+\uparrow\Sigma F_y = 0: \qquad -20 \text{ kN} - V_2 = 0 \qquad\qquad V_2 = -20 \text{ kN}$$
$$+\,\gamma\Sigma M_2 = 0: \qquad (20 \text{ kN})(2.5 \text{ m}) + M_2 = 0 \qquad M_2 = -50 \text{ kN} \cdot \text{m}$$

The shear and bending moment at sections 3, 4, 5, and 6 are determined in a similar way from the free-body diagrams shown. We obtain

$$V_3 = +26 \text{ kN} \qquad M_3 = -50 \text{ kN} \cdot \text{m}$$
$$V_4 = +26 \text{ kN} \qquad M_4 = +28 \text{ kN} \cdot \text{m}$$
$$V_5 = -14 \text{ kN} \qquad M_5 = +28 \text{ kN} \cdot \text{m}$$
$$V_6 = -14 \text{ kN} \qquad M_6 = 0$$

For several of the latter sections, the results may be more easily obtained by considering as a free body the portion of the beam to the right of the section. For example, considering the portion of the beam to the right of section 4, we write

$$+\uparrow\Sigma F_y = 0: \qquad V_4 - 40 \text{ kN} + 14 \text{ kN} = 0 \qquad V_4 = +26 \text{ kN}$$
$$+\,\gamma\Sigma M_4 = 0: \qquad -M_4 + (14 \text{ kN})(2 \text{ m}) = 0 \qquad M_4 = +28 \text{ kN} \cdot \text{m}$$

We may now plot the six points shown on the shear and bending-moment diagrams. As indicated in Sec. 7.4, the shear is of constant value between concentrated loads, and the bending moment varies linearly; we obtain therefore the shear and bending-moment diagrams shown.

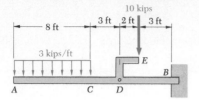

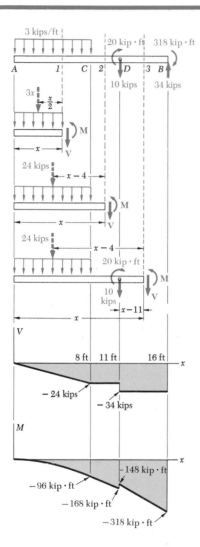

SAMPLE PROBLEM 7.3

Draw the shear and bending-moment diagrams for the cantilever beam *AB*. The distributed load of 3 kips/ft extends over 8 ft of the beam and the 10-kip load is applied at *E*.

Solution. The 10-kip load is replaced by an equivalent force-couple system acting on the beam at point *D*. The reaction at *B* is determined by considering the entire beam as a free body.

From A to C. We determine the internal forces at a distance x from point *A* by considering the portion of beam to the left of section *1*. That part of the distributed load acting on the free body is replaced by its resultant, and we write

$$+\uparrow\Sigma F_y = 0: \qquad -3x - V = 0 \qquad\qquad V = -3x \text{ kips}$$

$$+\gamma\Sigma M_1 = 0: \qquad 3x(\tfrac{1}{2}x) + M = 0 \qquad\qquad M = -1.5x^2 \text{ kip} \cdot \text{ft}$$

Since the free-body diagram shown may be used for all values of x smaller than 8 ft, the expressions obtained for V and M are valid in the region $0 < x < 8$ ft.

From C to D. Considering the portion of beam to the left of section *2* and again replacing the distributed load by its resultant, we obtain

$$+\uparrow\Sigma F_y = 0: \qquad -24 - V = 0 \qquad V = -24 \text{ kips}$$

$$+\gamma\Sigma M_2 = 0: \qquad 24(x - 4) + M = 0 \qquad M = 96 - 24x \qquad \text{kip} \cdot \text{ft}$$

These expressions are valid in the region 8 ft $< x <$ 11 ft.

From D to B. Using the portion of beam to the left of section *3*, we obtain for the region 11 ft $< x <$ 16 ft

$$V = -34 \text{ kips} \qquad M = 226 - 34x \qquad \text{kip} \cdot \text{ft}$$

The shear and bending-moment diagrams for the entire beam may now be plotted. We note that the couple of moment 20 kip · ft applied at point *D* introduces a discontinuity into the bending-moment diagram.

PROBLEMS

7.17 through 7.22 Draw the shear and bending-moment diagrams for the beam and loading shown.

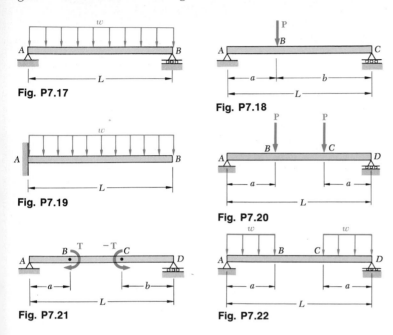

Fig. P7.17

Fig. P7.18

Fig. P7.19

Fig. P7.20

Fig. P7.21

Fig. P7.22

7.23 and 7.24 Draw the shear and bending-moment diagrams for the beam *AB*.

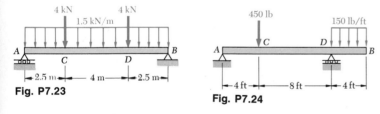

Fig. P7.23

Fig. P7.24

7.25 Draw the shear and bending-moment diagrams for the beam *AB* if $a = 6$ ft.

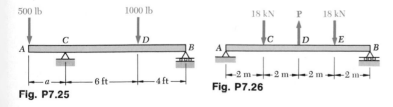

Fig. P7.25

Fig. P7.26

7.26 Draw the shear and bending-moment diagrams for the beam *AB* if the magnitude of the upward force **P** is 4 kN.

*7.27 Determine the distance *a* for which the maximum absolute value of the bending moment in the beam is as small as possible. (*Hint.* Draw the bending-moment diagram and then equate the maximum positive and negative bending moments obtained.)

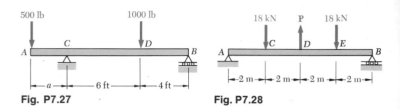

Fig. P7.27   **Fig. P7.28**

*7.28 Determine the magnitude of the upward force **P** for which the maximum absolute value of the bending moment in the beam is as small as possible. (See hint of Prob. 7.27.)

7.29 and 7.30 Assuming the upward reaction of the ground to be uniformly distributed, draw the shear and bending-moment diagrams for the beam *AB*.

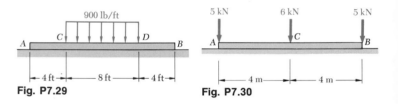

Fig. P7.29   **Fig. P7.30**

7.31 and 7.32 Draw the shear and bending-moment diagrams for the beam *AB*.

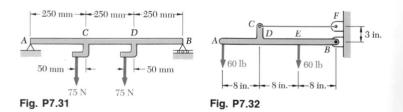

Fig. P7.31   **Fig. P7.32**

7.33 Draw the shear and bending-moment diagrams for the beam and loading of Prob. 6.59.

7.34 Draw the shear and bending-moment diagrams for the beam and loading of Prob. 6.60.

*7.35 In order to reduce the bending moment in the cantilever beam *AB*, a cable and counterweight are permanently attached at end *B*. Determine the magnitude of the counterweight for which the maximum absolute value of the bending moment in the beam is as small as possible. (*a*) Consider only the case when the force **P** is actually applied at *C*. (*b*) Consider the more general case when the force **P** may either be applied at *C* or removed.

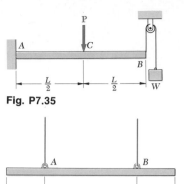

Fig. P7.35

*7.36 A uniform beam is to be picked up by crane cables attached at *A* and *B*. Determine the distance *a* from the ends of the beam to the points where the cables should be attached if the maximum absolute value of the bending moment in the beam is to be as small as possible. (*Hint.* Draw the bending-moment diagram in terms of *a*, *L*, and the weight *w* per unit length, and then equate the maximum positive and negative bending moments obtained.)

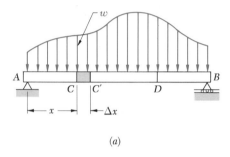

Fig. P7.36

*7.5. Relations between Load, Shear, and Bending Moment.

When a beam carries more than two or three concentrated loads, or when it carries distributed loads, the method outlined in Sec. 7.4 for plotting shear and bending moment may prove quite cumbersome. The construction of the shear diagram and, especially, of the bending-moment diagram will be greatly facilitated if certain relations existing between load, shear, and bending moment are taken into consideration.

Let us consider a simply supported beam *AB* carrying a distributed load *w* per unit length (Fig. 7.11*a*), and let *C* and *C'* be two points of the beam at a distance Δx from each other. The shear and bending moment at *C* will be denoted by *V* and *M*, respectively, and will be assumed positive; the shear and bending moment at *C'* will be denoted by $V + \Delta V$ and $M + \Delta M$.

We now shall detach the portion of beam *CC'* and draw its free-body diagram (Fig. 7.11*b*). The forces exerted on the free body include a load of magnitude $w\,\Delta x$ and internal forces and couples at *C* and *C'*. Since shear and bending moment have been assumed positive, the forces and couples will be directed as shown in the figure.

Relations between Load and Shear. Writing that the sum of the vertical components of the forces acting on the free body *CC'* is zero, we obtain

$$V - (V + \Delta V) - w\,\Delta x = 0$$
$$\Delta V = -w\,\Delta x$$

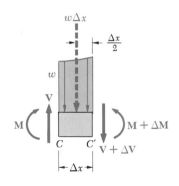

Fig. 7.11

Dividing both members of the equation by Δx and then letting Δx approach zero, we obtain

$$\frac{dV}{dx} = -w \qquad (7.1)$$

Formula (7.1) indicates that, for a beam loaded as shown in Fig. 7.11a, the slope dV/dx of the shear curve is negative; the numerical value of the slope at any point is equal to the load per unit length at that point.

Integrating (7.1) between points C and D, we obtain

$$V_D - V_C = -\int_{x_C}^{x_D} w\,dx \qquad (7.2)$$

$$V_D - V_C = -(\text{area under load curve between } C \text{ and } D) \quad (7.2')$$

Note that this result could also have been obtained by considering the equilibrium of the portion of beam CD, since the area under the load curve represents the total load applied between C and D.

It should be observed that formula (7.1) *is not valid* at a point where a concentrated load is applied; the shear curve is discontinuous at such a point, as seen in Sec. 7.4. Similarly, formulas (7.2) and (7.2') cease to be valid when concentrated loads are applied between C and D, since they do not take into account the sudden change in shear caused by a concentrated load. Formulas (7.2) and (7.2'), therefore, should be applied only between successive concentrated loads.

Relations between Shear and Bending Moment. Returning to the free-body diagram of Fig. 7.11b, and writing now that the sum of the moments about C' is zero, we obtain

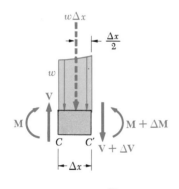

(a)

(b)

Fig. 7.11 (repeated)

$$(M + \Delta M) - M - V\,\Delta x + w\,\Delta x \frac{\Delta x}{2} = 0$$

$$\Delta M = V\,\Delta x - \tfrac{1}{2}w(\Delta x)^2$$

Dividing both members of the equation by Δx and then letting Δx approach zero, we obtain

$$\frac{dM}{dx} = V \qquad (7.3)$$

Formula (7.3) indicates that the slope dM/dx of the bending-moment curve is equal to the value of the shear. This is true at any point where the shear has a well-defined value, i.e., at any point where no concentrated load is applied. Formula (7.3) also shows that the shear is zero at points where the bending moment is maximum. This property facilitates the determination of the points where the beam is likely to fail under bending.

Integrating (7.3) between points C and D, we obtain

$$M_D - M_C = \int_{x_C}^{x_D} V\,dx \qquad (7.4)$$

$M_D - M_C =$ area under shear curve between C and D (7.4′)

Note that the area under the shear curve should be considered positive where the shear is positive and negative where the shear is negative. Formulas (7.4) and (7.4′) are valid even when concentrated loads are applied between C and D, as long as the shear curve has been correctly drawn. The formulas cease to be valid, however, if a *couple* is applied at a point between C and D, since they do not take into account the sudden change in bending moment caused by a couple (see Sample Prob. 7.7).

Example.   Let us consider a simply supported beam AB of span L carrying a uniformly distributed load w (Fig. 7.12a). From the free-body diagram of the entire beam we determine the magnitude of the reactions at the supports: $R_A = R_B = wL/2$ (Fig. 7.12b). Next, we draw the shear diagram. Close to the end A of the beam, the shear is equal to R_A, that is, to $wL/2$, as we may check by considering as a free body a very small portion of the beam. Using formula (7.2), we may then determine the shear V at any distance x from A; we write

$$V - V_A = -\int_0^x w\,dx = -wx$$

$$V = V_A - wx = \frac{wL}{2} - wx = w\left(\frac{L}{2} - x\right)$$

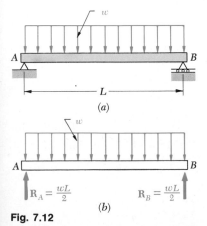

(a)

(b)

Fig. 7.12

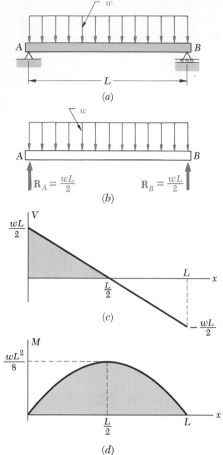

Fig. 7.12

The shear curve is thus an oblique straight line which crosses the x axis at $x = L/2$ (Fig. 7.12c). Considering, now, the bending moment, we first observe that $M_A = 0$. The value M of the bending moment at any distance x from A may then be obtained from formula (7.4); we have

$$M - M_A = \int_0^x V\,dx$$

$$M = \int_0^x w\left(\frac{L}{2} - x\right)dx = \frac{w}{2}(Lx - x^2)$$

The bending-moment curve is a parabola. The maximum value of the bending moment occurs when $x = L/2$, since V (and thus dM/dx) is zero for that value of x. Substituting $x = L/2$ in the last equation, we obtain $M_{\max} = wL^2/8$.

In most engineering applications, the value of the bending moment needs to be known only at a few specific points. Once the shear diagram has been drawn, and after M has been determined at one of the ends of the beam, the value of the bending moment may then be obtained at any given point by computing the area under the shear curve and using formula (7.4'). For instance, since $M_A = 0$ for the beam of Fig. 7.12a, the maximum value of the bending moment for that beam may be obtained simply by measuring the area of the shaded triangle in the shear diagram of Fig. 7.12c. We have

$$M_{\max} = \frac{1}{2}\frac{L}{2}\frac{wL}{2} = \frac{wL^2}{8}$$

We note that, in this example, the load curve is a horizontal straight line, the shear curve an oblique straight line, and the bending-moment curve a parabola. If the load curve had been an oblique straight line (first degree), the shear curve would have been a parabola (second degree) and the bending-moment curve a cubic (third degree). The shear and bending-moment curves will always be, respectively, one and two degrees higher than the load curve. With this in mind, we should be able to sketch the shear and bending-moment diagrams without actually determining the functions $V(x)$ and $M(x)$, once a few values of the shear and bending moment have been computed. The sketches obtained will be more accurate if we make use of the fact that, at any point where the curves are continuous, the slope of the shear curve is equal to $-w$ and the slope of the bending-moment curve is equal to V.

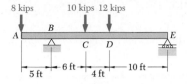

8 kips   10 kips   12 kips

A   B   C   D   E

5 ft   6 ft   4 ft   10 ft

SAMPLE PROBLEM 7.4

Draw the shear and bending-moment diagrams for the beam and loading shown.

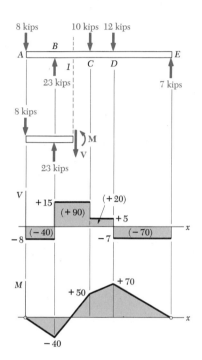

Solution. Considering the entire beam as a free body, we obtain the reactions

$$\mathbf{R}_B = 23 \text{ kips} \uparrow \qquad \mathbf{R}_E = 7 \text{ kips} \uparrow$$

We also note that at both A and E the bending moment is zero; thus two points (indicated by dots) are obtained on the bending-moment diagram.

Shear Diagram. Since $dV/dx = -w$, we find that between loads the slope of the shear diagram is zero (i.e., the shear is constant). The shear at any point is determined by dividing the beam into two parts and considering either part as a free body. For example, using the portion of beam to the left of section *1*, we obtain

$$+\uparrow \Sigma F_y = 0: \qquad -8 + 23 - V = 0 \qquad V = +15 \text{ kips}$$

Bending-Moment Diagram. We recall that the area under the shear curve between two points is equal to the change in bending moment between the same two points. For convenience, the area of each portion of the shear diagram is computed and is indicated on the diagram. Since the bending moment M_A at the free end is known to be zero, we write

$$
\begin{aligned}
M_B - M_A &= -40 & M_B &= -40 \text{ kip} \cdot \text{ft} \\
M_C - M_B &= +90 & M_C &= +50 \text{ kip} \cdot \text{ft} \\
M_D - M_C &= +20 & M_D &= +70 \text{ kip} \cdot \text{ft} \\
M_E - M_D &= -70 & M_E &= 0
\end{aligned}
$$

Since M_E is known to be zero, a check of the computations is obtained.

The shear being constant between successive loads, the slope dM/dx is constant and the bending-moment diagram is obtained by connecting the known points with straight lines. From the V and M diagrams we note that $V_{\max} = 15$ kips, and $M_{\max} = 70$ kip $\cdot$ ft.

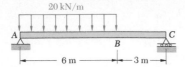

Draw the shear and bending-moment diagrams for the beam and loading shown.

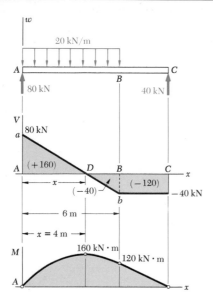

Solution. Considering the entire beam as a free body, we obtain the reactions

$$\mathbf{R}_A = 80 \text{ kN} \uparrow \qquad \mathbf{R}_C = 40 \text{ kN} \uparrow$$

Shear Diagram. The shear just to the right of A is $V_A = +80$ kN. Since the change in shear between two points is equal to *minus* the area under the load curve between the same two points, we obtain V_B by writing

$$V_B - V_A = -(20 \text{ kN/m})(6 \text{ m}) = -120 \text{ kN}$$
$$V_B = -120 + V_A = -120 + 80 = -40 \text{ kN}$$

The slope $dV/dx = -w$ being constant between A and B, the shear diagram between these two points is represented by a straight line. Between B and C, the area under the load curve is zero; therefore,

$$V_C - V_B = 0 \qquad V_C = V_B = -40 \text{ kN}$$

and the shear is constant between B and C.

Bending-Moment Diagram. We note that the bending moment at each end of the beam is zero. In order to determine the maximum bending moment, we locate the section D of the beam where $V = 0$. Considering the portion of the shear diagram between A and B, we note that the triangles DAa and DBb are similar; thus,

$$\frac{x}{80 \text{ kN}} = \frac{6 - x}{40 \text{ kN}} \qquad x = 4 \text{ m}$$

The maximum bending moment occurs at point D, where we have $dM/dx = V = 0$. The areas of the various portions of the shear diagram are computed and are given (in parentheses) on the diagram. Since the area of the shear diagram between two points is equal to the change in bending moment between the same two points, we write

$$M_D - M_A = +160 \text{ kN} \cdot \text{m} \qquad M_D = +160 \text{ kN} \cdot \text{m}$$
$$M_B - M_D = -\ 40 \text{ kN} \cdot \text{m} \qquad M_B = +120 \text{ kN} \cdot \text{m}$$
$$M_C - M_B = -120 \text{ kN} \cdot \text{m} \qquad M_C = 0$$

The bending-moment diagram consists of an arc of parabola followed by a segment of straight line; the slope of the parabola at A is equal to the value of V at that point.

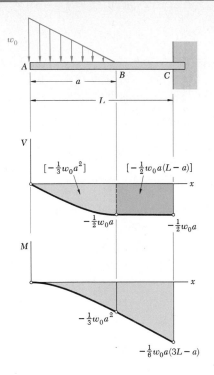

SAMPLE PROBLEM 7.6

Sketch the shear and bending-moment diagrams for the cantilever beam shown.

Solution. *Shear Diagram.* At the free end of the beam, we find $V_A = 0$. Between A and B, the area under the load curve is $\frac{1}{2}w_0 a$; we find V_B by writing

$$V_B - V_A = -\tfrac{1}{2}w_0 a \qquad V_B = -\tfrac{1}{2}w_0 a$$

Between B and C, the beam is not loaded; thus $V_C = V_B$. At A, we have $w = w_0$ and, according to Eq. (7.1), the slope of the shear curve is $dV/dx = -w_0$, while at B the slope is $dV/dx = 0$. Between A and B, the loading decreases linearly, and the shear diagram is parabolic. Between B and C, $w = 0$, and the shear diagram is a horizontal line.

Bending-Moment Diagram. The bending moment M_A at the free end of the beam is zero. We compute the area under the shear curve and write

$$M_B - M_A = -\tfrac{1}{3}w_0 a^2 \qquad M_B = -\tfrac{1}{3}w_0 a^2$$
$$M_C - M_B = -\tfrac{1}{2}w_0 a(L - a)$$
$$M_C = -\tfrac{1}{6}w_0 a(3L - a)$$

The sketch of the bending-moment diagram is completed by recalling that $dM/dx = V$. We find that between A and B the diagram is represented by a cubic curve with zero slope at A, and between B and C by a straight line.

SAMPLE PROBLEM 7.7

The simple beam AC is loaded by a couple of magnitude T applied at point B. Draw the shear and bending-moment diagrams of the beam.

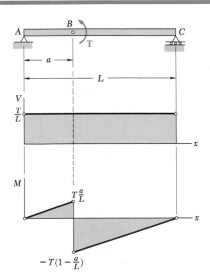

Solution. The entire beam is taken as a free body, and we obtain

$$\mathbf{R}_A = \frac{T}{L}\uparrow \qquad \mathbf{R}_C = \frac{T}{L}\downarrow$$

The shear at any section is constant and equal to T/L. Since a couple is applied at B, the bending-moment diagram is discontinuous at B; the bending moment decreases suddenly by an amount equal to T.

PROBLEMS

7.37 Using the methods of Sec. 7.5, solve Prob. 7.17.
7.38 Using the methods of Sec. 7.5, solve Prob. 7.18.
7.39 Using the methods of Sec. 7.5, solve Prob. 7.19.
7.40 Using the methods of Sec. 7.5, solve Prob. 7.22.
7.41 Using the methods of Sec. 7.5, solve Prob. 7.23.
7.42 Using the methods of Sec. 7.5, solve Prob. 7.24.
7.43 Using the methods of Sec. 7.5, solve Prob. 7.25.
7.44 Using the methods of Sec. 7.5, solve Prob. 7.26.
7.45 Using the methods of Sec. 7.5, solve Prob. 7.29.

7.46 through 7.49 Draw the shear and bending-moment diagrams for the beam and loading shown.

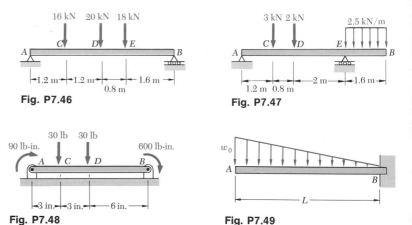

Fig. P7.46

Fig. P7.47

Fig. P7.48

Fig. P7.49

7.50 through 7.53 Draw the shear and bending-moment diagrams for the beam and loading shown, and determine the location and magnitude of the maximum bending moment.

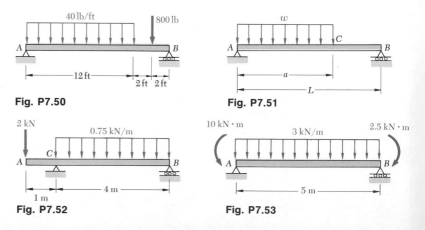

Fig. P7.50

Fig. P7.51

Fig. P7.52

Fig. P7.53

7.54 Determine the equations of the shear and bending-moment curves for the beam and loading of Prob. 7.17. (Place the origin at point A.)

7.55 Determine the equations of the shear and bending-moment curves for the beam and loading of Prob. 7.19. (Place the origin at point A.)

7.56 and 7.57 Determine the equations of the shear and bending-moment curves for the given beam and loading. Also determine the magnitude and location of the maximum bending moment in the beam.

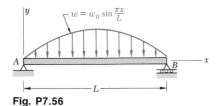

Fig. P7.56

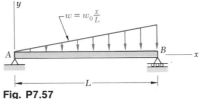

Fig. P7.57

∗7.58 The beam AB is acted upon by the uniformly distributed load of 100 lb/ft and by two forces **P** and **Q**. It has been experimentally determined that the bending moment is $+400$ lb · ft at point D and $+150$ lb · ft at point E. Draw the shear and bending-moment diagrams for the beam.

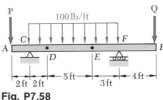

Fig. P7.58

∗7.59 The beam AB supports the uniformly distributed load of 400 N/m and two concentrated loads **P** and **Q**. The bending moment is known to be -17.7 kN · m at point A and -9.3 kN · m at point C. Draw the shear and bending-moment diagrams for the beam.

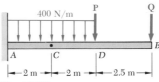

Fig. P7.59

7.60 A beam AB is loaded by couples spaced uniformly along its length. Assuming that the couples may be represented by a uniformly distributed couple loading of m lb · ft/ft, draw the shear and bending-moment diagrams for the beam when it is supported (*a*) as shown, (*b*) as a cantilever with a fixed support at A and no support at B.

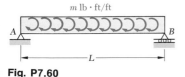

Fig. P7.60

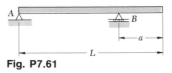

Fig. P7.61

∗7.61 A uniform beam of weight w per unit length is supported as shown. Determine the distance a if the maximum absolute value of the bending moment is to be as small as possible. (See hint of Prob. 7.36.)

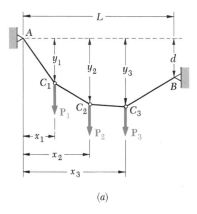

(a)

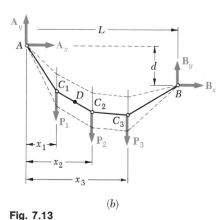

(b)

Fig. 7.13

CABLES

*7.6. Cables with Concentrated Loads. Cables are used in many engineering applications, such as suspension bridges, transmission lines, aerial tramways, guy wires for high towers, etc. Cables may be divided into two categories, according to their loading: (1) cables supporting concentrated loads, (2) cables supporting distributed loads. In this section, we shall examine cables of the first category.

Consider a cable attached to two fixed points A and B and supporting n given vertical concentrated loads $\mathbf{P}_1$, $\mathbf{P}_2$, ..., $\mathbf{P}_n$ (Fig. 7.13*a*). We assume that the cable is *flexible*, i.e., that its resistance to bending is small and may be neglected. We further assume that the *weight of the cable is negligible* compared with the loads supported by the cable. Any portion of cable between successive loads may therefore be considered as a two-force member, and the internal forces at any point in the cable reduce to a *force of tension directed along the cable*.

We assume that each of the loads lies in a given vertical line, i.e., that the horizontal distance from support A to each of the loads is known; we also assume that the horizontal and vertical distances between the supports are known. We propose to determine the shape of the cable, i.e., the vertical distance from A to each of the points C_1, C_2, ..., C_n, and also the tension T in each portion of the cable.

We first draw the free-body diagram of the entire cable (Fig. 7.13*b*). Since the slope of the portions of cable attached at A and B is not known, the reactions at A and B must be represented by two components each. Thus, four unknowns are involved, and the three equations of equilibrium are not sufficient to determine the reactions at A and B.† We must therefore obtain an additional equation by considering the equilibrium of a portion of the cable. This is possible if we know the coordinates x and y of a point D of the cable. Drawing the free-body diagram of the portion of cable AD (Fig. 7.14*a*) and writing $\Sigma M_D = 0$, we obtain an additional relation between the scalar components A_x and A_y and may determine the reactions at A and B. The problem would remain indeterminate, however, if we did not know the coordinates of D, unless some other relation between A_x and A_y (or between B_x and B_y) were given. The cable might hang in any of various possible ways, as indicated by the dashed lines in Fig. 7.13*b*.

Once A_x and A_y have been determined, the vertical distance from A to any point of the cable may be easily found. Considering point C_2, for example, we draw the free-body diagram of

† Clearly, the cable is not a rigid body; the equilibrium equations represent, therefore, *necessary but not sufficient conditions* (see Sec. 6.11).

the portion of cable AC_2 (Fig. 7.14b). Writing $\Sigma M_{C_2} = 0$, we obtain an equation which may be solved for y_2. Writing $\Sigma F_x = 0$ and $\Sigma F_y = 0$, we obtain the components of the force **T** representing the tension in the portion of cable to the right of C_2. We observe that $T \cos \theta = -A_x$; *the horizontal component of the tension force is the same at any point of the cable.* It follows that the tension T is maximum when $\cos \theta$ is minimum, i.e., in the portion of cable which has the largest angle of inclination θ. Clearly, this portion of cable must be adjacent to one of the two supports of the cable.

*7.7. Cables with Distributed Loads. Consider a cable attached to two fixed points A and B and carrying a *distributed load* (Fig. 7.15a). We saw in the preceding section that, for a cable supporting concentrated loads, the internal force at any point is a force of tension directed along the cable. In the case of a cable carrying a distributed load, the cable hangs in the shape of a curve, and the internal force at a point D is a force of tension **T** *directed along the tangent to the curve.* Given a certain distributed load, we propose in this section to determine the tension at any point of the cable. We shall also see in the following sections how the shape of the cable may be determined for two particular types of distributed loads.

Considering the most general case of distributed load, we draw the free-body diagram of the portion of cable extending from the lowest point C to a given point D of the cable (Fig. 7.15b). The forces acting on the free body are the tension force $\mathbf{T}_0$ at C, which is horizontal, the tension force **T** at D, directed along the tangent to the cable at D, and the resultant **W** of the distributed load supported by the portion of cable CD. Drawing the corresponding force triangle (Fig. 7.15c), we obtain the following relations:

$$T \cos \theta = T_0 \qquad T \sin \theta = W \qquad (7.5)$$

$$T = \sqrt{T_0^2 + W^2} \qquad \tan \theta = \frac{W}{T_0} \qquad (7.6)$$

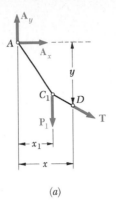

(a)

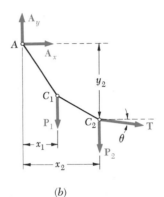

(b)

Fig. 7.14

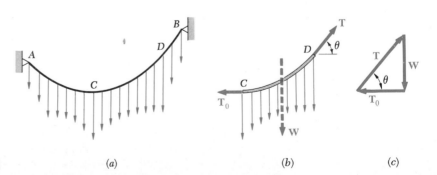

Fig. 7.15   (a)   (b)   (c)

From the relations (7.5), it appears that the horizontal component of the tension force **T** is the same at any point and that the vertical component of **T** is equal to the magnitude W of the load measured from the lowest point. Relations (7.6) show that the tension T is minimum at the lowest point and maximum at one of the two points of support.

7.8. Parabolic Cable. Let us assume, now, that the cable AB carries a load *uniformly distributed along the horizontal* (Fig. 7.16a). Cables of suspension bridges may be assumed loaded in this way, since the weight of the cables is small compared with the weight of the roadway. We denote by w the load per unit length (*measured horizontally*) and express it in N/m or in lb/ft. Choosing coordinate axes with origin at the

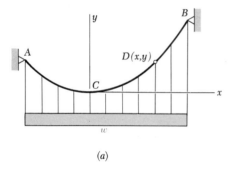

(a)

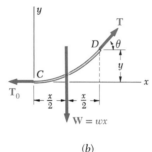

(b)   **Fig. 7.16**

lowest point C of the cable, we find that the magnitude W of the total load carried by the portion of cable extending from C to the point D of coordinates x and y is $W = wx$. The relations (7.6) defining the magnitude and direction of the tension force at D become

$$T = \sqrt{T_0^2 + w^2 x^2} \qquad \tan\theta = \frac{wx}{T_0} \qquad (7.7)$$

Moreover, the distance from D to the line of action of the resultant **W** is equal to half the horizontal distance from C to D (Fig. 7.16b). Summing moments about D, we write

$$+\curvearrowleft\Sigma M_D = 0: \qquad wx\frac{x}{2} - T_0 y = 0$$

$$y = \frac{wx^2}{2T_0} \qquad (7.8)$$

This is the equation of a *parabola* with a vertical axis and its vertex at the origin of coordinates. The curve formed by cables loaded uniformly along the horizontal is thus a parabola.†

When the supports A and B of the cable have the same elevation, the distance L between the supports is called the *span* of the cable and the vertical distance h from the supports to the lowest point is called the *sag* of the cable (Fig. 7.17a). If the span and sag of a cable are known, and if the load w per unit horizontal length is given, the minimum tension T_0 may be found by substituting $x = L/2$ and $y = h$ in formula (7.8). Formulas (7.7) and (7.8) will then define the tension at any point and the shape of the cable.

When the supports have different elevations, the position of the lowest point of the cable is not known and the coordinates x_A, y_A and x_B, y_B of the supports must be determined. To this effect, we express that the coordinates of A and B satisfy Eq. (7.8) and that $x_B - x_A = L$, $y_B - y_A = d$, where L and d denote, respectively, the horizontal and vertical distances between the two supports (Fig. 7.17b and c).

The length of the cable from its lowest point C to its support B may be obtained from the formula

$$s_B = \int_0^{x_B} \sqrt{1 + \left(\frac{dy}{dx}\right)^2}\, dx \qquad (7.9)$$

Differentiating (7.8), we obtain the derivative $dy/dx = wx/T_0$; substituting into (7.9) and using the binomial theorem to expand the radical in an infinite series, we have

$$s_B = \int_0^{x_B} \sqrt{1 + \frac{w^2x^2}{T_0^2}}\, dx$$

$$= \int_0^{x_B} \left(1 + \frac{w^2x^2}{2T_0^2} - \frac{w^4x^4}{8T_0^4} + \cdots\right) dx$$

$$= x_B\left(1 + \frac{w^2x_B^2}{6T_0^2} - \frac{w^4x_B^4}{40T_0^4} + \cdots\right)$$

and, since $wx_B^2/2T_0 = y_B$,

$$s_B = x_B\left[1 + \frac{2}{3}\left(\frac{y_B}{x_B}\right)^2 - \frac{2}{5}\left(\frac{y_B}{x_B}\right)^4 + \cdots\right] \qquad (7.10)$$

The series converges for values of the ratio y_B/x_B less than 0.5; in most cases, this ratio is much smaller, and only the first two terms of the series need be computed.

† Cables hanging under their own weight are not loaded uniformly along the horizontal, and they do not form a parabola. The error introduced by assuming a parabolic shape for cables hanging under their own weight, however, is small when the cable is sufficiently taut. A complete discussion of cables hanging under their own weight is given in the next section.

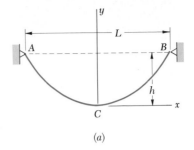

(a)

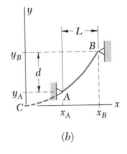

(b)

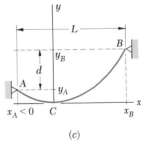

(c)

Fig. 7.17

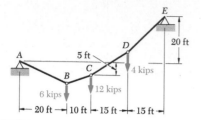

SAMPLE PROBLEM 7.8

The cable AE supports three vertical loads from the points indicated. If point C is 5 ft below the left support, determine (a) the elevations of points B and D, (b) the maximum slope and the maximum tension in the cable.

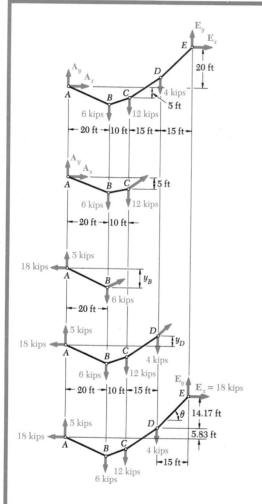

Solution. The reaction components $\mathbf{A}_x$ and $\mathbf{A}_y$ are determined as follows:

Free Body: Entire Cable

$+\,\text{↑}\Sigma M_E = 0:$
$$A_x(20\text{ ft}) - A_y(60\text{ ft}) + (6\text{ kips})(40\text{ ft})$$
$$+\ (12\text{ kips})(30\text{ ft}) + (4\text{ kips})(15\text{ ft}) = 0$$
$$20A_x - 60A_y + 660 = 0$$

Free Body: ABC

$+\,\text{↑}\Sigma M_C = 0:\qquad -A_x(5\text{ ft}) - A_y(30\text{ ft}) + (6\text{ kips})(10\text{ ft}) = 0$
$$-5A_x - 30A_y + 60 = 0$$

Solving the two equations simultaneously, we obtain

$$A_x = -18\text{ kips} \qquad \mathbf{A}_x = 18\text{ kips} \leftarrow$$
$$A_y = +5\text{ kips} \qquad \mathbf{A}_y = 5\text{ kips} \uparrow$$

a. Elevation of Point B. Considering the portion of cable AB as a free body, we write

$+\,\text{↑}\Sigma M_B = 0:\qquad (18\text{ kips})y_B - (5\text{ kips})(20\text{ ft}) = 0$
$$y_B = 5.56\text{ ft below }A \quad \blacktriangleleft$$

Elevation of Point D. Using the portion of cable $ABCD$ as a free body, we write

$+\,\text{↑}\Sigma M_D = 0:$
$$-(18\text{ kips})\,y_D - (5\text{ kips})(45\text{ ft}) + (6\text{ kips})(25\text{ ft}) + (12\text{ kips})(15\text{ ft}) = 0$$
$$y_D = 5.83\text{ ft above }A \quad \blacktriangleleft$$

b. Maximum Slope and Maximum Tension. We observe that the maximum slope occurs in portion DE. Since the horizontal component of the tension is constant and equal to 18 kips, we write

$$\tan\theta = \frac{14.17\text{ ft}}{15\text{ ft}} \qquad \theta = 43.4° \quad \blacktriangleleft$$

$$T_{\text{max}} = \frac{18\text{ kips}}{\cos\theta} \qquad T_{\text{max}} = 24.8\text{ kips} \quad \blacktriangleleft$$

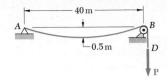

SAMPLE PROBLEM 7.9

A light cable is attached to a support at A, passes over a small pulley at B, and supports a load **P**. Knowing that the sag of the cable is 0.5 m and that the mass per unit length of the cable is 0.75 kg/m, determine (a) the magnitude of the load **P**, (b) the slope of the cable at B, and (c) the total length of the cable from A to B. Since the ratio of the sag to the span is small, assume the cable to be parabolic. Also, neglect the weight of the portion of cable from B to D.

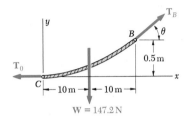

a. **Load P.** We denote by C the lowest point of the cable and draw the free-body diagram of the portion CB of cable. Assuming the load to be uniformly distributed along the horizontal, we write

$$w = (0.75 \text{ kg/m})(9.81 \text{ m/s}^2) = 7.36 \text{ N/m}$$

The total load for the portion CB of the cable is

$$W = wx_B = (7.36 \text{ N/m})(20 \text{ m}) = 147.2 \text{ N}$$

and is applied halfway between C and B. Summing moments about B, we write

$$+\gamma\Sigma M_B = 0: \qquad (147.2 \text{ N})(10 \text{ m}) - T_0(0.5 \text{ m}) = 0 \qquad T_0 = 2944 \text{ N}$$

From the force triangle we obtain

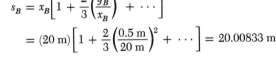

$$T_B = \sqrt{T_0^2 + W^2}$$
$$= \sqrt{(2944 \text{ N})^2 + (147.2 \text{ N})^2} = 2948 \text{ N}$$

Since the tension on each side of the pulley is the same, we find

$$P = T_B = 2948 \text{ N} \qquad \blacktriangleleft$$

b. **Slope of Cable at B.** We also obtain from the force triangle

$$\tan \theta = \frac{W}{T_0} = \frac{147.2 \text{ N}}{2944 \text{ N}} = 0.05$$

$$\theta = 2.9° \qquad \blacktriangleleft$$

c. **Length of Cable.** Applying Eq. (7.10) between C and B, we write

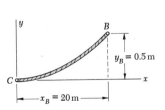

$$s_B = x_B\left[1 + \frac{2}{3}\left(\frac{y_B}{x_B}\right)^2 + \cdots\right]$$
$$= (20 \text{ m})\left[1 + \frac{2}{3}\left(\frac{0.5 \text{ m}}{20 \text{ m}}\right)^2 + \cdots\right] = 20.00833 \text{ m}$$

The total length of the cable between A and B is twice this value,

$$\text{Length} = 2s_B = 40.0167 \text{ m} \qquad \blacktriangleleft$$

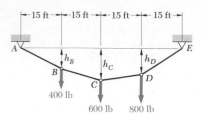

—15 ft—|—15 ft—|—15 ft—|—15 ft—

Fig. P7.62 and P7.63

PROBLEMS

7.62 Three loads are suspended as shown from the cable. Knowing that $h_C = 12$ ft, determine (a) the components of the reaction at E, (b) the maximum value of the tension in the cable.

7.63 Determine the sag at point C if the maximum tension in the cable is 2600 lb.

7.64 If $a = 2$ m and $b = 2.25$ m, determine the components of the reaction at E for the cable and loading shown.

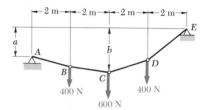

Fig. P7.64, P7.65, and P7.66

7.65 If $a = b = 1.25$ m, determine the components of the reaction at E and the maximum tension in the cable.

7.66 Determine the distance a if the portion BC of the cable is horizontal and if the maximum tension in the cable is 2600 N.

7.67 The center span of the George Washington Bridge, as originally constructed, consisted of a uniform roadway suspended from four cables. The uniform load supported by each cable was $w = 9.75$ kips/ft along the horizontal. Knowing that the span L is 3500 ft and that the sag h is 316 ft, determine the maximum and minimum tension in each cable.

7.68 Determine the length of each of the cables used in the center span of the George Washington Bridge. (See Prob. 7.67 for data.)

7.69 Two cables of the same gage are attached to a transmission tower at B. Since the tower is slender, the horizontal component of the resultant of the forces exerted by the cables at B is to be zero. Assuming the cables to be parabolic, determine the required sag h of cable AB.

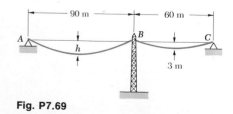

Fig. P7.69

7.70 An electric wire having a mass per unit length of 0.3 kg/m is strung between two insulators at the same elevation and 30 m apart. If the maximum tension in the wire is to be 400 N, determine the smallest value of the sag which may be used. (Assume the wire to be parabolic.)

7.71 Knowing that a 51-m length of wire was used in spanning a horizontal distance of 50 m, determine the approximate sag of the wire. Assume the wire to be parabolic. [*Hint.* Use only the first two terms of Eq. (7.10).]

7.72 A cable of length $L + \Delta$ is suspended between two points which are at the same elevation and a distance L apart. (*a*) Assuming that Δ is small compared to L and that the cable is parabolic, determine the approximate sag in terms of L and Δ. (*b*) If $L = 100$ ft and $\Delta = 4$ ft, determine the approximate sag. [*Hint.* Use only the first two terms of Eq. (7.10).]

7.73 Before being fed into a printing press located to the right of D, a continuous sheet of paper weighing 0.20 lb/ft passes over rollers at A and B. Assuming that the curve formed by the sheet is parabolic, determine the location of the lowest point C and the maximum tension in the sheet.

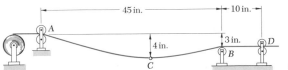

Fig. P7.73

7.74 A steam pipe has a mass per unit length of 70 kg/m. It passes between two buildings 20 m apart and is supported by a system of cables as shown. Assuming that the cable system causes the same loading as a single uniform cable having a mass per unit length of 5 kg/m, determine the location of the lowest point C of the cable and the maximum cable tension.

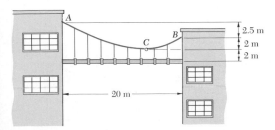

Fig. P7.74

*7.75 The total weight of cable AC is 60 lb. Assuming that the weight of the cable is distributed uniformly along the horizontal, determine the sag h and the slope of the cable at A and C.

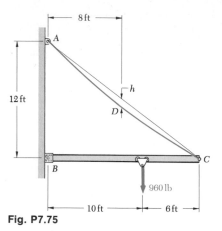

Fig. P7.75

*7.76 Solve Prob. 7.75, assuming that the total weight of cable AC is 120 lb.

7.77 Cable AB supports a load distributed uniformly along the horizontal as shown. The lowest point of the cable is 3 m below the support A. Determine the maximum and minimum values of the tension in the cable.

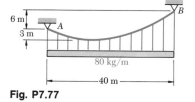

Fig. P7.77

*7.78 A cable AB of span L and a simple beam $A'B'$ of the same span are subjected to identical vertical loadings as shown. Show that the magnitude of the bending moment at a point C' in the beam is equal to the product T_0h, where T_0 is the magnitude of the horizontal component of the tension force in the cable and h is the vertical distance between point C of the cable and the chord joining the points of support A and B.

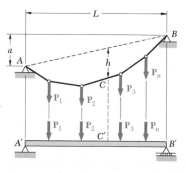

Fig. P7.78

✱7.79 Show that the curve assumed by a cable which carries a distributed load $w(x)$ is defined by the differential equation $d^2y/dx^2 = w(x)/T_0$, where T_0 is the tension at the lowest point.

✱7.80 Using the property indicated in Prob. 7.79, determine the curve assumed by a cable of span L and sag h carrying a distributed load $w = w_0 \cos(\pi x/L)$, where x is measured from mid-span. Also determine the maximum and minimum values of the tension.

✱7.81 A large number of ropes are tied to a light wire which is suspended from two points A and B at the same level. Show that, if the lower ends of the ropes have been cut so that they lie in the same horizontal line and if the ropes are kept uniformly spaced, the curve assumed by the wire ACB is

$$y + d = d \cosh\left(\frac{w_r}{T_0 e}\right)^{1/2} x$$

where w_r is the weight per unit length of the ropes, d is the length of the shortest rope, and e is the horizontal distance between adjacent ropes. (*Hint.* Use the property indicated in Prob. 7.79.)

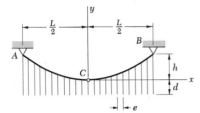

Fig. P7.81

✱7.82 If the weight per unit length of the cable AB is $w_0/\cos^2\theta$, prove that the curve formed by the cable is a circular arc. (*Hint.* Use the property indicated in Prob. 7.79.)

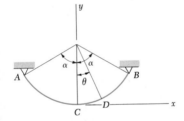

Fig. P7.82

✱7.9. Catenary. We shall consider now a cable AB carrying a load *uniformly distributed along the cable itself* (Fig. 7.18a). Cables hanging under their own weight are loaded in this way. We denote by w the load per unit length (*measured along the cable*) and express it in N/m or in lb/ft. The magnitude W of the total load carried by a portion of cable of length s extending from the lowest point C to a point D is $W = ws$. Substituting this value for W in formula (7.6), we obtain the tension at D,

$$T = \sqrt{T_0^2 + w^2 s^2}$$

In order to simplify the subsequent computations, we shall introduce the constant $c = T_0/w$. We thus write

$$T_0 = wc \qquad W = ws \qquad T = w\sqrt{c^2 + s^2} \qquad (7.11)$$

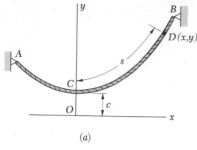

(a)

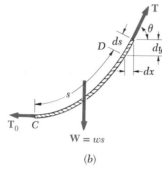

(b)

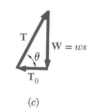

(c)

Fig. 7.18

The free-body diagram of the portion of cable CD is shown in Fig. 7.18b. This diagram, however, cannot be used to obtain directly the equation of the curve assumed by the cable, since we do not know the horizontal distance from D to the line of action of the resultant $\mathbf{W}$ of the load. To obtain this equation, we shall write first that the horizontal projection of a small element of cable of length ds is $dx = ds \cos \theta$. Observing from Fig. 7.18c that $\cos \theta = T_0/T$ and using (7.11), we write

$$dx = ds \cos \theta = \frac{T_0}{T} ds = \frac{wc \, ds}{w \sqrt{c^2 + s^2}} = \frac{ds}{\sqrt{1 + s^2/c^2}}$$

Selecting the origin O of the coordinates at a distance c directly below C (Fig. 7.18a) and integrating from $C(0,c)$ to $D(x,y)$, we obtain†

$$x = \int_0^s \frac{ds}{\sqrt{1 + s^2/c^2}} = c \left[\sinh^{-1} \frac{s}{c} \right]_0^s = c \sinh^{-1} \frac{s}{c}$$

This equation, which relates the length s of the portion of cable CD and the horizontal distance x, may be written in the form

$$s = c \sinh \frac{x}{c} \qquad (7.15)$$

The relation between the coordinates x and y may now be obtained by writing $dy = dx \tan \theta$. Observing from Fig. 7.18c that $\tan \theta = W/T_0$ and using (7.11) and (7.15), we write

$$dy = dx \tan \theta = \frac{W}{T_0} dx = \frac{s}{c} dx = \sinh \frac{x}{c} dx$$

Integrating from $C(0,c)$ to $D(x,y)$ and using (7.12) and (7.13), we obtain

† This integral may be found in all standard integral tables. The function

$$z = \sinh^{-1} u$$

(read "arc hyperbolic sine u") is the *inverse* of the function $u = \sinh z$ (read "hyperbolic sine z"). This function and the function $v = \cosh z$ (read "hyperbolic cosine z") are defined as follows:

$$u = \sinh z = \tfrac{1}{2}(e^z - e^{-z}) \qquad v = \cosh z = \tfrac{1}{2}(e^z + e^{-z})$$

Numerical values of the functions $\sinh z$ and $\cosh z$ are .found in *tables of hyperbolic functions*. They may also be computed on most calculators either directly or from the above definitions. The student is referred to any calculus text for a complete description of the properties of these functions. In this section, we shall make use only of the following properties, which may be easily derived from the above definitions:

$$\frac{d \sinh z}{dz} = \cosh z \qquad \frac{d \cosh z}{dz} = \sinh z \qquad (7.12)$$

$$\sinh 0 = 0 \qquad \cosh 0 = 1 \qquad (7.13)$$

$$\cosh^2 z - \sinh^2 z = 1 \qquad (7.14)$$

$$y - c = \int_0^x \sinh \frac{x}{c}\, dx = c\left[\cosh \frac{x}{c}\right]_0^x = c\left(\cosh \frac{x}{c} - 1\right)$$

$$y = c \cosh \frac{x}{c} \qquad (7.16)$$

This is the equation of a *catenary* with vertical axis. The ordinate c of the lowest point C is called the *parameter* of the catenary. Squaring both sides of Eqs. (7.15) and (7.16), subtracting, and taking (7.14) into account, we obtain the following relation between y and s:

$$y^2 - s^2 = c^2 \qquad (7.17)$$

Solving (7.17) for s^2 and carrying into the last of the relations (7.11), we write these relations as follows:

$$T_0 = wc \qquad W = ws \qquad T = wy \qquad (7.18)$$

The last relation indicates that the tension at any point D of the cable is proportional to the vertical distance from D to the horizontal line representing the x axis.

When the supports A and B of the cable have the same elevation, the distance L between the supports is called the *span* of the cable and the vertical distance h from the supports to the lowest point C is called the *sag* of the cable. These definitions are the same that were given in the case of parabolic cables, but it should be noted that, because of our choice of coordinate axes, the sag h is now

$$h = y_A - c \qquad (7.19)$$

It should also be observed that certain catenary problems involve transcendental equations which must be solved by successive approximations (see Sample Prob. 7.10). When the cable is fairly taut, however, the load may be assumed uniformly distributed *along the horizontal* and the catenary may be replaced by a parabola. The solution of the problem is thus greatly simplified, while the error introduced is small.

When the supports A and B have different elevations, the position of the lowest point of the cable is not known. The problem may be solved then in a manner similar to that indicated for parabolic cables, by expressing that the cable must pass through the supports and that $x_B - x_A = L$, $y_B - y_A = d$, where L and d denote, respectively, the horizontal and vertical distances between the two supports.

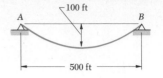

A uniform cable weighing 3 lb/ft is suspended between two points A and B as shown. Determine (a) the maximum and minimum values of the tension in the cable, (b) the length of the cable.

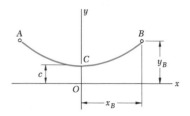

Solution. *Equation of Cable.* The origin of coordinates is placed at a distance c below the lowest point of the cable. The equation of the cable is given by Eq. (7.16),

$$y = c \cosh \frac{x}{c}$$

The coordinates of point B are

$$x_B = 250 \text{ ft} \qquad y_B = 100 + c$$

Substituting these coordinates into the equation of the cable, we obtain

$$100 + c = c \cosh \frac{250}{c}$$

$$\frac{100}{c} + 1 = \cosh \frac{250}{c}$$

The value of c is determined by assuming successive trial values, as shown in the following table:

c	$\dfrac{250}{c}$	$\dfrac{100}{c}$	$\dfrac{100}{c} + 1$	$\cosh \dfrac{250}{c}$
300	0.833	0.333	1.333	1.367
350	0.714	0.286	1.286	1.266
330	0.758	0.303	1.303	1.301
328	0.762	0.305	1.305	1.305

Taking $c = 328$, we have

$$y_B = 100 + c = 428 \text{ ft}$$

a. Maximum and Minimum Values of the Tension. Using Eqs. (7.18), we obtain

$$T_{\min} = T_0 = wc = (3 \text{ lb/ft})(328 \text{ ft}) \qquad T_{\min} = 984 \text{ lb} \blacktriangleleft$$
$$T_{\max} = T_B = wy_B = (3 \text{ lb/ft})(428 \text{ ft}) \qquad T_{\max} = 1284 \text{ lb} \blacktriangleleft$$

b. Length of Cable. One-half the length of the cable is found by solving Eq. (7.17),

$$y_B^2 - s_{CB}^2 = c^2 \qquad s_{CB}^2 = y_B^2 - c^2 = (428)^2 - (328)^2 \qquad s_{CB} = 275 \text{ ft}$$

The total length of the cable is therefore

$$s_{AB} = 2s_{CB} = 2(275 \text{ ft}) \qquad s_{AB} = 550 \text{ ft} \blacktriangleleft$$

PROBLEMS

7.83   An aerial tramway cable of length 200 m and mass 1000 kg is suspended between two points at the same elevation. Knowing that the sag is 50 m, find the horizontal distance between supports and the maximum tension.

7.84   A 100-ft rope is strung between the roofs of two buildings, each of height 30 ft. The maximum tension is found to be 50 lb and the lowest point of the cable is observed to be 10 ft above the ground. Determine the horizontal distance between the buildings and the total weight of the rope.

7.85   A 200-ft steel surveying tape weighs 4 lb. If the tape is stretched between two points at the same elevation and pulled until the tension at each end is 16 lb, determine the horizontal distance between the ends of the tape. Neglect the elongation of the tape due to the tension.

7.86   A copper transmission wire having a mass per unit length of 0.75 kg/m is attached to two insulators at the same elevation and 100 m apart. It has been established that the horizontal component of the tension in the wire must be 900 N if the insulators are not to be bent. Determine (a) the length of cable which should be used, (b) the resulting sag, (c) the resulting maximum tension.

7.87   A tramway wire rope weighing 5 lb/ft is suspended across a canyon; it is attached to two supports at the same elevation and 750 ft apart. Knowing that the sag is 100 ft, determine the total length of the cable and the maximum tension.

7.88   A counterweight of mass 200 kg is attached to a cable which passes over a small pulley at A and is attached to a support at B. If the sag is 3 m, determine (a) the length of the cable from A to B, (b) the mass per unit length of the cable. Neglect the mass of the cable from A to D.

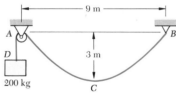

Fig. P7.88

7.89   A chain of length 10 m and total mass 20 kg is suspended between two points at the same elevation and 5 m apart. Determine the sag and the maximum tension.

7.90   A 75-ft wire is suspended between two points at the same elevation and 50 ft apart. Knowing that the maximum tension is 15 lb, determine the sag and the total weight of the wire.

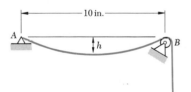

Fig. P7.91

7.91   A uniform cord 30 in. long passes over a frictionless pulley at B and is attached to a rigid support at A. Determine the smaller of the two values of h for which the cord is in equilibrium.

*7.92   The 4-m cable AB has a total mass of 10 kg and is attached to collars at A and B which may slide freely on the rods shown. Neglecting the weight of the collars, determine (a) the magnitude of the horizontal force $\mathbf{F}$ so that $h = a$, (b) the corresponding value of h and a, (c) the maximum tension in the cable.

7.93   Denoting by θ the angle formed by a uniform cable and the horizontal, show that at any point $y = c/\cos \theta$.

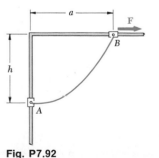

Fig. P7.92

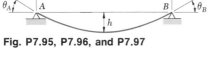

Fig. P7.95, P7.96, and P7.97

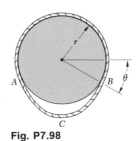

Fig. P7.98

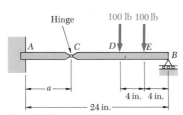

Fig. P7.100 and P7.101

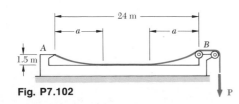

Fig. P7.102

*7.94 (a) Determine the maximum allowable horizontal span for a uniform cable of weight w per unit length if the tension in the cable is not to exceed the value T_m. (b) Using the result of part a, find the maximum span of a drawn-steel wire for which $w = 0.2$ lb/ft and $T_m = 6000$ lb.

*7.95 A chain has a mass per unit length of 6 kg/m and is supported as shown. Knowing that the span L is 8 m, determine the *two* values of the sag h for which the maximum tension is 750 N.

*7.96 Determine the sag-to-span ratio for which the maximum tension in the cable is equal to the total weight of the entire cable AB.

*7.97 A cable, of weight w per unit length, is suspended between two points at the same elevation and a distance L apart. Determine the sag-to-span ratio for which the maximum tension is as small as possible. What are the corresponding values of θ_B and T_m?

*7.98 A cable, of weight w per unit length, is looped over a cylinder and is in contact with the cylinder above points A and B. Knowing that $\theta = 30°$, determine (a) the length of the cable, (b) the tension in the cable at C. (*Hint.* Use the property indicated in Prob. 7.93.)

REVIEW PROBLEMS

7.99 A 12-mm-diameter wire rope has a mass per unit length of 0.5 kg/m and is suspended from two supports at the same elevation and 100 m apart. If the sag is 25 m, determine (a) the total length of the cable, (b) the maximum tension.

7.100 Draw the shear and bending-moment diagrams for the beam and loading shown when $a = 4$ in.

7.101 Determine the distance a for which the maximum absolute value of the bending moment is as small as possible.

7.102 The central portion of cable AB is supported by a frictionless horizontal surface. Knowing that the mass per unit length of the cable is 3 kg/m and assuming the cable to be parabolic, determine the magnitude of the load P when $a = 8$ m.

7.103 Solve Prob. 7.102 when (a) $a = 6$ m, (b) $a = 12$ m.

7.104 A 300-lb load is attached to a small pulley which may roll on the cable *ACB*. The pulley and load are held in the position shown by a second cable *DE* which is parallel to the portion *CB* of the main cable. Determine (*a*) the reactions at *A* and *B*, (*b*) the tension in cable *ACB*, (*c*) the tension in cable *DE*. Neglect the radius of the pulleys and the weight of the cables.

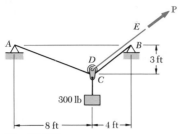

Fig. P7.104

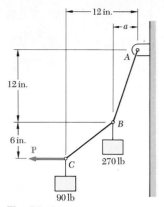

Fig. P7.105

7.105 Cable *ABC* supports two loads as shown. Determine (*a*) the required magnitude of the force **P**, (*b*) the corresponding distance *a*.

7.106 The axis of the curved member *ABC* is parabolic. If the tension in the cable *AC* is 1000 N and the magnitude of the force **P** is 500 N, determine the internal forces at a distance $\frac{1}{4}L$ from the left end. Assume $h = 80$ mm and $L = 480$ mm.

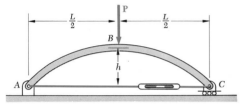

Fig. P7.106 and P7.107

7.107 Denoting by *T* the tension in the cable *AC*, determine (*a*) the internal forces just to the left of point *B*, (*b*) the value of *T* for which the bending moment at *B* is zero.

7.108 The axis of the curved member *AB* is a parabola with vertex at *A*. If a vertical load **P** of magnitude 3 kN is applied at *A*, determine the internal forces at *J* when $h = 225$ mm, $L = 750$ mm, and $a = 500$ mm.

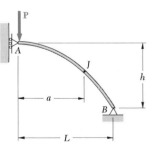

Fig. P7.108 and P7.109

7.109 The axis of the curved member *AB* is a parabola with vertex at *A*. (*a*) Determine the magnitude and location of the maximum bending moment. (*b*) Noting that *AB* is a two-force member, determine the maximum perpendicular distance from the chord *AB* to the curved member.

★7.110 A uniform semicircular rod of weight *W* is attached to a pin at *A* and bears against a frictionless surface at *B*. Determine (*a*) the bending moment at point *J* when $\theta = 90°$, (*b*) the magnitude and location of the maximum bending moment.

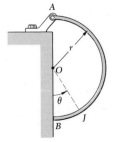

Fig. P7.110

CHAPTER 8 Friction

8.1. Introduction. In the preceding chapters, it was assumed that surfaces in contact were either *frictionless* or *rough*. If they were frictionless, the force each surface exerted on the other was normal to the surfaces and the two surfaces could move freely with respect to each other. If they were rough, it was assumed that tangential forces could develop to prevent the motion of one surface with respect to the other.

This view was a simplified one. Actually, no perfectly frictionless surface exists. When two surfaces are in contact, tangential forces, called *friction forces*, will always develop if one attempts to move one surface with respect to the other. On the other hand, these friction forces are limited in magnitude and will not prevent motion if sufficiently large forces are applied. The distinction between frictionless and rough surfaces is thus a matter of degree. This will be seen more clearly in the present chapter, which is devoted to the study of friction and of its applications to common engineering situations.

There are two types of friction: *dry friction*, sometimes called *Coulomb friction*, and *fluid friction*. Fluid friction develops between layers of fluid moving at different velocities. Fluid friction is of great importance in problems involving the flow of fluids through pipes and orifices or dealing with bodies immersed in moving fluids. It is also basic in the analysis of the motion of *lubricated mechanisms*. Such problems are considered in texts on fluid mechanics. We shall limit our present study to dry friction, i.e., to problems involving rigid bodies which are in contact along *nonlubricated* surfaces.

8.2. The Laws of Dry Friction. Coefficients of Friction. The laws of dry friction are best understood by the following experiment. A block of weight **W** is placed on a horizontal plane surface (Fig. 8.1a). The forces acting on the block are its weight **W** and the reaction of the surface. Since the weight has no horizontal component, the reaction of the surface also has no horizontal component; the reaction is therefore *normal* to the surface and is represented by **N** in Fig. 8.1a. Suppose, now, that a horizontal force **P** is applied to the block (Fig. 8.1b). If **P** is small, the block will not move; some other horizontal force must therefore exist, which balances **P**. This other force is the *static-friction force* **F**, which is actually the resultant of a great number of forces acting over the entire surface of contact between the block and the plane. The nature of these forces is not known exactly, but it is generally assumed that these forces are due to the irregularities of the surfaces in contact and also, to a certain extent, to molecular attraction.

If the force **P** is increased, the friction force **F** also increases, continuing to oppose **P**, until its magnitude reaches a certain *maximum value* F_m (Fig. 8.1c). If **P** is further increased, the friction force cannot balance it any more and the block starts sliding. As soon as the block has been set in motion, the magnitude of **F** drops from F_m to a lower value F_k. This is because there is less interpenetration between the irregularities of the surfaces in contact when these surfaces move with respect to each other. From then on, the block keeps sliding with increasing velocity while the friction force, denoted by **F**$_k$ and called the *kinetic-friction force*, remains approximately constant.

Experimental evidence shows that the maximum value F_m of the static-friction force is proportional to the normal component N of the reaction of the surface. We have

$$F_m = \mu_s N \tag{8.1}$$

where μ_s is a constant called the *coefficient of static friction*. Similarly, the magnitude F_k of the kinetic-friction force may be put in the form

$$F_k = \mu_k N \tag{8.2}$$

where μ_k is a constant called the *coefficient of kinetic friction*.

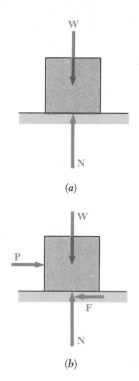

(a)

(b)

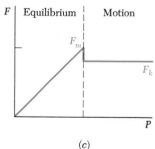

(c)

Fig. 8.1

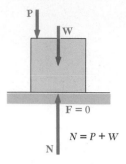

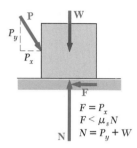

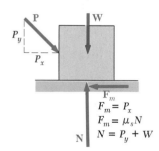

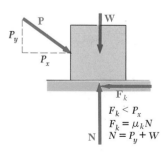

Fig. 8.2

The coefficients of friction μ_s and μ_k do not depend upon the area of the surfaces in contact. Both coefficients, however, depend strongly on the *nature* of the surfaces in contact. Since they also depend upon the exact condition of the surfaces, their value is seldom known with an accuracy greater than 5 percent. Approximate values of coefficients of static friction are given in Table 8.1 for various dry surfaces. The corresponding values

Table 8.1. Approximate Values of Coefficient of Static Friction for Dry Surfaces

Metal on metal	0.15–0.60
Metal on wood	0.20–0.60
Metal on stone	0.30–0.70
Metal on leather	0.30–0.60
Wood on wood	0.25–0.50
Wood on leather	0.25–0.50
Stone on stone	0.40–0.70
Earth on earth	0.20–1.00
Rubber on concrete	0.60–0.90

of the coefficient of kinetic friction would be about 25 percent smaller. Since coefficients of friction are dimensionless quantities, the values given in Table 8.1 may be used with both SI units and U.S. customary units.

From the description given above, it appears that four different situations may occur when a rigid body is in contact with a horizontal surface:

1. The forces applied to the body do not tend to move it along the surface of contact; there is no friction force (Fig. 8.2a).
2. The applied forces tend to move the body along the surface of contact but are not large enough to set it in motion. The friction force **F** which has developed may be found by solving the equations of equilibrium for the body. Since there is no evidence that the maximum value of the static-friction force has been reached, the equation $F_m = \mu_s N$ *cannot be used* to determine the friction force (Fig. 8.2b).

3. The applied forces are such that the body is just about to slide. We say that *motion is impending*. The friction force **F** has reached its maximum value F_m and, together with the normal force **N**, balances the applied forces. Both the equations of equilibrium and the equation $F_m = \mu_s N$ *may be used.* We also note that the friction force has a sense opposite to the sense of impending motion (Fig. 8.2c).

4. The body is sliding under the action of the applied forces, and the equations of equilibrium do not apply any more. However, **F** is now equal to the kinetic-friction force $\mathbf{F}_k$ and the equation $F_k = \mu_k N$ may be used. The sense of $\mathbf{F}_k$ is opposite to the sense of motion (Fig. 8.2d).

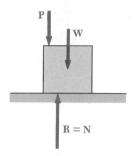

(a) No friction

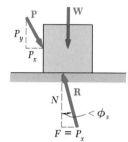

(b) No motion

8.3. Angles of Friction. It is sometimes found convenient to replace the normal force **N** and the friction force **F** by their resultant **R**. Let us consider again a block of weight **W** resting on a horizontal plane surface. If no horizontal force is applied to the block, the resultant **R** reduces to the normal force **N** (Fig. 8.3a). However, if the applied force **P** has a horizontal component $\mathbf{P}_x$ which tends to move the block, the force **R** will have a horizontal component **F** and, thus, will form a certain angle with the vertical (Fig. 8.3b). If $\mathbf{P}_x$ is increased until motion becomes impending, the angle between **R** and the vertical grows and reaches a maximum value (Fig. 8.3c). This value is called the *angle of static friction* and is denoted by ϕ_s. From the geometry of Fig. 8.3c, we note that

$$\tan \phi_s = \frac{F_m}{N} = \frac{\mu_s N}{N}$$

$$\tan \phi_s = \mu_s \tag{8.3}$$

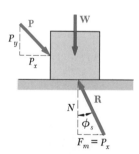

(c) Motion impending $\longrightarrow$

If motion actually takes place, the magnitude of the friction force drops to F_k; similarly, the angle between **R** and **N** drops to a lower value ϕ_k, called the *angle of kinetic friction* (Fig. 8.3d). From the geometry of Fig. 8.3d, we write

$$\tan \phi_k = \frac{F_k}{N} = \frac{\mu_k N}{N}$$

$$\tan \phi_k = \mu_k \tag{8.4}$$

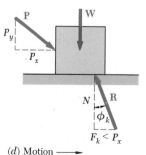

(d) Motion $\longrightarrow$

Fig 8.3

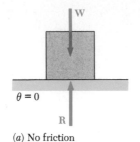

$\theta = 0$

(a) No friction

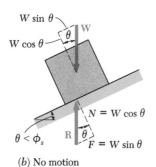

$W \sin \theta$

$W \cos \theta$

$N = W \cos \theta$

$\theta < \phi_s$ $\quad F = W \sin \theta$

(b) No motion

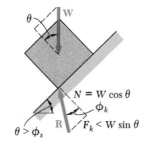

$\theta = \phi_s$

$N = W \cos \theta$

$F_m = W \sin \theta$

$\theta = \phi_s =$ angle of repose

(c) Motion impending

θ

$N = W \cos \theta$

ϕ_k

$\theta > \phi_s$ $\quad F_k < W \sin \theta$

(d) Motion

Fig. 8.4

Another example will show how the angle of friction may be used to advantage in the analysis of certain types of problems. Consider a block resting on a board which may be given any desired inclination; the block is subjected to no other force than its weight **W** and the reaction **R** of the board. If the board is horizontal, the force **R** exerted by the board on the block is perpendicular to the board and balances the weight **W** (Fig. 8.4a). If the board is given a small angle of inclination θ, the force **R** will deviate from the perpendicular to the board by the angle θ and will keep balancing **W** (Fig. 8.4b); it will then have a normal component **N** of magnitude $N = W \cos \theta$ and a tangential component **F** of magnitude $F = W \sin \theta$.

If we keep increasing the angle of inclination, motion will soon become impending. At that time, the angle between **R** and the normal will have reached its maximum value ϕ_s (Fig. 8.4c). The value of the angle of inclination corresponding to impending motion is called the *angle of repose*. Clearly, the angle of repose is equal to the angle of static friction ϕ_s. If the angle of inclination θ is further increased, motion starts and the angle between **R** and the normal drops to the lower value ϕ_k (Fig. 8.4d). The reaction **R** is not vertical any more, and the forces acting on the block are unbalanced.

8.4. Problems Involving Dry Friction. Problems involving dry friction are found in many engineering applications. Some deal with simple situations such as the block sliding on a plane described in the preceding sections. Others involve more complicated situations as in Sample Prob. 8.3; many deal with the stability of rigid bodies in accelerated motion and will be studied in dynamics. Also, a number of common machines and mechanisms may be analyzed by applying the laws of dry friction. These include wedges, screws, journal and thrust bearings, and belt transmissions. They will be studied in the following sections.

The *methods* which should be used to solve problems involving dry friction are the same that were used in the preceding chapters. If a problem involves only a motion of translation, with no possible rotation, the body under consideration may usually be treated as a particle, and the methods of Chap. 2 may be used. If the problem involves a possible rotation, the body must be considered as a rigid body and the methods of Chap. 4 should be used. If the structure considered is made of several parts, the principle of action and reaction must be used as was done in Chap. 6.

If the body considered is acted upon by more than three forces (including the reactions at the surfaces of contact), the reaction at each surface will be represented by its components **N** and

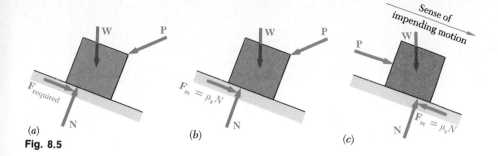

Fig. 8.5

F and the problem will be solved from the equations of equilibrium. If only three forces act on the body under consideration, it may be found more convenient to represent each reaction by the single force **R** and to solve the problem by drawing a force triangle.

Most problems involving friction fall into one of the following *three groups:* In the *first group* of problems, all applied forces are given, and the coefficients of friction are known; we are to determine whether the body considered will remain at rest or slide. The friction force **F** *required to maintain equilibrium* is unknown (its magnitude is *not* equal to $\mu_s N$) and should be determined, together with the normal force **N,** by drawing a free-body diagram and *solving the equations of equilibrium* (Fig. 8.5a). The value found for the magnitude F of the friction force is then compared with the maximum value $F_m = \mu_s N$. If F is smaller than or equal to F_m, the body remains at rest. If the value found for F is larger than F_m, equilibrium cannot be maintained and motion takes place; the actual magnitude of the friction force is then $F_k = \mu_k N$.

In problems of the *second group,* all applied forces are given, and the motion is known to be impending; we are to determine the value of the coefficient of static friction. Here again, we determine the friction force and the normal force by drawing a free-body diagram and solving the equations of equilibrium (Fig. 8.5b). Since we know that the value found for F is the maximum value F_m, the coefficient of friction may be found by writing and solving the equation $F_m = \mu_s N$.

In problems of the *third group,* the coefficient of static friction is given, and it is known that motion is impending in a given direction; we are to determine the magnitude or the direction of one of the applied forces. The friction force should be shown in the free-body diagram with a *sense opposite to that of the impending motion* and with a magnitude $F_m = \mu_s N$ (Fig. 8.5c). The equations of equilibrium may then be written, and the desired force may be determined.

As noted above, it may be more convenient, when only three forces are involved, to represent the reaction of the surface by a single force **R** and to solve the problem by drawing a force triangle. Such a solution is used in Sample Prob. 8.2.

When two bodies A and B are in contact (Fig. 8.6a), the forces of friction exerted, respectively, by A on B and by B on A are equal and opposite (Newton's third law). It is important, in drawing the free-body diagram of one of the bodies, to include the appropriate friction force with its correct sense. The following rule should then be observed: *The sense of the friction force acting on A is opposite to that of the motion (or impending motion) of A as observed from B* (Fig. 8.6b).† The sense of the friction force acting on B is determined in a similar way (Fig. 8.6c). Note that the motion of A as observed from B is a *relative motion*. Body A may be fixed; yet it will have a relative motion with respect to B if B itself moves. Also, A may actually move down yet be observed from B to move up if B moves down faster than A.

---

† It is therefore *the same as that of the motion of B as observed from A.*

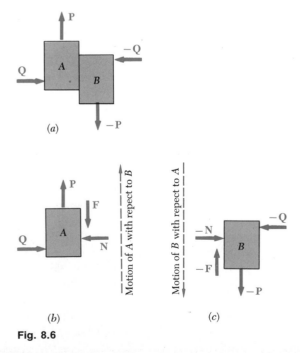

(a)

(b)   (c)

Fig. 8.6

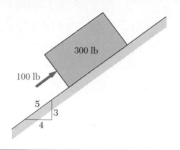

SAMPLE PROBLEM 8.1

A 100-lb force acts as shown on a 300-lb block placed on an inclined plane. The coefficients of friction between the block and the plane are $\mu_s = 0.25$ and $\mu_k = 0.20$. Determine whether the block is in equilibrium, and find the value of the friction force.

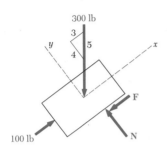

Force Required for Equilibrium. We first determine the value of the friction force *required to maintain equilibrium*. Assuming that **F** is directed down and to the left, we draw the free-body diagram of the block and write

$$+ \nearrow \Sigma F_x = 0: \qquad 100 \text{ lb} - \tfrac{3}{5}(300 \text{ lb}) - F = 0$$
$$F = -80 \text{ lb} \qquad \mathbf{F} = 80 \text{ lb} \nearrow$$

$$+ \nwarrow \Sigma F_y = 0: \qquad N - \tfrac{4}{5}(300 \text{ lb}) = 0$$
$$N = +240 \text{ lb} \qquad \mathbf{N} = 240 \text{ lb} \nwarrow$$

The force **F** required to maintain equilibrium is an 80-lb force directed up and to the right; the tendency of the block is thus to move down the plane.

Maximum Friction Force. The magnitude of the maximum friction force which may be developed is

$$F_{\max} = \mu_s N \qquad F_{\max} = 0.25(240 \text{ lb}) = 60 \text{ lb}$$

Since the value of the force required to maintain equilibrium (80 lb) is larger than the maximum value which may be obtained (60 lb), equilibrium will not be maintained and *the block will slide down the plane*.

Actual Value of Friction Force. The magnitude of the actual friction force is obtained as follows:

$$F_{\text{actual}} = F_k = \mu_k N$$
$$= 0.20(240 \text{ lb}) = 48 \text{ lb}$$

The sense of this force is opposite to the sense of motion; the force is thus directed up and to the right,

$$\mathbf{F}_{\text{actual}} = 48 \text{ lb} \nearrow \quad \blacktriangleleft$$

It should be noted that the forces acting on the block are not balanced; the resultant is

$$\tfrac{3}{5}(300 \text{ lb}) - 100 \text{ lb} - 48 \text{ lb} = 32 \text{ lb} \swarrow$$

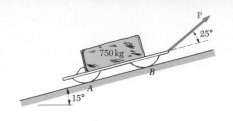

SAMPLE PROBLEM 8.2

A wooden sled supporting a large stone is pulled up a track inclined at 15°. The combined mass of the sled and stone is 750 kg, and the coefficients of friction between the sled runners and the track are $\mu_s = 0.40$ and $\mu_k = 0.30$. Determine the force P required (a) to start the sled up the track, (b) to keep the sled moving up after it has been started, (c) to keep the sled from sliding down.

Solution. The magnitude of the combined weight **W** of the sled and stone is

$$W = mg = (750 \text{ kg})(9.81 \text{ m/s}^2) = 7360 \text{ N} = 7.36 \text{ kN}$$

Noting that R_A and R_B have the same direction (same angle of friction ϕ), we draw a force triangle including the weight **W**, the force **P**, and the sum $\mathbf{R} = \mathbf{R}_A + \mathbf{R}_B$. The direction of **R** must be redetermined in each part of the problem. The law of sines is used to determine the magnitude of **P** in each part of the problem. If required, we could then determine the individual reactions R_A and R_B by summing moments about B and A, respectively.

a. Force P to Start Sled Moving Up

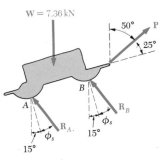

$$\tan \phi_s = \mu_s$$
$$= 0.40$$
$$\phi_s = 21.8°$$

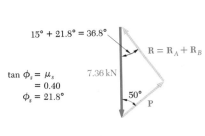

$$15° + 21.8° = 36.8°$$

$$\frac{P}{\sin 36.8°} = \frac{7.36 \text{ kN}}{\sin [180° - (50° + 36.8°)]}$$

$$P = 4.42 \text{ kN} \nearrow \quad \blacktriangleleft$$

b. Force P to Keep Sled Moving

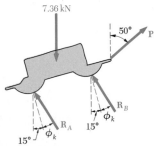

$$\tan \phi_k = \mu_k$$
$$= 0.30$$
$$\phi_k = 16.7°$$

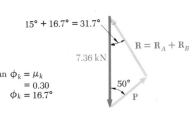

$$15° + 16.7° = 31.7°$$

$$\frac{P}{\sin 31.7°} = \frac{7.36 \text{ kN}}{\sin [180° - (50° + 31.7°)]}$$

$$P = 3.91 \text{ kN} \nearrow \quad \blacktriangleleft$$

c. Force P to Keep Sled from Sliding Down

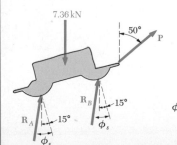

$$\phi_s = 21.8°$$

$$21.8° - 15° = 6.8°$$

$$\frac{P}{\sin 6.8°} = \frac{7.36 \text{ kN}}{\sin [180° - (130° + 6.8°)]}$$

$$P = 1.273 \text{ kN} \swarrow \quad \blacktriangleleft$$

Since the force **P** is directed downward, the sled will not slide down under its own weight.

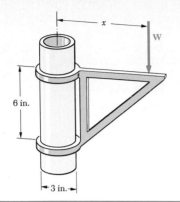

6 in.

3 in.

The movable bracket shown may be placed at any height on the 3-in.-diameter pipe. If the coefficient of static friction between the pipe and bracket is 0.25, determine the minimum distance x at which the load **W** can be supported. Neglect the weight of the bracket.

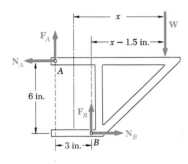

F_A

N_A

A

6 in.

F_B

3 in. B

N_B

x

$x - 1.5$ in.

W

Solution. We draw the free-body diagram of the bracket. When **W** is placed at the minimum distance x from the axis of the pipe, the bracket is just about to slip, and the forces of friction at A and B have reached their maximum values:

$$F_A = \mu N_A = 0.25 N_A$$
$$F_B = \mu N_B = 0.25 N_B$$

Equilibrium Equations

$\xrightarrow{+} \Sigma F_x = 0$: $N_B - N_A = 0$

$N_B = N_A$

$+\uparrow \Sigma F_y = 0$: $F_A + F_B - W = 0$

$0.25 N_A + 0.25 N_B = W$

And, since N_B has been found equal to N_A,

$$0.50 N_A = W$$
$$N_A = 2W$$

$+\circlearrowleft \Sigma M_B = 0$: $N_A(6 \text{ in.}) - F_A(3 \text{ in.}) - W(x - 1.5 \text{ in.}) = 0$

$6 N_A - 3(0.25 N_A) - Wx + 1.5W = 0$

$6(2W) - 0.75(2W) - Wx + 1.5W = 0$

Dividing through by W and solving for x:

$x = 12$ in. ◀

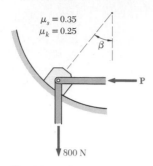

$\mu_s = 0.35$
$\mu_k = 0.25$

800 N

Fig. P8.1 and P8.2

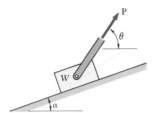

Fig. P8.3 and P8.4

PROBLEMS

8.1 The support block is acted upon by the two forces shown. For the position when $\beta = 25°$, determine (a) the force **P** required to start the block up the surface, (b) the smallest force **P** which will prevent the block from moving down the surface.

8.2 Solve Prob. 8.1 for the position when $\beta = 30°$.

8.3 Denoting by ϕ_s the angle of static friction between the block and the plane, determine the magnitude and direction of the smallest force **P** which will cause the block to move up the plane.

8.4 A block of weight $W = 40$ lb rests on a rough plane as shown. Knowing that $\alpha = 25°$ and $\mu_s = 0.20$, determine the magnitude and direction of the smallest force **P** required (a) to start the block up the plane, (b) to prevent the block from moving down the plane.

8.5 The two 25-lb boxes A and B may be attached to the lever by means of either one or two horizontal links. The coefficient of friction between all surfaces is 0.30. Determine the magnitude of the force **P** required to move the lever (a) if a single link EF is used, (b) if a single link CD is used, (c) if two links EF and CD are used at the same time.

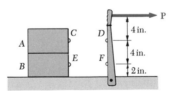

Fig. P8.5

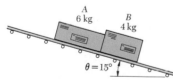

Fig. P8.6

8.6 Two packages are placed on a conveyor belt which is at rest. Between the belt and package A the coefficients of friction are $\mu_s = 0.2$ and $\mu_k = 0.15$; between package B and the belt the coefficients are $\mu_s = 0.3$ and $\mu_k = 0.25$. The packages are placed on the belt so that they are in contact with each other and at rest. Determine (a) whether either, or both, of the packages will move, (b) the friction force acting on each package.

8.7 Solve Prob. 8.6, assuming $\theta = 12°$.

8.8 The 20-lb block rests on a conveyor belt and is attached to the link AB. Knowing that the coefficient of friction between the block and the belt is 0.25 and neglecting the weight of the link, determine the magnitude of the force **P** required to move the belt to the left.

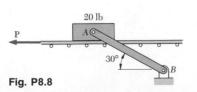

Fig. P8.8

8.9 A 120-lb cabinet is mounted on casters which can be locked to prevent their rotation. The coefficient of friction is 0.30. If $h = 32$ in., determine the magnitude of the force **P** required to move the cabinet to the right (*a*) if all casters are locked, (*b*) if the casters at *B* are locked and the casters at *A* are free to rotate, (*c*) if casters *A* are locked and casters *B* are free to rotate.

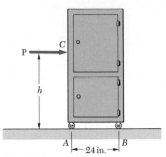

8.10 A 120-lb cabinet is mounted on casters which can be locked to prevent their rotation. The coefficient of friction between the floor and each caster is 0.30. Assuming that the casters at both *A* and *B* are locked, determine (*a*) the force **P** required to move the cabinet to the right, (*b*) the largest allowable value of h if the cabinet is not to tip over.

Fig. P8.9 and P8.10

8.11 A packing crate, of mass 30 kg, must be moved to the left along the floor without tipping. Knowing that the coefficient of friction between the crate and the floor is 0.35, determine (*a*) the largest allowable value of α, (*b*) the corresponding tension *T*.

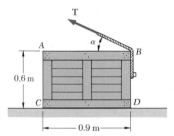

8.12 A packing crate of mass 30 kg is pulled by a rope as shown. The coefficient of friction between the crate and the floor is 0.35. If $\alpha = 30°$, determine (*a*) the tension *T* required to move the crate, (*b*) whether the crate will slide or tip.

Fig. P8.11 and P8.12

8.13 A 200-lb sliding door is mounted on a horizontal rail as shown. The coefficients of static friction between the rail and the door at *A* and *B* are 0.15 and 0.25, respectively. Determine the horizontal force which must be applied to the handle *C* in order to move the door to the right.

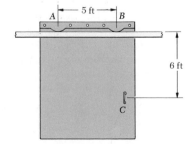

Fig. P8.13

8.14 Solve Prob. 8.13, assuming that the door is to be moved to the left.

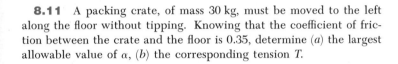

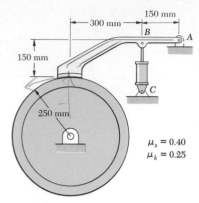

Fig. P8.15 and P8.16

8.15 A couple of magnitude 90 N · m is applied to the drum. Determine the smallest force which must be exerted by the hydraulic cylinder if the drum is not to rotate, when the applied couple is directed (*a*) clockwise, (*b*) counterclockwise.

8.16 The hydraulic cylinder exerts on point *B* a force of 2400 N directed downward. Determine the moment of the friction force about the axle of the drum when the drum is rotating (*a*) clockwise, (*b*) counterclockwise.

8.17 A uniform beam of weight *W* and length *L* is initially in position *AB*. As the cable is pulled over the pulley *C*, the beam first slides on the floor and is then raised, with its end *A* still sliding. Denoting by μ the coefficient of friction between the beam and the floor, determine the distance *x* that the beam moves before it begins to rise.

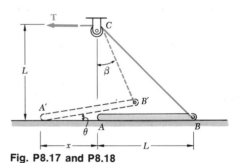

Fig. P8.17 and P8.18

8.18 The coefficient of friction between the beam of Prob. 8.17 and the floor is 0.40. The beam is slowly raised to a position *A'B'* where $\theta = 30°$. Determine the corresponding value of β.

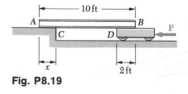

Fig. P8.19

8.19 A 10-ft beam, weighing 1200 lb, is to be moved to the left onto the platform. A horizontal force **P** is applied to the dolly, which is mounted on frictionless wheels. The coefficient of friction between all surfaces is 0.30. Knowing that the horizontal top surface of the dolly is slightly higher than the platform, determine the magnitude of **P** required to move the beam. (*Hint.* The beam is supported at *A* and *D*.)

8.20 (*a*) Show that the beam of Prob. 8.19 cannot be moved if the horizontal top surface of the dolly is slightly *lower* than the platform. (*b*) How far can the beam be moved to the left if two 175-lb men stand on the beam at *B*?

8.21 Two links, of negligible weight, are connected by frictionless pins to each other and to two blocks at *A* and *C*. The coefficient of friction is 0.25 at *A* and *C*. If neither block is to slip, determine the magnitude of the largest force **P** which can be applied at *B*.

8.22 In Prob. 8.21, determine the magnitude of the smallest force **P** which must be applied at *B* if neither block is to slip.

8.23 Two collars, each of weight *W*, are connected by a cord which passes over a frictionless pulley at *C*. Determine the smallest value of μ between the collars and the vertical rods for which the system will remain in equilibrium in the position shown.

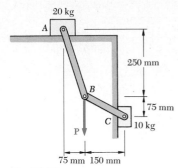

Fig. P8.21

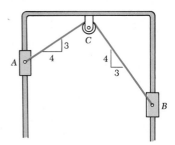

Fig. P8.23

8.24 The 10-lb uniform rod *AB* is held in the position shown by the force **P**. Knowing that the coefficient of friction is 0.20 at *A* and *B*, determine the smallest value of *P* for which equilibrium is maintained.

8.25 In Prob. 8.24, determine the largest value of **P** for which equilibrium is maintained.

8.26 The rod *AB* rests on a horizontal surface at *A* and against a sloping surface at *B*. Knowing that the coefficient of friction is 0.25 at both *A* and *B*, determine the maximum distance *a* at which the load **W** can be supported. Neglect the weight of the rod.

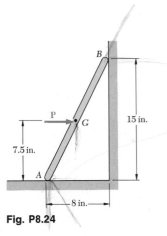

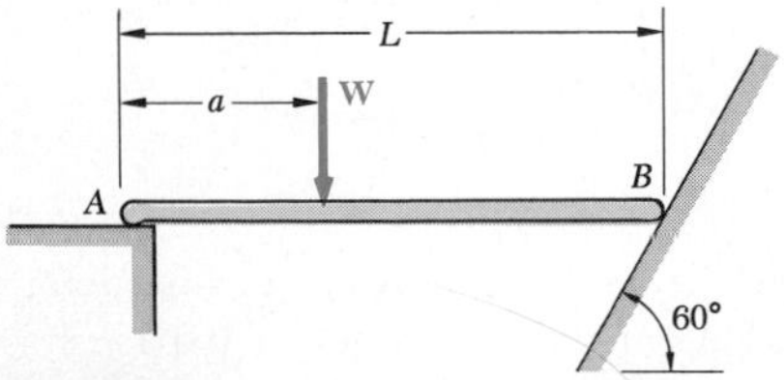

Fig. P8.24

Fig. P8.26

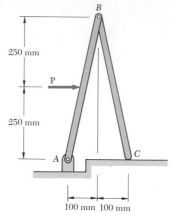

250 mm

250 mm

100 mm 100 mm

Fig. P8.27

8.27 Two uniform rods, each of mass 5 kg, are held by frictionless pins A and B. It is observed that if the magnitude of $\mathbf{P}$ is larger than 120 N, the rods will collapse. Determine the coefficient of friction at C.

8.28 A window sash of mass 6 kg is normally supported by two 3-kg sash weights. It is observed that the window remains open after one sash cord has broken. What is the smallest possible value of the coefficient of static friction? (Assume that the sash is slightly smaller than the frame and will bind only at points A and D.)

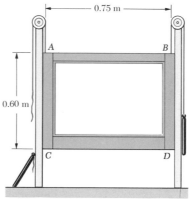

0.75 m

0.60 m

Fig. P8.28

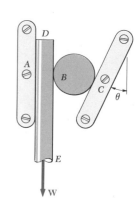

Fig. P8.29

8.29 A rod DE and a small cylinder are placed between two guides as shown. The rod is not to slip downward, however large the force $\mathbf{W}$ may be; i.e., the arrangement is to be *self-locking*. Determine the minimum allowable coefficients of friction at A, B, and C.

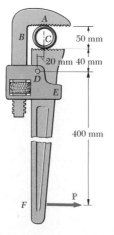

50 mm

20 mm 40 mm

400 mm

Fig. P8.30

8.30 A pipe of diameter 50 mm is gripped by the stillson wrench shown. Portions AB and DE of the wrench are rigidly attached to each other and portion CF is connected by a pin at D. If the wrench is to grip the pipe and be self-locking, determine the required minimum coefficients of friction at A and at C.

8.31 If the plumbing wrench of Prob. 6.105 is not to slip, determine the minimum allowable value of the coefficient of friction between (*a*) the nut and the jaw BC, (*b*) the nut and the rod A.

8.32 A 10-ft uniform plank of weight 45 lb rests on two joists as shown. The coefficient of friction between the joists and the plank is 0.40. (*a*) Determine the magnitude of the horizontal force **P** required to move the plank. (*b*) Solve part *a* assuming that a single nail driven into joist *A* prevents motion of the plank along joist *A*.

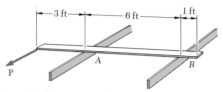

Fig. P8.32

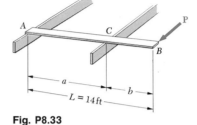

Fig. P8.33

8.33 Knowing that the coefficient of friction between the joists and the plank *AB* is 0.30, determine the smallest distance *a* for which the plank will slip at *C*.

8.34 The uniform slender rod *AB* rests on a horizontal surface at *A* and on a small wheel at *C*. Knowing that the wheel may rotate freely, determine the smallest coefficient of friction between the rod and the floor for which the rod will remain in the position shown.

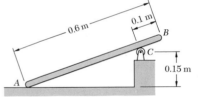

Fig. P8.34

8.35 The mathematical model shown has been developed for the analysis of a certain structure. It consists of 1-lb blocks connected in series by springs, each of constant $k = 0.20$ lb/in. The coefficient of friction between the base and each block is 0.40. Knowing that the initial tension in each spring is zero, construct a graph showing the magnitude of the force **P** versus the position of block *A* as *P* increases from zero to 1.00 lb and then decreases to zero.

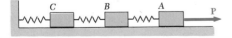

Fig. P8.35

8.36 A cylinder of weight *W* is placed in a V block as shown. Denoting by ϕ the angle of friction between the cylinder and the block and assuming $\phi < \theta$, determine (*a*) the axial force **P** required to move the cylinder, (*b*) the couple **M**, applied in the plane of the cross section of the cylinder, required to rotate the cylinder.

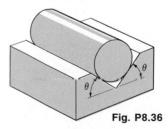

Fig. P8.36

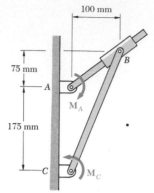

100 mm

75 mm

A

175 mm

M_A

B

C M_C

Fig. P8.37

8.37 Two rods are connected by a collar at *B*; a couple M_A of magnitude 40 N · m is applied to rod *AB*. Knowing that $\mu = 0.30$ between the collar and rod *AB*, determine the *maximum* couple M_C for which equilibrium will exist.

8.38 In Prob. 8.37, determine the *minimum* couple M_C required for equilibrium.

8.39 A 10-kg rod *AB* is placed between a movable block and a wall as shown. The coefficient of static friction at *A* and *B* is 0.25. If a 500-N force acts on the block as shown, determine the angle θ for which slipping impends.

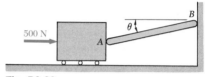

500 N

A

θ

B

Fig. P8.39

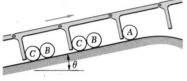

C *B*

C *B*

A

θ

Fig. P8.40

8.40 Identical cylindrical cans, each of weight *W*, are raised to the top of an incline by a series of moving arms. Either one or two cans are moved by each arm. The coefficient of friction between all surfaces is $\mu = 0.20$. If $W = 2.00$ lb and $\theta = 12°$, determine the force parallel to the incline which the arm must exert on can *A* to move it. Does can *A* roll or slide?

*8.41 Solve Prob. 8.40, considering can *C* instead of can *A*.

8.42 In Prob. 8.40, determine the range of values of θ for which can *A* will roll.

*8.43 In Prob. 8.40, determine the range of values of θ for which can *C* will roll.

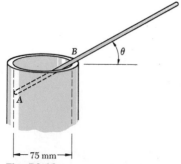

B

θ

A

75 mm

Fig. P8.44

*8.44 A small steel rod, of length 500 mm, is placed inside a pipe as shown. The coefficient of static friction between the rod and the pipe is 0.20. Determine the largest value of θ for which the rod will not fall into the pipe.

*8.45 In Prob. 8.44, determine the smallest value of θ for which the rod will not fall out of the pipe.

8.5. Wedges.

Wedges are simple machines used to raise large stone blocks and other heavy loads. These loads may be raised by applying to the wedge a force usually considerably smaller than the weight of the load. Besides, because of the friction existing between the surfaces in contact, a wedge, if properly shaped, will remain in place after being forced under the load. Wedges may thus be used advantageously to make small adjustments in the position of heavy pieces of machinery.

Consider the block A shown in Fig. 8.7a. This block rests against a vertical wall B and is to be raised slightly by forcing a wedge C between block A and a second wedge D. We want to find the minimum value of the force $\mathbf{P}$ which must be applied to the wedge C to move the block. We shall assume that the weight $\mathbf{W}$ of the block is known. It may have been given in pounds or determined in newtons from the mass of the block expressed in kilograms.

The free-body diagrams of block A and of wedge C have been drawn in Fig. 8.7b and c. The forces acting on the block include its weight and the normal and friction forces at the surfaces of contact with wall B and wedge C. The magnitudes of the friction forces $\mathbf{F}_1$ and $\mathbf{F}_2$ are equal, respectively, to $\mu_s N_1$ and $\mu_s N_2$ since the motion of the block must be started. It is important to show the friction forces with their correct sense. Since the block will move upward, the force $\mathbf{F}_1$ exerted by the wall on the block must be directed downward. On the other hand, since the wedge C moves to the right, the relative motion of A with respect to C is to the left and the force $\mathbf{F}_2$ exerted by C on A must be directed to the right.

Considering now the free body C in Fig. 8.7c, we note that the forces acting on C include the applied force $\mathbf{P}$ and the normal and friction forces at the surfaces of contact with A and D. The weight of the wedge is small compared with the other forces involved and may be neglected. The forces acting on C are equal and opposite to the forces $\mathbf{N}_2$ and $\mathbf{F}_2$ acting on A and are denoted, respectively, by $-\mathbf{N}_2$ and $-\mathbf{F}_2$; the friction force $-\mathbf{F}_2$ must therefore be directed to the left. We check that the force $\mathbf{F}_3$ is also directed to the left.

The total number of unknowns involved in the two free-body diagrams may be reduced to four if the friction forces are expressed in terms of the normal forces. Expressing that block A and wedge C are in equilibrium will provide four equations which may be solved to obtain the magnitude of $\mathbf{P}$. It should be noted that, in the example considered here, it will be more convenient to replace each pair of normal and friction forces

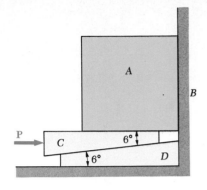

(a)

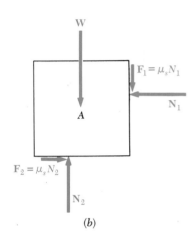

(b)

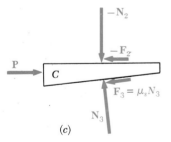

(c)

Fig. 8.7

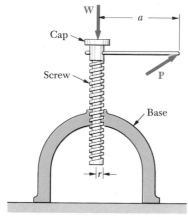

Fig. 8.8

by their resultant. Each free body is then subjected to only three forces, and the problem may be solved by drawing the corresponding force triangles (see Sample Prob. 8.4).

8.6. Square-threaded Screws. Square-threaded screws are frequently used in jacks, presses, and other mechanisms. Their analysis is similar to that of a block sliding along an inclined plane.

Consider the jack shown in Fig. 8.8. The screw carries a load **W** and is supported by the base of the jack. Contact between screw and base takes place along a portion of their threads. By applying a force **P** on the handle, the screw may be made to turn and to raise the load **W**.

The thread of the base has been unwrapped and shown as a straight line in Fig. 8.9*a*. The correct slope was obtained by plotting horizontally the product $2\pi r$, where r is the mean radius of the thread, and vertically the *lead L* of the screw, i.e., the distance through which the screw advances in one turn. The angle θ this line forms with the horizontal is the *lead angle*. Since the force of friction between two surfaces in contact does not depend upon the area of contact, the two threads may be assumed to be in contact over a much smaller area than they actually are and the screw may be represented by the block shown in Fig. 8.9*a*. It should be noted, however, that, in this analysis of the jack, the friction between cap and screw is neglected.

The free-body diagram of the block should include the load **W**, the reaction **R** of the base thread, and a horizontal force **Q** having the same effect as the force **P** exerted on the handle. The force **Q** should have the same moment as **P** about the axis of the screw and its magnitude should thus be $Q = Pa/r$. The force **Q**, and thus the force **P** required to raise the load **W**, may be obtained from the free-body diagram shown in Fig. 8.9*a*. The friction angle is taken equal to ϕ_s since the load will presumably be raised through a succession of short strokes. In mechanisms providing for the continuous rotation of a screw, it may be desirable to distinguish between the force required to start motion (using ϕ_s) and that required to maintain motion (using ϕ_k).

If the friction angle ϕ_s is larger than the lead angle θ, the screw is said to be *self-locking;* it will remain in place under the load. To lower the load, we must then apply the force shown in Fig. 8.9*b*. If ϕ_s is smaller than θ, the screw will unwind under the load; it is then necessary to apply the force shown in Fig. 8.9*c* to maintain equilibrium.

The lead of a screw should not be confused with its *pitch*. The lead was defined as the distance through which the screw advances in one turn; the pitch is the distance measured between two consecutive threads. While lead and pitch are equal in the case of *single-threaded* screws, they are different in the case of *multiple-threaded* screws, i.e., screws having several independent threads. It is easily verified that, for double-threaded screws, the lead is twice as large as the pitch; for triple-threaded screws, it is three times as large as the pitch; etc.

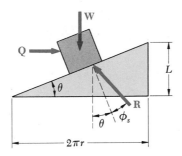

(*a*) Impending motion upward

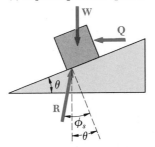

(*b*) Impending motion downward with $\phi_s > \theta$

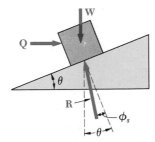

(*c*) Impending motion downward with $\phi_s < \theta$

Fig. 8.9 Block-and-incline analysis of a screw.

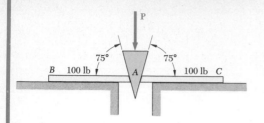

SAMPLE PROBLEM 8.4

A wedge A of negligible weight is to be driven between two 100-lb plates B and C. The coefficient of static friction between all surfaces of contact is 0.35. Determine the magnitude of the force $\mathbf{P}$ required to start moving the wedge (a) if the plates are equally free to move, (b) if plate C is securely bolted to the surface.

a. **With Both Plates Free to Move.** The free-body diagrams of wedge A and of plate B are shown, together with the corresponding force triangles. Because of symmetry it will not be necessary to consider the free-body diagram of plate C in the solution of the problem, and this diagram has been omitted.

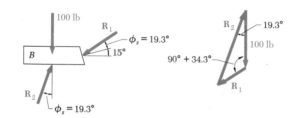

Free Body: Plate B

$$\phi_s = \tan^{-1} 0.35 = 19.3°$$

$$\frac{R_1}{\sin 19.3°} = \frac{100 \text{ lb}}{\sin[180° - (90° + 34.3°) - 19.3°]}$$

$$\frac{R_1}{\sin 19.3°} = \frac{100 \text{ lb}}{\sin 36.4°}$$

$$R_1 = 55.7 \text{ lb}$$

Free Body: Wedge A. We note that the force triangle is an isosceles triangle; thus

$$R_3 = R_1 = 55.7 \text{ lb}$$
$$P = 2R_1 \sin 34.3°$$
$$= 2(55.7 \text{ lb}) \sin 34.3°$$
$$P = 62.8 \text{ lb} \downarrow \quad \blacktriangleleft$$

b. **With Plate C Bolted.** The only free bodies remaining are those of the wedge A and of the plate B, and the corresponding diagrams will be the same as in part *a*. Thus, the answer is still

$$P = 62.8 \text{ lb} \downarrow \quad \blacktriangleleft$$

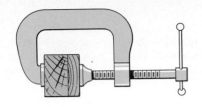

SAMPLE PROBLEM 8.5

A clamp is used to hold two pieces of wood together as shown. The clamp has a double square thread of mean diameter equal to 10 mm and with a pitch of 2 mm. The coefficient of friction between threads is $\mu_s = 0.30$. If a maximum torque of 40 N · m is applied in tightening the clamp, determine (a) the force exerted on the pieces of wood, (b) the torque required to loosen the clamp.

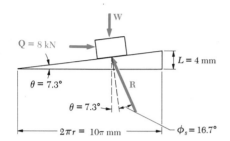

a. **Force Exerted by Clamp.** The mean radius of the screw is $r = 5$ mm. Since the screw is double-threaded, the lead L is equal to twice the pitch: $L = 2(2 \text{ mm}) = 4$ mm. The lead angle θ and the friction angle ϕ_s are obtained by writing

$$\tan \theta = \frac{L}{2\pi r} = \frac{4 \text{ mm}}{10\pi \text{ mm}} = 0.1273 \qquad \theta = 7.3°$$

$$\tan \phi_s = \mu_s = 0.30 \qquad \phi_s = 16.7°$$

The force **Q** which should be applied to the block representing the screw is obtained by expressing that its moment Qr about the axis of the screw is equal to the applied torque.

$$Q(5 \text{ mm}) = 40 \text{ N} \cdot \text{m}$$
$$Q = \frac{40 \text{ N} \cdot \text{m}}{5 \text{ mm}} = \frac{40 \text{ N} \cdot \text{m}}{5 \times 10^{-3} \text{ m}} = 8000 \text{ N} = 8 \text{ kN}$$

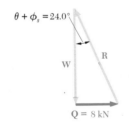

The free-body diagram and the corresponding force triangle may now be drawn for the block; the magnitude of the force **W** exerted on the pieces of wood is obtained by solving the triangle.

$$W = \frac{Q}{\tan (\theta + \phi_s)} = \frac{8 \text{ kN}}{\tan 24.0°}$$
$$W = 17.97 \text{ kN} \quad \blacktriangleleft$$

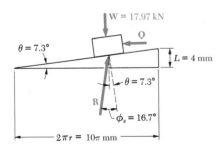

b. **Torque Required to Loosen Clamp.** The force **Q** required to loosen the clamp and the corresponding torque are obtained from the free-body diagram and force triangle shown.

$$Q = W \tan (\phi_s - \theta) = (17.97 \text{ kN}) \tan 9.4°$$
$$= 2.975 \text{ kN}$$

$$\text{Torque} = Qr = (2.975 \text{ kN})(5 \text{ mm})$$
$$= (2.975 \times 10^3 \text{ N})(5 \times 10^{-3} \text{ m}) = 14.87 \text{ N} \cdot \text{m}$$
$$\text{Torque} = 14.87 \text{ N} \cdot \text{m} \quad \blacktriangleleft$$

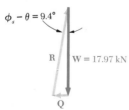

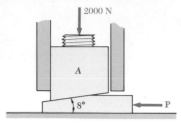

2000 N

A

8°

P

Fig. P8.46

PROBLEMS

8.46 The position of machine block A is adjusted by moving the wedge shown. Knowing that the coefficient of friction is 0.25 at all surfaces of contact, determine the force **P** required to raise the block.

8.47 In Prob. 8.46, determine the force **P** required to lower machine block A.

8.48 and 8.49 Determine the minimum value of P which must be applied to the wedge in order to move the 2000-lb block. The coefficient of static friction is 0.30 at all surfaces of contact.

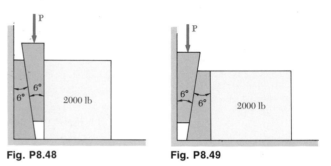

Fig. P8.48   **Fig. P8.49**

Fig. P8.50

8.50 A 12° wedge is used to split a log. The coefficient of friction between the wedge and the wood is 0.30. Knowing that a force **P** of magnitude 1500 N was required to insert the wedge, determine the magnitude of the forces exerted on the log by the wedge after it has been inserted.

8.51 The spring of the door latch has a constant of 1.5 lb/in. and in the position shown exerts a force of $\frac{1}{2}$ lb on the bolt. The coefficient of friction between the bolt and the guide plate is 0.40; all other surfaces are well lubricated and may be assumed to be frictionless. Determine the magnitude of the force **P** required to start closing the door.

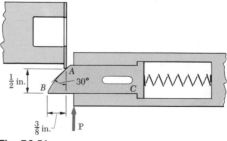

Fig. P8.51

8.52 In Prob. 8.51, determine the angle which the face of the bolt should form with the line *BC* if the force **P** required to close the door is to be the same for both the position shown and the position when *B* is almost at the guide plate.

8.53 The elevation of the end of the steel floor beam is adjusted by means of the steel wedges *E* and *F*. The steel base plate *CD* has been welded to the lower flange of the beam; the end reaction of the beam is known to be 100 kN. The coefficient of friction between two steel surfaces is 0.30 and between steel and concrete is 0.60. If horizontal motion of the base plate is prevented by the force **Q**, determine (*a*) the force **P** required to raise the beam, (*b*) the corresponding force **Q**.

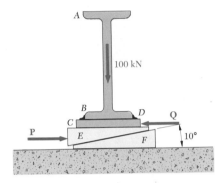

Fig. P8.53

8.54 Solve Prob. 8.53, assuming that the end of the beam is to be lowered.

8.55 A 5° wedge is to be forced under a machine base at *A*. Knowing that $\mu = 0.15$ at all surfaces, (*a*) determine the force **P** required to move the wedge, (*b*) indicate whether the machine will move.

8.56 Solve Prob. 8.55, assuming that the wedge is to be forced under the machine base at *B* instead of *A*.

8.57 A conical wedge is placed between two horizontal plates which are then slowly moved toward each other. Indicate what will happen to the wedge (*a*) if $\mu = 0.15$, (*b*) if $\mu = 0.25$.

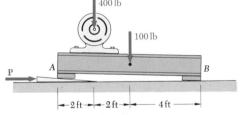

Fig. P8.55

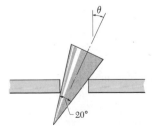

Fig. P8.57

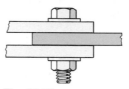

8.58 High-strength bolts are used in the construction of many modern steel structures. For a 1-in. nominal-diameter bolt the required minimum bolt tension is 47,250 lb. Assuming the coefficient of friction to be 0.40, determine the required torque which must be applied to the bolt and nut. The mean diameter of the thread is 0.94 in., and the lead is 0.125 in. Neglect friction between the nut and washer, and assume the bolt to be square-threaded.

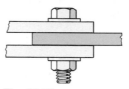

Fig. P8.58

8.59 Derive the following formulas relating the load W and the force $\mathbf{P}$ exerted on the handle of the jack discussed in Sec. 8.6: (a) $P = (Wr/a) \tan (\theta + \phi_s)$, to raise the load; (b) $P = (Wr/a) \tan (\phi_s - \theta)$, to lower the load if the screw is self-locking; (c) $P = (Wr/a) \tan (\theta - \phi_s)$, to hold the load if the screw is not self-locking.

8.60 The square-threaded worm gear shown has a mean radius of 40 mm and a pitch of 10 mm. The large gear is subjected to a constant clockwise torque of 900 N · m. Knowing that the coefficient of friction between gear teeth is 0.10, determine the torque which must be applied to shaft AB in order to rotate the large gear counter-clockwise. Neglect friction in the bearings at A, B, and C.

8.61 In Prob. 8.60, determine the torque which must be applied to shaft AB in order to rotate the large gear clockwise.

8.62 The position of the automobile jack shown is controlled by the screw ABC which is single-threaded at each end (right-handed thread at A, left-handed thread at C). Each thread has a pitch of 0.1 in. and a mean diameter of 0.375 in. If $\mu = 0.12$, determine the magnitude and sense of the couple $\mathbf{M}$ which must be applied to the screw (a) to raise the automobile, (b) to lower the automobile.

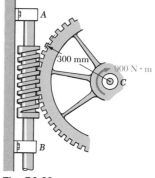

Fig. P8.60

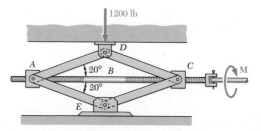

Fig. P8.62

8.63 The ends of two fixed rods A and B are each made in the form of a single-threaded screw of mean radius 6 mm and pitch 2 mm. The coefficient of friction between the rods and the threaded sleeve is 0.15. Determine the magnitude of the couple which must be applied to the sleeve in order to draw the rods closer together. Rod A has a left-handed thread and rod B a right-handed thread.

2 kN   2 kN

Fig. P8.63

8.64 In Prob. 8.63, a right-handed thread is used on *both* rods A and B. Determine the magnitude of the couple which must be applied to the sleeve in order to rotate it.

8.65 In the machinist's vise shown, the movable jaw D is rigidly attached to the tongue AB which fits loosely into the fixed body of the vise. The screw is single-threaded into the fixed base and has a mean diameter of 0.50 in. and a pitch of 0.20 in. The coefficient of friction is 0.25 between the threads and also between the tongue and the body. Neglecting bearing friction between the screw and the movable head, determine the torque which must be applied to the handle in order to produce a clamping force of 800 lb.

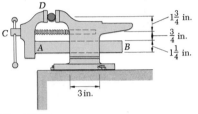

Fig. P8.65

8.66 In Prob. 8.65, a clamping force of 800 lb was obtained by tightening the vise. Determine the torque which must be applied to the screw to loosen the vise.

8.67 The vise shown consists of two members connected by two double-threaded screws of mean radius 5 mm and pitch 1 mm. The lower member is threaded at A and B ($\mu_s = 0.25$), but the upper member is not threaded. It is desired to apply two equal and opposite forces of 500 N on the blocks held between the jaws. (*a*) What screw should be adjusted first? (*b*) What is the maximum torque applied in tightening the second screw?

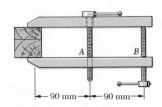

Fig. P8.67 |—90 mm—|—90 mm—|

∗ **8.7. Journal Bearings. Axle Friction.** Journal bearings are used to provide lateral support to rotating shafts and axles. Thrust bearings, which will be studied in the next section, are used to provide axial support to shafts and axles. If the journal bearing is fully lubricated, the frictional resistance depends upon the speed of rotation, the clearance between axle and bearing, and the viscosity of the lubricant. As indicated in Sec. 8.1, such problems are studied in fluid mechanics. The methods of this chapter, however, may be applied to the study of axle friction when the bearing is not lubricated or only partially lubricated. We may then assume that the axle and the bearing are in direct contact along a single straight line.

Consider two wheels, each of weight **W**, rigidly mounted on an axle supported symmetrically by two journal bearings (Fig. 8.10a). If the wheels rotate, we find that, to keep them rotating at constant speed, it is necessary to apply to each of them a couple **M**. A free-body diagram has been drawn in Fig. 8.10c, which represents one of the wheels and the corresponding half axle in projection on a plane perpendicular to the axle. The forces acting on the free body include the weight **W** of the wheel, the couple **M** required to maintain its motion, and a force **R** representing the reaction of the bearing. This force is vertical, equal, and opposite to **W** but does not pass through the center O of the axle; **R** is located to the right of O at a distance such that its moment about O balances the moment **M** of the couple. Contact between axle and bearing, therefore, does not take place at the lowest point A when the axle rotates. It takes place at point B (Fig. 8.10b) or, rather, along a straight line intersecting the plane of the figure at B. Physically, this is explained by the fact that, when the wheels are set in motion, the axle "climbs" in the bearings until slippage occurs. After sliding back slightly, the axle settles more or less in the position shown. This position is such that the angle between the reaction **R** and the normal to the surface of the bearing is equal to the angle of kinetic friction ϕ_k. The distance from O to the line of action of **R** is thus $r \sin \phi_k$, where r is the radius of the axle. Writing that $\Sigma M_O = 0$ for the forces acting on the free body considered, we obtain the magnitude of the couple **M** required to overcome the frictional resistance of one of the bearings:

$$M = Rr \sin \phi_k \tag{8.5}$$

Observing that, for small values of the angle of friction, $\sin \phi_k$

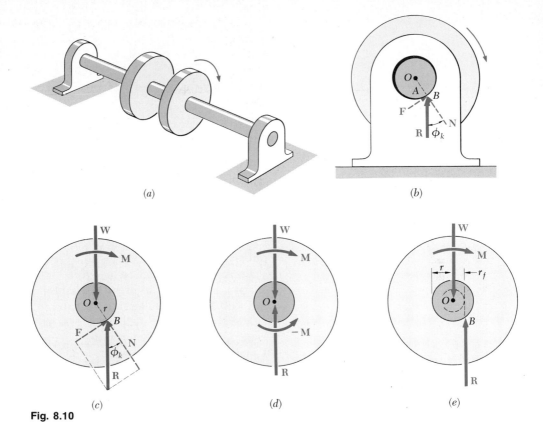

(a)

(b)

(c)

(d)

(e)

Fig. 8.10

may be replaced by $\tan \phi_k$, that is, by μ_k, we write the approximate formula

$$M \approx Rr\,\mu_k \qquad (8.6)$$

In the solution of certain problems, it may be more convenient to let the line of action of **R** pass through O, as it does when the axle does not rotate. A couple $-\mathbf{M}$ of the same magnitude as the couple **M** but of opposite sense must then be added to the reaction **R** (Fig. 8.10d). This couple represents the frictional resistance of the bearing.

In case a graphical solution is preferred, the line of action of **R** may be readily drawn (Fig. 8.10e) if we note that it must be tangent to a circle centered at O and of radius

$$r_f = r \sin \phi_k \approx r\,\mu_k \qquad (8.7)$$

This circle is called the *circle of friction* of the axle and bearing and is independent of the loading conditions of the axle.

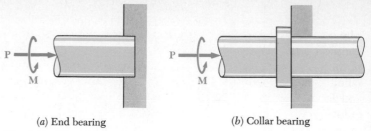

(a) End bearing   (b) Collar bearing

Fig. 8.11 Thrust bearings

∗8.8. Thrust Bearings. Disk Friction. Thrust bearings are used to provide axial support to rotating shafts and axles. They are of two types: (1) *end bearings* and (2) *collar bearings* (Fig. 8.11). In the case of collar bearings, friction forces develop between the two ring-shaped areas which are in contact. In the case of end bearings, friction takes place over full circular areas, or over ring-shaped areas when the end of the shaft is hollow. Friction between circular areas, called *disk friction*, also occurs in other mechanisms, such as *disk clutches*.

To obtain a formula which is valid in the most general case of disk friction, we shall consider a rotating hollow shaft. A couple **M** keeps the shaft rotating at constant speed while a force **P** maintains it in contact with a fixed bearing (Fig. 8.12). Contact between the shaft and the bearing takes place over a ring-shaped area of inner radius R_1 and outer radius R_2. Assuming that the pressure between the two surfaces in contact is uniform, we find that the magnitude of the normal force ΔN exerted on an element of area ΔA is $\Delta N = P\,\Delta A/A$, where $A = \pi(R_2^2 - R_1^2)$, and that the magnitude of the friction force $\Delta \mathbf{F}$ acting on ΔA is $\Delta F = \mu_k\,\Delta N$. Denoting by r the distance from the axis of the shaft to the element of area ΔA, we express as follows the moment ΔM of $\Delta \mathbf{F}$ about the axis of the shaft:

$$\Delta M = r\,\Delta F = \frac{r\,\mu_k P\,\Delta A}{\pi(R_2^2 - R_1^2)}$$

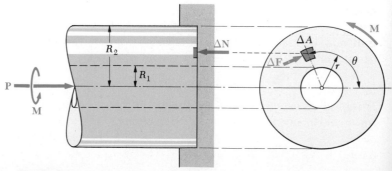

Fig. 8.12

The equilibrium of the shaft requires that the moment **M** of the couple applied to the shaft be equal in magnitude to the sum of the moments of the friction forces $\Delta \mathbf{F}$. Replacing ΔA by the infinitesimal element $dA = r \, d\theta \, dr$ used with polar coordinates, and integrating over the area of contact, we thus obtain the following expression for the magnitude of the couple **M** required to overcome the frictional resistance of the bearing:

$$M = \frac{\mu_k P}{\pi(R_2^2 - R_1^2)} \int_0^{2\pi} \int_{R_1}^{R_2} r^2 \, dr \, d\theta$$

$$= \frac{\mu_k P}{\pi(R_2^2 - R_1^2)} \int_0^{2\pi} \tfrac{1}{3}(R_2^3 - R_1^3) \, d\theta$$

$$M = \tfrac{2}{3} \mu_k P \frac{R_2^3 - R_1^3}{R_2^2 - R_1^2} \tag{8.8}$$

When contact takes place over a full circle of radius R, formula (8.8) reduces to

$$M = \tfrac{2}{3} \mu_k P R \tag{8.9}$$

The value of M is then the same as would be obtained if contact between shaft and bearing took place at a single point located at a distance $2R/3$ from the axis of the shaft.

The largest torque which may be transmitted by a disk clutch without causing slippage is given by a formula similar to (8.9), where μ_k has been replaced by the coefficient of static friction μ_s.

∗8.9. Wheel Friction. Rolling Resistance. The wheel is one of the most important inventions of our civilization. Its use makes it possible to move heavy loads with relatively little effort. Because the point of the wheel in contact with the ground at any given instant has no relative motion with respect to the ground, the wheel eliminates the large friction forces which would arise if the load were in direct contact with the ground. In practice, however, the wheel is not perfect, and some resistance to its motion exists. This resistance has two distinct causes. It is due (1) to a combined effect of axle friction and friction at the rim and (2) to the fact that the wheel and the ground deform, with the result that contact between wheel and ground takes place, not at a single point, but over a certain area.

To understand better the first cause of resistance to the motion of a wheel, we shall consider a railroad car supported by eight wheels mounted on axles and bearings. The car is assumed to be moving to the right at constant speed along a straight horizontal track. The free-body diagram of one of the wheels is shown in Fig. 8.13a. The forces acting on the free body include

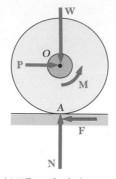

(a) Effect of axle friction

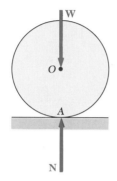

(b) Free wheel

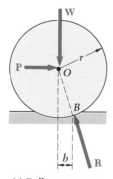

(c) Rolling resistance

Fig. 8.13

the load **W** supported by the wheel and the normal reaction **N** of the track. Since **W** is drawn through the center O of the axle, the frictional resistance of the bearing should be represented by a counterclockwise couple **M** (see Sec. 8.7). To keep the free body in equilibrium, we must add two equal and opposite forces **P** and **F**, forming a clockwise couple of moment $-\mathbf{M}$. The force **F** is the friction force exerted by the track on the wheel, and **P** represents the force which should be applied to the wheel to keep it rolling at constant speed. Note that the forces **P** and **F** would not exist if there were no friction between wheel and track. The couple **M** representing the axle friction would then be zero; the wheel would slide on the track without turning in its bearing.

The couple **M** and the forces **P** and **F** also reduce to zero when there is no axle friction. For example, a wheel which is not held in bearings and rolls freely and at constant speed on horizontal ground (Fig. 8.13b) will be subjected to only two forces: its own weight **W** and the normal reaction **N** of the ground. No friction force will act on the wheel, regardless of the value of the coefficient of friction between wheel and ground. A wheel rolling freely on horizontal ground should thus keep rolling indefinitely.

Experience, however, indicates that the wheel will slow down and eventually come to rest. This is due to the second type of resistance mentioned at the beginning of this section, known as the *rolling resistance*. Under the load **W**, both the wheel and the ground deform slightly, causing the contact between wheel and ground to take place over a certain area. Experimental evidence shows that the resultant of the forces exerted by the ground on the wheel over this area is a force **R** applied at a point B, which is not located directly under the center O of the wheel, but slightly in front of it (Fig. 8.13c). To balance the moment of **W** about B and to keep the wheel rolling at constant speed, it is necessary to apply a horizontal force **P** at the center of the wheel. Writing $\Sigma M_B = 0$, we obtain

$$Pr = Wb \tag{8.10}$$

where r = radius of wheel
$\quad\ b$ = horizontal distance between O and B
The distance b is commonly called the *coefficient of rolling resistance*. It should be noted that b is not a dimensionless coefficient since it represents a length; b is usually expressed in inches or in millimeters. The value of b depends upon several parameters in a manner which has not yet been clearly established. Values of the coefficient of rolling resistance vary from about 0.01 in. or 0.25 mm for a steel wheel on a steel rail to 5.0 in. or 125 mm for the same wheel on soft ground.

SAMPLE PROBLEM 8.6

A pulley of diameter 4 in. can rotate about a fixed shaft of diameter 2 in. The coefficients of static and kinetic friction between the pulley and shaft are both assumed equal to 0.20. Determine (a) the smallest vertical force **P** required to raise a 500-lb load, (b) the smallest vertical force **P** required to hold the load, (c) the smallest horizontal force **P** required to raise the same load.

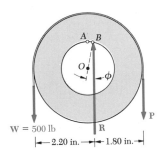

a. **Vertical Force P Required to Raise the Load.** When the forces in both parts of the rope are equal, contact between the pulley and shaft takes place at A. When **P** is increased, the pulley rolls around the shaft slightly and contact takes place at B. The free-body diagram of the pulley when motion is impending is drawn. The perpendicular distance from the center O of the pulley to the line of action of **R** is

$$r_f = r \sin \phi \approx r\mu \qquad r_f \approx (1 \text{ in.})0.20 = 0.20 \text{ in.}$$

Summing moments about B, we write

$$+\uparrow\Sigma M_B = 0: \qquad (2.20 \text{ in.})(500 \text{ lb}) - (1.80 \text{ in.})P = 0$$
$$P = 611 \text{ lb} \qquad\qquad P = 611 \text{ lb} \downarrow \quad \blacktriangleleft$$

b. **Vertical Force P to Hold the Load.** As the force **P** is decreased, the pulley rolls around the shaft and contact takes place at C. Considering the pulley as a free body and summing moments about C, we write

$$+\uparrow\Sigma M_C = 0: \qquad (1.80 \text{ in.})(500 \text{ lb}) - (2.20 \text{ in.})P = 0$$
$$P = 409 \text{ lb} \qquad\qquad P = 409 \text{ lb} \downarrow \quad \blacktriangleleft$$

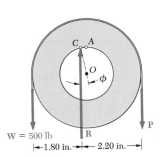

c. **Horizontal Force P to Raise the Load.** Since the three forces **W**, **P**, and **R** are not parallel, they must be concurrent. The direction of **R** is thus determined from the fact that its line of action must pass through the point of intersection D of **W** and **P**, and must be tangent to the circle of friction. Recalling that the radius of the circle of friction is $r_f = 0.20$ in., we write

$$\sin \theta = \frac{OE}{OD} = \frac{0.20 \text{ in.}}{(2 \text{ in.})\sqrt{2}} = 0.0707 \qquad \theta = 4.1°$$

From the force triangle, we obtain

$$P = W \cot (45° - \theta) = (500 \text{ lb}) \cot 40.9°$$
$$= 577 \text{ lb} \qquad\qquad P = 577 \text{ lb} \rightarrow \quad \blacktriangleleft$$

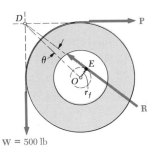

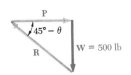

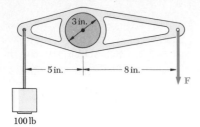

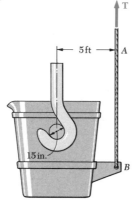

Fig. P8.68

Fig. P8.69

PROBLEMS

8.68 A lever of negligible weight is loosely fitted onto a 3-in.-diameter fixed shaft as shown. It is observed that a force **F** of magnitude 70 lb will just start rotating the lever clockwise. Determine (*a*) the coefficient of friction between the shaft and the lever, (*b*) the smallest force **F** for which the lever does not start rotating counterclockwise.

8.69 A hot-metal ladle and its contents weigh 60 tons. Knowing that the coefficient of friction between the hooks and the pinion is 0.30, determine the tension in the cable *AB* required to start tipping the ladle.

8.70 A bushing of outside diameter 75 mm fits loosely on a horizontal 50-mm-diameter shaft. A horizontal force **P** of magnitude 550 N is required to pull one end of the rope to the right when a 500-N force is applied to its other end. Determine the coefficient of friction between the shaft and the bushing. Assume that the rope does not slip on the bushing.

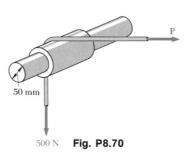

500 N   **Fig. P8.70**

8.71 A 600-N load is to be raised by the block and tackle shown. Each of the 60-mm-diameter pulleys rotates on a 10-mm-diameter axle. Knowing that $\mu = 0.20$, determine the tension in each portion of the rope as the load is slowly raised.

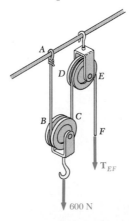

Fig. P8.71

8.72 In Prob. 8.71, determine the tension in each portion of the rope as the 600-N load is slowly lowered.

8.73 A certain railroad freight car has eight steel wheels of 32-in. diameter which are supported on 5-in.-diameter axles. Assuming $\mu = 0.015$, determine the horizontal force per ton of load required to move the car at a constant velocity.

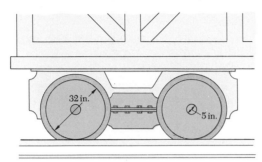

Fig. P8.73

8.74 A scooter is to be designed to roll down a 2 percent slope at a constant speed. Assuming that the coefficient of kinetic friction between the 1-in. axles and the bearings is 0.10, determine the required diameter of the wheels. Neglect the rolling resistance between the wheels and the ground.

8.75 A couple of magnitude 150 lb·ft is required to start the vertical shaft rotating. Determine the coefficient of static friction.

8.76 A 50-lb electric floor polisher is operated on a surface for which the coefficient of friction is 0.25. Assuming the normal force per unit area between the disk and the floor to be uniform, determine the magnitude Q of the horizontal forces required to prevent motion of the machine.

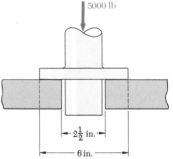

Fig. P8.75

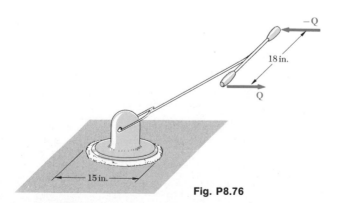

Fig. P8.76

8.77 In the clutch shown, disks A and B are keyed to the shaft but are free to slide along it. Disks C, D, and E are free to move parallel to the shaft but cannot rotate. The coefficient of static friction is 0.25 between all surfaces in contact. If disks C and E are pressed against the other disks as shown, determine the magnitude of the couple **M** required to rotate the shaft.

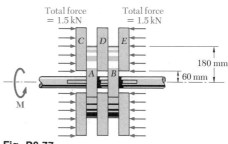

Fig. P8.77

8.78 In Prob. 8.77, determine the smallest forces with which disks C and E must be pressed against the other disks if the shaft is not to rotate when $M = 75 \text{ N} \cdot \text{m}$.

∗8.79 As the surfaces of shaft and bearing wear out, the frictional resistance of a thrust bearing decreases. It is generally assumed that the wear is directly proportional to the distance traveled by any given point of the shaft, and thus to the distance r from the point to the axis of the shaft. Assuming, then, that the normal force per unit area is inversely proportional to r, show that the magnitude M of the couple required to overcome the frictional resistance of a worn-out end bearing (with contact over the full circular area) is equal to 75 percent of the value given by formula (8.9) for a new bearing.

∗8.80 Assuming that bearings wear out as indicated in Prob. 8.79, show that the magnitude M of the couple required to overcome the frictional resistance of a worn-out collar bearing is

$$M = \tfrac{1}{2}\mu_k P(R_1 + R_2)$$

where P = magnitude of the total axial force
R_1, R_2 = inner and outer radii of collar

∗8.81 Assuming that the pressure between the surfaces of contact is uniform, show that the magnitude M of the couple required to overcome frictional resistance for the conical pivot shown is

$$M = \frac{2}{3}\frac{\mu_k P}{\sin\theta}\frac{R_2^3 - R_1^3}{R_2^2 - R_1^2}$$

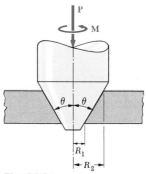

Fig. P8.81

8.82 Solve Prob. 8.76, assuming that between the disk and the surface the normal force per unit area varies uniformly from a maximum at the center to zero at the perimeter of the disk.

8.83 A circular disk of diameter 200 mm rolls at a constant velocity down a 2 percent incline. Determine the coefficient of rolling resistance.

8.84 Determine the horizontal force required to move a 1200 kg automobile along a horizontal road at a constant velocity. Neglect all forms of friction except rolling resistance, and assume the coefficient of rolling resistance to be 1.5 mm. The diameter of each tire is 600 mm.

8.85 Solve Prob. 8.73, including the effect of a coefficient of rolling resistance of 0.02 in.

8.86 Solve Prob. 8.74, including the effect of a coefficient of rolling resistance of 0.07 in.

8.10. Belt Friction.

Consider a flat belt passing over a fixed cylindrical drum (Fig. 8.14*a*). We propose to determine the relation existing between the values T_1 and T_2 of the tension in the two parts of the belt when the belt is just about to slide toward the right.

Let us detach from the belt a small element PP' subtending an angle $\Delta\theta$. Denoting by T the tension at P and by $T + \Delta T$ the tension at P', we draw the free-body diagram of the element of the belt (Fig. 8.14*b*). Besides the two forces of tension, the forces acting on the free body are the normal component $\Delta\mathbf{N}$ of the reaction of the drum and the friction force $\Delta\mathbf{F}$. Since motion is assumed to be impending, we have $\Delta F = \mu_s \, \Delta N$. It should be noted that if $\Delta\theta$ is made to approach zero, the magnitudes ΔN, ΔF, and the *difference* ΔT between the tension at P and the tension at P' will also approach zero; the value T of the tension at P, however, will remain unchanged. This observation helps in understanding our choice of notations.

Choosing the coordinate axes shown in Fig. 8.14*b*, we write the equations of equilibrium for the element PP':

$$\Sigma F_x = 0: \qquad (T + \Delta T)\cos\frac{\Delta\theta}{2} - T\cos\frac{\Delta\theta}{2} - \mu_s \, \Delta N = 0 \quad (8.11)$$

$$\Sigma F_y = 0: \qquad \Delta N - (T + \Delta T)\sin\frac{\Delta\theta}{2} - T\sin\frac{\Delta\theta}{2} = 0 \quad (8.12)$$

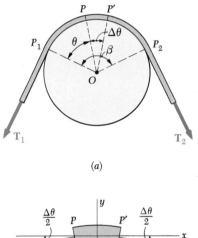

(*a*)

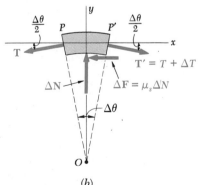

(*b*)

Fig. 8.14

Solving Eq. (8.12) for ΔN and substituting into (8.11), we obtain after reductions

$$\Delta T \cos \frac{\Delta\theta}{2} - \mu_s(2T + \Delta T)\sin\frac{\Delta\theta}{2} = 0$$

We shall now divide both terms by $\Delta\theta$; as far as the first term is concerned, this will be simply done by dividing ΔT by $\Delta\theta$. The division of the second term is carried out by dividing the terms in the parentheses by 2 and the sine by $\Delta\theta/2$. We write

$$\frac{\Delta T}{\Delta\theta}\cos\frac{\Delta\theta}{2} - \mu_s\left(T + \frac{\Delta T}{2}\right)\frac{\sin(\Delta\theta/2)}{\Delta\theta/2} = 0$$

If we now let $\Delta\theta$ approach 0, the cosine approaches 1 and $\Delta T/2$ approaches zero as noted above. On the other hand, the quotient of $\sin(\Delta\theta/2)$ over $\Delta\theta/2$ approaches 1, according to a lemma derived in all calculus textbooks. Since the limit of $\Delta T/\Delta\theta$ is by definition equal to the derivative $dT/d\theta$, we write

$$\frac{dT}{d\theta} - \mu_s T = 0 \qquad \frac{dT}{T} = \mu_s\,d\theta$$

We shall now integrate both members of the last equation obtained from P_1 to P_2 (Fig. 8.14a). At P_1, we have $\theta = 0$ and $T = T_1$; at P_2, we have $\theta = \beta$ and $T = T_2$. Integrating between these limits, we write

$$\int_{T_1}^{T_2}\frac{dT}{T} = \int_0^\beta \mu_s\,d\theta$$

$$\ln T_2 - \ln T_1 = \mu_s\beta$$

$$\ln\frac{T_2}{T_1} = \mu_s\beta \tag{8.13}$$

This relation may also be written in the form

$$\frac{T_2}{T_1} = e^{\mu_s\beta} \tag{8.14}$$

The formulas we have derived apply equally well to problems involving flat belts passing over fixed cylindrical drums and to problems involving ropes wrapped around a post or capstan. They may also be used to solve problems involving band brakes. In such problems, it is the drum which is about to rotate, while the band remains fixed. The formulas may also be applied to problems involving belt drives. In these problems, both the pulley and the belt rotate; our concern is then to find whether the belt will slip, i.e., whether it will move *with respect* to the pulley.

Formulas (8.13) and (8.14) should be used only if the belt, rope, or brake is *about to slip*. Formula (8.14) will be used if T_1 or T_2 is desired; formula (8.13) will be preferred if either μ_s or the angle of contact β is desired. We should note that T_2 is always larger than T_1; T_2 therefore represents the tension in that part of the belt or rope which *pulls*, while T_1 is the tension in the part which *resists*. We should also observe that the angle of contact β must be expressed in *radians*. The angle β may be larger than 2π; for example, if a rope is wrapped n times around a post, β is equal to $2\pi n$.

If the belt, rope, or brake is actually slipping, formulas similar to (8.13) and (8.14), but involving the coefficient of kinetic friction μ_k, should be used. If the belt, rope, or brake does not slip and is not about to slip, none of these formulas may be used.

The belts used in belt drives are often V-shaped. Such a belt, called a *V belt*, is shown in Fig. 8.15a. It is seen that contact between belt and pulley takes place along the sides of the groove. The relation existing between the values T_1 and T_2 of the tension in the two parts of the belt when the belt is just about to slip may again be obtained by drawing the free-body diagram of an element of belt (Fig. 8.15b and c). Equations similar to (8.11) and (8.12) are derived, but the magnitude of the total friction force acting on the element is now $2\,\Delta F$, and the sum of the y components of the normal forces is $2\,\Delta N \sin{(\alpha/2)}$. Proceeding as above, we obtain

$$\frac{T_2}{T_1} = e^{\mu_s \beta / \sin{(\alpha/2)}} \qquad\qquad (8.15)$$

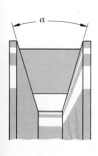
(a)

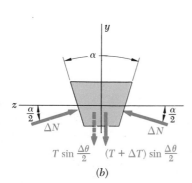
(b)

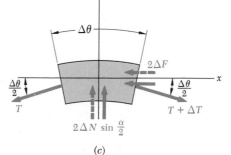
(c)

Fig. 8.15

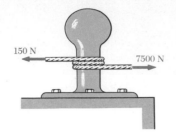

SAMPLE PROBLEM 8.7

A hawser thrown from a ship to a pier is wrapped two full turns around a capstan. The tension in the hawser is 7500 N; by exerting a force of 150 N on its free end, a longshoreman can just keep the hawser from slipping. (*a*) Determine the coefficient of friction between the hawser and the capstan. (*b*) Determine the tension in the hawser that could be resisted by the 150-N force if the hawser were wrapped three full turns around the capstan.

a. Coefficient of Friction. Since slipping of the hawser is impending, we use Eq. (8.13),

$$\ln \frac{T_2}{T_1} = \mu_s \beta$$

Since the hawser is wrapped two full turns around the capstan, we have

$$\beta = 2(2\pi \text{ rad}) = 12.6 \text{ rad}$$
$$T_1 = 150 \text{ N} \qquad T_2 = 7500 \text{ N}$$

Therefore,

$$\mu_s \beta = \ln \frac{T_2}{T_1}$$

$$\mu_s(12.6 \text{ rad}) = \ln \frac{7500 \text{ N}}{150 \text{ N}} = \ln 50 = 3.91$$

$$\mu_s = 0.31 \quad \blacktriangleleft$$

b. Hawser Wrapped Three Turns around Capstan. Using the value of μ_s obtained in part *a*, we have now

$$\beta = 3(2\pi \text{ rad}) = 18.9 \text{ rad}$$
$$T_1 = 150 \text{ N} \qquad \mu_s = 0.31$$

Substituting these values into Eq. (8.14), we obtain

$$\frac{T_2}{T_1} = e^{\mu_s \beta}$$

$$\frac{T_2}{150 \text{ N}} = e^{(0.31)(18.9)} = e^{5.86} = 350$$

$$T_2 = 52\,500 \text{ N} \qquad T_2 = 52.5 \text{ kN} \quad \blacktriangleleft$$

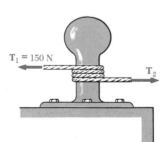

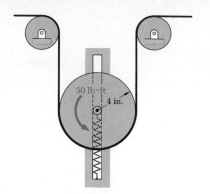

A flat belt passes over two idler pulleys and under a rotating drum of diameter 8 in. The axle of the drum is free to move vertically in a slot, and a spring keeps the drum in contact with the belt. What is the minimum force which should be exerted by the spring if slippage is not to occur when a 30-lb · ft torque is applied to the drum? The coefficient of static friction between belt and drum is 0.30.

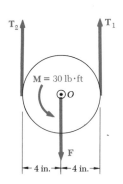

Solution. The free-body diagram of the drum is shown. The forces exerted on the drum by the spring and by each side of the belt are denoted by $\mathbf{F}$, $\mathbf{T}_1$, and $\mathbf{T}_2$, respectively. The force $\mathbf{T}_2$ is shown on the left since, if slippage occurs, the belt will be observed as moving clockwise *with respect to the drum.*

$$+\,\gamma\Sigma M_O = 0: \qquad T_1(4\text{ in.}) - T_2(4\text{ in.}) + 30\text{ lb} \cdot \text{ft} = 0$$
$$(T_2 - T_1)(4\text{ in.}) = 360\text{ lb} \cdot \text{in.}$$
$$T_2 - T_1 = 90\text{ lb} \tag{1}$$

The angle of contact between belt and drum is

$$\beta = 180° = 3.14\text{ rad}$$

Since the values of T_1 and T_2 correspond to impending slippage, we use Eq. (8.14):

$$\frac{T_2}{T_1} = e^{\mu_s \beta}$$
$$= e^{(0.30)(3.14)} = 2.57$$
$$T_2 = 2.57 T_1 \tag{2}$$

Substituting T_2 from Eq. (2) into Eq. (1), we obtain

$$2.57 T_1 - T_1 = 90\text{ lb}$$
$$1.57 T_1 = 90\text{ lb}$$
$$T_1 = 57.3\text{ lb}$$

$$T_2 = 2.57(57.3\text{ lb}) = 147.3\text{ lb}$$

Expressing that the sum of the vertical components of all forces must be zero:

$$+\uparrow\Sigma F_y = 0: \qquad T_1 + T_2 - F = 0$$
$$F = T_1 + T_2 = 57.3\text{ lb} + 147.3\text{ lb} = 204.6\text{ lb}$$
$$F = 205\text{ lb} \downarrow \qquad \blacktriangleleft$$

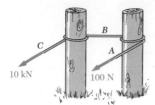

Fig. P8.87

PROBLEMS

8.87 A rope is wrapped around two posts as shown. If a 100-N force must be exerted at A to resist a 10-kN force at C, determine (*a*) the coefficient of static friction between the rope and the posts, (*b*) the corresponding tension in portion B of the rope.

8.88 A hawser is wrapped two full turns around a capstan head. By exerting a force of 160 lb on the free end of the hawser, a seaman can resist a force of 10,000 lb on the other end of the hawser. Determine (*a*) the coefficient of friction, (*b*) the number of times the hawser should be wrapped around the capstan if a 40,000-lb force is to be resisted by the same 160-lb force.

8.89 Knowing that the coefficient of friction between the rope and the rods is 0.30, determine the range of values of P for which the rope will not slip.

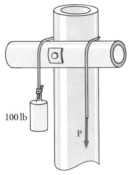

8.90 Assume that the bushing of Prob. 8.70 has become frozen to the shaft and cannot rotate. Determine the coefficient of friction between the bushing and the rope if a force P of magnitude 600 N is required to pull the rope to the right.

Fig. P8.89

8.91 A flat belt is used to deliver a torque to pulley A. Determine the largest torque which can be delivered to pulley A if the maximum allowable belt tension is 3 kN. Will the belt slip first on pulley A or on pulley B if the torque is increased?

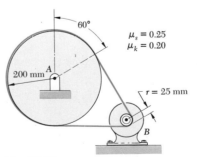

Fig. P8.91

8.92 A band brake is used to control the speed of a flywheel as shown. What torque must be applied to the flywheel in order to keep it rotating at a constant speed, when $P = 12$ lb? Assume that the flywheel rotates clockwise.

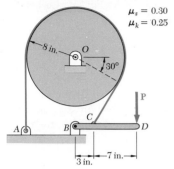

$\mu_s = 0.30$
$\mu_k = 0.25$

Fig. P8.92

8.93 Solve Prob. 8.92, assuming that the flywheel rotates counterclockwise.

8.94 A brake drum of radius $r = 150$ mm is rotating clockwise when a force **P** of magnitude 75 N is applied at D. Knowing that $\mu = 0.25$, determine the moment about O of the friction forces applied to the drum when $a = 75$ mm and $b = 400$ mm.

8.95 Knowing that $r = 150$ mm and $a = 75$ mm, determine the maximum value of μ for which the brake is not self-locking. The brake drum rotates clockwise.

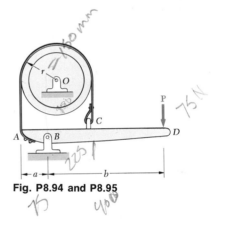

Fig. P8.94 and P8.95

8.96 A cord is placed over two cylinders, each of 75-mm diameter. Knowing that $\mu = 0.25$, determine the largest mass m which can be raised if cylinder B is rotated slowly and cylinder A is fixed.

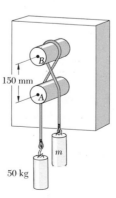

150 mm

50 kg

Fig. P8.96

Fig. P8.97

8.97 A 50-ft rope passes over a small horizontal shaft; one end of the rope is attached to a bucket which weighs 5 lb. The excess rope is coiled inside the bucket. The coefficient of static friction between the rope and the shaft is 0.30, and the rope weighs 0.50 lb/ft. (a) If the shaft is held fixed, show that the system is in equilibrium. (b) If the shaft is slowly rotated, how far will the bucket rise before slipping? Neglect the diameter of the shaft.

8.98 The shaft of Prob. 8.97 is slowly rotated. Determine how far the bucket can be lowered before the rope slips on the shaft.

8.99 A cable is placed around three pipes, each of 100-mm outside diameter, located in the same horizontal plane. Two of the pipes are fixed and do not rotate; the third pipe is rotated slowly. Knowing that $\mu = 0.30$ for each pipe, determine the largest mass m which can be raised (a) if only pipe A is rotated, (b) if only pipe B is rotated, (c) if only pipe C is rotated.

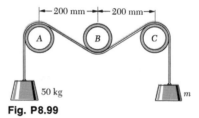

Fig. P8.99

Fig. P8.100 and P8.101

8.100 The strap wrench shown is used to grip the pipe firmly and at the same time not mar the external surface of the pipe. Knowing that $a = 10$ in. and $r = 2$ in., determine the minimum coefficient of friction for which the wrench will be self-locking.

8.101 Denoting by μ the coefficient of friction between the strap and the pipe, determine the minimum value of the ratio r/a for which the wrench will be self-locking.

8.102 Solve Prob. 8.91, assuming that the flat belt and pulley are replaced by a V belt and V pulley with $\alpha = 28°$. (The angle α is as shown in Fig. 8.15a.)

8.103 Complete the derivation of Eq. (8.15), which relates the tension in both parts of a V belt.

8.104 Prove that Eqs. (8.13) and (8.14) are valid for any shape of surface provided that the coefficient of friction is the same at all points of contact.

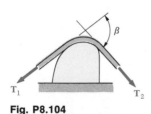

Fig. P8.104

REVIEW PROBLEMS

8.105 Two uniform rods, each of weight W, are held by frictionless pins A and B. It is observed that if the value of θ is greater than $10°$, the rods will not remain in equilibrium. Determine the coefficient of friction at C.

8.106 Determine the smallest value of the coefficient of friction for which three identical cylindrical rods may be placed as shown.

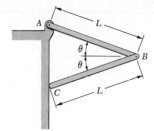

Fig. P8.105

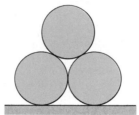

Fig. P8.106

8.107 A flat belt is used to transmit a torque from pulley A to pulley B. The radius of each pulley is 2 in. and $\mu_s = 0.30$. Determine the largest torque which can be transmitted if the maximum allowable tension is 600 lb.

8.108 Solve Prob. 8.107, assuming that the belt is looped around the pulleys in a figure 8.

8.109 The drive disk of the Scotch crosshead mechanism shown is slowly rotated counterclockwise by the couple **M**. A constant force **P** is applied to rod BC which slides in the frictionless bearings D and E. Denoting by μ the coefficient of friction between the pin A and the crosshead, determine the magnitude of the couple **M** in terms of P, r, μ, and θ.

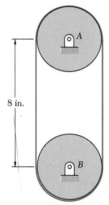

Fig. P8.107

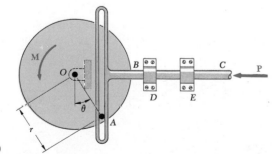

Fig. P8.109

8.110 A windlass, of diameter 150 mm, is used to raise a 40-kg load. It is supported by two axles and bearings of diameter 50 mm, poorly lubricated ($\mu = 0.40$). Determine the magnitude of the force **P** required to raise the load for each of the two positions shown.

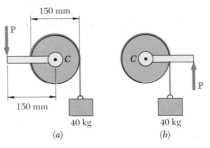

Fig. P8.110

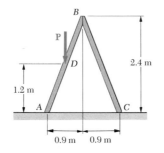

Fig. P8.111

8.111 Two identical uniform boards, each of mass 20 kg, are temporarily leaned against each other as shown. Knowing that the coefficient of friction between all surfaces is 0.40, determine (a) the largest magnitude of the force **P** for which equilibrium will be maintained, (b) the surface at which motion will impend.

8.112 Solve Prob. 8.111, assuming that the force **P** is applied at D and is directed horizontally to the right.

8.113 Knowing that $\mu = 0.20$ at all surfaces of contact, determine the magnitude of the force **P** required to move the 10-kg plate B to the left. (Neglect bearing friction in the pulley.)

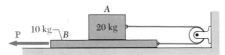

Fig. P8.113

8.114 A couple $\mathbf{M}_0$, of magnitude 150 lb · ft, must be applied to the drive pulley B in order to maintain a constant motion in the conveyor shown. Knowing that $\mu = 0.20$, determine the minimum tension in the lower portion of the belt if no slipping is to occur.

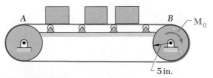

Fig. P8.114

8.115 A uniform slender rod is attached to a collar at B and rests on a circular cylindrical surface of radius r. The collar may slide without friction along a vertical guide. Denoting by ϕ the angle of friction between the rod and the surface, show that the largest and smallest values of θ corresponding to equilibrium must satisfy the equation

$$\tan \theta + \tan^2 \theta \tan (\theta \mp \phi) = 1$$

8.116 A block of weight W rests on an incline which forms an angle θ with the horizontal plane. A horizontal force **P**, parallel to the incline, is applied to the block. Denoting by μ the coefficient of friction, determine (a) the magnitude of **P** required to move the block, (b) the direction in which the block moves. (c) Obtain numerical values for parts a and b, when $\theta = 15°$, $\mu = 0.30$, and $W = 50$ lb.

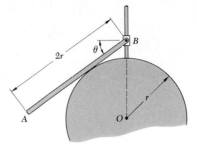

Fig. P8.115

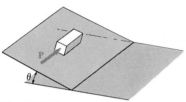

Fig. P8.116

CHAPTER 9

Distributed Forces: Moments of Inertia

MOMENTS OF INERTIA OF AREAS

9.1. Second Moment, or Moment of Inertia, of an Area. In Chap. 5, we analyzed various systems of forces distributed over an area. The two main types of forces considered were (1) weights of homogeneous plates of uniform thickness (Secs. 5.2 to 5.4) and (2) distributed loads on beams and hydrostatic forces (Secs. 5.6 and 5.7). In the case of homogeneous plates, the magnitude ΔW of the weight of an element of plate was proportional to the area ΔA of the element. In the case of distributed loads on beams, the magnitude ΔW of each elementary weight was represented by an element of area $\Delta A = \Delta W$ under the load curve; in the case of hydrostatic forces on submerged rectangular surfaces, a similar procedure was followed. Thus, in all cases considered in Chap. 5, the distributed forces were proportional to the elementary areas associated with them. The resultant of these forces, therefore, could be obtained by summing the corresponding areas, and the moment of the resultant about any given axis could be determined by computing the first moments of the areas about that axis.

In this chapter, we shall consider distributed forces $\Delta \mathbf{F}$ whose magnitudes depend not only upon the element of area ΔA on which they act but also upon the distance from ΔA to some given axis. More precisely, the magnitude of the force per unit area

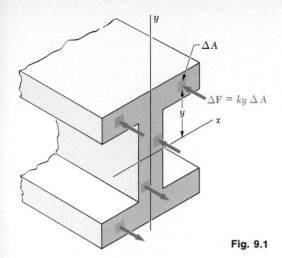

Fig. 9.1

$\Delta F/\Delta A$ will vary linearly with the distance to the axis.

Consider, for example, a beam of uniform cross section, subjected to two equal and opposite couples applied at each end of the beam. Such a beam is said to be in *pure bending,* and it is shown in mechanics of materials that the internal forces in any section of the beam are distributed forces whose magnitudes $\Delta F = ky\,\Delta A$ vary linearly with the distance y from an axis passing through the centroid of the section. This axis, represented by the x axis in Fig. 9.1, is known as the *neutral axis* of the section. The forces on one side of the neutral axis are forces of compression, and on the other side forces of tension, while on the neutral axis itself the forces are zero.

The magnitude of the resultant $\mathbf{R}$ of the elementary forces $\Delta \mathbf{F}$ over the entire section is

$$R = \int ky\,dA = k\int y\,dA$$

The last integral obtained is recognized as the *first moment* of the section about the x axis; it is equal to $\bar{y}A$ and to zero, since the centroid of the section is located on the x axis. The system of the forces $\Delta \mathbf{F}$ thus reduces to a couple. The magnitude M of this couple (bending moment) must be equal to the sum of the moments $\Delta M_x = y\,\Delta F = ky^2\,\Delta A$ of the elementary forces. Integrating over the entire section, we obtain

$$M = \int ky^2\,dA = k\int y^2\,dA$$

The last integral is known as the *second moment,* or *moment of inertia,*† of the beam section with respect to the x axis and

† The term second moment is more proper than the term moment of inertia, since, logically, the latter should be used only to denote integrals of mass (see Sec. 9.10). In common engineering practice, however, moment of inertia is used in connection with areas as well as masses.

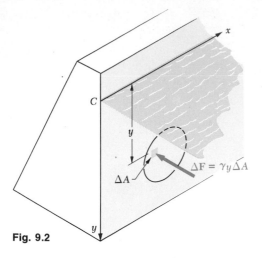

Fig. 9.2

is denoted by I_x. It is obtained by multiplying each element of area dA by the *square of its distance* from the x axis and integrating over the beam section. Since each product $y^2\, dA$ is positive, whether y is itself positive or negative (or zero if y is zero), the integral I_x will always be different from zero and positive.

Another example of second moment, or moment of inertia, of an area is provided by the following problem of hydrostatics: A vertical circular gate used to close the outlet of a large reservoir is submerged under water as shown in Fig. 9.2. What is the resultant of the forces exerted by the water on the gate, and what is the moment of the resultant about the line of intersection of the plane of the gate with the water surface (x axis)?

If the gate were rectangular, the resultant of the forces of pressure could be determined from the pressure curve, as was done in Sec. 5.7. Since the gate is circular, however, a more general method must be used. Denoting by y the depth of an element of area ΔA and by γ the specific weight of water, the pressure at the element is $p = \gamma y$, and the magnitude of the elementary force exerted on ΔA is $\Delta F = p\, \Delta A = \gamma y\, \Delta A$. The magnitude of the resultant of the elementary forces is thus

$$R = \int \gamma y\, dA = \gamma \int y\, dA$$

and may be obtained by computing the first moment of the area of the gate with respect to the x axis. The moment M_x of the resultant must be equal to the sum of the moments $\Delta M_x = y\, \Delta F = \gamma y^2\, \Delta A$ of the elementary forces. Integrating over the area of the gate, we have

$$M_x = \int \gamma y^2\, dA = \gamma \int y^2\, dA$$

Here again, the integral obtained represents the second moment, or moment of inertia, I_x of the area with respect to the x axis.

9.2. Determination of the Moment of Inertia of an Area by Integration. We have defined in the preceding section the second moment, or moment of inertia, of an area A with respect to the x axis. Defining in a similar way the moment of inertia I_y of the area A with respect to the y axis, we write (Fig. 9.3a)

$$I_x = \int y^2\, dA \qquad I_y = \int x^2\, dA \tag{9.1}$$

These integrals, known as the *rectangular moments of inertia* of the area A, may be more easily computed if we choose for dA a thin strip parallel to one of the axes of coordinates. To compute I_x, the strip is chosen parallel to the x axis, so that all the points forming the strip are at the same distance y from the x axis (Fig. 9.3b); the moment of inertia dI_x of the strip is

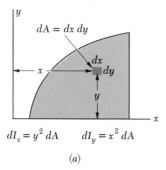

$dI_x = y^2\, dA \qquad dI_y = x^2\, dA$

(a)

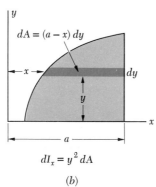

$dI_x = y^2\, dA$

(b)

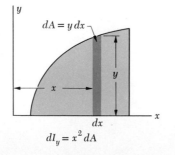

$dI_y = x^2\, dA$

Fig. 9.3   (c)

then obtained by multiplying the area dA of the strip by y^2. To compute I_y, the strip is chosen parallel to the y axis so that all the points forming the strip are at the same distance x from the y axis (Fig. 9.3c); the moment of inertia dI_y of the strip is $x^2\, dA$.

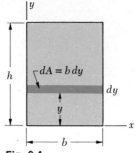

Fig. 9.4

Moment of Inertia of a Rectangular Area. As an example, we shall determine the moment of inertia of a rectangle with respect to its base (Fig. 9.4). Dividing the rectangle into strips parallel to the x axis, we obtain

$$dA = b\, dy \qquad dI_x = y^2 b\, dy \qquad I_x = \int_0^h b y^2\, dy = \tfrac{1}{3}bh^3 \qquad (9.2)$$

Computing I_x and I_y from the Same Elementary Strips. The formula just derived may be used to determine the moment of inertia dI_x with respect to the x axis of a rectangular strip parallel to the y axis such as the one shown in Fig. 9.3c. Making $b = dx$ and $h = y$ in formula (9.2), we write

$$dI_x = \tfrac{1}{3}y^3\, dx$$

On the other hand, we have

$$dI_y = x^2\, dA = x^2 y\, dx$$

The same element may thus be used to compute the moments of inertia I_x and I_y of a given area (Fig. 9.5).

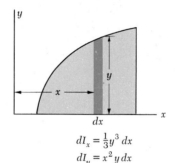

$$dI_x = \tfrac{1}{3}y^3\, dx$$
$$dI_y = x^2 y\, dx$$

Fig. 9.5

9.3. Polar Moment of Inertia. An integral of great importance in problems concerning the torsion of cylindrical shafts and in problems dealing with the rotation of slabs is

$$J_O = \int r^2\, dA \qquad (9.3)$$

where r is the distance from the element of area dA to the pole O (Fig. 9.6). This integral is the *polar moment of inertia* of the area A with respect to O.

The polar moment of inertia of a given area may be computed from the rectangular moments of inertia I_x and I_y of the area if these integrals are already known. Indeed, noting that $r^2 = x^2 + y^2$, we write

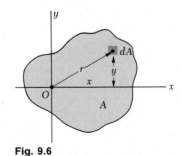

Fig. 9.6

$$J_O = \int r^2\, dA = \int (x^2 + y^2)\, dA = \int y^2\, dA + \int x^2\, dA$$

that is,

$$J_O = I_x + I_y \qquad (9.4)$$

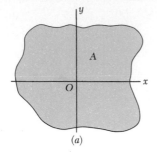

(a)

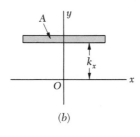

(b)

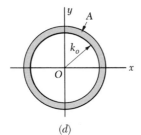

(c)

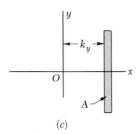

(d)

Fig. 9.7

9.4. Radius of Gyration of an Area. Consider an area A which has a moment of inertia I_x with respect to the x axis (Fig. 9.7a). Let us imagine that we concentrate this area into a thin strip parallel to the x axis (Fig. 9.7b). If the area A, thus concentrated, is to have the same moment of inertia with respect to the x axis, the strip should be placed at a distance k_x from the x axis, defined by the relation

$$I_x = k_x^2 A$$

Solving for k_x, we write

$$k_x = \sqrt{\frac{I_x}{A}} \tag{9.5}$$

The distance k_x is referred to as the *radius of gyration* of the area with respect to the x axis. We may define in a similar way the radii of gyration k_y and k_O (Fig. 9.7c and d); we write

$$I_y = k_y^2 A \qquad k_y = \sqrt{\frac{I_y}{A}} \tag{9.6}$$

$$J_O = k_O^2 A \qquad k_O = \sqrt{\frac{J_O}{A}} \tag{9.7}$$

Substituting for J_O, I_x, and I_y in terms of the radii of gyration in the relation (9.4), we observe that

$$k_O^2 = k_x^2 + k_y^2 \tag{9.8}$$

Example. As an example, let us compute the radius of gyration k_x of the rectangle shown in Fig. 9.4. Using formulas (9.5) and (9.2), we write

$$k_x^2 = \frac{I_x}{A} = \frac{\frac{1}{3}bh^3}{bh} = \frac{h^2}{3} \qquad k_x = \frac{h}{\sqrt{3}}$$

The radius of gyration k_x of the rectangle is shown in Fig. 9.8. It should not be confused with the ordinate $\bar{y} = h/2$ of the centroid of the area. While k_x depends upon the *second moment,* or moment of inertia, of the area, the ordinate $\bar{y}$ is related to the *first moment* of the area.

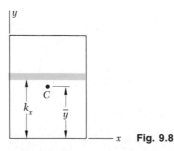

Fig. 9.8

Determine the moment of inertia of a triangle with respect to its base.

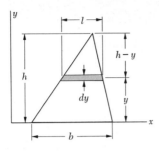

Solution. A triangle of base b and height h is drawn; the x axis is chosen to coincide with the base. A differential strip parallel to the x axis is chosen. Since all portions of the strip are at the same distance from the x axis, we write

$$dI_x = y^2\, dA \qquad dA = l\, dy$$

From the similar triangles, we have

$$\frac{l}{b} = \frac{h-y}{h} \qquad l = b\frac{h-y}{h} \qquad dA = b\frac{h-y}{h}\, dy$$

Integrating dI_x from $y = 0$ to $y = h$, we obtain

$$I_x = \int y^2\, dA = \int_0^h y^2 b\frac{h-y}{h}\, dy = \frac{b}{h}\int_0^h (hy^2 - y^3)\, dy$$

$$= \frac{b}{h}\left[h\frac{y^3}{3} - \frac{y^4}{4}\right]_0^h \qquad\qquad I_x = \frac{bh^3}{12} \quad \blacktriangleleft$$

(a) Determine the centroidal polar moment of inertia of a circular area by direct integration. (b) Using the result of part a, determine the moment of inertia of a circular area with respect to a diameter.

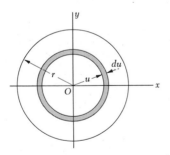

a. Polar Moment of Inertia. An annular differential element of area is chosen. Since all portions of this differential area are at the same distance from the origin, we write

$$dJ_0 = u^2\, dA \qquad dA = 2\pi u\, du$$

$$J_0 = \int dJ_0 = \int_0^r u^2(2\pi u\, du) = 2\pi \int_0^r u^3\, du$$

$$J_0 = \frac{\pi}{2}r^4 \quad \blacktriangleleft$$

b. Moment of Inertia. Because of the symmetry of the circular area we have $I_x = I_y$. We then write

$$J_0 = I_x + I_y = 2I_x \qquad \frac{\pi}{2}r^4 = 2I_x \qquad I_{\text{diameter}} = I_x = \frac{\pi}{4}r^4 \quad \blacktriangleleft$$

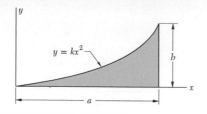

SAMPLE PROBLEM 9.3

(a) Determine the moment of inertia of the shaded area shown with respect to each of the coordinate axes. This area has also been considered in Sample Prob. 5.4. (b) Using the results of part a, determine the radius of gyration of the shaded area with respect to each of the coordinate axes.

Solution. Referring to Sample Prob. 5.4, we obtain the following expressions for the equation of the curve and the total area:

$$y = \frac{b}{a^2}x^2 \qquad A = \tfrac{1}{3}ab$$

Moment of Inertia I_x. A vertical differential element of area is chosen. Since all portions of this element are *not* at the same distance from the x axis, we must treat the element as a thin rectangle. The moment of inertia of the element with respect to the x axis is

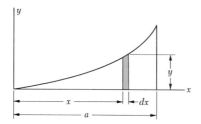

$$dI_x = \tfrac{1}{3}y^3\, dx = \frac{1}{3}\left(\frac{b}{a^2}x^2\right)^3 dx = \frac{1}{3}\frac{b^3}{a^6}x^6\, dx$$

$$I_x = \int dI_x = \int_0^a \frac{1}{3}\frac{b^3}{a^6}x^6\, dx = \left[\frac{1}{3}\frac{b^3}{a^6}\frac{x^7}{7}\right]_0^a$$

$$I_x = \frac{ab^3}{21} \qquad \blacktriangleleft$$

Moment of Inertia I_y. The same vertical differential element of area is used. Since all portions of the element are at the same distance from the y axis, we write

$$dI_y = x^2\, dA = x^2(y\, dx) = x^2\left(\frac{b}{a^2}x^2\right)dx = \frac{b}{a^2}x^4\, dx$$

$$I_y = \int dI_y = \int_0^a \frac{b}{a^2}x^4\, dx = \left[\frac{b}{a^2}\frac{x^5}{5}\right]_0^a$$

$$I_y = \frac{a^3 b}{5} \qquad \blacktriangleleft$$

Radii of Gyration k_x and k_y

$$k_x^2 = \frac{I_x}{A} = \frac{ab^3/21}{ab/3} = \frac{b^2}{7} \qquad k_x = \sqrt{\tfrac{1}{7}}\,b \qquad \blacktriangleleft$$

$$k_y^2 = \frac{I_y}{A} = \frac{a^3 b/5}{ab/3} = \tfrac{3}{5}a^2 \qquad k_y = \sqrt{\tfrac{3}{5}}\,a \qquad \blacktriangleleft$$

PROBLEMS

9.1 through 9.4 Determine by direct integration the moment of inertia of the shaded area with respect to the y axis.

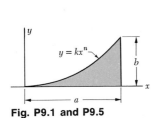

Fig. P9.1 and P9.5

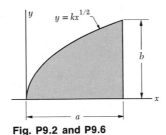

Fig. P9.2 and P9.6

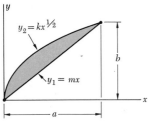

Fig. P9.3 and P9.7

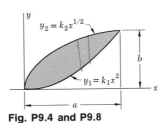

Fig. P9.4 and P9.8

9.5 through 9.8 Determine by direct integration the moment of inertia of the shaded area with respect to the x axis.

9.9 and 9.10 Determine the moment of inertia and radius of gyration of the shaded area shown with respect to the x axis.

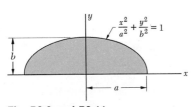

Fig. P9.9 and P9.11

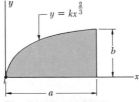

Fig. P9.10 and P9.12

9.11 and 9.12 Determine the moment of inertia and radius of gyration of the shaded area shown with respect to the y axis.

9.13 Determine the polar moment of inertia and the polar radius of gyration of an equilateral triangle of side a with respect to one of its vertices.

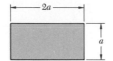

Fig. P9.14

9.14 Determine the polar moment of inertia and the polar radius of gyration of the rectangle shown with respect to the midpoint of one of its (*a*) longer sides, (*b*) shorter sides.

9.15 (*a*) Determine by direct integration the polar moment of inertia of the annular area shown. (*b*) Using the results of part *a*, determine the moment of inertia of the given area with respect to the *x* axis.

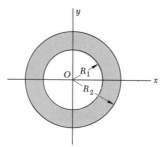

Fig. P9.15 and P9.16

9.16 (*a*) Show that the polar radius of gyration k_O of the annular area shown is approximately equal to the mean radius $R_m = (R_1 + R_2)/2$ for small values of the thickness $t = R_2 - R_1$. (*b*) Determine the percentage error introduced by using R_m in place of k_O for values of t/R_m respectively equal to 1, $\frac{1}{2}$, and $\frac{1}{10}$.

*9.17 Determine the moment of inertia of the shaded area with respect to the *x* axis.

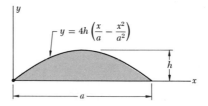

$$y = 4h\left(\frac{x}{a} - \frac{x^2}{a^2}\right)$$

Fig. P9.17 and P9.18

*9.18 Determine the moment of inertia of the shaded area with respect to the *y* axis.

*9.19 Prove that the centroidal polar moment of inertia of a given area A cannot be smaller than $A^2/2\pi$. (*Hint.* Compare the moment of inertia of the given area with the moment of inertia of a circle of the same area and same centroid.)

9.5. Parallel-Axis Theorem. Consider the moment of inertia I of an area A with respect to an axis AA' (Fig. 9.9). Denoting by y the distance from an element of area dA to AA', we write

$$I = \int y^2 \, dA$$

Let us now draw an axis BB' parallel to AA' through the centroid C of the area; this axis is called a *centroidal axis*. Denoting by

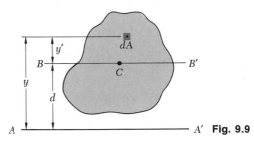

Fig. 9.9

y' the distance from the element dA to BB', we write $y = y' + d$, where d is the distance between the axes AA' and BB'. Substituting for y in the integral representing I, we write

$$\begin{aligned} I = \int y^2 \, dA &= \int (y' + d)^2 \, dA \\ &= \int y'^2 \, dA + 2d \int y' \, dA + d^2 \int dA \end{aligned}$$

The first integral represents the moment of inertia $\bar{I}$ of the area with respect to the centroidal axis BB'. The second integral represents the first moment of the area with respect to BB'; since the centroid C of the area is located on that axis, the second integral must be zero. Finally, we observe that the last integral is equal to the total area A. We write therefore

$$I = \bar{I} + Ad^2 \tag{9.9}$$

This formula expresses that the moment of inertia I of an area with respect to any given axis AA' is equal to the moment of inertia $\bar{I}$ of the area with respect to a centroidal axis BB' parallel to AA' *plus* the product Ad^2 of the area A and of the square of the distance d between the two axes. This theorem is known as the *parallel-axis theorem*. Substituting k^2A for I and $\bar{k}^2A$ for $\bar{I}$, the theorem may also be expressed in the following way:

$$k^2 = \bar{k}^2 + d^2 \tag{9.10}$$

A similar theorem may be used to relate the polar moment of inertia J_O of an area about a point O and the polar moment

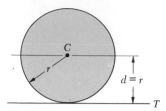

Fig. 9.10

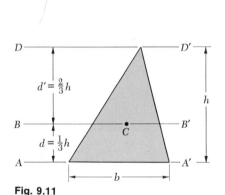

Fig. 9.11

of inertia $\bar{I}_C$ of the same area about its centroid C. Denoting by d the distance between O and C, we write

$$J_O = \bar{J}_C + Ad^2 \qquad \text{or} \qquad k_O^2 = \bar{k}_C^2 + d^2 \qquad (9.11)$$

Example 1. As an application of the parallel-axis theorem, we shall determine the moment of inertia I_T of a circular area with respect to a line tangent to the circle (Fig. 9.10). We found in Sample Prob. 9.2 that the moment of inertia of a circular area about a centroidal axis is $\bar{I} = \frac{1}{4}\pi r^4$. We may write, therefore,

$$I_T = \bar{I} + Ad^2 = \frac{1}{4}\pi r^4 + \pi r^2 r^2 = \frac{5}{4}\pi r^4$$

Example 2. The parallel-axis theorem may also be used to determine the centroidal moment of inertia of an area when the moment of inertia of this area with respect to some parallel axis is known. Consider, for instance, a triangular area (Fig. 9.11). We found in Sample Prob. 9.1 that the moment of inertia of a triangle with respect to its base AA' is equal to $\frac{1}{12}bh^3$. Using the parallel-axis theorem, we write

$$I_{AA'} = \bar{I}_{BB'} + Ad^2$$
$$\bar{I}_{BB'} = I_{AA'} - Ad^2 = \frac{1}{12}bh^3 - \frac{1}{2}bh(\tfrac{1}{3}h)^2 = \frac{1}{36}bh^3$$

It should be observed that the product Ad^2 was *subtracted* from the given moment of inertia in order to obtain the centroidal moment of inertia of the triangle. While this product is *added* in transferring *from* a centroidal axis to a parallel axis, it should be *subtracted* in transferring *to* a centroidal axis. In other words, the moment of inertia of an area is always smaller with respect to a centroidal axis than with respect to any other parallel axis.

Returning to Fig. 9.11, we observe that the moment of inertia of the triangle with respect to a line DD' drawn through a vertex may be obtained by writing

$$I_{DD'} = \bar{I}_{BB'} + Ad'^2 = \frac{1}{36}bh^3 + \frac{1}{2}bh(\tfrac{2}{3}h)^2 = \frac{1}{4}bh^3$$

Note that $I_{DD'}$ *could not* have been obtained directly from $I_{AA'}$. The parallel-axis theorem can be applied only if one of the two parallel axes passes through the centroid of the area.

9.6. Moments of Inertia of Composite Areas. Consider a composite area A made of several component areas A_1, A_2, etc. Since the integral representing the moment of inertia of A may be subdivided into integrals computed over A_1, A_2, etc., the moment of inertia of A with respect to a given axis will be obtained by adding the moments of inertia of the areas A_1, A_2, etc., with respect to the same axis. The moment of inertia of an area made of several of the common shapes shown in Fig. 9.12 may thus be obtained from the formulas given in that figure. Before adding the moments of inertia of the component areas, however, the parallel-axis theorem should be used to transfer each moment of inertia to the desired axis. This is shown in Sample Probs. 9.4 and 9.5.

Rectangle	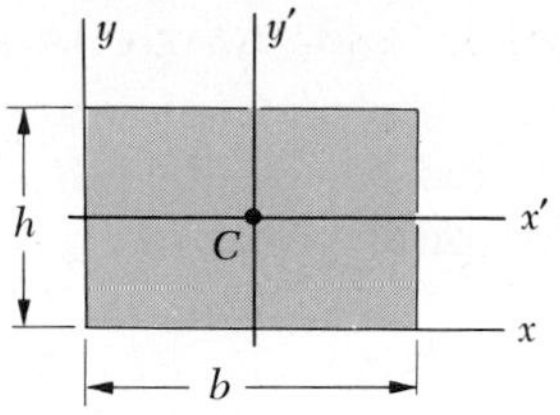	$\bar{I}_{x'} = \frac{1}{12}bh^3$ $\bar{I}_{y'} = \frac{1}{12}b^3h$ $I_x = \frac{1}{3}bh^3$ $I_y = \frac{1}{3}b^3h$ $J_C = \frac{1}{12}bh(b^2 + h^2)$
Triangle	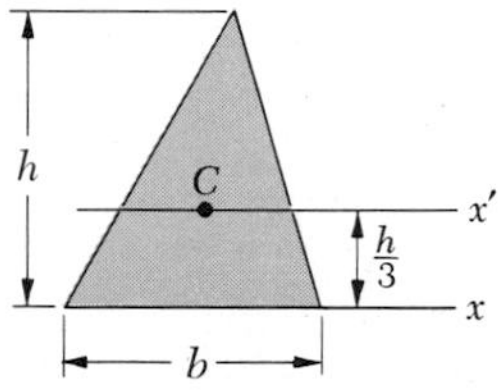	$\bar{I}_{x'} = \frac{1}{36}bh^3$ $I_x = \frac{1}{12}bh^3$
Circle	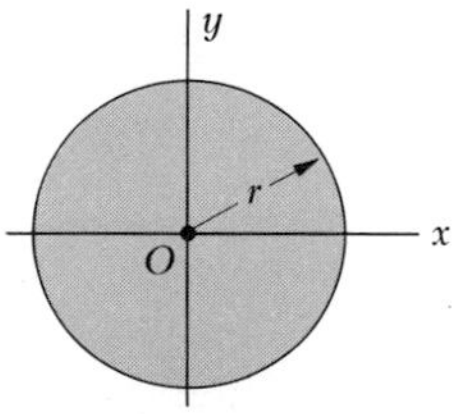	$\bar{I}_x = \bar{I}_y = \frac{1}{4}\pi r^4$ $J_O = \frac{1}{2}\pi r^4$
Semicircle	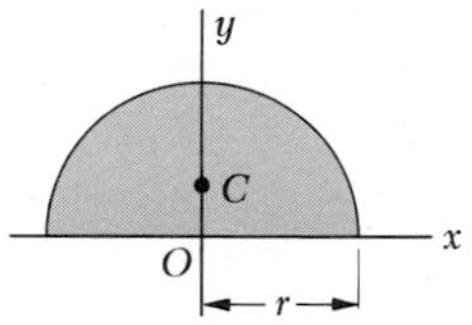	$I_x = I_y = \frac{1}{8}\pi r^4$ $J_O = \frac{1}{4}\pi r^4$
Quarter circle	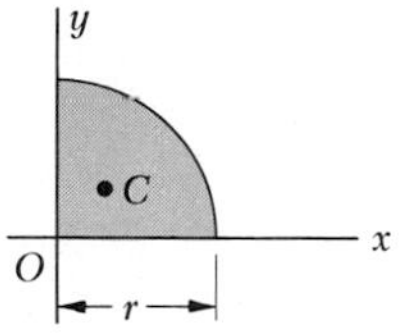	$I_x = I_y = \frac{1}{16}\pi r^4$ $J_O = \frac{1}{8}\pi r^4$
Ellipse	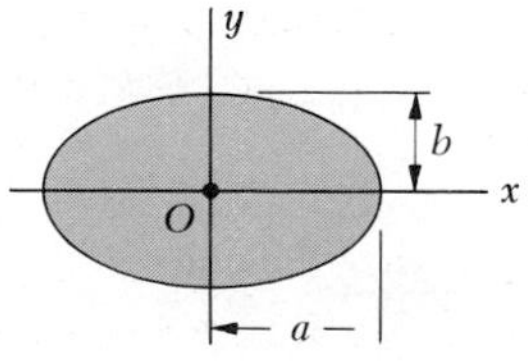	$\bar{I}_x = \frac{1}{4}\pi ab^3$ $\bar{I}_y = \frac{1}{4}\pi a^3b$ $J_O = \frac{1}{4}\pi ab(a^2 + b^2)$

Fig. 9.12 Moments of inertia of common geometric shapes.

The properties of the cross sections of various structural shapes are given in Fig. 9.13. As noted in Sec. 9.1, the moment of inertia of a beam section about its neutral axis is closely related to the value of the internal forces. The determination of moments of inertia is thus a prerequisite to the analysis and design of structural members.

It should be noted that the radius of gyration of a composite area is *not* equal to the sum of the radii of gyration of the component areas. In order to determine the radius of gyration of a composite area, it is necessary first to compute the moment of inertia of the area.

Fig. 9.13 Properties of rolled-steel structural shapes.

Shape		Designation	Width, in.
Wide-flange section	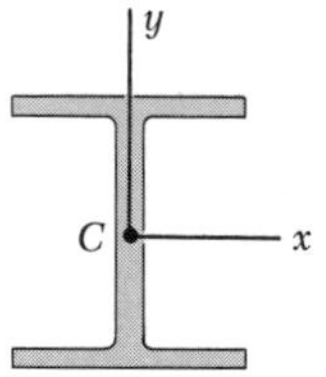	W16 × 64† W14 × 43 W8 × 31	$8\frac{1}{2}$ 8 8
American Standard beam	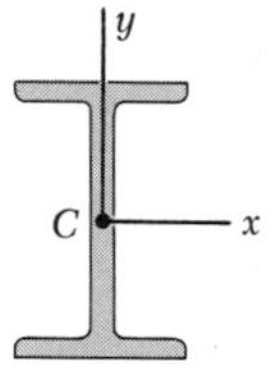	S18 × 70† S12 × 35 S6 × 12.5	$6\frac{1}{4}$ $5\frac{1}{8}$ $3\frac{3}{8}$
American Standard channel	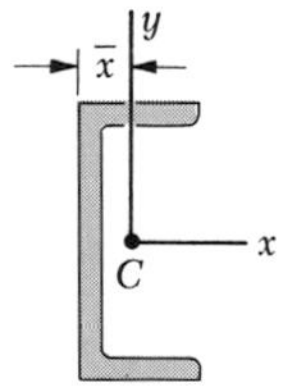	C10 × 25† C8 × 11.5 C6 × 8.2	$2\frac{7}{8}$ $2\frac{1}{4}$ $1\frac{7}{8}$
Angles	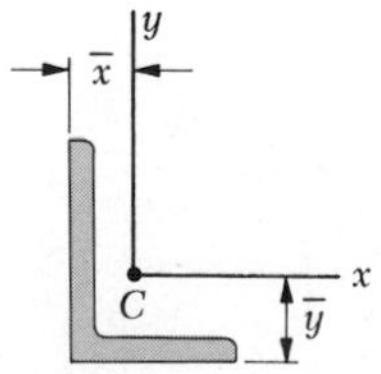	L6 × 6 × 1‡ L4 × 4 × $\frac{1}{2}$ L8 × 6 × 1 L6 × 4 × $\frac{3}{4}$ L5 × $3\frac{1}{2}$ × $\frac{1}{2}$	

†Depth in inches and weight per unit length in lb/ft.
‡Depth, width, and thickness in inches.

Area, in^2	$\bar{I}_x$, in^4	$\bar{k}_x$, in.	$\bar{y}$, in.	$\bar{I}_y$, in^4	$\bar{k}_y$, in.	$\bar{x}$, in.
18.80	836	6.66	· · ·	73.3	1.97	
12.60	429	5.82	· · ·	45.1	1.89	
9.12	110	3.47	· · ·	37.0	2.01	
20.6	926	6.71	· · ·	24.1	1.08	
10.3	229	4.72	· · ·	9.87	0.98	
3.67	21.1	2.45	· · ·	1.8	0.71	
7.35	91.2	3.52	· · ·	3.4	0.68	0.62
3.38	32.6	3.11	· · ·	1.3	0.63	0.57
2.40	13.1	2.34	· · ·	0.7	0.54	0.51
11.00	35.5	1.80	1.86	35.5	1.80	1.86
3.75	5.6	1.22	1.18	5.6	1.22	1.18
13.00	80.8	2.49	2.65	38.8	1.73	1.65
6.94	24.5	1.88	2.08	8.7	1.12	1.08
4.00	9.99	1.58	1.66	4.1	1.01	0.91

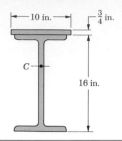

The strength of a 16-in., 64-lb wide-flange beam is increased by attaching a 10- by $\frac{3}{4}$-in. plate to its upper flange as shown. Determine the moment of inertia and the radius of gyration of the composite section with respect to an axis through its centroid C and parallel to the plate.

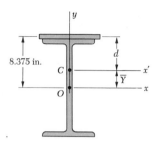

Solution. The origin of coordinates O is placed at the centroid of the wide-flange section, and the distance $\overline{Y}$ to the centroid of the composite section is computed by the methods of Chap. 5. The area of the wide-flange section is found by referring to Fig. 9.13.

Section	Area, in^2	$\overline{y}$, in.	$\overline{y}A$, in^3
Plate	7.5	8.375	62.8
Wide-flange section	18.8	0	0
	26.3	. . .	62.8

$$\overline{Y}\,\Sigma A = \Sigma\,\overline{y}A \qquad \overline{Y}(26.3) = 62.8 \qquad \overline{Y} = 2.39 \text{ in.}$$

Moment of Inertia. The parallel-axis theorem is used to determine the moments of inertia of the wide-flange section and of the plate with respect to the x' axis. This axis is a centroidal axis for the composite section, but *not* for either of the elements considered separately. The value of $\overline{I}_x$ for the wide-flange section is obtained from Fig. 9.13.

For the wide-flange section

$$I_{x'} = \overline{I}_x + A\overline{Y}^2 = 836 + (18.8)(2.39)^2 = 943 \text{ in}^4$$

For the plate

$$I_{x'} = \overline{I}_x + Ad^2 = (\tfrac{1}{12})(10)(\tfrac{3}{4})^3 + (7.5)(8.375 - 2.39)^2 = 269 \text{ in}^4$$

For the composite area

$$I_{x'} = 943 + 269 = 1212 \text{ in}^4$$

$$I_{x'} = 1212 \text{ in}^4 \quad \blacktriangleleft$$

Radius of Gyration

$$k_{x'}^2 = \frac{I_{x'}}{A} = \frac{1212 \text{ in}^4}{26.3 \text{ in}^2}$$

$$k_{x'} = 6.79 \text{ in.} \quad \blacktriangleleft$$

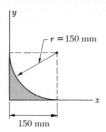

SAMPLE PROBLEM 9.5

r = 150 mm

Determine the moment of inertia of the shaded area with respect to the *y* axis.

150 mm

Solution. The given area may be obtained by subtracting a quarter circle from a square. The moments of inertia of the square and of the quarter circle are computed separately. To avoid excessively large numerical values, they will be expressed in cm⁴ rather than in mm⁴. We shall convert, therefore, the given data into centimeters.

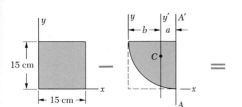

I_y *for Square.* Referring to Fig. 9.12, we obtain

$$I_y = \tfrac{1}{3}b^3h = \tfrac{1}{3}(15\text{ cm})^3(15\text{ cm}) = 16\,875\text{ cm}^4$$

I_y *for the Quarter Circle.* Referring to Fig. 5.8, we locate the centroid C of the quarter circle with respect to side AA'.

$$a = \frac{4r}{3\pi} = \frac{4(15\text{ cm})}{3\pi} = 6.366\text{ cm}$$

The distance b from the centroid C to the y axis is

$$b = 15\text{ cm} - a = 15\text{ cm} - 6.366\text{ cm} = 8.634\text{ cm}$$

Referring now to Fig. 9.12, we compute the moment of inertia of the quarter circle with respect to side AA'; we also compute the area of the quarter circle.

$$I_{AA'} = \tfrac{1}{16}\pi r^4 = \tfrac{1}{16}\pi(15\text{ cm})^4 = 9940\text{ cm}^4$$
$$A = \tfrac{1}{4}\pi r^2 = \tfrac{1}{4}\pi(15\text{ cm})^2 = 176.7\text{ cm}^2$$

Using the parallel-axis theorem, we obtain the value of $\overline{I}_{y'}$,

$$I_{AA'} = \overline{I}_{y'} + Aa^2$$
$$\overline{I}_{y'} = I_{AA'} - Aa^2$$
$$= 9940\text{ cm}^4 - (176.7\text{ cm}^2)(6.366\text{ cm})^2 = 2779\text{ cm}^4$$

Again using the parallel-axis theorem, we obtain the value of I_y,

$$I_y = \overline{I}_{y'} + Ab^2 = 2779\text{ cm}^4 + (176.7\text{ cm}^2)(8.634\text{ cm})^2 = 15\,950\text{ cm}^4$$

I_y *for Given Area.* Subtracting the moment of inertia of the quarter circle from that of the square, we obtain

$$I_y = 16\,875\text{ cm}^4 - 15\,950\text{ cm}^4 = 925\text{ cm}^4$$

or, since $1\text{ cm}^4 = (10\text{ mm})^4 = 10^4\text{ mm}^4$,

$$I_y = 925 \times 10^4\text{ mm}^4 \qquad I_y = 9.25 \times 10^6\text{ mm}^4 \quad \blacktriangleleft$$

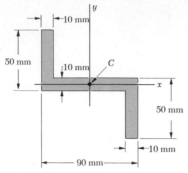

Fig. P9.20 and P9.21

PROBLEMS

9.20 and 9.22 Determine the moment of inertia and the radius of gyration of the shaded area with respect to the x axis.

9.21 and 9.23 Determine the moment of inertia and the radius of gyration of the shaded area with respect to the y axis.

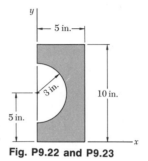

Fig. P9.22 and P9.23

9.24 Determine the shaded area and its moment of inertia with respect to a centroidal axis parallel to AA', knowing that its moments of inertia with respect to AA' and BB' are respectively 2000 in⁴ and 4000 in⁴, and that $d_1 = 8$ in. and $d_2 = 4$ in.

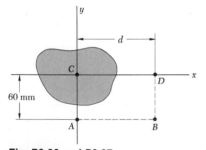

Fig. P9.24 and P9.25

9.25 Knowing that the shaded area is equal to 25 in² and that its moment of inertia with respect to AA' is 800 in⁴, determine its moment of inertia with respect to BB', for $d_1 = 5$ in. and $d_2 = 2$ in.

9.26 The shaded area is equal to 5000 mm². Determine its centroidal moments of inertia $\bar{I}_x$ and $\bar{I}_y$, knowing that $2\bar{I}_x = \bar{I}_y$ and that the polar moment of inertia of the area about point A is $J_A = 22.5 \times 10^6$ mm⁴.

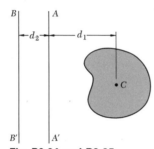

Fig. P9.26 and P9.27

9.27 The polar moments of inertia of the shaded area with respect to points A, B, and D are, respectively, $J_A = 28.8 \times 10^6$ mm⁴, $J_B = 67.2 \times 10^6$ mm⁴, and $J_D = 45.6 \times 10^6$ mm⁴. Determine the shaded area, its centroidal moment of inertia $\bar{J}_C$, and the distance d from C to D.

9.28 Determine the centroidal polar moment of inertia of the area shown.

9.29 Determine the moments of inertia $\bar{I}_x$ and $\bar{I}_y$ of the area shown with respect to centroidal axes respectively parallel and perpendicular to the side AB.

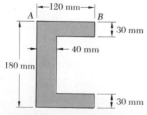

Fig. P9.28 and P9.29

9.30 Determine the polar moment of inertia of the area shown with respect to (*a*) point *O*, (*b*) the centroid of the area.

9.31 Determine the moments of inertia $\bar{I}_x$ and $\bar{I}_y$ of the area shown with respect to centroidal axes respectively parallel and perpendicular to the side *AB*.

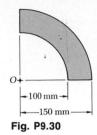

Fig. P9.30

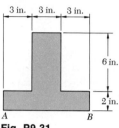

Fig. P9.31

9.32 Two 10-in., 25-lb American Standard channels and a 14- by $\frac{1}{2}$-in. plate are used to form the column section shown. For *b* = 7 in., determine the moments of inertia and the radii of gyration of the total section with respect to the centroidal axes shown.

9.33 In Prob. 9.32 determine the distance *b* for which the centroidal moments of inertia $\bar{I}_x$ and $\bar{I}_y$ of the column section are equal.

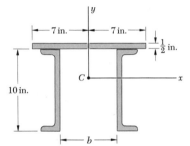

Fig. P9.32

9.34 A plate-girder section consists of a $31\frac{1}{2}$- by $\frac{5}{16}$-in. plate and of four 5- by $3\frac{1}{2}$-in. angles each $\frac{1}{2}$ in. thick. Determine the moments of inertia of the section with respect to the centroidal axes shown.

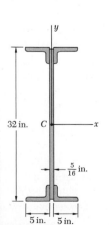

Fig. P9.34

9.35 Two 4- by 4- by $\frac{1}{2}$-in. angles are welded to a $\frac{1}{2}$- by 10-in. steel plate as shown. Determine the moments of inertia and the radii of gyration of the combined section with respect to centroidal axes respectively parallel and perpendicular to the plate.

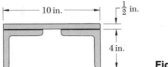

Fig. P9.35

9.36 through 9.38 The panel shown forms the end of a trough which is filled with water to the line AA'. Referring to Sec. 9.1, determine the depth of the point of application of the resultant of the hydrostatic forces acting on the panel (center of pressure).

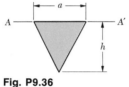

Fig. P9.36

Fig. P9.37

Fig. P9.38

*9.39 Determine the x coordinate of the centroid of the volume shown; this volume was obtained by intersecting a circular cylinder by the oblique plane. (*Hint.* The height of the volume is proportional to the x coordinate; consider an analogy between this height and the water pressure on a submerged surface.)

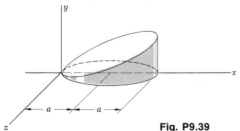

Fig. P9.39

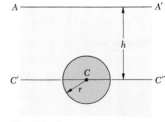

Fig. P9.40

9.40 A vertical circular gate of radius r is hinged about its diameter $C'C''$. Denoting the specific weight of water by γ, determine (a) the reaction at each hinge, (b) the moment of the couple required to keep the gate closed.

9.41 The center of a vertical circular gate, 2 m in diameter, is located 3 m below the water surface. The gate is held by three bolts equally spaced as shown. Determine the force on each bolt.

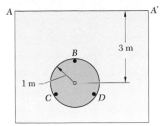

Fig. P9.41

∗9.42 Show that the system of hydrostatic forces acting on a submerged plane area A may be reduced to a force **P** at the centroid C of the area and two couples. The force **P** is perpendicular to the area and of magnitude $P = \gamma A\bar{y} \sin \theta$, where γ is the specific weight of the liquid, and the couples are represented by vectors directed as shown and of magnitude $M_{x'} = \gamma \bar{I}_{x'} \sin \theta$ and $M_{y'} = \gamma \bar{P}_{x'y'} \sin \theta$, where $\bar{P}_{x'y'} = \int x' y' \, dA$ (see Sec. 9.7). Note that the couples are independent of the depth at which the area is submerged.

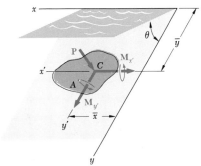

Fig. P9.42

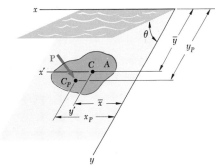

Fig. P9.43

∗9.43 Show that the resultant of the hydrostatic forces acting on a submerged plane area A is a force **P** perpendicular to the area and of magnitude $P = \gamma A\bar{y} \sin \theta = \bar{p}A$, where γ is the specific weight of the liquid and $\bar{p}$ the pressure at the centroid C of the area. Show that **P** is applied at a point C_P, called the center of pressure, of coordinates $x_P = P_{xy}/A\bar{y}$ and $y_P = I_x/A\bar{y}$, where $P_{xy} = \int xy \, dA$ (see Sec. 9.7). Show also that the difference of ordinates $y_P - \bar{y}$ is equal to $\bar{k}_{x'}^2/\bar{y}$ and thus depends upon the depth at which the area is submerged.

∗9.7. Product of Inertia.   The integral

$$P_{xy} = \int xy \, dA \tag{9.12}$$

obtained by multiplying each element dA of an area A by its coordinates x and y and integrating over the area (Fig. 9.14), is known as the *product of inertia* of the area A with respect to the x and y axes. Unlike the moments of inertia I_x and I_y, the product of inertia P_{xy} may be either positive or negative.

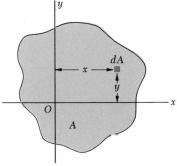

Fig. 9.14

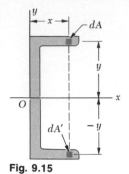

Fig. 9.15

When one or both of the x and y axes are axes of symmetry for the area A, the product of inertia P_{xy} is zero. Consider, for example, the channel section shown in Fig. 9.15. Since this section is symmetrical with respect to the x axis, we can associate to each element dA of coordinates x and y an element dA' of coordinates x and $-y$. Clearly, the contributions of any pair of elements chosen in this way cancel out, and the integral (9.12) reduces to zero.

A parallel-axis theorem similar to the one established in Sec. 9.5 for moments of inertia may be derived for products of inertia. Consider an area A and a system of rectangular coordinates x and y (Fig. 9.16). Through the centroid C of the area, of coordinates $\bar{x}$ and $\bar{y}$, we draw two *centroidal axes* x' and y'

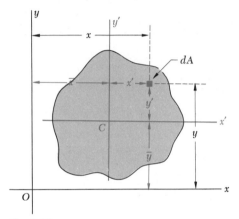

Fig. 9.16

parallel, respectively, to the x and y axes. Denoting by x and y the coordinates of an element of area dA with respect to the original axes, and by x' and y' the coordinates of the same element with respect to the centroidal axes, we write $x = x' + \bar{x}$ and $y = y' + \bar{y}$. Substituting into (9.12), we obtain the following expression for the product of inertia P_{xy}:

$$P_{xy} = \int xy \, dA = \int (x' + \bar{x})(y' + \bar{y}) \, dA$$
$$= \int x'y' \, dA + \bar{y}\int x' \, dA + \bar{x}\int y' \, dA + \bar{x}\bar{y}\int dA$$

The first integral represents the product of inertia $\bar{P}_{x'y'}$ of the area A with respect to the centroidal axes x' and y'. The next two integrals represent first moments of the area with respect to the centroidal axes; they reduce to zero, since the centroid C is located on these axes. Finally, we observe that the last integral is equal to the total area A. We write therefore

$$P_{xy} = \bar{P}_{x'y'} + \bar{x}\bar{y}A \tag{9.13}$$

*9.8. Principal Axes and Principal Moments of Inertia.

Consider the area A and the coordinate axes x and y (Fig. 9.17). We assume that the moments and product of inertia

$$I_x = \int y^2\, dA \qquad I_y = \int x^2\, dA \qquad P_{xy} = \int xy\, dA \qquad (9.14)$$

of the area A are known, and we propose to determine the moments and product of inertia I_u, I_v, and P_{uv} of A with respect

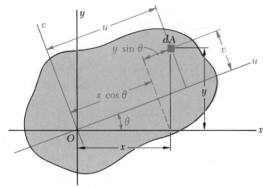

Fig. 9.17

to new axes u and v obtained by rotating the original axes about the origin through an angle θ.

We first note the following relations between the coordinates u, v and x, y of an element of area dA:

$$u = x \cos \theta + y \sin \theta \qquad v = y \cos \theta - x \sin \theta$$

Substituting for v in the expression for I_u, we write

$$I_u = \int v^2\, dA = \int (y \cos \theta - x \sin \theta)^2\, dA$$
$$= \cos^2 \theta \int y^2\, dA - 2 \sin \theta \cos \theta \int xy\, dA + \sin^2 \theta \int x^2\, dA$$

Taking the relations (9.14) into account, we write

$$I_u = I_x \cos^2 \theta - 2P_{xy} \sin \theta \cos \theta + I_y \sin^2 \theta \qquad (9.15)$$

Similarly, we obtain for I_v and P_{uv} the expressions

$$I_v = I_x \sin^2 \theta + 2P_{xy} \sin \theta \cos \theta + I_y \cos^2 \theta \qquad (9.16)$$
$$P_{uv} = I_x \sin \theta \cos \theta + P_{xy} (\cos^2 \theta - \sin^2 \theta) - I_y \sin \theta \cos \theta \qquad (9.17)$$

We observe, by adding (9.15) and (9.16) member by member, that

$$I_u + I_v = I_x + I_y \qquad (9.18)$$

This result could have been anticipated, since both members of (9.18) are equal to the polar moment of inertia J_O.

Using the trigonometric relations $\sin 2\theta = 2 \sin \theta \cos \theta$ and

$\cos 2\theta = \cos^2 \theta - \sin^2 \theta$, we may write (9.15), (9.16), and (9.17) as follows:

$$I_u = \frac{I_x + I_y}{2} + \frac{I_x - I_y}{2} \cos 2\theta - P_{xy} \sin 2\theta \qquad (9.19)$$

$$I_v = \frac{I_x + I_y}{2} - \frac{I_x - I_y}{2} \cos 2\theta + P_{xy} \sin 2\theta \qquad (9.20)$$

$$P_{uv} = \frac{I_x - I_y}{2} \sin 2\theta + P_{xy} \cos 2\theta \qquad (9.21)$$

Equations (9.19) and (9.21) are the parametric equations of a circle. This means that, if we choose a set of rectangular axes and plot a point M of abscissa I_u and ordinate P_{uv} for any given value of the parameter θ, all the points thus obtained will lie on a circle. To establish this property we shall eliminate θ from Eqs. (9.19) and (9.21); this is done by transposing $(I_x + I_y)/2$ in Eq. (9.19), squaring both members of Eqs. (9.19) and (9.21), and adding. We write

$$\left(I_u - \frac{I_x + I_y}{2}\right)^2 + P_{uv}^2 = \left(\frac{I_x - I_y}{2}\right)^2 + P_{xy}^2 \qquad (9.22)$$

Setting

$$I_{av} = \frac{I_x + I_y}{2} \qquad \text{and} \qquad R = \sqrt{\left(\frac{I_x - I_y}{2}\right)^2 + P_{xy}^2} \qquad (9.23)$$

we write the identity (9.22) in the form

$$(I_u - I_{av})^2 + P_{uv}^2 = R^2 \qquad (9.24)$$

which is the equation of a circle of radius R centered at the point C of abscissa I_{av} and ordinate 0 (Fig. 9.18).

The two points A and B where the circle obtained intersects

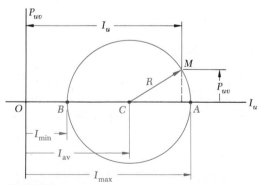

Fig. 9.18

the axis of abscissas are of special interest: Point A corresponds to the maximum value of the moment of inertia I_u, while point B corresponds to its minimum value. Besides, both points correspond to a zero value of the product of inertia P_{uv}. Thus, the values θ_m of the parameter θ which correspond to the points A and B may be obtained by setting $P_{uv} = 0$ in Eq. (9.21). We obtain†

$$\tan 2\theta_m = -\frac{2P_{xy}}{I_x - I_y} \qquad (9.25)$$

This equation defines two values $2\theta_m$ which are 180° apart and thus two values θ_m which are 90° apart. One of them corresponds to point A in Fig. 9.18 and to an axis through O in Fig. 9.17 with respect to which the moment of inertia of the given area is maximum; the other value corresponds to point B and an axis through O with respect to which the moment of inertia of the area is minimum. The two axes thus defined, which are perpendicular to each other, are called the *principal axes of the area about O*, and the corresponding values I_{max} and I_{min} of the moment of inertia are called the *principal moments of inertia of the area about O*. We check from Fig. 9.18 that

$$I_{max} = I_{av} + R \qquad \text{and} \qquad I_{min} = I_{av} - R \qquad (9.26)$$

Substituting for I_{av} and R from formulas (9.23), we write

$$I_{max,min} = \frac{I_x + I_y}{2} \pm \sqrt{\left(\frac{I_x - I_y}{2}\right)^2 + P_{xy}^2} \qquad (9.27)$$

Since the two values θ_m defined by Eq. (9.25) were obtained by setting $P_{uv} = 0$ in Eq. (9.21), it is clear that the product of inertia of the given area with respect to its principal axes is zero. Referring to Sec. 9.7, we note that, if an area possesses an axis of symmetry through a point O, this axis must be a principal axis of the area about O. On the other hand, a principal axis does not need to be an axis of symmetry; whether or not an area possesses properties of symmetry, it will have two principal axes of inertia about any point O.

The properties established here hold for any point O located inside or outside the given area. If the point O is chosen to coincide with the centroid of the area, any axis through O is a centroidal axis; the two principal axes of the area about its centroid are referred to as the *principal centroidal axes of the area*.

---

† This relation may also be obtained by differentiating I_u in Eq. (9.19) and setting $dI_u/d\theta = 0$.

*9.9. Mohr's Circle for Moments and Products of Inertia.

The circle used in the preceding section to illustrate the relations existing between the moments and products of inertia of a given area with respect to axes passing through a fixed point O was first introduced by the German engineer Otto Mohr (1835–1918) and is known as *Mohr's circle*. We shall see that, if the moments and product of inertia of an area A are known with respect to two rectangular x and y axes through a point O, Mohr's circle may be used to determine graphically (*a*) the principal axes and principal moments of inertia of the area about O, or (*b*) the moments and product of inertia of the area with respect to any other pair of rectangular axes u and v through O.

Consider a given area A and two rectangular coordinate axes x and y (Fig. 9.19*a*). We shall assume that the moments of inertia I_x and I_y and the product of inertia P_{xy} are known, and we shall represent them on a diagram by plotting a point X of coordinates I_x and P_{xy} and a point Y of coordinates I_y and $-P_{xy}$ (Fig. 9.19*b*). Joining X and Y by a straight line, we define the point C of intersection of line XY with the I axis and draw the circle of center C and diameter XY. Noting that the abscissa of C and the radius of the circle are respectively equal to the quantities I_{av} and R defined by the formulas (9.23), we conclude that the circle obtained is Mohr's circle for the given area about point O. Thus the abscissas of the points A and B where the circle intersects the I axis represent respectively the principal moments of inertia $I_{\max}$ and $I_{\min}$ of the area.

We also note that, since $\tan(XCA) = 2P_{xy}/(I_x - I_y)$, the angle XCA is equal in magnitude to one of the angles $2\theta_m$ which satisfy Eq. (9.25); thus the angle θ_m which defines in Fig. 9.19*a* the principal axis Oa corresponding to point A in Fig. 9.19*b* may be obtained by dividing in half the angle XCA measured on Mohr's circle. We further observe that, if $I_x > I_y$ and $P_{xy} > 0$, as in the case considered here, the rotation which brings CX into CA is clockwise. But, in that case, the angle θ_m obtained from Eq. (9.25) and defining the principal axis Oa in Fig. 9.19*a* is negative; thus the rotation bringing Ox into Oa is also clockwise. We conclude that the senses of rotation in both parts of Fig. 9.19 are the same; if a clockwise rotation through $2\theta_m$ is required to bring CX into CA on Mohr's circle, a clockwise rotation through θ_m will bring Ox into the corresponding principal axis Oa in Fig. 9.19*a*.

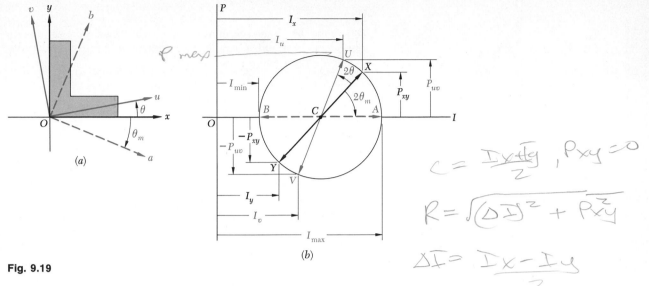

Fig. 9.19

Since Mohr's circle is uniquely defined, the same circle may be obtained by considering the moments and product of inertia of the area A with respect to rectangular axes u and v (Fig. 9.19a). The point U of coordinates I_u and P_{uv} and the point V of coordinates I_v and $-P_{uv}$ are therefore located on Mohr's circle, and the angle UCA in Fig. 9.19b must be equal to twice the angle uOa in Fig. 9.19a. Since, as noted above, the angle XCA is twice the angle xOa, it follows that the angle XCU in Fig. 9.19b is twice the angle xOu in Fig. 9.19a. Thus the diameter UV defining the moments and product of inertia I_u, I_v, and P_{uv} of the given area with respect to rectangular axes u and v forming an angle θ with the x and y axes may be obtained by rotating through an angle 2θ the diameter XY corresponding to the moments and product of inertia I_x, I_y, and P_{xy}. We note that the rotation which brings the diameter XY into the diameter UV in Fig. 9.19b has the same sense as the rotation which brings the xy axes into the uv axes in Fig. 9.19a.

It should be noted that the use of Mohr's circle is not limited to graphical solutions, i.e., to solutions based on the careful drawing and measuring of the various parameters involved. By merely sketching Mohr's circle and using trigonometry, one may easily derive the various relations required for a numerical solution of a given problem. Actual computations may then be carried out on a calculator (see Sample Prob. 9.8).

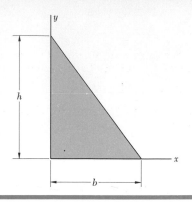

Determine the product of inertia of the right triangle shown (*a*) with respect to the *x* and *y* axes and (*b*) with respect to centroidal axes parallel to the *x* and *y* axes.

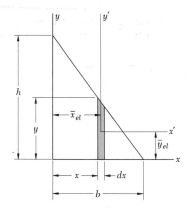

a. **Product of Inertia** P_{xy}. A vertical rectangular strip is chosen as the differential element of area. Using the parallel-axis theorem, we write

$$dP_{xy} = dP_{x'y'} + \bar{x}_{el}\bar{y}_{el}\, dA$$

Since the element is symmetrical with respect to the x' and y' axes, we note that $dP_{x'y'} = 0$. From the geometry of the triangle, we obtain

$$y = h\left(1 - \frac{x}{b}\right) \qquad dA = y\, dx = h\left(1 - \frac{x}{b}\right) dx$$

$$\bar{x}_{el} = x \qquad \bar{y}_{el} = \tfrac{1}{2}y = \tfrac{1}{2}h\left(1 - \frac{x}{b}\right)$$

Integrating dP_{xy} from $x = 0$ to $x = b$, we obtain

$$P_{xy} = \int dP_{xy} = \int \bar{x}_{el}\bar{y}_{el}\, dA = \int_0^b x(\tfrac{1}{2})h^2\left(1 - \frac{x}{b}\right)^2 dx$$

$$= h^2 \int_0^b \left(\frac{x}{2} - \frac{x^2}{b} + \frac{x^3}{2b^2}\right) dx = h^2\left[\frac{x^2}{4} - \frac{x^3}{3b} + \frac{x^4}{8b^2}\right]_0^b$$

$$P_{xy} = \tfrac{1}{24}b^2h^2 \cdot \blacktriangleleft$$

b. **Product of Inertia** $\bar{P}_{x''y''}$. The coordinates of the centroid of the triangle are

$$\bar{x} = \tfrac{1}{3}b \qquad \bar{y} = \tfrac{1}{3}h$$

Using the expression for P_{xy} obtained in part *a*, we apply the parallel-axis theorem and write

$$P_{xy} = \bar{P}_{x''y''} + \bar{x}\bar{y}A$$
$$\tfrac{1}{24}b^2h^2 = \bar{P}_{x''y''} + (\tfrac{1}{3}b)(\tfrac{1}{3}h)(\tfrac{1}{2}bh)$$
$$\bar{P}_{x''y''} = \tfrac{1}{24}b^2h^2 - \tfrac{1}{18}b^2h^2$$

$$\bar{P}_{x''y''} = -\tfrac{1}{72}b^2h^2 \quad \blacktriangleleft$$

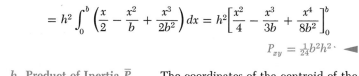

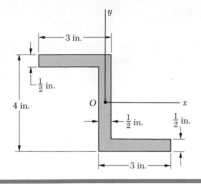

For the section shown, the moments of inertia with respect to the x and y axes have been computed and are known to be

$$I_x = 10.38 \text{ in}^4 \qquad I_y = 6.97 \text{ in}^4$$

Determine (a) the principal axes of the section about O, (b) the values of the principal moments of inertia of the section about O.

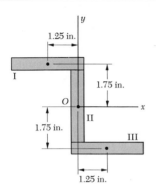

Solution. We first compute the product of inertia with respect to the x and y axes. The area is divided into three rectangles as shown. We note that the product of inertia $\bar{P}_{x'y'}$ with respect to centroidal axes parallel to the x and y axes is zero for each rectangle. Using the parallel-axis theorem $P_{xy} = \bar{P}_{x'y'} + \bar{x}\bar{y}A$, we thus find that, for each rectangle, P_{xy} reduces to $\bar{x}\bar{y}A$.

Rectangle	Area, in^2	$\bar{x}$, in.	$\bar{y}$, in.	$\bar{x}\bar{y}A$, in^4
I	1.5	-1.25	$+1.75$	-3.28
II	1.5	0	0	0
III	1.5	$+1.25$	-1.75	-3.28
				-6.56

$$P_{xy} = \Sigma \bar{x}\bar{y}A = -6.56 \text{ in}^4$$

a. Principal Axes. Since the magnitudes of I_x, I_y, and P_{xy} are known, Eq. (9.25) is used to determine the values of θ_m,

$$\tan 2\theta_m = -\frac{2P_{xy}}{I_x - I_y} = -\frac{2(-6.56)}{10.38 - 6.97} = +3.85$$

$$2\theta_m = 75.4° \text{ and } 255.4°$$

$$\theta_m = 37.7° \qquad \text{and} \qquad \theta_m = 127.7° \quad \blacktriangleleft$$

b. Principal Moments of Inertia. Using Eq. (9.27), we write

$$I_{max,min} = \frac{I_x + I_y}{2} \pm \sqrt{\left(\frac{I_x - I_y}{2}\right)^2 + P_{xy}^2}$$

$$= \frac{10.38 + 6.97}{2} \pm \sqrt{\left(\frac{10.38 - 6.97}{2}\right)^2 + (-6.56)^2}$$

$$I_{max} = 15.45 \text{ in}^4 \qquad I_{min} = 1.897 \text{ in}^4 \quad \blacktriangleleft$$

Noting that the area of the section is farther away from the a axis than from the b axis, we conclude that $I_a = I_{max} = 15.45 \text{ in}^4$ and $I_b = I_{min} = 1.897 \text{ in}^4$. This conclusion may be verified by substituting $\theta = 37.7°$ into Eqs. (9.19) and (9.20).

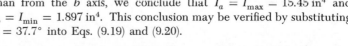

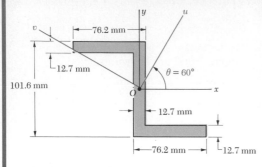

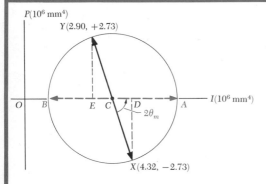

For the section shown, the moments and product of inertia with respect to the x and y axes have been computed and are known to be $I_x = 4.32 \times 10^6$ mm^4, $I_y = 2.90 \times 10^6$ mm^4, $P_{xy} = -2.73 \times 10^6$ mm^4. Using Mohr's circle, determine (a) the principal axes of the section about O, (b) the values of the principal moments of inertia of the section about O, (c) the moments and product of inertia of the section with respect to the u and v axes forming an angle of 60° with the x and y axes. (*Note:* This section is the same as that of Sample Prob. 9.7.)

Solution. We first plot point X of coordinates $I_x = 4.32$, $P_{xy} = -2.73$, and point Y of coordinates $I_y = 2.90$, $-P_{xy} = +2.73$. Joining X and Y by a straight line, we define the center C of Mohr's circle. The abscissa of C, which represents I_{av}, and the radius R of the circle may be measured directly or calculated as follows:

$$I_{av} = OC = \tfrac{1}{2}(I_x + I_y) = \tfrac{1}{2}(4.32 \times 10^6 \text{ mm}^4 + 2.90 \times 10^6 \text{ mm}^4)$$
$$= 3.61 \times 10^6 \text{ mm}^4$$

$$R = \sqrt{(CD)^2 + (DX)^2} = \sqrt{(0.71 \times 10^6 \text{ mm}^4)^2 + (2.73 \times 10^6 \text{ mm}^4)^2}$$
$$= 2.82 \times 10^6 \text{ mm}^4$$

a. Principal Axes. The principal axes of the section correspond to points A and B on Mohr's circle and the angle through which we should rotate CX to bring it into CA defines $2\theta_m$. We have

$$\tan 2\theta_m = \frac{DX}{CD} = \frac{2.73}{0.71} = 3.85 \qquad 2\theta_m = 75.4° \text{↺} \quad \theta_m = 37.7° \text{↺} \quad \blacktriangleleft$$

Thus the principal axis Oa corresponding to the maximum value of the moment of inertia is obtained by rotating the x axis through 37.7° counterclockwise; the principal axis corresponding to the minimum value of the moment of inertia may be obtained by rotating the y axis through the same angle.

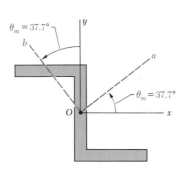

b. Principal Moments of Inertia. The principal moments of inertia are represented by the abscissas of A and B. We have

$$I_{max} = OA = OC + CA = I_{av} + R = (3.61 + 2.82)10^6 \text{ mm}^4$$
$$I_{max} = 6.43 \times 10^6 \text{ mm}^4 \quad \blacktriangleleft$$
$$I_{min} = OB = OC - BC = I_{av} - R = (3.61 - 2.82)10^6 \text{ mm}^4$$
$$I_{min} = 0.79 \times 10^6 \text{ mm}^4 \quad \blacktriangleleft$$

c. Moments and Product of Inertia with Respect to uv Axes. The points U and V on Mohr's circle which correspond to the u and v axes are obtained by rotating CX and CY through an angle $2\theta = 2(60°) = 120°$ counterclockwise. The coordinates of U and V yield the desired moments and product of inertia. Noting that the angle that CU forms with the I axis is $\phi = 120° - 75.4° = 44.6°$, we write

$$I_u = OF = OC + CF = 3.61 \times 10^6 \text{ mm}^4 + (2.82 \times 10^6 \text{ mm}^4) \cos 44.6°$$
$$I_u = 5.62 \times 10^6 \text{ mm}^4 \quad \blacktriangleleft$$
$$I_v = OG = OC - GC = 3.61 \times 10^6 \text{ mm}^4 - (2.82 \times 10^6 \text{ mm}^4) \cos 44.6°$$
$$I_v = 1.60 \times 10^6 \text{ mm}^4 \quad \blacktriangleleft$$
$$P_{uv} = FU = (2.82 \times 10^6 \text{ mm}^4) \sin 44.6°$$
$$P_{uv} = +1.98 \times 10^6 \text{ mm}^4 \quad \blacktriangleleft$$

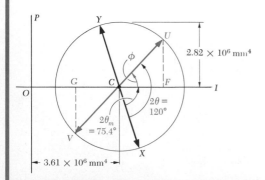

PROBLEMS

9.44 through 9.47 Determine by direct integration the product of inertia of the given area with respect to the x and y axes.

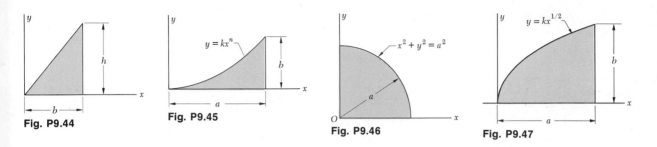

Fig. P9.44

Fig. P9.45

Fig. P9.46

Fig. P9.47

9.48 through 9.51 Using the parallel-axis theorem, determine the product of inertia of the area shown with respect to the centroidal x and y axes.

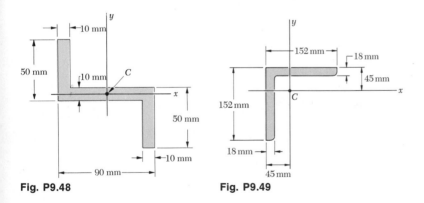

Fig. P9.48

Fig. P9.49

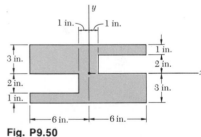

Fig. P9.50

9.52 Determine the moments of inertia and the product of inertia of the area of Prob. 9.48 with respect to new centroidal axes obtained by rotating the x and y axes through 30° counterclockwise.

9.53 Determine the moments of inertia and the product of inertia of the quarter circle of Prob. 9.46 with respect to new axes obtained by rotating the x and y axes about O (a) through 30° counterclockwise, (b) through 45° counterclockwise.

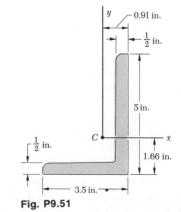

Fig. P9.51

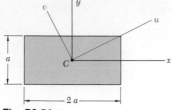

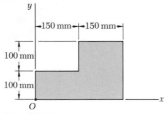

Fig. P9.54

Fig. P9.57

9.54 Determine the moments of inertia and the product of inertia of the rectangle shown with respect to the u and v axes.

9.55 Determine the moments of inertia and the product of inertia of the angle cross section of Prob. 9.51 with respect to new centroidal axes obtained by rotating the x and y axes through 45° clockwise. (The moments of inertia $\bar{I}_x$ and $\bar{I}_y$ are given in Fig. 9.13.)

9.56 Determine the orientation of the principal axes through the centroid and the corresponding values of the moment of inertia for the area of Prob. 9.48.

9.57 Determine the orientation of the principal axes through O and the corresponding values of the moment of inertia for the area shown.

9.58 Determine the orientation of the principal axes through the centroid and the corresponding values of the moment of inertia for the angle cross section shown. Neglect the effect of fillets in computing $\bar{P}_{xy}$. (The moments of inertia $\bar{I}_x$ and $\bar{I}_y$ of the section are given in Fig. 9.13.)

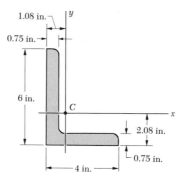

Fig. P9.58

9.59 Determine the orientation of the principal axes through the centroid and the corresponding values of the moment of inertia for the angle cross section of Prob. 9.51. (The moments of inertia $\bar{I}_x$ and $\bar{I}_y$ of the section are given in Fig. 9.13.)

9.60 Solve Prob. 9.54, using Mohr's circle.

9.61 Using Mohr's circle, determine the moments of inertia and the product of inertia of the angle cross section of Prob. 9.51 with respect to new centroidal axes obtained by rotating the x and y axes through 45° clockwise. (The moments of inertia $\bar{I}_x$ and $\bar{I}_y$ of the section are given in Fig. 9.13.)

9.62 Using Mohr's circle, determine the moments of inertia and the product of inertia of the area of Prob. 9.48 with respect to new centroidal axes obtained by rotating the x and y axes through 30° counterclockwise.

9.63 Using Mohr's circle, determine the moments of inertia and the product of inertia of the quarter circle of Prob. 9.46 with respect to new axes obtained by rotating the x and y axes about O (*a*) through 30° counterclockwise, (*b*) through 45° counterclockwise.

9.64 Solve Prob. 9.58, using Mohr's circle.

9.65 Using Mohr's circle, determine the orientation of the principal axes through the centroid and the corresponding values of the moment of inertia for the angle cross section of Prob. 9.51. (The moments of inertia $\bar{I}_x$ and $\bar{I}_y$ of the section are given in Fig. 9.13.)

9.66 Using Mohr's circle, determine the orientation of the principal axes through the centroid and the corresponding values of the moment of inertia for the area of Prob. 9.48.

9.67 Solve Prob. 9.57, using Mohr's circle.

9.68 Using Mohr's circle, show that, for any regular polygon (such as a pentagon), (*a*) the moment of inertia with respect to every axis through the centroid is the same, (*b*) the product of inertia with respect to any pair of rectangular axes through the centroid is zero.

9.69 The moments and product of inertia of a given area with respect to two rectangular axes x and y through O are, respectively, $I_x = 14.4 \times 10^6$ mm^4, $I_y = 8 \times 10^6$ mm^4, and $P_{xy} > 0$, while the minimum value of the moment of inertia of the area with respect to any axis through O is $I_{\min} = 7.2 \times 10^6$ mm^4. Using Mohr's circle, determine (*a*) the product of inertia P_{xy} of the area, (*b*) the orientation of the principal axes, (*c*) the value of $I_{\max}$.

*9.70 Prove that the expression $I_u I_v - P_{uv}^2$, where I_u, I_v, and P_{uv} represent, respectively, the moments and product of inertia of a given area with respect to two rectangular axes u and v through a given point O, is independent of the orientation of the u and v axes. Considering the particular case when the u and v axes correspond to the maximum value of P_{uv}, show that the given expression represents the square of the tangent drawn from the origin of the coordinates to Mohr's circle.

*9.71 Using the invariance property established in the preceding problem, express the product of inertia P_{xy} of an area A with respect to two rectangular axes through O in terms of the moments of inertia I_x and I_y of A and of the principal moments of inertia $I_{\min}$ and $I_{\max}$ of A about O. Apply the formula obtained to calculate the product of inertia $\bar{P}_{xy}$ of the 8- by 6- by 1-in.-angle cross section shown in Fig. 9.13, knowing that its minimum moment of inertia is 21.3 in^4.

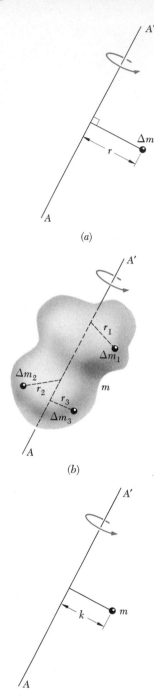

(a)

(b)

(c)

Fig. 9.20

MOMENTS OF INERTIA OF MASSES

9.10. Moment of Inertia of a Mass. Consider a small mass Δm mounted on a rod of negligible mass which may rotate freely about an axis AA' (Fig. 9.20a). If a couple is applied to the system, the rod and mass, assumed initially at rest, will start rotating about AA'. The details of this motion will be studied later in dynamics. At present, we wish only to indicate that the time required for the system to reach a given speed of rotation is proportional to the mass Δm and to the square of the distance r. The product $r^2 \Delta m$ provides, therefore, a measure of the *inertia* of the system, i.e., of the resistance the system offers when we try to set it in motion. For this reason, the product $r^2 \Delta m$ is called the *moment of inertia* of the mass Δm with respect to the axis AA'.

Consider now a body of mass m which is to be rotated about an axis AA' (Fig. 9.20b). Dividing the body into elements of mass Δm_1, Δm_2, etc., we find that the resistance offered by the body is measured by the sum $r_1^2 \Delta m_1 + r_2^2 \Delta m_2 + \cdots$. This sum defines, therefore, the moment of inertia of the body with respect to the axis AA'. Increasing the number of elements, we find that the moment of inertia is equal, at the limit, to the integral

$$I = \int r^2 \, dm \qquad (9.28)$$

The *radius of gyration* k of the body with respect to the axis AA' is defined by the relation

$$I = k^2 m \qquad \text{or} \qquad k = \sqrt{\frac{I}{m}} \qquad (9.29)$$

The radius of gyration k represents, therefore, the distance at which the entire mass of the body should be concentrated if its moment of inertia with respect to AA' is to remain unchanged (Fig. 9.20c). Whether it is kept in its original shape (Fig. 9.20b) or whether it is concentrated as shown in Fig. 9.20c, the mass m will react in the same way to a rotation, or *gyration*, about AA'.

If SI units are used, the radius of gyration k is expressed in meters and the mass m in kilograms. The moment of inertia of a mass, therefore, will be expressed in kg · m². If U.S. customary units are used, the radius of gyration is expressed in feet and the mass in slugs, i.e., in lb·s²/ft. The moment of inertia of a mass, then, will be expressed in lb·ft·s².†

The moment of inertia of a body with respect to a coordinate axis may easily be expressed in terms of the coordinates x, y, z of the element of mass dm (Fig. 9.21). Noting, for example, that the square of the distance r from the element dm to the y axis is $z^2 + x^2$, we express the moment of inertia of the body with respect to the y axis as

$$I_y = \int r^2\, dm = \int (z^2 + x^2)\, dm$$

Similar expressions may be obtained for the moments of inertia with respect to the x and z axes. We write

$$
\begin{aligned}
I_x &= \int (y^2 + z^2)\, dm \\
I_y &= \int (z^2 + x^2)\, dm \\
I_z &= \int (x^2 + y^2)\, dm
\end{aligned}
\qquad (9.30)
$$

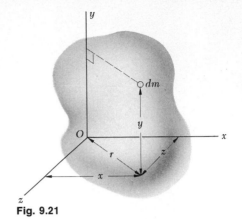

Fig. 9.21

9.11. Parallel-Axis Theorem. Consider a body of mass m. Let $Oxyz$ be a system of rectangular coordinates with origin at an arbitrary point O, and $Gx'y'z'$ a system of parallel *centroidal axes*, i.e., a system with origin at the center of gravity G of the body‡ and with axes x', y', z', respectively parallel to x, y, z (Fig. 9.22). Denoting by $\bar{x}$, $\bar{y}$, $\bar{z}$ the coordinates of G with respect to $Oxyz$, we write the following relations between the coordinates x, y, z of the element dm with respect to $Oxyz$ and its coordinates x', y', z' with respect to the centroidal axes $Gx'y'z'$:

$$x = x' + \bar{x} \qquad y = y' + \bar{y} \qquad z = z' + \bar{z} \qquad (9.31)$$

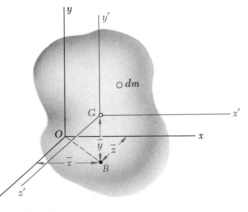

Fig. 9.22

---

† It should be kept in mind, when converting the moment of inertia of a mass from U.S. customary units to SI units, that the base unit pound used in the derived unit lb·ft·s² is a unit of force (*not* of mass) and, therefore, should be converted into newtons. We have

$$1\ \text{lb·ft·s}^2 = (4.45\ \text{N})(0.3048\ \text{m})(1\ \text{s})^2 = 1.356\ \text{N·m·s}^2$$

or, since $N = kg \cdot m/s^2$,

$$1\ \text{lb·ft·s}^2 = 1.356\ \text{kg·m}^2$$

‡ Note that the term centroidal is used to define an axis passing through the center of gravity G of the body, whether or not G coincides with the centroid of the volume of the body.

Referring to Eqs. (9.30), we may express the moment of inertia of the body with respect to the x axis as follows:

$$I_x = \int(y^2 + z^2)\,dm = \int[(y' + \overline{y})^2 + (z' + \overline{z})^2]\,dm$$
$$= \int(y'^2 + z'^2)\,dm + 2\overline{y}\int y'\,dm + 2\overline{z}\int z'\,dm + (\overline{y}^2 + \overline{z}^2)\int dm$$

The first integral in the expression obtained represents the moment of inertia $\overline{I}_{x'}$ of the body with respect to the centroidal axis x'; the second and third integrals represent the first moment of the body with respect to the $z'x'$ and $x'y'$ planes, respectively, and, since both planes contain G, the two integrals are zero; the last integral is equal to the total mass m of the body. We write, therefore,

$$I_x = \overline{I}_{x'} + m(\overline{y}^2 + \overline{z}^2) \tag{9.32}$$

and, similarly,

$$I_y = \overline{I}_{y'} + m(\overline{z}^2 + \overline{x}^2) \qquad I_z = \overline{I}_{z'} + m(\overline{x}^2 + \overline{y}^2)$$

$$\tag{9.32'}$$

We easily verify from Fig. 9.22 that the sum $\overline{z}^2 + \overline{x}^2$ represents the square of the distance OB between the y and y' axis. Similarly, $\overline{y}^2 + \overline{z}^2$ and $\overline{x}^2 + \overline{y}^2$ represent the squares of the distances between the x and x' axes, and the z and z' axes, respectively. Denoting by d the distance between an arbitrary axis AA' and a parallel centroidal axis BB' (Fig. 9.23), we may, therefore, write the following general relation between the moment of inertia I of the body with respect to AA' and its moment of inertia $\overline{I}$ with respect to BB':

$$I = \overline{I} + md^2 \tag{9.33}$$

Expressing the moments of inertia in terms of the corresponding radii of gyration, we may also write

$$k^2 = \overline{k}^2 + d^2 \tag{9.34}$$

where k and $\overline{k}$ represent the radii of gyration about AA' and BB', respectively.

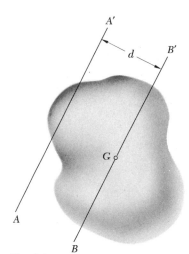

Fig. 9.23

9.12. Moments of Inertia of Thin Plates. Consider a thin plate of uniform thickness t, made of a homogeneous material of density ρ (density = mass per unit volume). The mass moment of inertia of the plate with respect to an axis AA' *contained in the plane* of the plate (Fig. 9.24a) is

$$I_{AA',\text{mass}} = \int r^2 \, dm$$

Since $dm = \rho t \, dA$, we write

$$I_{AA',\text{mass}} = \rho t \int r^2 \, dA$$

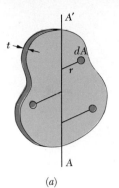

(a)

But r represents the distance of the element of area dA to the axis AA'; the integral is therefore equal to the moment of inertia of the area of the plate with respect to AA'. We have

$$I_{AA',\text{mass}} = \rho t I_{AA',\text{area}} \qquad (9.35)$$

Similarly, we have with respect to an axis BB' perpendicular to AA' (Fig. 9.24b)

$$I_{BB',\text{mass}} = \rho t I_{BB',\text{area}} \qquad (9.36)$$

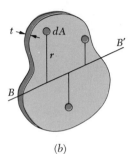

(b)

Considering now the axis CC' *perpendicular* to the plate through the point of intersection C of AA' and BB' (Fig. 9.24c), we write

$$I_{CC',\text{ mass}} = \rho t J_{C,\text{ area}} \qquad (9.37)$$

where J_C is the *polar* moment of inertia of the area of the plate with respect to point C.

Recalling the relation $J_C = I_{AA'} + I_{BB'}$ existing between polar and rectangular moments of inertia of an area, we write the following relation between the mass moments of inertia of a thin plate:

$$I_{CC'} = I_{AA'} + I_{BB'} \qquad (9.38)$$

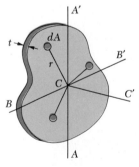

(c) **Fig. 9.24**

Rectangular Plate. In the case of a rectangular plate of sides a and b (Fig. 9.25), we obtain the following mass moments of inertia with respect to axes through the center of gravity of the plate:

$$I_{AA',\text{mass}} = \rho t I_{AA',\text{area}} = \rho t (\tfrac{1}{12} a^3 b)$$
$$I_{BB',\text{mass}} = \rho t I_{BB',\text{area}} = \rho t (\tfrac{1}{12} a b^3)$$

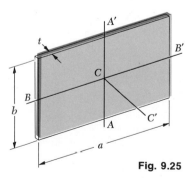

Fig. 9.25

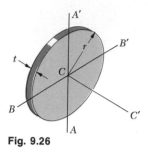

Fig. 9.26

Observing that the product ρabt is equal to the mass m of the plate, we write the mass moments of inertia of a thin rectangular plate as follows:

$$I_{AA'} = \tfrac{1}{12}ma^2 \qquad I_{BB'} = \tfrac{1}{12}mb^2 \tag{9.39}$$
$$I_{CC'} = I_{AA'} + I_{BB'} = \tfrac{1}{12}m(a^2 + b^2) \tag{9.40}$$

Circular Plate. In the case of a circular plate, or disk, of radius r (Fig. 9.26), we write

$$I_{AA',\text{mass}} = \rho t I_{AA',\text{area}} = \rho t(\tfrac{1}{4}\pi r^4)$$

Observing that the product $\rho \pi r^2 t$ is equal to the mass m of the plate and that $I_{AA'} = I_{BB'}$, we write the mass moments of inertia of a circular plate as follows:

$$I_{AA'} = I_{BB'} = \tfrac{1}{4}mr^2 \tag{9.41}$$
$$I_{CC'} = I_{AA'} + I_{BB'} = \tfrac{1}{2}mr^2 \tag{9.42}$$

9.13. Determination of the Moment of Inertia of a Three-dimensional Body by Integration.

The moment of inertia of a three-dimensional body is obtained by computing the integral $I = \int r^2\, dm$. If the body is made of a homogeneous material of density ρ, we have $dm = \rho\, dV$ and write $I = \rho\int r^2\, dV$. This integral depends only upon the shape of the body. In order to compute it, it will generally be necessary to perform a triple, or at least a double, integration.

However, if the body possesses two planes of symmetry, it is usually possible to determine its moment of inertia through a single integration by choosing as an element of mass dm the mass of a thin slab perpendicular to the planes of symmetry. In the case of bodies of revolution, for example, the element of mass should be a thin disk (Fig. 9.27). Using formula (9.42), the moment of inertia of the disk with respect to the axis of revolution may be readily expressed as indicated in Fig. 9.27. Its moment of inertia with respect to each of the other two axes of coordinates will be obtained by using formula (9.41) and the parallel-axis theorem. Integration of the expressions obtained will yield the desired moments of inertia of the body of revolution.

9.14. Moments of Inertia of Composite Bodies.

The moments of inertia of a few common shapes are shown in Fig. 9.28. The moment of inertia with respect to a given axis of a body made of several of these simple shapes may be obtained by computing the moments of inertia of its component parts about the desired axis and adding them together. We should note, as we already have noted in the case of areas, that the radius of gyration of a composite body *cannot* be obtained by adding the radii of gyration of its component parts.

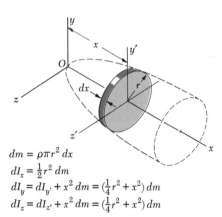

$$dm = \rho\pi r^2\, dx$$
$$dI_x = \tfrac{1}{2}r^2\, dm$$
$$dI_y = dI_{y'} + x^2\, dm = (\tfrac{1}{4}r^2 + x^2)\, dm$$
$$dI_z = dI_{z'} + x^2\, dm = (\tfrac{1}{4}r^2 + x^2)\, dm$$

Fig. 9.27 Determination of the moment of inertia of a body of revolution.

Slender rod	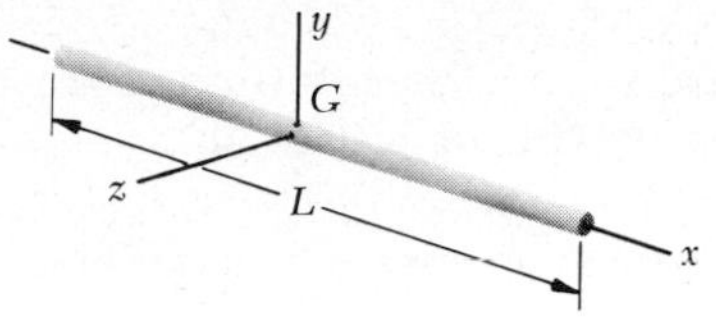	$I_y = I_z = \frac{1}{12}mL^2$
Thin rectangular plate	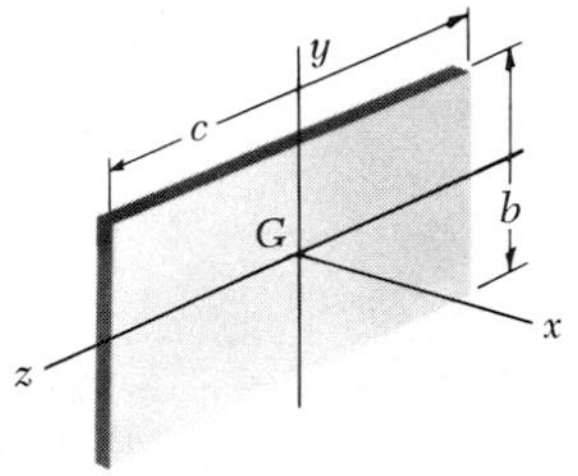	$I_x = \frac{1}{12}m(b^2 + c^2)$ $I_y = \frac{1}{12}mc^2$ $I_z = \frac{1}{12}mb^2$
Rectangular prism	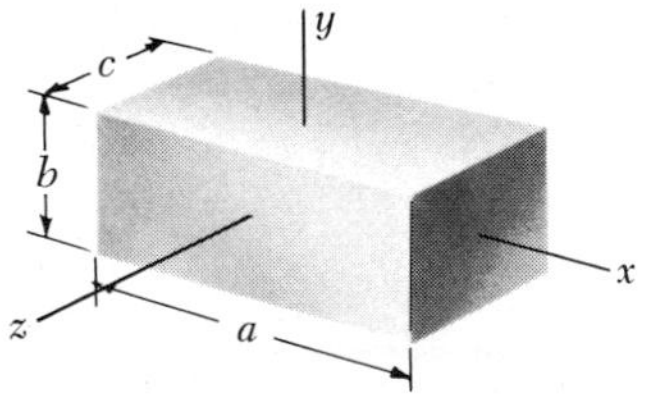	$I_x = \frac{1}{12}m(b^2 + c^2)$ $I_y = \frac{1}{12}m(c^2 + a^2)$ $I_z = \frac{1}{12}m(a^2 + b^2)$
Thin disk	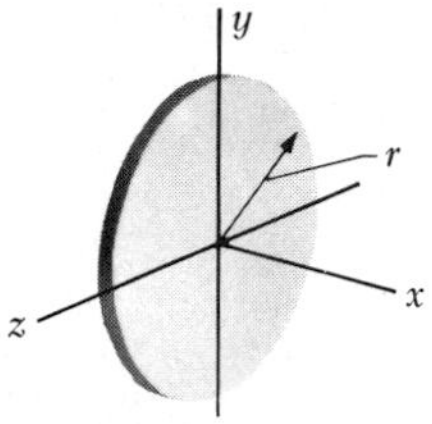	$I_x = \frac{1}{2}mr^2$ $I_y = I_z = \frac{1}{4}mr^2$
Circular cylinder	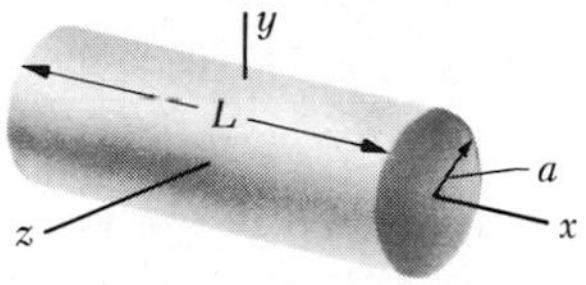	$I_x = \frac{1}{2}ma^2$ $I_y = I_z = \frac{1}{12}m(3a^2 + L^2)$
Circular cone	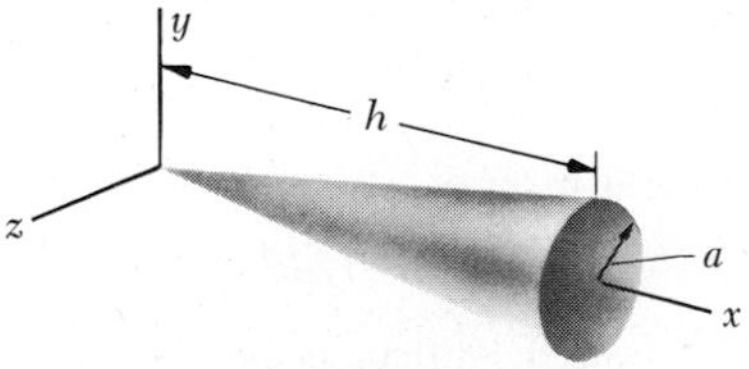	$I_x = \frac{3}{10}ma^2$ $I_y = I_z = \frac{3}{5}m(\frac{1}{4}a^2 + h^2)$
Sphere	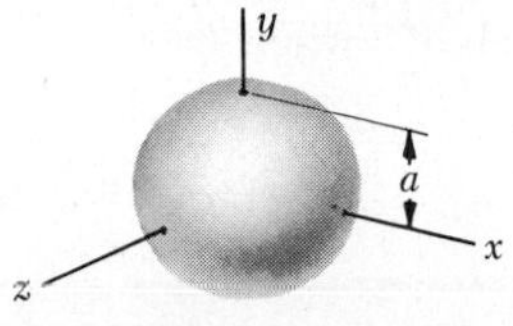	$I_x = I_y = I_z = \frac{2}{5}ma^2$

Fig. 9.28 Mass moments of inertia of common geometric shapes

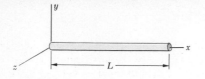

SAMPLE PROBLEM 9.9

Determine the mass moment of inertia of a slender rod of length L and mass m with respect to an axis perpendicular to the rod and passing through one end of the rod.

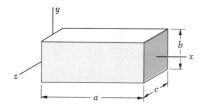

Solution. Choosing the differential element of mass shown, we write

$$dm = \frac{m}{L}\,dx$$

$$I_y = \int x^2\,dm = \int_0^L x^2\frac{m}{L}\,dx = \left[\frac{m}{L}\frac{x^3}{3}\right]_0^L \qquad I_y = \frac{mL^2}{3} \quad \blacktriangleleft$$

SAMPLE PROBLEM 9.10

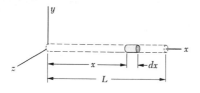

Determine the mass moment of inertia of the homogeneous rectangular prism shown with respect to the z axis.

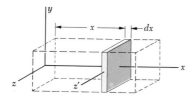

Solution. We choose as a differential element of mass the thin slab shown for which

$$dm = \rho bc\,dx$$

Referring to Sec. 9.12, we find that the moment of inertia of the element with respect to the z' axis is

$$dI_{z'} = \tfrac{1}{12}b^2\,dm$$

Applying the parallel-axis theorem, we obtain the mass moment of inertia of the slab with respect to the z axis.

$$dI_z = dI_{z'} + x^2\,dm = \tfrac{1}{12}b^2\,dm + x^2\,dm = (\tfrac{1}{12}b^2 + x^2)\,\rho bc\,dx$$

Integrating from $x = 0$ to $x = a$, we obtain

$$I_z = \int dI_z = \int_0^a (\tfrac{1}{12}b^2 + x^2)\,\rho bc\,dx = \rho abc\,(\tfrac{1}{12}b^2 + \tfrac{1}{3}a^2)$$

Since the total mass of the prism is $m = \rho abc$, we may write

$$I_z = m(\tfrac{1}{12}b^2 + \tfrac{1}{3}a^2) \qquad I_z = \tfrac{1}{12}m(4a^2 + b^2) \quad \blacktriangleleft$$

We note that if the prism is slender, b is small compared to a and the expression for I_z reduces to $ma^2/3$, which is the result obtained in Sample Prob. 9.9 when $L = a$.

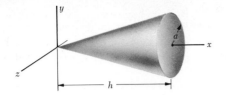

SAMPLE PROBLEM 9.11

Determine the mass moment of inertia of a right circular cone with respect to (a) its longitudinal axis, (b) an axis through the apex of the cone and perpendicular to its longitudinal axis, (c) an axis through the centroid of the cone and perpendicular to its longitudinal axis.

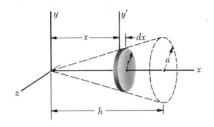

Solution. We choose the differential element of mass shown.

$$r = a\frac{x}{h} \qquad dm = \rho\pi r^2\, dx = \rho\pi \frac{a^2}{h^2}x^2\, dx$$

a. Moment of Inertia I_x. Using the expression derived in Sec. 9.12 for a thin disk, we compute the mass moment of inertia of the differential element with respect to the x axis.

$$dI_x = \tfrac{1}{2}r^2\, dm = \tfrac{1}{2}\left(a\frac{x}{h}\right)^2\left(\rho\pi\frac{a^2}{h^2}x^2\, dx\right) = \tfrac{1}{2}\rho\pi\frac{a^4}{h^4}x^4\, dx$$

Integrating from $x = 0$ to $x = h$, we obtain

$$I_x = \int dI_x = \int_0^h \tfrac{1}{2}\rho\pi\frac{a^4}{h^4}x^4\, dx = \tfrac{1}{2}\rho\pi\frac{a^4}{h^4}\frac{h^5}{5} = \tfrac{1}{10}\rho\pi a^4 h$$

Since the total mass of the cone is $m = \tfrac{1}{3}\rho\pi a^2 h$, we may write

$$I_x = \tfrac{1}{10}\rho\pi a^4 h = \tfrac{3}{10}a^2(\tfrac{1}{3}\rho\pi a^2 h) = \tfrac{3}{10}ma^2 \qquad I_x = \tfrac{3}{10}ma^2 \quad \blacktriangleleft$$

b. Moment of Inertia I_y. The same differential element will be used. Applying the parallel-axis theorem and using the expression derived in Sec. 9.12 for a thin disk, we write

$$dI_y = dI_{y'} + x^2\, dm = \tfrac{1}{4}r^2\, dm + x^2\, dm = (\tfrac{1}{4}r^2 + x^2)\, dm$$

Substituting the expressions for r and dm, we obtain

$$dI_y = \left(\frac{1}{4}\frac{a^2}{h^2}x^2 + x^2\right)\left(\rho\pi\frac{a^2}{h^2}x^2\, dx\right) = \rho\pi\frac{a^2}{h^2}\left(\frac{a^2}{4h^2} + 1\right)x^4\, dx$$

$$I_y = \int dI_y = \int_0^h \rho\pi\frac{a^2}{h^2}\left(\frac{a^2}{4h^2} + 1\right)x^4\, dx = \rho\pi\frac{a^2}{h^2}\left(\frac{a^2}{4h^2} + 1\right)\frac{h^5}{5}$$

Introducing the total mass of the cone m, we rewrite I_y as follows:

$$I_y = \tfrac{3}{5}(\tfrac{1}{4}a^2 + h^2)\tfrac{1}{3}\rho\pi a^2 h \qquad I_y = \tfrac{3}{5}m(\tfrac{1}{4}a^2 + h^2) \quad \blacktriangleleft$$

c. Moment of Inertia $\bar{I}_{y''}$. We apply the parallel-axis theorem and write

$$I_y = \bar{I}_{y''} + m\bar{x}^2$$

Solving for $\bar{I}_{y''}$ and recalling that $\bar{x} = \tfrac{3}{4}h$, we have

$$\bar{I}_{y''} = I_y - m\bar{x}^2 = \tfrac{3}{5}m(\tfrac{1}{4}a^2 + h^2) - m(\tfrac{3}{4}h)^2$$

$$\bar{I}_{y''} = \tfrac{3}{20}m(a^2 + \tfrac{1}{4}h^2) \quad \blacktriangleleft$$

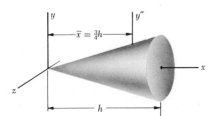

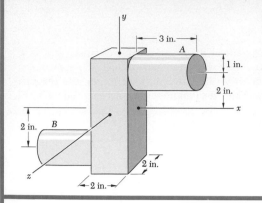

A steel forging consists of a rectangular prism 6 by 2 by 2 in. and of two cylinders of diameter 2 in. and length 3 in., as shown. Determine the mass moments of inertia with respect to the coordinate axes. (Specific weight of steel = 490 lb/ft³.)

Computation of Masses
Prism

$$V = 24 \text{ in}^3 \qquad W = \frac{(24 \text{ in}^3)(490 \text{ lb/ft}^3)}{1728 \text{ in}^3/\text{ft}^3} = 6.81 \text{ lb}$$

$$m = \frac{6.81 \text{ lb}}{32.2 \text{ ft/s}^2} = 0.211 \text{ lb} \cdot \text{s}^2/\text{ft}$$

Each Cylinder

$$V = \pi(1 \text{ in.})^2(3 \text{ in.}) = 9.42 \text{ in}^3$$

$$W = \frac{(9.42 \text{ in}^3)(490 \text{ lb/ft}^3)}{1728 \text{ in}^3/\text{ft}^3} = 2.67 \text{ lb}$$

$$m = \frac{2.67 \text{ lb}}{32.2 \text{ ft/s}^2} = 0.0829 \text{ lb} \cdot \text{s}^2/\text{ft}$$

Mass Moments of Inertia.   The mass moments of inertia of each component are computed from Fig. 9.28, using the parallel-axis theorem when necessary. Note that all lengths should be expressed in feet.

Prism

$I_x = I_z = \frac{1}{12}(0.211 \text{ lb} \cdot \text{s}^2/\text{ft})[(\frac{6}{12} \text{ ft})^2 + (\frac{2}{12} \text{ ft})^2] = 4.88 \times 10^{-3} \text{ lb} \cdot \text{ft} \cdot \text{s}^2$
$I_y = \frac{1}{12}(0.211 \text{ lb} \cdot \text{s}^2/\text{ft})[(\frac{2}{12} \text{ ft})^2 + (\frac{2}{12} \text{ ft})^2] = 0.977 \times 10^{-3} \text{ lb} \cdot \text{ft} \cdot \text{s}^2$

Each cylinder

$I_x = \frac{1}{2}ma^2 + m\bar{y}^2 = \frac{1}{2}(0.0829 \text{ lb} \cdot \text{s}^2/\text{ft})(\frac{1}{12} \text{ ft})^2$
$\qquad\qquad + (0.0829 \text{ lb} \cdot \text{s}^2/\text{ft})(\frac{2}{12} \text{ ft})^2 = 2.59 \times 10^{-3} \text{ lb} \cdot \text{ft} \cdot \text{s}^2$
$I_y = \frac{1}{12}m(3a^2 + L^2) + m\bar{x}^2 = \frac{1}{12}(0.0829 \text{ lb} \cdot \text{s}^2/\text{ft})[3(\frac{1}{12} \text{ ft})^2 + (\frac{3}{12} \text{ ft})^2]$
$\qquad\qquad + (0.0829 \text{ lb} \cdot \text{s}^2/\text{ft})(\frac{2.5}{12} \text{ ft})^2 = 4.17 \times 10^{-3} \text{ lb} \cdot \text{ft} \cdot \text{s}^2$
$I_z = \frac{1}{12}m(3a^2 + L^2) + m(\bar{x}^2 + \bar{y}^2) = \frac{1}{12}(0.0829)[3(\frac{1}{12})^2 + (\frac{3}{12})^2]$
$\qquad\qquad + (0.0829)[(\frac{2.5}{12})^2 + (\frac{2}{12})^2] = 6.48 \times 10^{-3} \text{ lb} \cdot \text{ft} \cdot \text{s}^2$

Entire Body.   Adding the values obtained:

$I_x = 4.88 \times 10^{-3} + 2(2.59 \times 10^{-3})$

$\qquad\qquad\qquad I_x = 10.06 \times 10^{-3} \text{ lb} \cdot \text{ft} \cdot \text{s}^2$ ◀

$I_y = 0.977 \times 10^{-3} + 2(4.17 \times 10^{-3})$

$\qquad\qquad\qquad I_y = 9.32 \times 10^{-3} \text{ lb} \cdot \text{ft} \cdot \text{s}^2$ ◀

$I_z = 4.88 \times 10^{-3} + 2(6.48 \times 10^{-3})$

$\qquad\qquad\qquad I_z = 17.84 \times 10^{-3} \text{ lb} \cdot \text{ft} \cdot \text{s}^2$ ◀

SAMPLE PROBLEM 9.13

Solve Sample Prob. 9.12 using SI units.

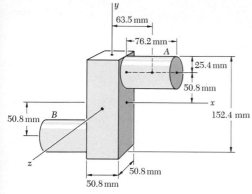

Solution. First, the dimensions are converted into millimeters (1 in. = 25.4 mm). Next, the density of steel ρ (mass per unit volume) is determined in SI units. Recalling that 1 ft = 0.3048 m and that the mass of a block weighing 1 lb is 0.454 kg, we have

$$\rho = (490 \text{ lb/ft}^3)\left(\frac{0.454 \text{ kg}}{1 \text{ lb}}\right)\left(\frac{1 \text{ ft}}{0.3048 \text{ m}}\right)^3 = 7850 \text{ kg/m}^3$$

Computation of Masses

Prism. $V = (50.8 \text{ mm})^2(152.4 \text{ mm}) = 0.393 \times 10^6 \text{ mm}^3$

or, since $1 \text{ mm}^3 = (10^{-3} \text{ m})^3 = 10^{-9} \text{ m}^3$,

$$V = 0.393 \times 10^6 \times 10^{-9} \text{ m}^3 = 0.393 \times 10^{-3} \text{ m}^3$$
$$m = \rho V = (7.85 \times 10^3 \text{ kg/m}^3)(0.393 \times 10^{-3} \text{ m}^3) = 3.09 \text{ kg}$$

Each Cylinder

$$V = \pi r^2 h = \pi(25.4 \text{ mm})^2(76.2 \text{ mm}) = 0.1544 \times 10^6 \text{ mm}^3$$
$$= 0.1544 \times 10^{-3} \text{ m}^3$$
$$m = \rho V = (7.85 \times 10^3 \text{ kg/m}^3)(0.1544 \times 10^{-3} \text{ m}^3) = 1.212 \text{ kg}$$

Mass Moments of Inertia. The mass moments of inertia of each component are computed from Fig. 9.28, using the parallel-axis theorem when necessary. Note that all lengths should be expressed in millimeters.

Prism

$$I_x = I_z = \tfrac{1}{12}(3.09 \text{ kg})[(152.4 \text{ mm})^2 + (50.8 \text{ mm})^2] = 6640 \text{ kg} \cdot \text{mm}^2$$
$$I_y = \tfrac{1}{12}(3.09 \text{ kg})[(50.8 \text{ mm})^2 + (50.8 \text{ mm})^2] = 1329 \text{ kg} \cdot \text{mm}^2$$

Each Cylinder

$$I_x = \tfrac{1}{2}ma^2 + m\bar{y}^2 = \tfrac{1}{2}(1.212 \text{ kg})(25.4 \text{ mm})^2 + (1.212 \text{ kg})(50.8 \text{ mm})^2$$
$$= 3520 \text{ kg} \cdot \text{mm}^2$$
$$I_y = \tfrac{1}{12}m(3a^2 + L^2) + m\bar{x}^2 = \tfrac{1}{12}(1.212 \text{ kg})[3(25.4 \text{ mm})^2 + (76.2 \text{ mm})^2]$$
$$+ (1.212 \text{ kg})(63.5 \text{ mm})^2 = 5670 \text{ kg} \cdot \text{mm}^2$$
$$I_z = \tfrac{1}{12}m(3a^2 + L^2) + m(\bar{x}^2 + \bar{y}^2)$$
$$= \tfrac{1}{12}(1.212 \text{ kg})[3(25.4 \text{ mm})^2 + (76.2 \text{ mm})^2]$$
$$+ (1.212 \text{ kg})[(63.5 \text{ mm})^2 + (50.8 \text{ mm})^2] = 8800 \text{ kg} \cdot \text{mm}^2$$

Entire Body. Adding the values obtained, and observing that $1 \text{ mm}^2 = (10^{-3} \text{ m})^2 = 10^{-6} \text{ m}^2$, we have

$$I_x = 6640 \text{ kg} \cdot \text{mm}^2 + 2(3520 \text{ kg} \cdot \text{mm}^2) = 13.68 \times 10^3 \text{ kg} \cdot \text{mm}^2$$
$$I_x = 13.68 \times 10^{-3} \text{ kg} \cdot \text{m}^2 \quad \blacktriangleleft$$
$$I_y = 1329 \text{ kg} \cdot \text{mm}^2 + 2(5670 \text{ kg} \cdot \text{mm}^2) = 12.67 \times 10^3 \text{ kg} \cdot \text{mm}^2$$
$$I_y = 12.67 \times 10^{-3} \text{ kg} \cdot \text{m}^2 \quad \blacktriangleleft$$
$$I_z = 6640 \text{ kg} \cdot \text{mm}^2 + 2(8800 \text{ kg} \cdot \text{mm}^2) = 24.2 \times 10^3 \text{ kg} \cdot \text{mm}^2$$
$$I_z = 24.2 \times 10^{-3} \text{ kg} \cdot \text{m}^2 \quad \blacktriangleleft$$

Recalling that $1 \text{ lb} \cdot \text{ft} \cdot \text{s}^2 = 1.356 \text{ kg} \cdot \text{m}^2$ (see footnote, page 383) we may check these answers against the values obtained in Sample Prob. 9.12.

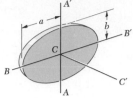

Fig. P9.72

PROBLEMS

9.72 Determine the mass moment of inertia of a thin elliptical plate of mass m with respect to (a) the axes AA' and BB' of the ellipse, (b) the axis CC' perpendicular to the plate.

9.73 Determine the mass moment of inertia of a ring of mass m, cut from a thin uniform plate, with respect to (a) the diameter AA' of the ring, (b) the axis CC' perpendicular to the plane of the ring.

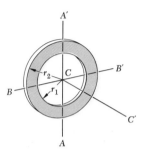

Fig. P9.73

9.74 A thin plate of mass m is cut in the shape of an isosceles triangle of base b and height h. Determine the mass moment of inertia of the plate with respect to (a) the centroidal axes AA' and BB' in the plane of the plate, (b) the centroidal axis CC' perpendicular to the plate.

9.75 Determine the mass moments of inertia of the plate of Prob. 9.74 with respect to the axes DD' and EE' parallel to the centroidal axes AA' and BB' respectively.

9.76 Determine by direct integration the mass moment of inertia with respect to the y axis of the right circular cylinder shown, assuming a uniform density and a mass m.

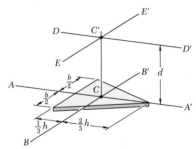

Fig. P9.74

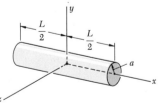

Fig. P9.76

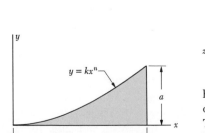

Fig. P9.77

9.77 The area shown is revolved about the x axis to form a homogeneous solid of revolution of mass m. Express the mass moment of inertia of the solid with respect to the x axis in terms of m, a, and n. The expression obtained may be used to verify (a) the value given in Fig. 9.28 for a cone (with $n = 1$), (b) the answer to Prob. 9.78 (with $n = \frac{1}{2}$), (c) the answer to Prob. 9.80 (with $n = 2$).

9.78 Determine by direct integration the mass moment of inertia and the radius of gyration with respect to the x axis of the paraboloid shown, assuming a uniform density and a mass m.

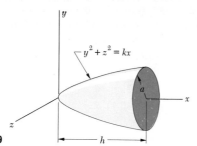

$y^2 + z^2 = kx$

Fig. P9.78 and P9.79

9.79 Determine by direct integration the mass moment of inertia and the radius of gyration with respect to the y axis of the paraboloid shown, assuming a uniform density and a mass m.

9.80 The homogeneous solid shown was obtained by rotating the area of Prob. 9.77, with $n = 2$, through $360°$ about the x axis. Determine the mass moment of inertia $\bar{I}_x$ in terms of m and a.

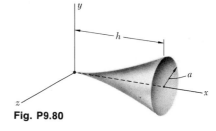

Fig. P9.80

9.81 Determine in terms of m and a the mass moment of inertia and the radius of gyration of the homogeneous solid of Prob. 9.80 with respect to the y axis.

9.82 Determine by direct integration the mass moment of inertia with respect to the x axis of the pyramid shown, assuming a uniform density and a mass m.

9.83 Determine by direct integration the mass moment of inertia with respect to the y axis of the pyramid shown, assuming a uniform density and a mass m.

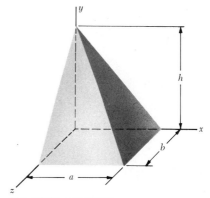

Fig. P9.82 and P9.83

9.84 Knowing that the thin hemispherical shell shown is of mass m and thickness t, determine the mass moment of inertia of the shell with respect to the x axis. (*Hint.* Consider the shell as formed by removing a hemisphere of radius r from a hemisphere of radius $r + t$; then neglect the terms containing t^2 and t^3, and keep those terms containing t.)

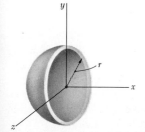

Fig. P9.84

Fig. P9.85

9.85 Determine the mass moment of inertia of the frustum of a right circular cone of mass m with respect to its axis of symmetry.

9.86 Determine the mass moment of inertia and the radius of gyration of the steel flywheel shown with respect to the axis of rotation. The web of the flywheel consists of a solid plate 25 mm thick. (Density of steel = 7850 kg/m³.)

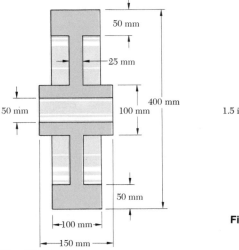

Fig. P9.86

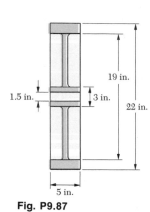

Fig. P9.87

9.87 The cross section of a small flywheel is shown. The rim and hub are connected by eight spokes (two of which are shown in the cross section). Each spoke has a cross-sectional area of 0.400 in². Determine the mass moment of inertia and radius of gyration of the flywheel with respect to the axis of rotation. (Specific weight of steel = 490 lb/ft³.)

9.88 Three slender homogeneous rods are welded together as shown. Denoting the mass of each rod by m, determine the mass moment of inertia and the radius of gyration of the assembly with respect to (a) the x axis, (b) the y axis, (c) the z axis.

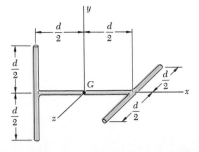

Fig. P9.88

9.89 In using the parallel-axis theorem, the error introduced by neglecting the centroidal moment of inertia is sometimes small. For a homogeneous sphere of radius a and mass m, (a) determine the mass moment of inertia with respect to an axis AA' at a distance R from the center of the sphere, (b) express as a function of a/R the relative error introduced by neglecting the centroidal moment of inertia, (c) determine the distance R in terms of a for which the relative error is 0.4 percent.

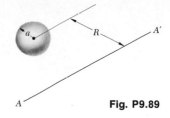

Fig. P9.89

9.90 A section of sheet steel, 2 mm thick, is cut and bent into the machine component shown. Knowing that the density of steel is 7850 kg/m³, determine the mass moment of inertia of the component with respect to (a) the x axis, (b) the y axis, (c) the z axis.

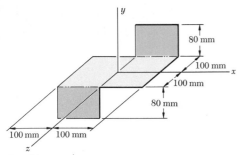

Fig. P9.90

9.91 Twelve uniform slender rods, each of length l, are welded together to form the cubical figure shown. Denoting by m the total mass of the twelve rods, determine the mass moment of inertia of the figure about the x axis.

9.92 and 9.93 Determine the mass moment of inertia and the radius of gyration of the steel machine element shown with respect to the x axis. (Specific weight of steel = 490 lb/ft³; density of steel = 7850 kg/m³.)

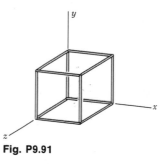

Fig. P9.91

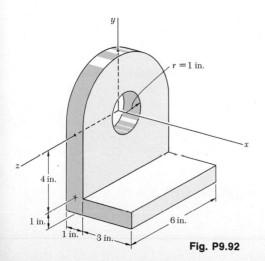

Fig. P9.92

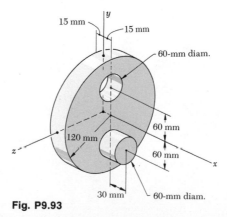

Fig. P9.93

9.94 A homogeneous wire, of weight 2 lb/ft, is used to form the figure shown. Determine the mass moment of inertia of the wire figure with respect to (*a*) the *x* axis, (*b*) the *y* axis, (*c*) the *z* axis.

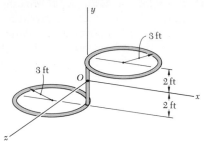

Fig. P9.94

9.95 Two holes, each of diameter 50 mm, are drilled through the steel block shown. Determine the mass moment of inertia of the body with respect to the axis of either of the holes. (Density of steel = 7850 kg/m³.)

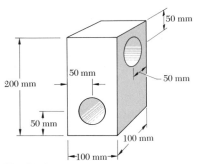

Fig. P9.95 .

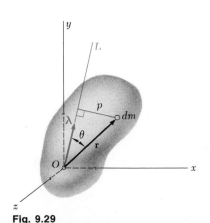

Fig. 9.29

*9.15. Moment of Inertia of a Body with Respect to an Arbitrary Axis through *O*. Mass Products of Inertia.

We shall see in this section how the moment of inertia of a body may be determined with respect to an arbitrary axis *OL* through the origin (Fig. 9.29) if we have computed beforehand its moments of inertia with respect to the three coordinate axes, as well as certain other quantities to be defined below.

The moment of inertia of the body with respect to *OL* is represented by the integral $I_{OL} = \int p^2 \, dm$, where *p* denotes the perpendicular distance from the element of mass *dm* to the axis *OL*. But, denoting by **λ** the unit vector along *OL* and by **r** the position vector of the element *dm*, we observe that the perpendicular distance *p* is equal to the magnitude $r \sin \theta$ of the vector product **λ** × **r**. We write therefore

$$I_{OL} = \int p^2 \, dm = \int (\boldsymbol{\lambda} \times \mathbf{r})^2 \, dm \tag{9.43}$$

Expressing the square of the vector product in terms of its rectangular components, we have

$$I_{OL} = \int [(\lambda_x y - \lambda_y x)^2 + (\lambda_y z - \lambda_z y)^2 + (\lambda_z x - \lambda_x z)^2]\, dm$$

where the components λ_x, λ_y, λ_z of the unit vector $\boldsymbol{\lambda}$ represent the direction cosines of the axis OL, and the components x, y, z of $\mathbf{r}$ represent the coordinates of the element of mass dm. Expanding the squares in the expression obtained and rearranging the terms, we write

$$I_{OL} = \lambda_x^2 \int (y^2 + z^2)\, dm + \lambda_y^2 \int (z^2 + x^2)\, dm + \lambda_z^2 \int (x^2 + y^2)\, dm$$
$$- 2\lambda_x \lambda_y \int xy\, dm - 2\lambda_y \lambda_z \int yz\, dm - 2\lambda_z \lambda_x \int zx\, dm \quad (9.44)$$

Referring to Eqs. (9.30), we note that the first three integrals in (9.44) represent, respectively, the moments of inertia I_x, I_y, and I_z of the body with respect to the coordinate axes. The last three integrals in (9.44), which involve products of coordinates, are called the *products of inertia* of the body with respect to the x and y axes, the y and z axes, and the z and x axes, respectively. We write

$$P_{xy} = \int xy\, dm \qquad P_{yz} = \int yz\, dm \qquad P_{zx} = \int zx\, dm \quad (9.45)$$

Substituting for the various integrals from (9.30) and (9.45) into (9.44), we have

$$I_{OL} = I_x \lambda_x^2 + I_y \lambda_y^2 + I_z \lambda_z^2 - 2P_{xy}\lambda_x\lambda_y - 2P_{yz}\lambda_y\lambda_z - 2P_{zx}\lambda_z\lambda_x$$

$$(9.46)$$

We note that the definition of the products of inertia of a mass given in Eqs. (9.45) is an extension of the definition of the product of inertia of an area (Sec. 9.7). Mass products of inertia reduce to zero under the same conditions of symmetry as products of inertia of areas, and the parallel-axis theorem for mass products of inertia is expressed by relations similar to the formula derived for the product of inertia of an area. Substituting for x, y, z from Eqs. (9.31) into Eqs. (9.45), we verify that

$$\begin{aligned} P_{xy} &= \bar{P}_{x'y'} + m\bar{x}\bar{y} \\ P_{yz} &= \bar{P}_{y'z'} + m\bar{y}\bar{z} \\ P_{zx} &= \bar{P}_{z'x'} + m\bar{z}\bar{x} \end{aligned} \qquad (9.47)$$

where $\bar{x}$, $\bar{y}$, $\bar{z}$ are the coordinates of the center of gravity G of the body, and $\bar{P}_{x'y'}$, $\bar{P}_{y'z'}$, $\bar{P}_{z'x'}$ denote the products of inertia with respect to the centroidal axes x', y', z' (Fig. 9.22).

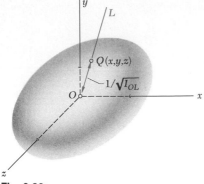

Fig. 9.30

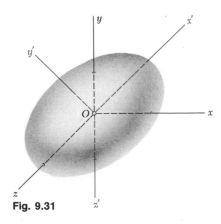

Fig. 9.31

∗9.16. Ellipsoid of Inertia. Principal Axes of Inertia. Let us assume that the moment of inertia of the body considered in the preceding section has been determined with respect to a large number of axes OL through the fixed point O, and that a point Q has been plotted on each axis OL at a distance $OQ = 1/\sqrt{I_{OL}}$ from O. The locus of the points Q thus obtained forms a surface (Fig. 9.30). The equation of that surface may be obtained by substituting $1/(OQ)^2$ for I_{OL} in (9.46) and multiplying both sides of the equation by $(OQ)^2$. Observing that

$$(OQ)\lambda_x = x \qquad (OQ)\lambda_y = y \qquad (OQ)\lambda_z = z$$

where x, y, z denote the rectangular coordinates of a point Q of the surface, we write

$$I_x x^2 + I_y y^2 + I_z z^2 - 2P_{xy}xy - 2P_{yz}yz - 2P_{zx}zx = 1 \quad (9.48)$$

The equation obtained is that of a *quadric*. Since the moment of inertia I_{OL} is different from zero for every axis OL, no point Q may be at an infinite distance from O. Thus, the quadric obtained is an *ellipsoid*. This ellipsoid, which defines the moment of inertia of the body with respect to any axis through O, is known as the *ellipsoid of inertia* of the body at O.

We observe that, if the axes in Fig. 9.30 are rotated, the coefficients of the equation defining the ellipsoid change, since these are equal to the moments and products of inertia of the body with respect to the rotated coordinate axes. However, the *ellipsoid itself remains unaffected*, since its shape depends only upon the distribution of mass in the body considered. Suppose now that we choose as coordinate axes the principal axes x', y', z' of the ellipsoid of inertia (Fig. 9.31). The equation of the ellipsoid with respect to these coordinate axes will be of the form

$$I_{x'}x'^2 + I_{y'}y'^2 + I_{z'}z'^2 = 1 \qquad (9.49)$$

which does not contain any product of coordinates. Thus, the products of inertia of the body with respect to the x', y', z' axes are zero. The x', y', z' axes are known as the *principal axes of inertia* of the body at O, and the coefficients $I_{x'}$, $I_{y'}$, $I_{z'}$ as the *principal moments of inertia* of the body at O. Note that, given a body of arbitrary shape and a point O, it is always possible to find axes which are the principal axes of inertia of the body at O, i.e., axes with respect to which the products of inertia of the body are zero. Indeed, no matter how odd or irregular the shape of the body may be, the moments of inertia of the body with respect to axes through O will define an ellipsoid, and this ellipsoid will have principal axes which, by definition, are the principal axes of the body at O.

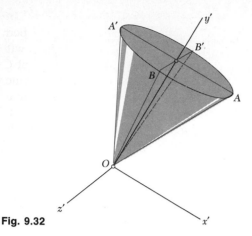

Fig. 9.32

If the principal axes of inertia x', y', z' are used as coordinate axes, the expression obtained in Eq. (9.46) for the moment of inertia of a body with respect to an arbitrary axis through O reduces to

$$I_{OL} = I_{x'}\lambda_{x'}^2 + I_{y'}\lambda_{y'}^2 + I_{z'}\lambda_{z'}^2 \qquad (9.50)$$

While the determination of the principal axes of inertia of a body of arbitrary shape is somewhat involved and requires solving a cubic equation,[†] there are many cases when these axes may be spotted immediately. Consider, for instance, the homogeneous cone of elliptical base shown in Fig. 9.32; this cone possesses two mutually perpendicular planes of symmetry OAA' and OBB'. We check from the definition (9.45) that, if the $x'y'$ and $y'z'$ planes are chosen to coincide with the two planes of symmetry, all the products of inertia are zero. The x', y', and z' axes thus selected are therefore the principal axes of inertia of the cone at O. In the case of the homogeneous regular tetrahedron $OABC$ shown in Fig. 9.33, the line joining the corner O to the center D of the opposite face is a principal axis of inertia at O and any line through O perpendicular to OD is also a principal axis of inertia at O. This property may be recognized if we observe that a rotation through 120° about OD leaves the shape and the mass distribution of the tetrahedron unchanged. It follows that the ellipsoid of inertia at O also remains unchanged under this rotation. The ellipsoid, therefore, is of revolution about OD, and the line OD, as well as any perpendicular line through O, must be a principal axis of the ellipsoid.

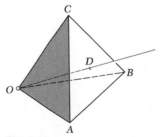

Fig. 9.33

---

† Cf. Synge and Griffith, *Principles of Mechanics*, McGraw-Hill Book Company, sec. 11.3.

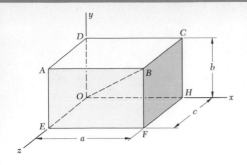

Consider a rectangular prism of mass m and sides a, b, c. Determine (a) the mass moments and products of inertia of the prism with respect to the coordinate axes shown, (b) its moment of inertia with respect to the diagonal OB.

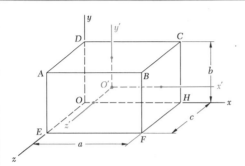

a. Moments and Products of Inertia with Respect to the Coordinate Axes. *Moments of Inertia.* Introducing the centroidal axes x', y', z', with respect to which the moments of inertia are given in Fig. 9.28, we apply the parallel-axis theorem:

$$I_x = \bar{I}_{x'} + m(\bar{y}^2 + \bar{z}^2) = \tfrac{1}{12}m(b^2 + c^2) + m(\tfrac{1}{4}b^2 + \tfrac{1}{4}c^2)$$

$$I_x = \tfrac{1}{3}m(b^2 + c^2) \quad \blacktriangleleft$$

Similarly: $\qquad I_y = \tfrac{1}{3}m(c^2 + a^2) \qquad I_z = \tfrac{1}{3}m(a^2 + b^2) \quad \blacktriangleleft$

Products of Inertia. Because of symmetry, the products of inertia with respect to the centroidal axes x', y', z' are zero and these axes are principal axes of inertia. Using the parallel-axis theorem, we have

$$P_{xy} = \bar{P}_{x'y'} + m\bar{x}\bar{y} = 0 + m(\tfrac{1}{2}a)(\tfrac{1}{2}b) \qquad P_{xy} = \tfrac{1}{4}mab \quad \blacktriangleleft$$

Similarly: $\qquad\qquad\qquad\qquad\qquad\quad P_{yz} = \tfrac{1}{4}mbc \qquad P_{zx} = \tfrac{1}{4}mca \quad \blacktriangleleft$

b. Moment of Inertia with Respect to OB. We recall Eq. (9.46):

$$I_{OB} = I_x\lambda_x^2 + I_y\lambda_y^2 + I_z\lambda_z^2 - 2P_{xy}\lambda_x\lambda_y - 2P_{yz}\lambda_y\lambda_z - 2P_{zx}\lambda_z\lambda_x$$

where the direction cosines of OB are

$$\lambda_x = \cos\theta_x = (OH)/(OB) = a/(a^2 + b^2 + c^2)^{1/2}$$
$$\lambda_y = b/(a^2 + b^2 + c^2)^{1/2} \qquad \lambda_z = c/(a^2 + b^2 + c^2)^{1/2}$$

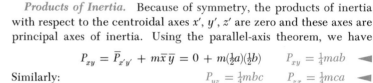

Substituting the values obtained for the moments and products of inertia and for the direction cosines:

$$I_{OB} = \frac{1}{a^2 + b^2 + c^2}[\tfrac{1}{3}m(b^2 + c^2)a^2 + \tfrac{1}{3}m(c^2 + a^2)b^2 + \tfrac{1}{3}m(a^2 + b^2)c^2$$
$$- \tfrac{1}{2}ma^2b^2 - \tfrac{1}{2}mb^2c^2 - \tfrac{1}{2}mc^2a^2]$$

$$I_{OB} = \frac{m}{6}\frac{a^2b^2 + b^2c^2 + c^2a^2}{a^2 + b^2 + c^2} \quad \blacktriangleleft$$

Alternate Solution. The moment of inertia I_{OB} may be obtained directly from the principal moments of inertia $\bar{I}_{x'}$, $\bar{I}_{y'}$, $\bar{I}_{z'}$, since the line OB passes through the centroid O'. Since the x', y', z' axes are principal axes of inertia, we use Eq. (9.50) and write

$$I_{OB} = \bar{I}_{x'}\lambda_x^2 + \bar{I}_{y'}\lambda_y^2 + \bar{I}_{z'}\lambda_z^2$$
$$= \frac{1}{a^2 + b^2 + c^2}\left[\frac{m}{12}(b^2 + c^2)a^2 + \frac{m}{12}(c^2 + a^2)b^2 + \frac{m}{12}(a^2 + b^2)c^2\right]$$

$$I_{OB} = \frac{m}{6}\frac{a^2b^2 + b^2c^2 + c^2a^2}{a^2 + b^2 + c^2} \quad \blacktriangleleft$$

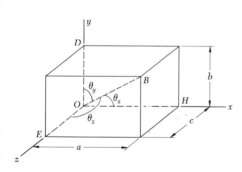

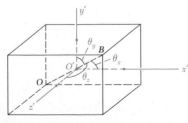

PROBLEMS

9.96 and 9.97 Determine the mass products of inertia P_{xy}, P_{yz}, and P_{zx} of the steel machine element shown. (Specific weight of steel = 490 lb/ft³; density of steel = 7850 kg/m³.)

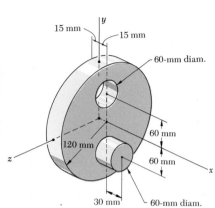

15 mm
15 mm
60-mm diam.
60 mm
120 mm
60 mm
30 mm
60-mm diam.

Fig. P9.96

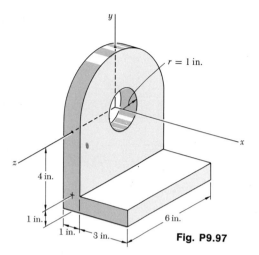

$r = 1$ in.
4 in.
1 in.
1 in.
3 in.
6 in.

Fig. P9.97

9.98 A homogeneous wire, of weight 2 lb/ft, is used to form the figure shown. Determine the mass products of inertia P_{xy}, P_{yz}, and P_{zx} of the wire figure.

9.99 A section of sheet steel, 2 mm thick, is cut and bent into the machine component shown. Knowing that the density of steel is 7850 kg/m³, determine the mass products of inertia P_{xy}, P_{yz}, and P_{zx} of the component.

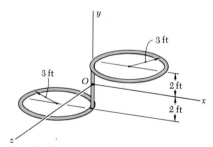

3 ft
3 ft
O
2 ft
2 ft

Fig. P9.98

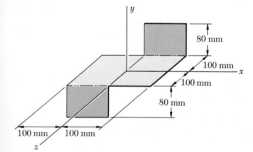

80 mm
100 mm
100 mm
80 mm
100 mm   100 mm

Fig. P9.99

9.100 For the homogeneous tetrahedron of mass m which is shown, (a) determine by direct integration the mass product of inertia P_{zx}, (b) deduce P_{yz} and P_{xy} from the result obtained in part a.

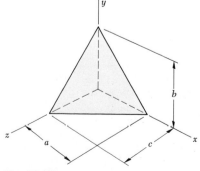

b
a
c

Fig. P9.100

9.101 Complete the derivation of Eqs. (9.47), which express the parallel-axis theorem for mass products of inertia.

9.102 Determine the mass moment of inertia of the right circular cone of Sample Prob. 9.11 with respect to a generator of the cone.

9.103 Determine the mass moment of inertia of the rectangular prism of Sample Prob. 9.14 with respect to the diagonal OF of its base.

9.104 Determine the mass moment of inertia of the bent wire of Probs. 9.94 and 9.98 with respect to the axis through O which forms equal angles with the x, y, and z axes.

9.105 Determine the mass moment of inertia of the forging of Sample Prob. 9.12 with respect to an axis through O characterized by the unit vector $\boldsymbol{\lambda} = \frac{2}{3}\mathbf{i} + \frac{1}{3}\mathbf{j} + \frac{2}{3}\mathbf{k}$.

9.106 Three uniform rods, each of mass m, are welded together as shown. Determine (a) the mass moments of inertia and the mass products of inertia with respect to the coordinate axes, (b) the mass moment of inertia with respect to a line joining the origin O and point D.

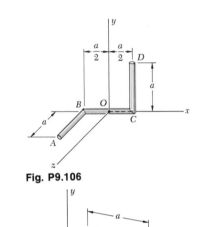

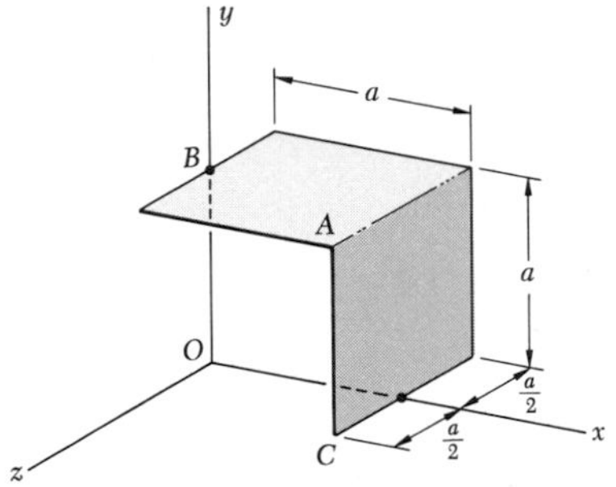

Fig. P9.106

9.107 The thin bent plate shown is of uniform density and mass m. Determine its mass moment of inertia with respect to a line joining the origin O and point A.

9.108 The thin bent plate shown is of uniform density and mass m. Determine its mass moment of inertia with respect to a line joining points B and C.

Fig. P9.107 and P9.108

9.109 Consider a homogeneous circular cylinder of radius a and length L. Determine the value of the ratio a/L for which the ellipsoid of inertia of the cylinder is a sphere when computed (a) at the centroid of the cylinder, (b) at the center of one of its bases.

9.110 Determine the value of the ratio a/h for which the ellipsoid of inertia of the right circular cone of Sample Prob. 9.11 is a sphere when computed (a) at the apex of the cone, (b) at the centroid of the cone.

9.111 Given an arbitrary solid and three rectangular axes x, y, and z, prove that the mass moment of inertia of the solid with respect to any one of the three axes cannot be larger than the sum of the moments of inertia of the solid with respect to the other two axes; i.e., prove that the inequality $I_x \leq I_y + I_z$ is satisfied, as well as two similar inequalities. Further prove that, if the solid is homogeneous and of revolution, and if x is the axis of revolution and y a transverse axis, then $I_y \geq \frac{1}{2}I_x$.

9.112 Given a homogeneous solid of mass m and of arbitrary shape, and three rectangular axes x, y, and z of origin O, prove that the sum $I_x + I_y + I_z$ of the mass moments of inertia of the solid cannot be smaller than the similar sum computed for a sphere of the same mass and same material centered at O. Further prove, using the result of Prob. 9.111, that, if the solid is of revolution and if x is the axis of revolution, then its moment of inertia I_y about a transverse axis y must satisfy the inequality

$$I_y \geq \frac{3}{10}ma^2$$

where a is the radius of the sphere of the same mass and same material.

9.113 Consider a cube of mass m and side a. (*a*) Show that the ellipsoid of inertia at the center of the cube is a sphere, and use this property to determine the mass moment of inertia of the cube with respect to one of its diagonals. (*b*) Show that the ellipsoid of inertia at one of the corners of the cube is an ellipsoid of revolution, and determine the principal moments of inertia of the cube at that point.

REVIEW PROBLEMS

9.114 Sheet metal of thickness t is used to form the conical shell shown. Denoting by m the total mass of the shell, determine the mass moment of inertia of the shell with respect to the y axis.

9.115 Determine the mass moment of inertia of the shell of Prob. 9.114 with respect to the x axis.

9.116 For the shaded area shown, determine (*a*) I_x, (*b*) I_y, (*c*) P_{xy}.

9.117 Determine the radius of gyration of an equilateral triangle of side a with respect to one of its sides.

9.118 For the homogeneous ring shown, which is of density ρ, determine (*a*) the mass moment of inertia with respect to the axis BB', (*b*) the value of a_1 for which, given a_2 and h, $I_{BB'}$ is maximum, (*c*) the corresponding value of $I_{BB'}$.

9.119 Determine the moment of inertia of the triangular area shown with respect to (*a*) the x axis, (*b*) the y axis, (*c*) side AB.

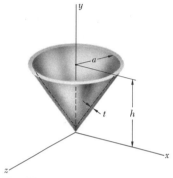

Fig. P9.114

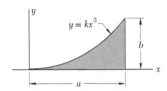

Fig. P9.116

Fig. P9.118

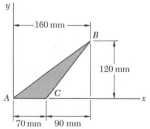

Fig. P9.119

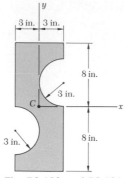

Fig. P9.120 and P9.121

9.120 Determine the centroidal moments of inertia $\bar{I}_x$ and $\bar{I}_y$ and the centroidal radii of gyration $\bar{k}_x$ and $\bar{k}_y$ for the shaded area shown.

9.121 Determine the orientation of the principal axes through the centroid C and the corresponding values of the moments of inertia for the shaded area shown.

9.122 (a) Determine by direct integration the mass moment of inertia of the homogeneous regular tetrahedron of mass m and side a with respect to an axis through A perpendicular to the base BCD. (b) Show that the ellipsoid of inertia at the center of a regular tetrahedron is a sphere. (c) Using the results of parts a and b, determine the principal moments of inertia of the tetrahedron at one of its corners.

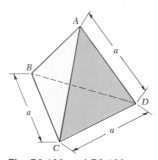

Fig. P9.122 and P9.123

9.123 Using the result of Prob. 9.122, determine the mass moment of inertia of a homogeneous regular tetrahedron with respect to one of its edges.

9.124 The assembly shown consists of two thin slabs, each of mass 3 kg, and four slender rods, each of mass 0.4 kg. Determine the mass moments of inertia of the assembly with respect to centroidal axes parallel to the coordinate axes.

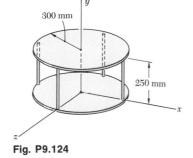

Fig. P9.124

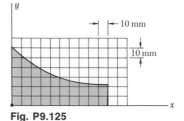

Fig. P9.125

9.125 Determine by approximate means the moment of inertia of the shaded area with respect to (a) the x axis, (b) the y axis.

Method of Virtual Work

∗10.1. Work of a Force. In the preceding chapters, problems involving the equilibrium of rigid bodies were solved by expressing that the external forces acting on the bodies were balanced. The equations of equilibrium $\Sigma F_x = 0$, $\Sigma F_y = 0$, $\Sigma M_A = 0$ were written and solved for the desired unknowns. We shall now consider a different method, which will prove more effective for solving certain types of equilibrium problems. This method is based on the concept of the *work of a force* and was first formally used by the Swiss mathematician Jean Bernoulli in the eighteenth century.

We shall first define the terms *displacement* and *work* as they are used in mechanics. Consider a particle which moves from a point A to a neighboring point A' (Fig. 10.1). If $\mathbf{r}$ denotes the position vector corresponding to point A, the small vector joining A and A' may be denoted by the differential $d\mathbf{r}$; the vector $d\mathbf{r}$ is called the *displacement* of the particle. Now let us assume that a force $\mathbf{F}$ is acting on the particle. The *work of the force* $\mathbf{F}$ *corresponding to the displacement* $d\mathbf{r}$ is defined as the quantity

$$dU = \mathbf{F} \cdot d\mathbf{r} \qquad (10.1)$$

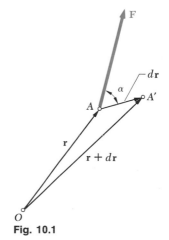

Fig. 10.1

obtained by forming the scalar product of the force $\mathbf{F}$ and of the displacement $d\mathbf{r}$. Denoting respectively by F and ds the magnitudes of the force and of the displacement, and by α the angle formed by $\mathbf{F}$ and $d\mathbf{r}$, and recalling the definition of the scalar product of two vectors (Sec. 3.8), we write

$$dU = F \, ds \cos \alpha \qquad (10.1')$$

405

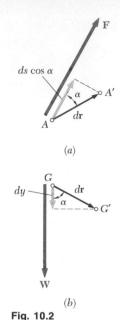

(a)

(b)

Fig. 10.2

Being a *scalar quantity*, work has a magnitude and a sign, but no direction. We also note that work should be expressed in units obtained by multiplying units of length by units of force. Thus, if U.S. customary units are used, work should be expressed in ft · lb or in · lb. If SI units are used, work should be expressed in N · m. The unit of work N · m is called a *joule* (J).†

It follows from (10.1′) that the work dU is positive if the angle α is acute and negative if α is obtuse. Three particular cases are of special interest. If the force **F** has the same direction as $d\mathbf{r}$, the work dU reduces to $F\,ds$. If **F** has a direction opposite to that of $d\mathbf{r}$, the work is $dU = -F\,ds$. Finally, if **F** is perpendicular to $d\mathbf{r}$, the work dU is zero.

The work dU of a force **F** during a displacement $d\mathbf{r}$ may also be considered as the product of F and of the component $ds \cos \alpha$ of the displacement $d\mathbf{r}$ along **F** (Fig. 10.2a). This view is particularly useful in the computation of the work done by the weight **W** of a body (Fig. 10.2b). The work of **W** is equal to the product of W and of the vertical displacement dy of the center of gravity G of the body. If the displacement is downward, the work is positive; if it is upward, the work is negative.

A number of forces frequently encountered in statics *do no work*. They are forces applied to fixed points ($ds = 0$) or acting in a direction perpendicular to the displacement ($\cos \alpha = 0$). Among the forces which do no work are the following: the reaction at a frictionless pin when the body supported rotates about the pin, the reaction at a frictionless surface when the body in contact moves along the surface, the reaction at a roller moving along its track, the weight of a body when its center of gravity moves horizontally, the friction force acting on a wheel rolling without slipping (since at any instant the point of contact does not move). Examples of forces which *do work* are the weight of a body (except in the case considered above), the friction force acting on a body sliding on a rough surface, and most forces applied on a moving body.

In certain cases, the sum of the work done by several forces is zero. Consider, for example, two rigid bodies AC and BC connected at C by a *frictionless pin* (Fig. 10.3a). Among the forces acting on AC is the force **F** exerted at C by BC. In general, the work of this force will not be zero, but it will be equal in

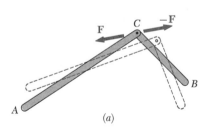

(a)

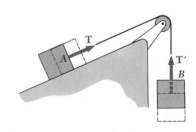

Fig. 10.3   *(b)*

---

†The joule is the SI unit of *energy*, whether in mechanical form (work, potential energy, kinetic energy) or in chemical, electrical, or thermal form. We should note that even though N · m = J, the moment of a force must be expressed in N · m, and not in joules, since the moment of a force is not a form of energy.

magnitude and opposite in sign to the work of the force $-\mathbf{F}$ exerted by AC on BC, since these forces are equal and opposite and are applied to the same particle. Thus, when the total work done by all the forces acting on AB and BC is considered, the work of the two internal forces at C cancels out. A similar result is obtained if we consider a system consisting of two blocks connected by an *inextensible cord AB* (Fig. 10.3*b*). The work of the tension force $\mathbf{T}$ at A is equal in magnitude to the work of the tension force $\mathbf{T}'$ at B, since these forces have the same magnitude and the points A and B move through the same distance; but in one case the work is positive, and in the other it is negative. Thus, the work of the internal forces again cancels out.

It may be shown that the total work of the internal forces holding together the particles of a rigid body is zero. Consider two particles A and B of a rigid body and the two equal and opposite forces $\mathbf{F}$ and $-\mathbf{F}$ they exert on each other (Fig. 10.4). While, in general, small displacements $d\mathbf{r}$ and $d\mathbf{r}'$ of the two particles are different, the components of these displacements along AB must be equal; otherwise, the particles would not remain at the same distance from each other, and the body would not be rigid. Therefore, the work of $\mathbf{F}$ is equal in magnitude and opposite in sign to the work of $-\mathbf{F}$, and their sum is zero.

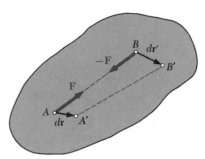

Fig. 10.4

In computing the work of the external forces acting on a rigid body, it is often convenient to determine the work of a couple without considering separately the work of each of the two forces forming the couple. Consider the two forces $\mathbf{F}$ and $-\mathbf{F}$ forming a couple of moment $\mathbf{M}$ and acting on a rigid body (Fig. 10.5). Any small displacement of the rigid body bringing A and B, respectively, into A' and B'' may be divided into two parts, one in which points A and B undergo equal displacements $d\mathbf{r}_1$, the other in which A' remains fixed while B' moves into B'' through a displacement $d\mathbf{r}_2$ of magnitude $ds_2 = r\,d\theta$. In the first part of the motion, the work of $\mathbf{F}$ is equal in magnitude and opposite in sign to the work of $-\mathbf{F}$, and their sum is zero. In the second part of the motion, only force $\mathbf{F}$ works, and its work is $dU = F\,ds_2 = Fr\,d\theta$. But the product Fr is equal to the magnitude M of the moment of the couple. Thus, the work of a couple of moment $\mathbf{M}$ acting on a rigid body is

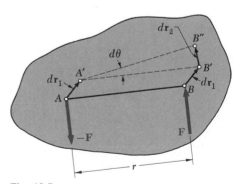

Fig. 10.5

$$dU = M\,d\theta \qquad (10.2)$$

where $d\theta$ is the small angle expressed in radians through which the body rotates. We again note that work should be expressed in units obtained by multiplying units of force by units of length.

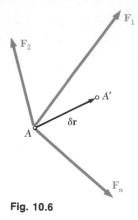

Fig. 10.6

*10.2. Principle of Virtual Work. Consider a particle acted upon by several forces $\mathbf{F}_1, \mathbf{F}_2, \ldots, \mathbf{F}_n$ (Fig. 10.6). We shall assume that the particle undergoes a small displacement from A to A'. This displacement is possible, but it will not necessarily take place. The forces may be balanced and the particle at rest, or the particle may move under the action of the given forces in a direction different from that of AA'. The displacement considered is therefore an imaginary displacement; it is called a *virtual displacement* and is denoted by $\delta\mathbf{r}$. The symbol $\delta\mathbf{r}$ represents a differential of the first order; it is used to distinguish the virtual displacement from the displacement $d\mathbf{r}$ which would take place under actual motion. As we shall see, virtual displacements may be used to determine whether the conditions of equilibrium of a particle are satisfied.

The work of each of the forces $\mathbf{F}_1, \mathbf{F}_2, \ldots, \mathbf{F}_n$ during the virtual displacement $\delta\mathbf{r}$ is called *virtual work*. The virtual work of all the forces acting on the particle of Fig. 10.6 is

$$\begin{aligned}
\delta U &= \mathbf{F}_1 \cdot \delta\mathbf{r} + \mathbf{F}_2 \cdot \delta\mathbf{r} + \cdots + \mathbf{F}_n \cdot \delta\mathbf{r} \\
&= (\mathbf{F}_1 + \mathbf{F}_2 + \cdots + \mathbf{F}_n) \cdot \delta\mathbf{r} \\
&= \mathbf{R} \cdot \delta\mathbf{r} \tag{10.3}
\end{aligned}$$

where $\mathbf{R}$ is the resultant of the given forces. Thus, the total virtual work of the forces $\mathbf{F}_1, \mathbf{F}_2, \ldots, \mathbf{F}_n$ is equal to the virtual work of their resultant $\mathbf{R}$.

The principle of virtual work for a particle states that, *if a particle is in equilibrium, the total virtual work of the forces acting on the particle is zero for any virtual displacement of the particle.* This condition is necessary: if the particle is in equilibrium, the resultant $\mathbf{R}$ of the forces is zero, and it follows from (10.3) that the total virtual work δU is zero. The condition is also sufficient: if the total virtual work δU is zero for any virtual displacement, the scalar product $\mathbf{R} \cdot \delta\mathbf{r}$ is zero for any $\delta\mathbf{r}$, and the resultant $\mathbf{R}$ must be zero.

In the case of a rigid body, the principle of virtual work states that, *if a rigid body is in equilibrium, the total virtual work of the external forces acting on the rigid body is zero for any virtual displacement of the body.* The condition is necessary: if the body is in equilibrium, all the particles forming the body are in equilibrium and the total virtual work of the forces acting on all the particles must be zero; but we have seen in the preceding section that the total work of the internal forces is zero; the total work of the external forces must therefore also be zero. The condition may also be proved to be sufficient.

The principle of virtual work may be extended to the case of a *system of connected rigid bodies.* If the system remains connected during the virtual displacement, *only the work of the forces external to the system need be considered,* since the total work of the internal forces at the various connections is zero.

∗10.3. Applications of the Principle of Virtual Work.   The principle of virtual work is particularly effective when applied to the solution of problems involving machines or mechanisms consisting of several connected rigid bodies. Consider, for instance, the toggle vise *ACB* of Fig. 10.7*a*, used to compress a wooden block. We wish to determine the force exerted by the vise on the block when a given force **P** is applied at *C*, assuming that there is no friction. Denoting by **Q** the reaction of the block on the vise, we draw the free-body diagram of the vise and consider the virtual displacement obtained by giving to the angle *θ* a positive increment *δθ* (Fig. 10.7*b*).

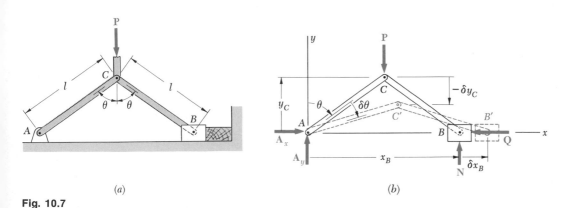

(a)   (b)

Fig. 10.7

Choosing a system of coordinate axes with origin at *A*, we note that x_B increases while y_C decreases. This is indicated in the figure by means of the positive increment δx_B and the negative increment $-\delta y_C$. The reactions $\mathbf{A}_x$, $\mathbf{A}_y$, and **N** will do no work during the virtual displacement considered, and we need only compute the work of **P** and **Q**. Since **Q** and δx_B have opposite senses, the virtual work of **Q** is $\delta U_Q = -Q\,\delta x_B$. Since **P** and the increment shown $(-\delta y_C)$ have the same sense, the virtual work of **P** is $\delta U_P = +P(-\delta y_C) = -P\,\delta y_C$. The minus signs obtained could have been predicted by simply noting that the forces **Q** and **P** are directed opposite to the positive *x* and *y* axes, respectively. Expressing the coordinates x_B and y_C in terms of the angle *θ* and differentiating, we obtain

$$x_B = 2l \sin \theta \qquad y_C = l \cos \theta$$
$$\delta x_B = 2l \cos \theta \, \delta \theta \qquad \delta y_C = -l \sin \theta \, \delta \theta \tag{10.4}$$

The total virtual work of the forces **Q** and **P** is thus

$$\delta U = \delta U_Q + \delta U_P = -Q\,\delta x_B - P\,\delta y_C$$
$$= -2Ql\cos\theta\,\delta\theta + Pl\sin\theta\,\delta\theta$$

Making $\delta U = 0$, we obtain

$$2Ql\cos\theta\,\delta\theta = Pl\sin\theta\,\delta\theta \qquad (10.5)$$
$$Q = \tfrac{1}{2}P\tan\theta \qquad (10.6)$$

The superiority of the method of virtual work over the conventional equilibrium equations in the problem considered here is clear: by using the method of virtual work, we were able to eliminate all unknown reactions, while the equation $\Sigma M_A = 0$ would have eliminated only two of the unknown reactions. We may take advantage of this characteristic of the method of virtual work to solve many problems involving machines and mechanisms. *If the virtual displacement considered is consistent with the constraints imposed by the supports and connections, all reactions and internal forces are eliminated and only the work of the loads, applied forces, and friction forces need be considered.*

We shall observe that the method of virtual work may also be used to solve problems involving completely constrained structures, although the virtual displacements considered will never actually take place. Consider, for example, the frame *ACB* shown in Fig. 10.8*a*. If point *A* is kept fixed, while *B* is given a horizontal virtual displacement (Fig. 10.8*b*), we need consider only the work of **P** and **B**$_x$. We may thus determine the reaction component **B**$_x$ in the same way as the force **Q** of the preceding example (Fig. 10.7*b*); we have

$$B_x = -\tfrac{1}{2}P\tan\theta$$

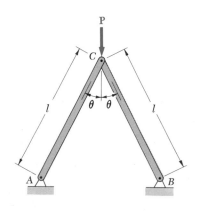

(a)

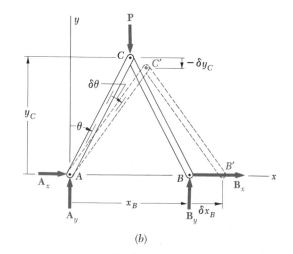

(b)

Fig. 10.8

Keeping B fixed and giving to A a horizontal virtual displacement, we may similarly determine the reaction component $\mathbf{A}_x$. The components $\mathbf{A}_y$ and $\mathbf{B}_y$ may be determined by rotating the frame ACB as a rigid body about B and A, respectively.

The method of virtual work may also be used to determine the configuration of a system in equilibrium under given forces. For example, the value of the angle θ for which the linkage of Fig. 10.7 is in equilibrium under two given forces $\mathbf{P}$ and $\mathbf{Q}$ may be obtained by solving Eq. (10.6) for $\tan \theta$.

It should be noted, however, that the attractiveness of the method of virtual work depends to a large extent upon the existence of simple geometric relations between the various virtual displacements involved in the solution of a given problem. When no such simple relations exist, it is usually advisable to revert to the conventional method of Chap. 6.

*10.4. Real Machines. Mechanical Efficiency. In

analyzing the toggle vise in the preceding section, we assumed that no friction forces were involved. Thus, the virtual work consisted only of the work of the applied force $\mathbf{P}$ and of the reaction $\mathbf{Q}$. But the work of the reaction $\mathbf{Q}$ is equal in magnitude and opposite in sign to the work of the force exerted by the vise on the block. Equation (10.5), therefore, expresses that the *output work* $2Ql \cos \theta \; \delta\theta$ is equal to the *input work* $Pl \sin \theta \; \delta\theta$. A machine in which input and output work are equal is said to be an "ideal" machine. In a "real" machine, friction forces will always do some work, and the output work will be smaller than the input work.

Consider, for example, the toggle vise of Fig. 10.7a, and assume now that a friction force $\mathbf{F}$ develops between the sliding block B and the horizontal plane (Fig. 10.9). Using the conventional methods of statics and summing moments about A, we find $N = P/2$. Denoting by μ the coefficient of friction between

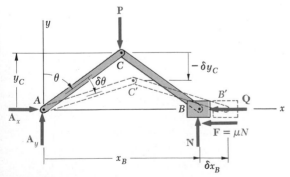

Fig. 10.9

block B and the horizontal plane, we have $F = \mu N = \mu P/2$. Recalling formulas (10.4), we find that the total virtual work of the forces $\mathbf{Q}$, $\mathbf{P}$, and $\mathbf{F}$ during the virtual displacement shown in Fig. 10.9 is

$$
\begin{aligned}
\delta U &= -Q\,\delta x_B - P\,\delta y_C - F\,\delta x_B \\
&= -2Ql\cos\theta\,\delta\theta + Pl\sin\theta\,\delta\theta - \mu Pl\cos\theta\,\delta\theta
\end{aligned}
$$

Making $\delta U = 0$, we obtain

$$
2Ql\cos\theta\,\delta\theta = Pl\sin\theta\,\delta\theta - \mu Pl\cos\theta\,\delta\theta \tag{10.7}
$$

which expresses that the output work is equal to the input work minus the work of the friction force. Solving for Q, we have

$$
Q = \tfrac{1}{2}P(\tan\theta - \mu) \tag{10.8}
$$

We note that $Q = 0$ when $\tan\theta = \mu$, that is, when θ is equal to the angle of friction ϕ, and that $Q < 0$ when $\theta < \phi$. The toggle vise may thus be used only for values θ larger than the angle of friction.

The *mechanical efficiency* of a machine is defined as the ratio

$$
\eta = \frac{\text{output work}}{\text{input work}} \tag{10.9}
$$

Clearly, the mechanical efficiency of an ideal machine is $\eta = 1$, since input and output work are then equal, while the mechanical efficiency of a real machine will always be less than 1.

In the case of the toggle vise we have just analyzed, we write

$$
\eta = \frac{\text{output work}}{\text{input work}} = \frac{2Ql\cos\theta\,\delta\theta}{Pl\sin\theta\,\delta\theta}
$$

Substituting from (10.8) for Q, we obtain

$$
\eta = \frac{P(\tan\theta - \mu)l\cos\theta\,\delta\theta}{Pl\sin\theta\,\delta\theta} = 1 - \mu\cot\theta \tag{10.10}
$$

We check that, in the absence of friction forces, we would have $\mu = 0$ and $\eta = 1$. In the general case, when μ is different from zero, the efficiency η becomes zero for $\mu\cot\theta = 1$, that is, for $\tan\theta = \mu$, or $\theta = \tan^{-1}\mu = \phi$. We check again that the toggle vise may be used only for values of θ larger than the angle of friction ϕ.

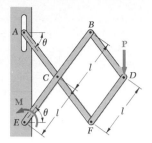

SAMPLE PROBLEM 10.1

Using the method of virtual work, determine the moment of the couple **M** required to maintain the equilibrium of the mechanism shown.

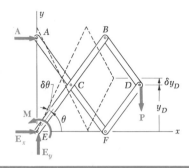

Solution. Choosing a coordinate system with origin at E, we write

$$y_D = l \sin \theta \qquad \delta y_D = l \cos \theta \, \delta \theta$$

Principle of Virtual Work. Since the reactions **A**, $\mathbf{E}_x$, and $\mathbf{E}_y$ will do no work during the virtual displacement, the total virtual work done by **M** and **P** must be zero.

$$\delta U = 0: \qquad + M \, \delta\theta - P \, \delta y_D = 0$$
$$+ M \, \delta\theta - P(l \cos \theta \, \delta\theta) = 0 \qquad\qquad M = Pl \cos \theta \quad \blacktriangleleft$$

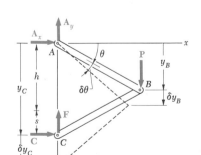

SAMPLE PROBLEM 10.2

Determine the expressions for θ and for the tension in the spring which correspond to the equilibrium position of the mechanism. The unstretched length of the spring is h, and the constant of the spring is k. Neglect the weight of the mechanism.

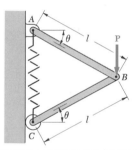

Solution. With the coordinate system shown

$$y_B = l \sin \theta \qquad\qquad y_C = 2l \sin \theta$$
$$\delta y_B = l \cos \theta \, \delta \theta \qquad \delta y_C = 2l \cos \theta \, \delta \theta$$

The elongation of the spring is

$$s = y_C - h = 2l \sin \theta - h$$

The magnitude of the force exerted at C by the spring is

$$F = ks = k(2l \sin \theta - h) \qquad\qquad (1)$$

Principle of Virtual Work. Since the reactions $\mathbf{A}_x$, $\mathbf{A}_y$, and **C** do no work, the total virtual work done by **P** and **F** must be zero.

$$\delta U = 0: \qquad P \, \delta y_B - F \, \delta y_C = 0$$
$$P(l \cos \theta \, \delta\theta) - k(2l \sin \theta - h)(2l \cos \theta \, \delta\theta) = 0$$

$$\sin \theta = \frac{P + 2kh}{4kl} \quad \blacktriangleleft$$

Substituting this expression into (1), we obtain

$$F = \tfrac{1}{2}P \quad \blacktriangleleft$$

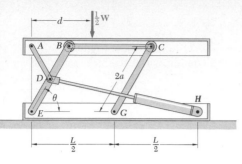

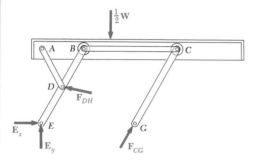

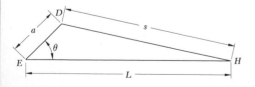

SAMPLE PROBLEM 10.3

A hydraulic-lift table is used to raise a 1000-kg crate. It consists of a platform and of two identical linkages on which hydraulic cylinders exert equal forces. (Only one linkage and one cylinder are shown.) Members EDB and CG are each of length $2a$, and member AD is pinned to the midpoint of EDB. If the crate is placed on the table, so that half of its weight is supported by the system shown, determine the force exerted by each cylinder in raising the crate for $\theta = 60°$, $a = 0.70$ m, and $L = 3.20$ m. This mechanism has been previously considered in Sample Prob. 6.7.

Solution. The machine considered consists of the platform and of the linkage, with an input force $\mathbf{F}_{DH}$ exerted by the cylinder and an output force equal and opposite to $\frac{1}{2}\mathbf{W}$.

Principle of Virtual Work. We first observe that the reactions at E and G do no work. Denoting by y the elevation of the platform about the base, and by s the length DH of the cylinder-and-piston assembly, we write

$$\delta U = 0: \qquad -\tfrac{1}{2}W\,\delta y + F_{DH}\,\delta s = 0 \qquad (1)$$

The vertical displacement δy of the platform is expressed in terms of the angular displacement $\delta\theta$ of EDB as follows:

$$y = (EB)\sin\theta = 2a\sin\theta$$
$$\delta y = 2a\cos\theta\,\delta\theta$$

To express δs similarly in terms of $\delta\theta$, we first note that, by the law of cosines,

$$s^2 = a^2 + L^2 - 2aL\cos\theta$$

Differentiating:

$$2s\,\delta s = -2aL(-\sin\theta)\,\delta\theta$$
$$\delta s = \frac{aL\sin\theta}{s}\,\delta\theta$$

Substituting for δy and δs into (1), we write

$$(-\tfrac{1}{2}W)2a\cos\theta\,\delta\theta + F_{DH}\frac{aL\sin\theta}{s}\,\delta\theta = 0$$

$$F_{DH} = W\frac{s}{L}\cot\theta$$

With the given numerical data, we have

$$W = mg = (1000\text{ kg})(9.81\text{ m/s}^2) = 9810\text{ N} = 9.81\text{ kN}$$

$$s^2 = a^2 + L^2 - 2aL\cos\theta$$
$$= (0.70)^2 + (3.20)^2 - 2(0.70)(3.20)\cos 60° = 8.49$$

$$s = 2.91\text{ m}$$

$$F_{DH} = W\frac{s}{L}\cot\theta = (9.81\text{ kN})\frac{2.91\text{ m}}{3.20\text{ m}}\cot 60°$$

$$F_{DH} = 5.15\text{ kN} \quad \blacktriangleleft$$

PROBLEMS

10.1 Determine the magnitude of the force **P** required to maintain the equilibrium of the linkage shown.

10.2 Determine the weight *W* which balances the 10-lb load.

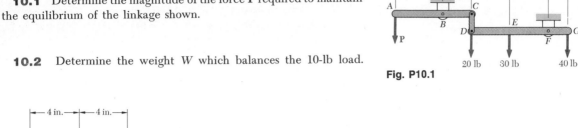

Fig. P10.1

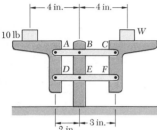

Fig. P10.2

10.3 Determine the horizontal force **P** which must be applied at *A* to maintain the equilibrium of the linkage.

10.4 Determine the force **P** required to maintain the equilibrium of the linkage shown. All members are of the same length and the wheels at *A* and *B* roll freely on the horizontal rod.

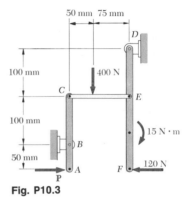

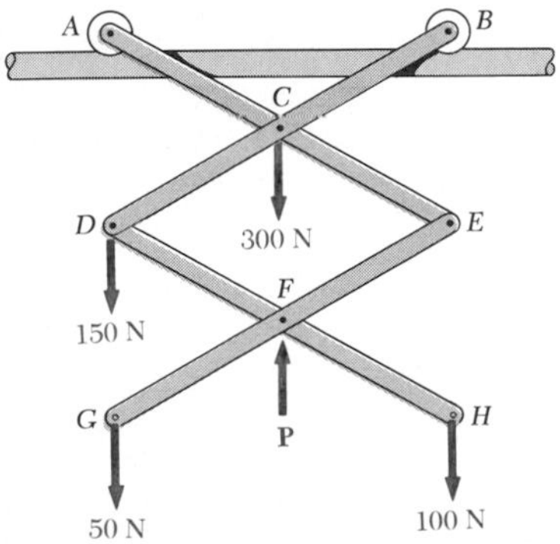

Fig. P10.3

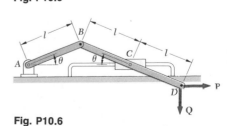

Fig. P10.4

10.5 and 10.6 The mechanism shown is acted upon by the force **P**; derive an expression for the magnitude of the force **Q** required to maintain equilibrium.

Fig. P10.5

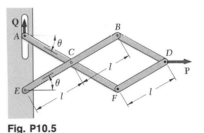

Fig. P10.6

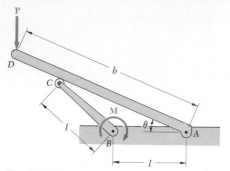

Fig. P10.7

10.7 Derive an expression for the magnitude of the couple **M** required to maintain equilibrium.

10.8 The slender rod AB is attached to a collar at A and rests on a wheel at C. Neglecting the effect of friction, derive an expression for the magnitude of the force **Q** required to maintain equilibrium.

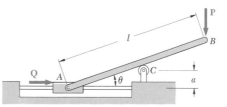

Fig. P10.8

10.9 The mechanism shown is acted upon by the force **P**; determine an expression for the magnitude of the force **Q** required to maintain equilibrium.

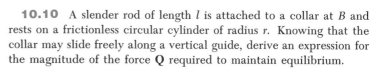

Fig. P10.9

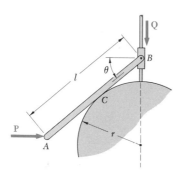

Fig. P10.10

10.10 A slender rod of length l is attached to a collar at B and rests on a frictionless circular cylinder of radius r. Knowing that the collar may slide freely along a vertical guide, derive an expression for the magnitude of the force **Q** required to maintain equilibrium.

10.11 Solve Prob. 10.10, assuming that the force **P** is removed and that a couple **M**, directed counterclockwise, is applied to rod AB.

10.12 Neglecting the effect of friction, derive an expression for the magnitude of the couple **M** required to maintain equilibrium.

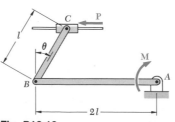

Fig. P10.12

10.13 Collar A weighs 5.60 lb and may slide freely on the frictionless vertical rod. Knowing that the length of the wire AB is 18 in., determine the magnitude of **P** required for equilibrium when (*a*) $c = 2$ in., (*b*) $c = 8$ in.

10.14 Solve Prob. 10.13 when (*a*) $c = 14$ in., (*b*) $c = 16$ in.

10.15 In Prob. 10.7, determine the magnitude of the couple **M** required for equilibrium when $l = 80$ mm, $b = 200$ mm, $P = 300$ N, and $\theta = 30°$.

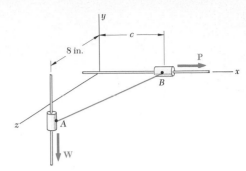

Fig. P10.13

10.16 In Prob. 10.8, determine the magnitude of the force **Q** required for equilibrium when $l = 500$ mm, $a = 100$ mm, $\theta = 30°$ and $P = 500$ N.

10.17 Determine the value of θ corresponding to the equilibrium position of the mechanism of Prob. 10.5 when $P = 30$ lb and $Q = 100$ lb.

10.18 Determine the value of θ corresponding to the equilibrium position of the mechanism of Prob. 10.6 when $P = 50$ lb and $Q = 80$ lb.

10.19 Determine the value of θ corresponding to the equilibrium position of the mechanism of Prob. 10.7 when $P = 400$ N, $M = 50$ N $\cdot$ m, $b = 300$ mm, and $l = 90$ mm.

10.20 Determine the value of θ corresponding to the equilibrium position of the mechanism of Prob. 10.8 when $P = 500$ N, $Q = 400$ N, $a = 100$ mm, and $l = 500$ mm.

10.21 A vertical load **W** is applied to the linkage at B. The constant of the spring is k, and the spring is unstretched when AB and BC are horizontal. Neglecting the weight of the linkage, derive an equation in θ, W, l, and k, which must be satisfied when the linkage is in equilibrium.

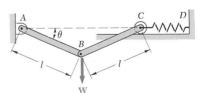

Fig. P10.21 and P10.22

10.22 A load **W** of magnitude 150 lb is applied to the linkage at B. Neglecting the weight of the linkage and knowing that $l = 8$ in., determine the value of θ corresponding to equilibrium. The constant of the spring is $k = 10$ lb/in., and the spring is unstretched when AB and BC are horizontal. (*Hint.* Obtain the approximate value of θ by solving by trial and error the equation obtained.)

10.23 Two bars AB and BC, each of mass 4 kg, are connected by a pin at B and by a spring DE. When unstretched, the spring is 150 mm long; the constant of the spring is 800 N/m. Determine the value of x corresponding to equilibrium.

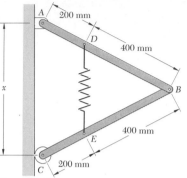

Fig. P10.23

10.24 A load **W** of magnitude 800 N is applied to the mechanism at B. Neglecting the weight of the mechanism, determine the value of θ corresponding to equilibrium. The constant of the spring is $k = 10$ kN/m, and the spring is unstretched when bar AC is horizontal.

10.25 The elevation of the overhead platform is controlled by two identical mechanisms, only one of which is shown. A load of 5 kN is applied to the mechanism shown. Neglecting the weight of the mechanism, determine the force exerted by the hydraulic cylinder on pin E when $\theta = 30°$.

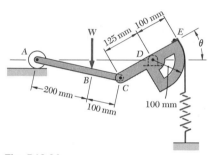

Fig. P10.24

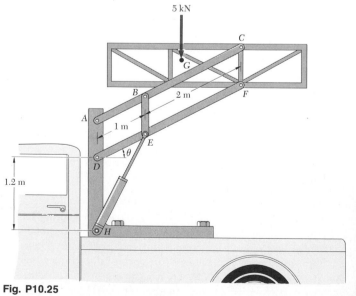

Fig. P10.25

10.26 For the mechanism of Prob. 10.25, (*a*) express the force exerted by the hydraulic cylinder on pin *E* as a function of the length *HE*, (*b*) determine the largest possible value of the angle θ if the maximum force that the cylinder can exert on pin *E* is 25 kN.

10.27 The lever *AB* is attached to the horizontal shaft *BC* which passes through a bearing and is welded to a fixed support at *C*. The torsional spring constant of the shaft *BC* is *K*; that is, a couple of magnitude *K* is required to rotate end *B* through one radian. Knowing that the shaft is untwisted when *AB* is horizontal, determine the value of θ corresponding to the position of equilibrium, if $P = 20$ lb, $l = 10$ in., and $K = 100$ lb $\cdot$ in./rad.

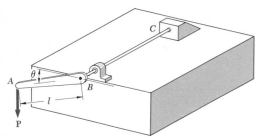

Fig. P.10.27

10.28 Solve Prob. 10.27, if $P = 70$ lb, $l = 10$ in., and $K = 100$ lb $\cdot$ in./rad. Obtain an answer in each of the following quadrants: $0 < \theta < 90°$, $270° < \theta < 360°$, and $360° < \theta < 450°$. (It is assumed that *K* remains constant for the angles of twist considered.)

10.29 Derive an expression for the mechanical efficiency of the jack discussed in Sec. 8.6. Show that, if the jack is to be self-locking, the mechanical efficiency cannot exceed $\frac{1}{2}$.

10.30 A block of weight *W* is pulled up a plane forming an angle α with the horizontal by a force **P** directed along the plane. If μ is the coefficient of friction between the block and the plane, derive an expression for the mechanical efficiency of the system. Show that the mechanical efficiency cannot exceed $\frac{1}{2}$ if the block is to remain in place when the force **P** is removed.

10.31 Denoting by μ the coefficient of friction between the collar *C* and the horizontal rod, determine the largest magnitude of the couple **M** for which equilibrium is maintained. Explain what happens if $\mu \geq \tan \theta$.

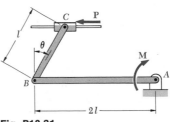

Fig. P10.31

10.32 In Prob. 10.5, assume that the coefficient of friction between the pin and the slot at *A* is $\mu = 0.30$, and determine the smallest and the largest magnitudes of the force **Q** for which equilibrium is maintained when $\theta = 30°$ and $P = 300$ lb.

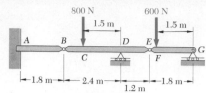

Fig. P10.33 and P10.34

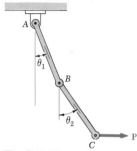

Fig. P10.36

10.33 Using the method of virtual work, determine separately the force and the couple representing the reaction at A.

10.34 Using the method of virtual work, determine the reaction at D.

10.35 In Prob. 10.4 the force P is removed and the linkage is maintained in equilibrium by a cord which is attached to pins E and H. Determine the tension in the cord.

10.36 Two uniform rods, each of weight W and length l, are connected as shown. Using the method of virtual work, determine θ_1 and θ_2 corresponding to equilibrium.

10.37 Determine the vertical movement of joint G if member AB is shortened 0.8 in. (*Hint.* Apply a vertical load at joint G, and, using the methods of Chap. 6, compute the force exerted by member AB on joints A and B. Then apply the method of virtual work for a virtual displacement making member AB shorter. This method should be used only for small changes in the length of members.)

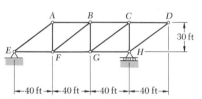

Fig. P10.37, P10.38 and P10.39

10.38 Determine the vertical movement of joint D if the length of member BF is changed to 50 ft $1\frac{1}{2}$ in. (See hint of Prob. 10.37).

10.39 Determine the horizontal movement of joint D if the length of member BF is changed to 50 ft $1\frac{1}{2}$ in. (See hint of Prob. 10.37.)

*10.5. Work of a Force during a Finite Displacement. Consider a force F acting on a particle. The work of F corresponding to an infinitesimal displacement $d\mathbf{r}$ of the particle was defined in Sec. 10.1 as

$$dU = \mathbf{F} \cdot d\mathbf{r} \qquad (10.1)$$

The work of F corresponding to a finite displacement of the particle from A_1 to A_2 (Fig. 10.10a) is denoted by $U_{1\rightarrow2}$ and is obtained by integrating (10.1) along the curve described by the particle:

$$U_{1 \to 2} = \int_{A_1}^{A_2} \mathbf{F} \cdot d\mathbf{r} \qquad (10.11)$$

Using the alternate expression

$$dU = F \, ds \cos \alpha \qquad (10.1')$$

given in Sec. 10.1 for the elementary work dU, we may also express the work $U_{1 \to 2}$ as

$$U_{1 \to 2} = \int_{s_1}^{s_2} (F \cos \alpha) \, ds \qquad (10.11')$$

where the variable of integration s measures the distance along the path traveled by the particle. The work $U_{1 \to 2}$ is represented by the area under the curve obtained by plotting $F \cos \alpha$ against s (Fig. 10.10b). In the case of a force $\mathbf{F}$ of constant magnitude acting in the direction of motion, formula (10.11′) yields $U_{1 \to 2} = F(s_2 - s_1)$.

Recalling from Sec. 10.1 that the work of a couple of moment $\mathbf{M}$ during an infinitesimal rotation $d\theta$ of a rigid body is

$$dU = M \, d\theta \qquad (10.2)$$

we express as follows the work of the couple during a finite rotation of the body:

$$U_{1 \to 2} = \int_{\theta_1}^{\theta_2} M \, d\theta \qquad (10.12)$$

In the case of a constant couple, formula (10.12) yields

$$U_{1 \to 2} = M(\theta_2 - \theta_1)$$

Work of a Weight. It was stated in Sec. 10.1 that the work of the weight $\mathbf{W}$ of a body during an infinitesimal displacement of the body is equal to the product of W and of the vertical displacement of the center of gravity of the body. With the y axis pointing upward, the work of $\mathbf{W}$ during a finite displacement of the body (Fig. 10.11) is obtained by writing

$$dU = -W \, dy$$

$$U_{1 \to 2} = -\int_{y_1}^{y_2} W \, dy = Wy_1 - Wy_2 \qquad (10.13)$$

or

$$U_{1 \to 2} = -W(y_2 - y_1) = -W \, \Delta y \qquad (10.13')$$

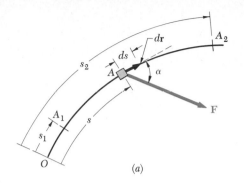

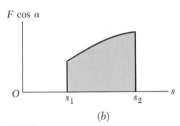

(b)

Fig. 10.10

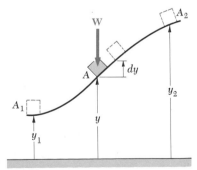

Fig. 10.11

where Δy is the vertical displacement from A_1 to A_2. The work of the weight $\mathbf{W}$ is thus equal to *the product of W and of the vertical displacement of the center of gravity of the body.* The work is *positive* when $\Delta y < 0$, that is, *when the body moves down.*

Work of the Force Exerted by a Spring. Consider a body A attached to a fixed point B by a spring; it is assumed that the spring is undeformed when the body is at A_0 (Fig. 10.12a). Experimental evidence shows that the magnitude of the force $\mathbf{F}$ exerted by the spring on a body A is proportional to the deflection x of the spring measured from the position A_0. We have

$$F = kx \tag{10.14}$$

where k is the *spring constant,* expressed in N/m if SI units are used, and in lb/ft or lb/in. if U.S. customary units are used. The work of the force $\mathbf{F}$ exerted by the spring during a finite displacement of the body from $A_1(x = x_1)$ to $A_2(x = x_2)$ is obtained by writing

$$dU = -F\,dx = -kx\,dx$$

$$U_{1 \to 2} = -\int_{x_1}^{x_2} kx\,dx = \tfrac{1}{2}kx_1^2 - \tfrac{1}{2}kx_2^2 \tag{10.15}$$

Care should be taken to express k and x in consistent units. For example, if U.S. customary units are used, k should be expressed in lb/ft and x in feet, or k in lb/in. and x in inches; in the first case, the work is obtained in ft · lb; in the second case, in in · lb. We note that the work of the force $\mathbf{F}$ exerted by the spring on the body is *positive* when $x_2 < x_1$, that is, *when the spring is returning to its undeformed position.*

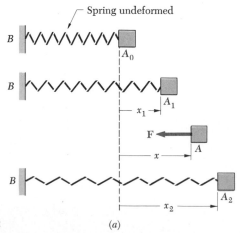

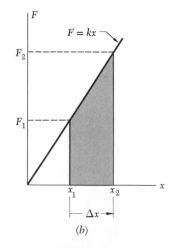

Fig. 10.12 (a) (b)

Since Eq. (10.14) is the equation of a straight line of slope k passing through the origin, the work $U_{1 \to 2}$ of $\mathbf{F}$ during the displacement from A_1 to A_2 may be obtained by evaluating the area of the trapezoid shown in Fig. 10.12b. This is done by computing the values F_1 and F_2 and multiplying the base Δx of the trapezoid by its mean height $\frac{1}{2}(F_1 + F_2)$. Since the work of the force $\mathbf{F}$ exerted by the spring is positive for a negative value of Δx, we write

$$U_{1 \to 2} = -\tfrac{1}{2}(F_1 + F_2)\,\Delta x \qquad (10.16)$$

Formula (10.16) is usually more convenient to use than (10.15) and affords fewer chances of confusing the units involved.

∗10.6. Potential Energy. Considering again the body of Fig. 10.11, we note from (10.13) that the work of the weight $\mathbf{W}$ during a finite displacement is obtained by subtracting the value of the function Wy corresponding to the second position of the body from its value corresponding to the first position. The work of $\mathbf{W}$ is thus independent of the actual path followed; it depends only upon the initial and final values of the function Wy. This function is called the *potential energy* of the body with respect to the *force of gravity* $\mathbf{W}$ and is denoted by V_g. We write

$$U_{1 \to 2} = (V_g)_1 - (V_g)_2 \qquad \text{with } V_g = Wy \qquad (10.17)$$

We note that if $(V_g)_2 > (V_g)_1$, that is, *if the potential energy increases* during the displacement (as in the case considered here), *the work $U_{1 \to 2}$ is negative.* If, on the other hand, the work of $\mathbf{W}$ is positive, the potential energy decreases. Therefore, the potential energy V_g of the body provides a measure of *the work which may be done* by its weight $\mathbf{W}$. Since only the *change* in potential energy, and not the actual value of V_g, is involved in formula (10.17), an arbitrary constant may be added to the expression obtained for V_g. In other words, the level from which the elevation y is measured may be chosen arbitrarily. Note that potential energy is expressed in the same units as work, i.e., in joules (J) if SI units are used,† and in ft · lb or in · lb if U.S. customary units are used.

Considering now the body of Fig. 10.12a, we note from formula (10.15) that the work of the elastic force $\mathbf{F}$ is obtained by subtracting the value of the function $\frac{1}{2}kx^2$ corresponding to the second position of the body from its value corresponding to the first position. This function is denoted by V_e and is called the *potential energy* of the body with respect to the *elastic force* $\mathbf{F}$. We write

† See footnote, page 406.

$$U_{1 \rightarrow 2} = (V_e)_1 - (V_e)_2 \qquad \text{with } V_e = \tfrac{1}{2}kx^2 \qquad (10.18)$$

and observe that, during the displacement considered, the work of the force **F** exerted by the spring on the body is negative and the potential energy V_e increases. We should note that the expression obtained for V_e is valid only if the deflection of the spring is measured from its undeformed position.

The concept of potential energy may be used when forces other than gravity forces and elastic forces are involved. It remains valid as long as the elementary work dU of the force considered is an *exact differential*. It is then possible to find a function V, called potential energy, such that

$$dU = -dV \qquad (10.19)$$

Integrating (10.19) over a finite displacement, we obtain the general formula

$$U_{1 \rightarrow 2} = V_1 - V_2 \qquad (10.20)$$

which expresses that *the work of the force is independent of the path followed and is equal to minus the change in potential energy*. A force which satisfies Eq. (10.20) is said to be a *conservative force*.†

*10.7. Potential Energy and Equilibrium.** The application of the principle of virtual work is considerably simplified when the potential energy of a system is known. In the case of a virtual displacement, formula (10.19) becomes $\delta U = -\delta V$. Besides, if the position of the system is defined by a single independent variable θ, we may write $\delta V = (dV/d\theta)\,\delta\theta$. Since $\delta\theta$ must be different from zero, the condition $\delta U = 0$ for the equilibrium of the system becomes

$$\frac{dV}{d\theta} = 0 \qquad (10.21)$$

In terms of potential energy, the principle of virtual work states therefore that, *if a system is in equilibrium, the derivative of its total potential energy is zero*. If the position of the system depends upon several independent variables (the system is then said to possess *several degrees of freedom*), the partial derivatives of V with respect to each of the independent variables must be zero.

Consider, for example, a structure made of two members AC and CB and carrying a load W at C. The structure is supported by a pin at A and a roller at B, and a spring BD connects B

† A detailed discussion of conservative forces is given in Sec. 13.7.

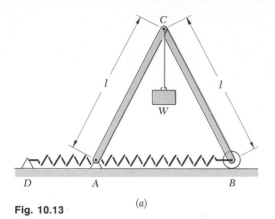

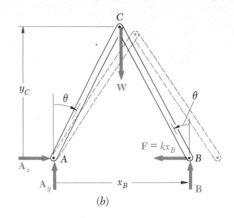

Fig. 10.13

(a)　　　　　(b)

to a fixed point D (Fig. 10.13a). The constant of the spring is k, and it is assumed that the natural length of the spring is equal to AD and thus that the spring is undeformed when B coincides with A. Neglecting the friction forces and the weight of the members, we find that the only forces which work during a displacement of the structure are the weight $\mathbf{W}$ and the force $\mathbf{F}$ exerted by the spring at point B (Fig. 10.13b). The total potential energy of the system will thus be obtained by adding the potential energy V_g corresponding to the gravity force $\mathbf{W}$ and the potential energy V_e corresponding to the elastic force $\mathbf{F}$.

Choosing a coordinate system with origin at A and noting that the deflection of the spring, measured from its undeformed position, is $AB = x_B$, we write

$$V_e = \tfrac{1}{2}kx_B^2 \qquad V_g = Wy_C$$

Expressing the coordinates x_B and y_C in terms of the angle θ, we have

$$
\begin{aligned}
x_B &= 2l \sin \theta & y_C &= l \cos \theta \\
V_e &= \tfrac{1}{2}k(2l \sin \theta)^2 & V_g &= W(l \cos \theta) \\
V &= V_e + V_g = 2kl^2 \sin^2 \theta + Wl \cos \theta & & (10.22)
\end{aligned}
$$

The positions of equilibrium of the system are obtained by equating to zero the derivative of the potential energy V,

$$\frac{dV}{d\theta} = 4kl^2 \sin \theta \cos \theta - Wl \sin \theta = 0$$

$$\sin \theta = 0 \qquad 4kl \cos \theta - W = 0$$

There are therefore two positions of equilibrium, corresponding to the values $\theta = 0$ and $\theta = \cos^{-1}(W/4kl)$.†

---

† The second position does not exist if $W > 4kl$ (see Prob. 10.63 for a further discussion of the equilibrium of this system).

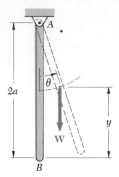

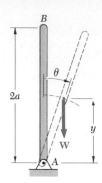

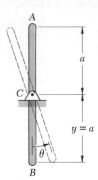

Fig. 10.14   (*a*) Stable equilibrium   (*b*) Unstable equilibrium   (*c*) Neutral equilibrium

∗10.8. Stability of Equilibrium.   Consider the three uniform rods of length $2a$ and weight **W** shown in Fig. 10.14. While each rod is in equilibrium, there is an important difference between the three cases considered. Suppose that each rod is slightly disturbed from its position of equilibrium and then released: rod *a* will move back toward its original position, rod *b* will keep moving away from its original position, and rod *c* will remain in its new position. In case *a*, the equilibrium of the rod is said to be *stable;* in case *b*, the equilibrium is said to be *unstable;* and, in case *c*, to be *neutral.*

Recalling from Sec. 10.6 that the potential energy V_g with respect to gravity is equal to Wy, where y is the elevation of the point of application of **W** measured from an arbitrary level, we observe that the potential energy of rod *a* is minimum in the position of equilibrium considered, that the potential energy of rod *b* is maximum, and that the potential energy of rod *c* is constant. Equilibrium is thus *stable, unstable,* or *neutral* according to whether the potential energy is *minimum, maximum,* or *constant* (Fig. 10.15).

That the result obtained is quite general may be seen as follows: We first observe that a force always tends to do positive

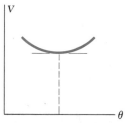

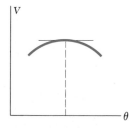

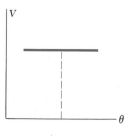

Fig. 10.15   (*a*) Stable equilibrium   (*b*) Unstable equilibrium   (*c*) Neutral equilibrium

work and thus to decrease the potential energy of the system on which it is applied. Therefore, when a system is disturbed from its position of equilibrium, the forces acting on the system will tend to bring it back to its original position if V is minimum (Fig. 10.15a) and to move it farther away if V is maximum (Fig. 10.15b). If V is constant (Fig. 10.15c), the forces will not tend to move the system either way.

Recalling from calculus that a function is minimum or maximum according to whether its second derivative is positive or negative, we may summarize as follows the conditions for the equilibrium of a system with one degree of freedom (i.e., a system the position of which is defined by a single independent variable θ):

$$\frac{dV}{d\theta} = 0 \qquad \frac{d^2V}{d\theta^2} > 0: \text{ stable equilibrium}$$

$$\frac{dV}{d\theta} = 0 \qquad \frac{d^2V}{d\theta^2} < 0: \text{ unstable equilibrium}$$

$$(10.23)$$

If both the first and the second derivatives of V are zero, it is necessary to examine derivatives of a higher order to determine whether the equilibrium is stable, unstable, or neutral. The equilibrium will be neutral if all derivatives are zero, since the potential energy V is then a constant. The equilibrium will be stable if the first derivative found to be different from zero is of even order, and if that derivative is positive. In all other cases the equilibrium will be unstable.

If the system considered possesses *several degrees of freedom,* the potential energy V depends upon several variables and it is thus necessary to apply the theory of functions of several variables to determine whether V is minimum. It may be verified that a system with two degrees of freedom will be stable, and the corresponding potential energy $V(\theta_1, \theta_2)$ will be minimum, if the following relations are satisfied simultaneously:

$$\frac{\partial V}{\partial \theta_1} = \frac{\partial V}{\partial \theta_2} = 0$$

$$\left(\frac{\partial^2 V}{\partial \theta_1 \, \partial \theta_2}\right)^2 - \frac{\partial^2 V}{\partial \theta_1^2}\frac{\partial^2 V}{\partial \theta_2^2} < 0 \qquad (10.24)$$

$$\frac{\partial^2 V}{\partial \theta_1^2} > 0 \qquad \text{or} \qquad \frac{\partial^2 V}{\partial \theta_2^2} > 0$$

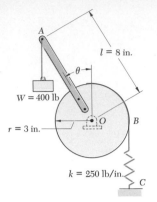

W = 400 lb

l = 8 in.

r = 3 in.

k = 250 lb/in.

SAMPLE PROBLEM 10.4

A 400-lb weight is attached to the lever AO as shown. The constant of the spring BC is $k = 250$ lb/in., and the spring is unstretched when $\theta = 0$. Determine the position or positions of equilibrium, and state in each case whether the equilibrium is stable, unstable, or neutral.

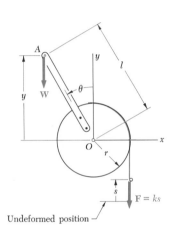

Undeformed position

Potential Energy. Denoting by s the deflection of the spring from its undeformed position and placing the origin of coordinates at O, the potential energy of the system is

$$V_e = \tfrac{1}{2}ks^2 \qquad V_g = Wy$$

Measuring θ in radians, we have

$$s = r\theta \qquad\qquad y = l\cos\theta$$
$$V_e = \tfrac{1}{2}kr^2\theta^2 \qquad V_g = Wl\cos\theta$$
$$V = V_e + V_g = \tfrac{1}{2}kr^2\theta^2 + Wl\cos\theta$$

Positions of Equilibrium. Setting $dV/d\theta = 0$, we write

$$\frac{dV}{d\theta} = kr^2\theta - Wl\sin\theta = 0$$

$$\sin\theta = \frac{kr^2}{Wl}\theta$$

Substituting the given data, we obtain

$$\sin\theta = \frac{(250 \text{ lb/in.})(3 \text{ in.})^2}{(400 \text{ lb})(8 \text{ in.})}\theta = 0.703\,\theta$$

Solving by trial and error for θ, we find

$$\theta = 0 \qquad \text{and} \qquad \theta = 80.4° \quad \blacktriangleleft$$

Stability of Equilibrium. The second derivative of the potential energy V with respect to θ is

$$\frac{d^2V}{d\theta^2} = kr^2 - Wl\cos\theta = (250 \text{ lb/in.})(3 \text{ in.})^2 - (400 \text{ lb})(8 \text{ in.})\cos\theta$$

$$= 2250 - 3200\cos\theta$$

For $\theta = 0$:
$$\frac{d^2V}{d\theta^2} = 2250 - 3200\cos 0° = -950 < 0$$

The equilibrium is unstable for $\theta = 0°$ $\quad\blacktriangleleft$

For $\theta = 80.4°$:
$$\frac{d^2V}{d\theta^2} = 2250 - 3200\cos 80.4° = +1716 > 0$$

The equilibrium is stable for $\theta = 80.4°$ $\quad\blacktriangleleft$

PROBLEMS

10.40 Using the method of Sec. 10.7, solve Prob. 10.22.

10.41 Using the method of Sec. 10.7, solve Prob. 10.21.

10.42 Using the method of Sec. 10.7, solve Prob. 10.24.

10.43 Using the method of Sec. 10.7, solve Prob. 10.23.

10.44 Show that the equilibrium is neutral in Prob. 10.4.

10.45 Show that the equilibrium is neutral in Prob. 10.1.

10.46 Two uniform rods, each of weight W, are attached to gears of equal radii as shown. Determine the positions of equilibrium of the system and state in each case whether the equilibrium is stable, unstable, or neutral.

10.47 A vertical force P of magnitude 10 lb is applied to rod CD at D. Knowing that the uniform rods AB and CD weigh 5 lb each, determine the positions of equilibrium of the system and state in each case whether the equilibrium is stable, unstable, or neutral.

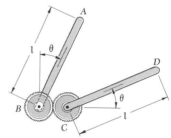

Fig. P10.46 and P10.47

10.48 Using the method of Sec. 10.7, solve Prob. 10.27. Determine whether the equilibrium is stable, unstable, or neutral. (*Hint*. The potential energy corresponding to the couple exerted by a torsion spring is $\frac{1}{2}K\theta^2$, where K is the torsional spring constant and θ is the angle of twist.)

10.49 In Prob. 10.28, determine whether each of the positions of equilibrium is stable, unstable, or neutral. (See hint of Prob. 10.48.)

10.50 A collar B, of weight W, may move freely along the vertical rod shown. Knowing that the constant of the spring is k and that the spring is unstretched when $y = 0$, derive an equation in y, W, a, and k which must be satisfied when the collar is in equilibrium.

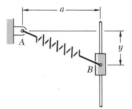

Fig. P10.50

10.51 A slender rod AB, of weight W, is attached to blocks A and B which may move freely in the guides shown. The constant of the spring is k and the spring is unstretched when AB is horizontal. Neglecting the weight of the blocks, derive an equation in θ, W, l, and k which must be satisfied when the rod is in equilibrium.

10.52 In Prob. 10.51, determine three values of θ corresponding to equilibrium when $W = 40$ lb, $k = 10$ lb/in., and $l = 15$ in. State in each case whether the equilibrium is stable, unstable, or neutral.

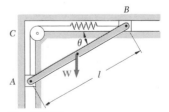

Fig. P10.51

10.53 For the mechanism considered in Prob. 10.21, assuming $0° \le \theta \le 90°$, show that for any values of W and k there is only one equilibrium position. Further show that the equilibrium position is stable.

10.54 The constant of the spring is k and the spring is unstretched when $\theta = 0$. (a) Derive an equation defining the angle θ corresponding to the equilibrium position. (b) Determine the angle θ corresponding to equilibrium if $P = 2kl$.

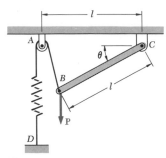

Fig. P10.55

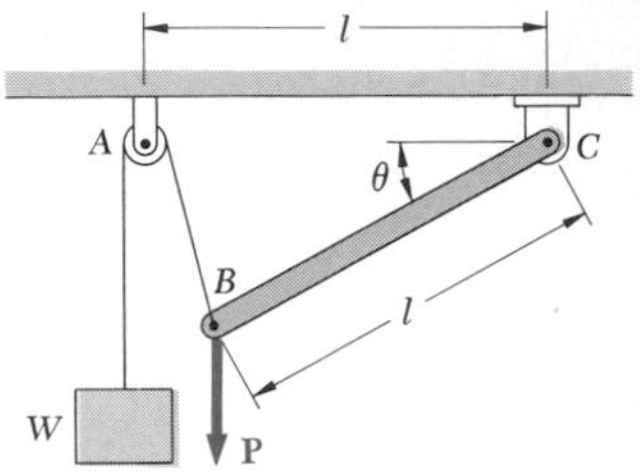

Fig. P10.54

10.55 (a) Derive an equation defining the angle θ corresponding to the equilibrium position. (b) Determine the angle θ corresponding to the equilibrium position $P = 2W$.

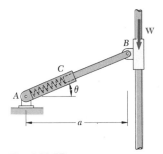

Fig. P10.57

10.56 For the mechanism of Sample Prob. 10.4, determine the range of values of W for which stable equilibrium exists when lever AO is vertical. Express the result in terms of l, r, and k.

10.57 The internal spring AC is of constant k and is undeformed when $\theta = 45°$. Derive an equation defining the values of θ corresponding to equilibrium positions.

10.58 In Prob. 10.57, determine the values of θ corresponding to equilibrium positions when $W = 75$ N, $k = 2$ kN/m, and $a = 500$ mm.

10.59 The rod AB is attached to a hinge at A and to two springs each of constant k. If $h = 450$ mm, $d = 300$ mm, and $m = 200$ kg, determine the range of values of k for which the equilibrium of rod AB is stable in the position shown. Each spring can act in either tension or compression.

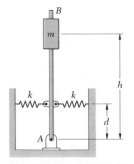

Fig. P10.59 and P10.60

10.60 If $m = 100$ kg, $h = 600$ mm, and the constant of each spring is $k = 3$ kN/m, determine the range of values of the distance d for which the equilibrium of the rod AB is stable in the position shown. Each spring can act in either tension or compression.

*10.61 Determine (*a*) the largest ratio P/W for which $\theta = 0$ is a stable position of equilibrium, (*b*) the smallest ratio P/W for which $\theta = 180°$ is a stable position of equilibrium, (*c*) the range of values of P/W for which both $\theta = 0$ and $\theta = 180°$ are stable positions of equilibrium.

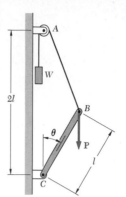

Fig. P10.61

*10.62 In Prob. 10.61 show that the value of θ defined by the equation $(5 - 4 \cos \theta)^{1/2} = 2W/P$ corresponds to a position of equilibrium. Further show that this position of equilibrium is always unstable.

10.63 In Sec. 10.7, two positions of equilibrium were obtained for the system shown in Fig. 10.13, namely, $\theta = 0$ and $\theta = \cos^{-1}(W/4kl)$. Show that (*a*) if $W < 4kl$, the equilibrium is stable in the first position ($\theta = 0$) and unstable in the second, (*b*) if $W = 4kl$, the two positions coincide and the equilibrium is unstable, (*c*) if $W > 4kl$, the equilibrium is unstable in the first position ($\theta = 0$) and the second position does not exist. (*Note.* It is assumed that the system must deform as shown and that the system cannot rotate as a single rigid body about A when A and B coincide.)

10.64 and 10.65 Two bars AB and BC are attached to a single spring of constant k which is unstretched when the bars are horizontal. Determine the range of values of the magnitude P of the two equal and opposite forces $\mathbf{P}$ and $-\mathbf{P}$ for which the equilibrium of the system is stable in the position shown.

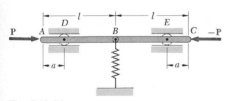

Fig. P10.64

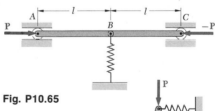

Fig. P10.65

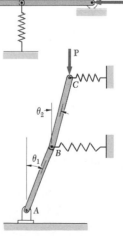

Fig. P10.66

*10.66 The bars shown, each of length l and of negligible weight, are attached to springs each of constant k. The springs are undeformed and the system is in equilibrium when $\theta_1 = \theta_2 = 0$. Determine the range of values of P for which the equilibrium position is stable.

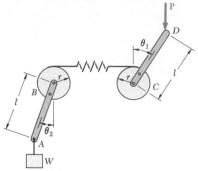

Fig. P10.67

*10.67 Two rods, of negligible weight, are attached to drums of radius r which are connected by a belt and a spring of constant k. The spring is undeformed when $\theta_1 = \theta_2 = 0$. Determine the largest value of the force **P** for which the equilibrium position $\theta_1 = \theta_2 = 0$ is stable.

*10.68 Solve Prob. 10.67 knowing that $k = 20$ lb/in., $r = 3$ in., $l = 6$ in., and $W = 10$ lb.

REVIEW PROBLEMS

10.69 A spring AB of constant k is attached to two identical gears as shown. A uniform bar CD of weight W is supported by cords wrapped around drums of radius b which are attached to the gears. If the spring is undeformed when $\theta = 0$, obtain an equation defining the angle θ corresponding to the equilibrium position.

10.70 For the mechanism of Prob. 10.69 the following numerical values are given: $k = 50$ lb/in., $a = 4$ in., $b = 3$ in., $r = 6$ in., and $W = 200$ lb. Determine the values of θ corresponding to equilibrium positions and state in each case whether the equilibrium is stable, unstable, or neutral.

10.71 The top surface of the work platform is maintained at a given elevation by the single hydraulic cylinder shown. Knowing that members AF and BD are each of length l, determine the magnitude of the force exerted by the hydraulic cylinder when $\theta = 30°$.

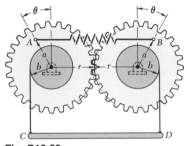

Fig. P10.69

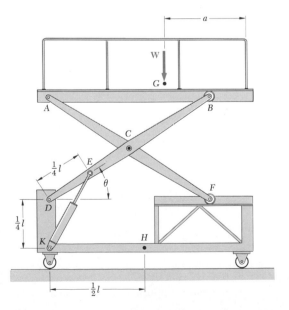

Fig. P10.71

10.72 Solve Prob. 10.71, assuming that the hydraulic cylinder has been moved and attached to points E and H.

10.73 Knowing that rod AB may slide freely along the floor and the inclined plane, derive an expression for the magnitude of the force $\mathbf{Q}$ required to maintain equilibrium. Neglect the weight of the rod.

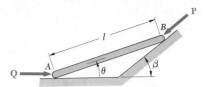

Fig. P10.73

10.74 Solve Prob. 10.73, taking into account the weight W of the uniform rod.

10.75 Three links, each of length l, are connected as shown. Knowing that the line of action of the force $\mathbf{Q}$ passes through point A, derive an expression for the magnitude of $\mathbf{Q}$ required to maintain equilibrium.

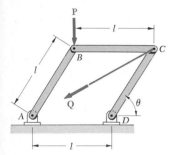

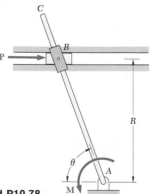

Fig. P10.75

10.76 Each of the uniform rods AB and BC is of weight W. Derive an expression for the magnitude of the couple $\mathbf{M}$ required to maintain equilibrium.

10.77 Collar B may slide along rod AC and is attached by a pin to a block which may slide in the horizontal slot. Derive an expression for the magnitude of the couple $\mathbf{M}$ required to maintain equilibrium.

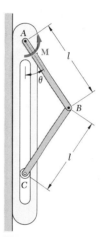

Fig. P10.76

Fig. P10.77 and P10.78

10.78 Determine the value of θ corresponding to the equilibrium position of rod AC when $R = 400$ mm, $P = 100$ N, and $M = 50$ N $\cdot$ m.

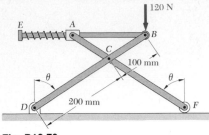

Fig. P10.79

10.79 Two rods AF and BD, each of length 300 mm, and a third rod BE are connected as shown. The spring EA is of constant $k = 1.8$ kN/m and is undeformed when $\theta = 0$. Determine the values of θ corresponding to equilibrium and state in each case whether the equilibrium is stable, unstable, or neutral.

10.80 Denoting by μ the coefficient of friction between end C of rod BC and the horizontal surface, determine the largest magnitude of the couple $\mathbf{M}$ for which equilibrium is maintained.

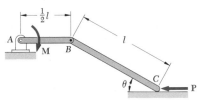

Fig. P10.80

Index

Answers to Even-Numbered Problems

2.2   3240 N ⟱49.1°.

2.4   14.3 kN ⟰19.9°.

2.6   123.4°.

2.8   (a) 2830 lb. (b) 5460 lb.
SI: (a) 12.59 kN. (b) 24.3 kN.

2.10   5380 lb ⟰8.9°. SI: 23.9 kN ⟰8.9°.

2.12 $P = 2990$ N; $\alpha = 72.8°$.

2.14 -2.35 kN, $+0.855$ kN.

2.16   (20 lb) $+17.32$ lb, -10 lb;
(25 lb) $+8.55$ lb, -23.5 lb;
(30 lb) -23.0 lb, -19.28 lb.
SI: (20 lb) $+77.0$ N, -44.5 N;
(25 lb) $+38.0$ N, -104.5 N;
(30 lb) -102.3 N, -85.8 N.

2.18 $+250$ N, -600 N.

2.20   200 lb ⟱4.3°. SI: 890 N ⟱4.3°.

2.22   52.9 lb ⟱86.9° SI: 235 N ⟱86.9°.

2.26   14.04 kips. SI: 62.5 kN.

2.28   6.3° and 133.7°.

2.30 $T_{AC} = 120.1$ lb; $T_{BC} = 156.3$ lb.
SI: $T_{AC} = 534$ N; $T_{BC} = 695$ N.

2.32 $T_{AC} = 981$ N; $T_{BC} = 2350$ N.

2.34   (a) 30°. (b) $T_{AC} = 300$ lb; $T_{BC} = 520$ lb.
SI: (b) $T_{AC} = 1334$ N; $T_{BC} = 2310$ N.

2.36 $F = 2.87$ kN; $\alpha = 75°$.

2.38 $C = 2170$ N; $D = 3750$ N.

2.40 $T_{AC} = 46.2$ lb; $T_{BC} = 36.9$ lb.
SI: $T_{AC} = 206$ N; $T_{BC} = 164.1$ N.

2.42   (a) 1560 lb. (b) 1377 lb.
SI: (a) 6.94 kN. (b) 6.13 kN.

2.44   (b) 654 N. (d) 491 N.

2.46   1.252 in. SI: 31.8 mm.

2.48   (a) 18 lb. (b) 24 lb.
SI: (a) 80.1 N. (b) 106.8 N.

2.50   (a) $+113.3$ N, $+217$ N, -52.8 N.
(b) 63.1°, 30.0°, 102.2°.

2.52   (a) -78.6 lb, $+282$ lb, -66.0 lb.
(b) 105.2°, 20.0°, 102.7°.
SI: (a) -350 N, $+1254$ N, -294 N.

2.54   61.0°; $+105.7$ lb, $+191.5$ lb, $+121.2$ lb.
SI: 61.0°; $+470$ N, $+852$ N, $+539$ N.

2.56   1444 N; 61.0°, 124.6°, 48.3°.

2.58 -200 lb, $+200$ lb, -350 lb.
SI: -890 N, $+890$ N, -1557 N.

2.60   116.4°, 63.6°, 141.1°.

2.62   (a) 54.7°, 125.3°. (b) 60°, 120°.

2.64 $T_{AC} = 21$ kN; $T_{AD} = 64.3$ kN.

2.66   2775 lb. SI: 12.34 kN.

2.68 $T_{AD} = 262.5$ lb; $T_{BD} = 195.0$ lb;
$T_{CD} = 187.5$ lb. SI: $T_{AD} = 1168$ N;
$T_{BD} = 867$ N; $T_{CD} = 834$ N.

2.70   4.80 kN, down.

2.72 $T_{AD} = 139.9$ N, $T_{BD} = T_{CD} = 115.7$ N.

2.74 (a) 6.3 lb. (b) 7.2 lb.
SI: (a) 28.0 N. (b) 32.0 N.

2.76 $P = 101.8$ N; $T_{AB} = 211$ N.

2.78 $T_{AC} = 5950$ N; $T_{BC} = 4490$ N.

2.80 1223 N $\leqslant P \leqslant 3173$ N.

2.82 (a) 500 lb. (b) 375 lb.
SI: (a) 2220 N. (b) 1668 N.

2.84 $T_{AD} = 350$ lb, $T_{BD} = 350$ lb,
$T_{CD} = 600$ lb. SI: $T_{AD} = 1557$ N,
$T_{BD} = 1557$ N, $T_{CD} = 2670$ N.

2.86 $T_{AC} = 547$ N; $T_{BC} = 800$ N.

2.88 418 N $\measuredangle 32.3°$ or 767 N $\measuredangle 32.3°$.

2.90 $T_{AD} = 31.7$ lb; $T_{BD} = 27.5$ lb;
$T_{CD} = 15.85$ lb. SI: $T_{AD} = 141.0$ N;
$T_{BD} = 122.3$ N; $T_{CD} = 70.5$ N.

CHAPTER 3

3.2 37.5 N $\cdot$ m $\downarrow$; $\alpha = 20°$.

3.4 (a) 88.8 N $\cdot$ m $\downarrow$. (b) 395 N $\leftarrow$.
(c) 280 N $\nearrow 45°$.

3.6 1077 lb $\cdot$ in. $\downarrow$. SI: 121.7 N $\cdot$ m $\downarrow$.

3.8 547 lb $\cdot$ in. $\downarrow$. SI: 61.8 N $\cdot$ m $\downarrow$.

3.12 $\mathbf{M}_O = \dfrac{P(x_1 y_2 - x_2 y_1)\mathbf{k}}{[(x_2 - x_1)^2 + (y_2 - y_1)^2]^{1/2}}$.

3.14 (a) $\mathbf{M}_O = -58\mathbf{i} + 4\mathbf{j} + 32\mathbf{k}$.
(b) $\mathbf{M}_O = +6\mathbf{i} - 4\mathbf{k}$.
(c) $\mathbf{M}_O = -30\mathbf{i} + 12\mathbf{j}$.

3.16 $(3600$ lb $\cdot$ ft$)\mathbf{i} + (9600$ lb $\cdot$ ft$)\mathbf{j}$
$- (3600$ lb $\cdot$ ft$)\mathbf{k}$. SI: $(4.88$ kN $\cdot$ m$)\mathbf{i}$
$+ (13.02$ kN $\cdot$ m$)\mathbf{j} - (4.88$ kN $\cdot$ m$)\mathbf{k}$.

3.18 (a) $(40$ N $\cdot$ m$)\mathbf{i} - (90$ N $\cdot$ m$)\mathbf{j} - (40$ N $\cdot$ m$)\mathbf{k}$.
(b) $(9$ N $\cdot$ m$)\mathbf{i} - (86.25$ N $\cdot$ m$)\mathbf{j} - (42$ N $\cdot$ m$)\mathbf{k}$.

3.20 $(36$ N $\cdot$ m$)\mathbf{i} + (24$ N $\cdot$ m$)\mathbf{j} + (32$ N $\cdot$ m$)\mathbf{k}$.

3.22 11.53 in. SI: 293 mm.

3.24 5 m.

3.26 $\mathbf{P} \cdot \mathbf{Q} = 0$; $\mathbf{P} \cdot \mathbf{S} = -11$; $\mathbf{Q} \cdot \mathbf{S} = 2$.

3.28 36.9°.

3.30 (a) 23.5°. (b) 413 N.

3.32 (a) 70.5°. (b) 60 lb. SI: (b) 267 N.

3.34 4.

3.36 4.88 kN.

3.38 40 lb; $\theta = 30°$. SI: 177.9 N.

3.40 -280 N $\cdot$ m.

3.42 $+309$ lb $\cdot$ in. SI: $+34.9$ N $\cdot$ m.

3.44 $+124.2$ N $\cdot$ m.

3.48 229 mm.

3.50 (a) 271 N. (b) 390 N. (c) 250 N.

3.52 (a) 20 lb. (b) 16 lb. (c) 12 lb.
SI: (a) 89.0 N. (b) 71.2 N. (c) 53.4 N.

3.54 $M_x = -540$ lb $\cdot$ in., $M_y = 240$ lb $\cdot$ in.,
$M_z = 180$ lb $\cdot$ in. SI: $M_x = -61.0$ N $\cdot$ m,
$M_y = 27.1$ N $\cdot$ m, $M_z = 20.3$ N $\cdot$ m.

3.56 $M_x = 3.6$ kN $\cdot$ m, $M_y = 7.72$ kN $\cdot$ m,
$M_z = 0$.

3.58 (a) 400 N $\measuredangle 60°$, 77.8 N $\cdot$ m $\downarrow$.
(b) 400 N $\measuredangle 60°$, 55.1 N $\cdot$ m $\downarrow$.

3.60 14.5°.

3.62 (a) $\mathbf{F} = 50$ lb $\measuredangle 60°$, $\mathbf{M} = 233$ lb $\cdot$ in. $\downarrow$.
(b) $\mathbf{A} = 77.7$ lb $\leftarrow$; $\mathbf{B} = 77.7$ lb $\rightarrow$.
SI: (a) $\mathbf{F} = 222$ N $\measuredangle 60°$,
$\mathbf{M} = 26.3$ N $\cdot$ m $\downarrow$.
(b) $\mathbf{A} = 346$ N $\leftarrow$; $\mathbf{B} = 346$ N $\rightarrow$.

3.64 $\mathbf{B} = 4\mathbf{P}$; $\mathbf{C} = -3\mathbf{P}$.

3.66 $\mathbf{F} = -(173.2$ N$)\mathbf{j} + (100$ N$)\mathbf{k}$,
$\mathbf{M} = (7.5$ N $\cdot$ m$)\mathbf{i} - (6$ N $\cdot$ m$)\mathbf{j}$
$- (10.39$ N $\cdot$ m$)\mathbf{k}$.

3.68 (a) $\mathbf{F} = (600$ lb$)\mathbf{i} - (300$ lb$)\mathbf{j} + (600$ lb$)\mathbf{k}$,
$\mathbf{M} = (3600$ lb $\cdot$ ft$)\mathbf{i} - (3600$ lb $\cdot$ ft$)\mathbf{k}$.
(b) $\mathbf{F} = (600$ lb$)\mathbf{i} - (300$ lb$)\mathbf{j} + (600$ lb$)\mathbf{k}$,
$\mathbf{M} = -(1200$ lb $\cdot$ ft$)\mathbf{i} - (9600$ lb $\cdot$ ft$)\mathbf{j}$
$- (3600$ lb $\cdot$ ft$)\mathbf{k}$. SI: (a) $\mathbf{F} = (2.67$ kN$)\mathbf{i}$
$- (1.334$ kN$)\mathbf{j} + (2.67$ kN$)\mathbf{k}$,
$\mathbf{M} = (4.88$ kN $\cdot$ m$)\mathbf{i} - (4.88$ kN $\cdot$ m$)\mathbf{k}$.
(b) $\mathbf{F} = (2.67$ kN$)\mathbf{i} - (1.334$ kN$)\mathbf{j}$
$+ (2.67$ kN$)\mathbf{k}$, $\mathbf{M} = -(1.627$ kN $\cdot$ m$)\mathbf{i}$
$- (13.02$ kN $\cdot$ m$)\mathbf{j} - (4.88$ kN $\cdot$ m$)\mathbf{k}$.

3.70 Force-couple system at C.

3.72 $\mathbf{F} = (30.7$ lb$)\mathbf{i} + (15.36$ lb$)\mathbf{j} - (20.5$ lb$)\mathbf{k}$,
$\mathbf{M} = -(277$ lb $\cdot$ in.$)\mathbf{i} + (76.8$ lb $\cdot$ in.$)\mathbf{j}$
$- (287$ lb $\cdot$ in.$)\mathbf{k}$.
SI: $\mathbf{F} = (136.6$ N$)\mathbf{i} + (68.3$ N$)\mathbf{j} - (91.2$ N$)\mathbf{k}$,
$\mathbf{M} = -(31.3$ N $\cdot$ m$)\mathbf{i} + (8.68$ N $\cdot$ m$)\mathbf{j}$
$- (32.4$ N $\cdot$ m$)\mathbf{k}$.

3.74 e.

3.76 (a) 0.6 m. (b) 1 m. (c) 1.8 m.

3.78 12 kips $\downarrow$; 17.33 ft to the right of A.
SI: 53.4 kN $\downarrow$; 5.28 m to the right of A.

3.80 1050 lb; 9.33 ft to the right of A.
SI: 4670 N; 2.84 m to the right of A.

3.82 400 N $\measuredangle 36.9°$ (a) 75 mm to the left
of C.
(b) 56.3 mm below C.

3.84 100 lb $\measuredangle 36.9°$ (a) At A. (b) 8 in. to the
right of B.
(c) 3 in. below C. SI: 445 N $\measuredangle 36.9°$
(a) At A.
(b) 203 mm to the right of B.
(c) 76.2 mm below C.

3.86 (a) $\mathbf{F} = P \measuredangle \theta$, $\mathbf{M} = Pa \sin 2\theta \cos \theta \downarrow$.
(b) 35.3°.

3.88 $\mathbf{R} = -(120 \text{ lb})\mathbf{i} - (60 \text{ lb})\mathbf{j} - (60 \text{ lb})\mathbf{k}$,
$\mathbf{M} = (720 \text{ lb} \cdot \text{ft})\mathbf{j} - (120 \text{ lb} \cdot \text{ft})\mathbf{k}$.
SI: $\mathbf{R} = -(534 \text{ N})\mathbf{i} - (267 \text{ N})\mathbf{j} - (267 \text{ N})\mathbf{k}$,
$\mathbf{M} = (976 \text{ N} \cdot \text{m})\mathbf{j} - (162.7 \text{ N} \cdot \text{m})\mathbf{k}$.

3.90 $\mathbf{R} = -(400 \text{ N})\mathbf{j} - (200 \text{ N})\mathbf{k}$,
$\mathbf{M} = (120 \text{ N} \cdot \text{m})\mathbf{i} + (40 \text{ N} \cdot \text{m})\mathbf{j}$
$- (105 \text{ N} \cdot \text{m})\mathbf{k}$.
(*a*) Tightens. (*b*) Tightens.

3.92 200 N; $y = 63.4$ mm, $z = 200$ mm.

3.94 $10W$; $x = 0.6a$, $z = 0.3a$.

3.96 $R = 10$ lb, $M_1 = 360$ lb $\cdot$ in.; parallel
to the x axis (opposite sense) at
$y = -14$ in., $z = 28$ in.;
pitch = 36 in. SI: $R = 44.5$ N,
$M_1 = 40.7$ N $\cdot$ m; parallel to the x axis
(opposite sense) at $y = -356$ mm,
$z = 711$ mm; pitch = 914 mm.

3.98 $R = 300$ N, $M_1 = -5$ N $\cdot$ m; parallel to
the y axis (opposite sense) at
$x = 125$ mm, $z = 0$;
pitch = -16.67 mm.

3.100 (*a*) $\mathbf{R} = \sqrt{2}P\,\mathbf{i}$; $\mathbf{M} = (Pa/\sqrt{2})(\mathbf{i} + \mathbf{j} - \mathbf{k})$.
(*b*) $R = \sqrt{2}P$, $M_1 = Pa/\sqrt{2}$; parallel to
the x axis at $y = z = \frac{1}{2}a$; pitch = $\frac{1}{2}a$.

3.102 $R = \sqrt{3}P$, $M_1 = -\sqrt{3}Pa$; at origin,
direction of axis, $\theta_x = \theta_y = \theta_z = 54.7°$.

3.104 $\mathbf{F}_B = -(20 \text{ lb})\mathbf{i} + (30 \text{ lb})\mathbf{j} + (60 \text{ lb})\mathbf{k}$,
$\mathbf{F}_{xz} = (80 \text{ lb})\mathbf{i} - (40 \text{ lb})\mathbf{k}$, at $x = 1.50$ in.,
$y = z = 0$.
SI: $\mathbf{F}_B = -(89.0 \text{ N})\mathbf{i} + (133.4 \text{ N})\mathbf{j}$
$+ (267 \text{ N})\mathbf{k}$, $\mathbf{F}_{xz} = (356 \text{ N})\mathbf{i} - (177.9 \text{ N})\mathbf{k}$,
at $x = 38.1$ mm, $y = z = 0$.

3.110 (*a*) 300 lb $\rightarrow$; 10.67 in. above A.
(*b*) 150 lb $\leftarrow$; 50.7 in. above A.
(*c*) System reduces to 4000-lb $\cdot$ in. $\mathbf{\jmath}$
couple.
SI: (*a*) 1334 N $\rightarrow$; 271 mm above A.
(*b*) 667 N $\leftarrow$; 1288 mm above A.
(*c*) System reduces to 425-N $\cdot$ m $\mathbf{\jmath}$
couple.

3.112 (*a*) $\mathbf{F} = 600 \text{ N} \, \diagdown 45°$, $\mathbf{M} = 16.97 \text{ N} \cdot \text{m} \, \mathbf{\jmath}$.
(*b*) $\mathbf{B} = 150 \text{ N} \, \diagdown 45°$, $\mathbf{C} = 450 \text{ N} \, \diagdown 45°$.

3.114 32 kips $\downarrow$, on AB, 8.75 ft from A.
SI: 142.3 kN $\downarrow$, on AB, 2.67 m from A.

3.116 (*a*) $(70 \text{ lb} \cdot \text{ft})\mathbf{i} + (1540 \text{ lb} \cdot \text{ft})\mathbf{j}$
$- (700 \text{ lb} \cdot \text{ft})\mathbf{k}$. (*b*) $+80$ lb $\cdot$ ft.
SI: (*a*) $(94.9 \text{ N} \cdot \text{m})\mathbf{i} + (2088 \text{ N} \cdot \text{m})\mathbf{j}$
$- (949 \text{ N} \cdot \text{m})\mathbf{k}$. (*b*) $+108.5$ N $\cdot$ m.

3.118 (*a*) $R = P$, $M_1 = Qr \sin \theta$; parallel to the
x axis at $x = 0$, $y = (rQ/P) \cos \theta$,
$z = r \sin \theta$. (*b*) $(P/Q)^2 y^2 + z^2 = r^2$

3.120 $\mathbf{F} = -(1 \text{ kN})\mathbf{i} - (9 \text{ kN})\mathbf{j} - (2 \text{ kN})\mathbf{k}$,
$\mathbf{M} = -(12 \text{ kN} \cdot \text{m})\mathbf{i} - (6 \text{ kN} \cdot \text{m})\mathbf{k}$.

CHAPTER 4

4.2 (*a*) 1.672 kN $\uparrow$. (*b*) 4.01 kN $\uparrow$.

4.4 (*a*) $T = (W \cos \theta)/(2 \cos \frac{1}{2}\theta)$.
(*b*) 11.74 lb. SI: (*b*) 52.2 N.

4.6 $\pm 60°$.

4.8 2.4 kN.

4.10 $\mathbf{A} = 346 \text{ N} \, \measuredangle 60.6°$; $\mathbf{B} = 196.2 \text{ N} \, \diagdown 30°$.

4.12 $31.0°$; $\mathbf{A} = 268 \text{ N} \, \measuredangle 70.8°$,
$\mathbf{B} = 171.5 \text{ N} \, \diagdown 59.0°$.

4.14 (*a*) $\mathbf{B} = 360 \text{ lb} \, \nearrow$; $\mathbf{C} = 360 \text{ lb} \, \swarrow$;
$\mathbf{D} = 200 \text{ lb} \, \uparrow$. (*b*) Rollers 2 and 3.
SI: (*a*) $\mathbf{B} = 1601 \text{ N} \, \nearrow$; $\mathbf{C} = 1601 \text{ N} \, \swarrow$;
$\mathbf{D} = 890 \text{ N} \, \uparrow$.

4.16 (*a*) 706 N. (*b*) $\mathbf{A} = 441 \text{ N} \, \uparrow$;
$\mathbf{B} = 353 \text{ N} \rightarrow$.

4.18 $\mathbf{A} = 200 \text{ lb} \, \downarrow$; $\mathbf{B} = 200 \text{ lb} \, \uparrow$.
SI: $\mathbf{A} = 890 \text{ N} \, \downarrow$; $\mathbf{B} = 890 \text{ N} \, \uparrow$.

4.20 (*a*) $\mathbf{B} = 2P/\sqrt{3} \, \measuredangle 60°$;
$\mathbf{C} = 2P/\sqrt{3} \, \diagdown 60°$; $\mathbf{D} = P \, \downarrow$.
(*b*) $\mathbf{B} = \frac{1}{2}P \, \measuredangle 60°$; $\mathbf{C} = \frac{3}{2}P \, \diagdown 60°$;
$\mathbf{D} = \sqrt{3}\,P/2 \, \downarrow$.

4.22 $30 \text{ kN} \leqslant P \leqslant 210 \text{ kN}$.

4.24 $60 \text{ lb} \leqslant P \leqslant 560 \text{ lb}$.
SI: $267 \text{ N} \leqslant P \leqslant 2490 \text{ N}$.

4.26 (*a*) $\mathbf{A} = 5540 \text{ N} \, \measuredangle 87.3°$;
$\mathbf{C} = 683 \text{ N} \, \measuredangle\!\!\!\diagup 67.4°$.
(*b*) $\mathbf{A} = 4900 \text{ N} \, \uparrow$; $\mathbf{M}_A = 1890 \text{ N} \cdot \text{m} \, \mathbf{\jmath}$.
(*c*) $\mathbf{A} = 6740 \text{ N} \, \measuredangle 83.6°$;
$\mathbf{M}_A = 3510 \text{ N} \cdot \text{m} \, \mathbf{\jmath}$;
$\mathbf{C} = 1950 \text{ N} \, \measuredangle\!\!\!\diagup 67.4°$.

4.28 $\mathbf{K} = 147.6 \text{ N} \, \measuredangle 28.3°$;
$\mathbf{L} = 183.8 \text{ N} \, \diagdown 45°$.

4.30 (*a*) $\mathbf{E} = 47.5 \text{ kips} \, \diagdown 67.8°$;
$\mathbf{M}_E = 60 \text{ kip} \cdot \text{ft} \, \mathbf{\jmath}$. (*b*) 25 kips.
SI: (*a*) $\mathbf{E} = 211 \text{ kN} \, \diagdown 67.8°$;
$\mathbf{M}_E = 81.3 \text{ kN} \cdot \text{m} \, \mathbf{\jmath}$. (*b*) 111.2 kN.

4.32 (*a*) 6.77 kips. (*b*) 1.734 kips $\rightarrow$;
4.17 kips $\uparrow$. SI: (*a*) 30.1 kN.
(*b*) 7.71 kN $\rightarrow$; 18.55 kN $\uparrow$.

4.34 $18.4°$.

4.36 (*a*) $T = \frac{1}{2} W/(1 - \tan \theta)$. (*b*) $36.9°$.

4.38 (*a*) $\cos \frac{1}{2}\theta = \frac{1}{4}\left(\dfrac{W}{P} + \sqrt{\dfrac{W^2}{P^2} + 8}\right)$.
(*b*) $65.0°$.

4.40 (1) Completely constrained; det.;
$\mathbf{A} = 481 \text{ N} \, \measuredangle 56.3°$; $\mathbf{B} = 267 \text{ N} \leftarrow$.

(2) Improperly constrained; indet.; no equil. (3) Partially constrained; det.; equil. maintained; $\mathbf{A} = \mathbf{C} = 200$ N $\uparrow$. (4) Completely constrained; det.; $\mathbf{A} = 200$ N $\uparrow$; $\mathbf{B} = 333$ N $\diagdown 36.9°$; $\mathbf{C} = 267$ N $\rightarrow$. (5) Completely constrained; indet.; $\mathbf{A}_y = 200$ N $\uparrow$. (6) Completely constrained; indet.; $\mathbf{A}_x = 267$ N $\rightarrow$; $\mathbf{B}_x = 267$ N $\leftarrow$. (7) Completely constrained; det.; $\mathbf{A} = \mathbf{C} = 200$ N $\uparrow$. (8) Improperly constrained; indet.; no equil.

4.44 (a) $32.6°$. (b) $\mathbf{C} = 211$ lb $\diagup 36.9°$; $\mathbf{D} = 234$ N $\uparrow$.
SI: (b) $\mathbf{C} = 937$ N $\diagup 36.9°$; $\mathbf{D} = 1041$ N $\uparrow$.

4.46 $\mathbf{A} = 534$ N $\measuredangle 69.4°$; $\mathbf{E} = 187.5$ N $\leftarrow$.

4.48 $\mathbf{A} = 7.07$ lb $\rightarrow$; $\mathbf{B} = 40.6$ lb $\diagdown 80.0°$.
SI: $\mathbf{A} = 31.5$ N $\rightarrow$;
$\mathbf{B} = 180.6$ N $\diagdown 80.0°$.

4.50 (a) 1500 N $\diagup 30°$. (b) 593 N $\diagdown 30°$.

4.52 $\mathbf{A} = 269$ lb $\measuredangle 21.8°$;
$\mathbf{E} = 269$ lb $\diagdown 21.8°$.
SI: $\mathbf{A} = 1198$ N $\measuredangle 21.8°$;
$\mathbf{E} = 1198$ N $\diagdown 21.8°$.

4.54 (a) $30°$. (b) $40°$. (c) $60°$.

4.56 $\sin^3 \theta = 2a/L$.

4.58 $23.2°$.

4.60 $T_{BD} = T_{BE} = 11$ kN;
$\mathbf{A} = -(3.6$ kN$)\mathbf{i} + (14$ kN$)\mathbf{j}$.

4.62 $T_{BC} = 560$ lb; $T_{BD} = 360$ lb;
$\mathbf{A} = 900$ lb $\measuredangle 36.9°$ (along AB).
SI: $T_{BC} = 2490$ N; $T_{BD} = 1601$ N;
$\mathbf{A} = 4000$ N $\measuredangle 36.9°$ (along AB).

4.64 $T_{EBF} = 35.4$ kN; $T_{CD} = 24.6$ kN.

4.66 $T_{AD} = T_{BD} = 1.717$ kN; $T_{CD} = 3.10$ kN.

4.68 $T_A = 5$ lb; $T_B = 15$ lb; $T_C = 20$ lb.
SI: $T_A = 22.2$ N; $T_B = 66.7$ N;
$T_C = 89.0$ N.

4.70 20 lb at $x = 20$ in., $z = 15$ in.
SI: 89.0 N at $x = 508$ mm, $z = 381$ mm.

4.72 (a) $T_B = 27$ lb; $T'_B = 9$ lb.
(b) $A_y = 27.4$ lb, $A_z = -4.18$ lb;
$E_y = 5.47$ lb, $E_z = 37.6$ lb; A_x and E_x indeterminate $(A_x + E_x = 0)$.
SI: (a) $T_B = 120.1$ N; $T'_B = 40.0$ N.
(b) $A_y = 121.9$ N, $A_z = -18.59$ N;
$E_y = 24.3$ N, $E_z = 167.3$ N.

4.74 $P = 400$ N; $\mathbf{A} = (192$ N$)\mathbf{j} - (120$ N$)\mathbf{k}$,
$\mathbf{B} = (954$ N$)\mathbf{j} + (320$ N$)\mathbf{k}$.

4.76 (a) 45 lb. (b) 60 lb; 2.5 ft.
SI: (a) 200 N. (b) 267 N; 762 mm.

4.78 $\mathbf{B} = (60$ N$)\mathbf{k}$; $\mathbf{C} = (30$ N$)\mathbf{j} - (16$ N$)\mathbf{k}$.
$\mathbf{D} = -(30$ N$)\mathbf{j} + (4$ N$)\mathbf{k}$.

4.80 $P = 118.9$ N; $\mathbf{A} = (42.9$ N$)\mathbf{i} - (69.9$ N$)\mathbf{k}$;
$\mathbf{B} = (61.1$ N$)\mathbf{i} + (196.2$ N$)\mathbf{j} + (84.7$ N$)\mathbf{k}$.

4.82 $F_{CD} = 19.62$ N;
$\mathbf{B} = -(19.22$ N$)\mathbf{i} + (94.2$ N$)\mathbf{j}$;
$\mathbf{M}_B = -(40.6$ N $\cdot$ m$)\mathbf{i} - (17.30$ N $\cdot$ m$)\mathbf{j}$.

4.84 $T = 846$ lb; $A_x = 0$, $A_y = -83.3$ lb;
$B_x = 400$ lb, $B_y = 250$ lb; A_z and B_z indeterminate $(A_z + B_z = 667$ lb$)$.
SI: $T = 3760$ N; $A_x = 0$, $A_y = -371$ N;
$B_x = 1780$ N, $B_y = 1112$ N; A_z and B_z indeterminate $(A_z + B_z = 2970$ N$)$.

4.86 $F_{CE} = 168.3$ lb;
$\mathbf{B} = -(93.8$ lb$)\mathbf{i} + (125$ lb$)\mathbf{j} - (62.5$ lb$)\mathbf{k}$;
$\mathbf{M}_B = -(1875$ lb $\cdot$ in.$)\mathbf{j}$. SI: $F_{CE} = 749$ N;
$\mathbf{B} = -(417$ N$)\mathbf{i} + (556$ N$)\mathbf{j} - (278$ N$)\mathbf{k}$;
$\mathbf{M}_B = -(212$ N $\cdot$ m$)\mathbf{j}$.

4.88 $T_{BE} = 6.5$ kN, $T_{CE} = 7.16$ kN, $T_{CF} = 0$;
$\mathbf{A} = (12$ kN$)\mathbf{i} - (3$ kN$)\mathbf{k}$.

4.90 $\mathbf{A} = (50$ lb$)\mathbf{j} + (62.5$ lb$)\mathbf{k}$;
$\mathbf{B} = (75$ lb$)\mathbf{i} - (62.5$ lb$)\mathbf{k}$;
$\mathbf{C} = -(75$ lb$)\mathbf{i} - (50$ lb$)\mathbf{j}$.
SI: $\mathbf{A} = (222$ N$)\mathbf{j} + (278$ N$)\mathbf{k}$;
$\mathbf{B} = (334$ N$)\mathbf{i} - (278$ N$)\mathbf{k}$;
$\mathbf{C} = -(334$ N$)\mathbf{i} - (222$ N$)\mathbf{j}$.

4.92 420 lb. SI: 1868 N.

4.94 206 N.

4.96 $\mathbf{D} = (250$ N$)\mathbf{i}$.

4.98 $\mathbf{A} = 166.3$ lb $\measuredangle 30°$;
$\mathbf{B} = 166.3$ lb $\diagup 30°$.
SI: $\mathbf{A} = 740$ N $\measuredangle 30°$;
$\mathbf{B} = 740$ N $\diagup 30°$.

4.100 (a) $\mathbf{A} = 300$ N $\uparrow$; $\mathbf{B} = 680$ N $\rightarrow$;
$\mathbf{C} = 160$ N $\leftarrow$. (b) $\mathbf{A} = 0$;
$\mathbf{B} = 600$ N $\leftarrow$; $\mathbf{C} = 1200$ N $\rightarrow$.

4.102 (a) $0°$ or $180°$. (b) $26.6°$. (c) $73.3°$.
(d) $90°$. (e) 8 in. SI: (e) 203 mm.

4.104 $T = 289$ lb; $\mathbf{A} = 577$ lb $\measuredangle 60°$.
SI: $T = 1284$ N; $\mathbf{A} = 2570$ N $\measuredangle 60°$.

4.106 $T_{BG} = 441$ N, $T_{EF} = 1373$ N;
$\mathbf{A} = (1472$ N$)\mathbf{i} + (441$ N$)\mathbf{j} - (98.1$ N$)\mathbf{k}$.

4.108 (a) $\mathbf{E} = (8$ lb$)\mathbf{i} + (10.67$ lb$)\mathbf{j}$.
(b) $\mathbf{A} = -(8$ lb$)\mathbf{i} + (14.33$ lb$)\mathbf{j} - (7.50$ lb$)\mathbf{k}$;
$\mathbf{B} = (7.50$ lb$)\mathbf{k}$.
SI: (a) $\mathbf{E} = (35.6$ N$)\mathbf{i} + (47.5$ N$)\mathbf{j}$.
(b) $\mathbf{A} = -(35.6$ N$)\mathbf{i} + (63.7$ N$)\mathbf{j} - (33.4$ N$)\mathbf{k}$;
$\mathbf{B} = (33.4$ N$)\mathbf{k}$.

CHAPTER 5

5.2 $\bar{x} = 55.4$ mm, $\bar{y} = 66.2$ mm.

5.4 $\bar{x} = 2.17$ in., $\bar{y} = 2.67$ in.
     SI: $\bar{x} = 55.2$ mm, $\bar{y} = 67.9$ mm.

5.6 $\bar{x} = 1.613$ in., $\bar{y} = 0$.
     SI: $\bar{x} = 41.0$ mm, $\bar{y} = 0$.

5.8 $\bar{x} = 0$, $\bar{y} = 9.26$ mm.

5.10 $\bar{x} = 6$ in., $\bar{y} = 1.633$ in.
     SI: $\bar{x} = 152.4$ mm, $\bar{y} = 41.5$ mm.

5.12 $\bar{x} = 152.5$ mm, $\bar{y} = 17.2$ mm.

5.14 $\bar{x} = \dfrac{4r}{3} \dfrac{\sin^3 \alpha}{2\alpha - \sin 2\alpha}$.

5.16 $\bar{x} = 53.0$ mm, $\bar{y} = 68.6$ mm.

5.18 $\bar{x} = 1.468$ in., $\bar{y} = 0$.
     SI: $\bar{x} = 37.3$ mm, $\bar{y} = 0$.

5.20 $\bar{x} = \dfrac{4}{3\pi} \dfrac{r_2^2 + r_1 r_2 + r_1^2}{r_2 + r_1}$, $\bar{y} = 0$.

5.22 $26.6°$.

5.24   27.3 in. SI: 693 mm.

5.26 $\alpha = 43.9°$; $\theta_{max} = 46.1°$.

5.28 $\bar{x} = 0$, $\bar{y} = 59$ mm.

5.30 $\bar{x} = 0.7424a$; -1.01%.

5.32 $\bar{x} = \frac{1}{2}a$, $\bar{y} = \frac{2}{5}h$.

5.34 $\bar{x} = \bar{y} = \frac{9}{20} a$.

5.42 $\bar{x} = 0.300a$.

5.44 $\bar{x} = 0.549R$, $\bar{y} = 0$.

5.46 $\bar{x} = \bar{y} = \dfrac{(n + 1)^2}{(n + 2)(2n + 1)}$.

5.48 $\bar{x} = 1.820a$, $\bar{y} = 0.303a$.

5.50   (a) 6.48×10^6 mm^3.
     (b) 5.43×10^6 mm^3.

5.52 $4\pi R^2 \sin^2 (\phi/2)$.

5.54 0.655 m^3.

5.56   (a) 867 ft^3. (b) 1093 ft^3.
     SI: (a) 24.6 m^3. (b) 31.0 m^3.

5.58 $V = 1.590$ in^3, $A = 11.04$ in^2.
     SI: $V = 26.1 \times 10^3$ mm^3;
     $A = 7.12 \times 10^3$ mm^2.

5.60 $V = 248 \times 10^3$ mm^3;
     $A = 34.2 \times 10^3$ mm^2.

5.62   (a) $\pi R^2 h$. (b) $\frac{2}{3}\pi R^2 h$. (c) $\frac{1}{2}\pi R^2 h$.
     (d) $\frac{1}{3}\pi R^2 h$. (e) $\frac{1}{6}\pi R^2 h$.

5.66 $R = 9.45$ kN $\downarrow$, 2.57 m to the right of A;
     $A = 4.05$ kN $\uparrow$, $B = 5.40$ kN $\uparrow$.

5.68 $B = 550$ lb $\uparrow$; $M_B = 2300$ lb $\cdot$ in. $\downarrow$.
     SI: $B = 2450$ N $\uparrow$; $M_B = 260$ N $\cdot$ m $\downarrow$.

5.70 $B = 1200$ N $\uparrow$; $M_B = 800$ N $\cdot$ m $\downarrow$.

5.72   (a) $R = \frac{1}{2}(w_A + w_B)L \downarrow$,
     $\bar{X} = (w_A + 2w_B)L/3(w_A + w_B)$.

(b) $A = \frac{1}{6}L(2w_A + w_B) \uparrow$,
$B = \frac{1}{6}L(w_A + 2w_B) \uparrow$.

5.74   (a) 0.500; $B = \frac{1}{4}w_B L \uparrow$, $M_B = 0$.
     (b) 1.000; $B = 0$, $M_B = \frac{1}{6}w_B L^2 \downarrow$.

5.76   (a) $H = 17,970$ lb $\rightarrow$, $V = 60,100$ lb $\uparrow$;
     12.63 ft to the right of A.
     (b) $R = 19,190$ lb $\nearrow 20.6°$.
     SI: (a) $H = 79.9$ kN $\rightarrow$, $V = 267$ kN $\uparrow$;
     3.85 m to the right of A.
     (b) $R = 85.4$ kN $\nearrow 20.6°$.

5.78   8.12 ft. SI: 2.47 m.

5.80   90 mm.

5.82   5.18 ft. SI: 1.580 m.

5.84 $A_x = 1.635$ kN $\rightarrow$, $A_y = 9.81$ kN $\downarrow$;
     $D = 8.18$ kN $\rightarrow$.

5.86   1.491 m.

5.88   2.58 m.

5.90   0.792.

5.92   1.732.

5.94   0.707.

5.96 $-3a/2\pi$.

5.98 $\bar{x} = 0.610$ in., $\bar{y} = -2.34$ in., $\bar{z} = 0$.
     SI: $\bar{x} = 15.50$ mm, $\bar{y} = -59.4$ mm, $\bar{z} = 0$.

5.100   82.2 mm above the base.

5.102 $\bar{x} = 93.9$ mm, $\bar{y} = 3.40$ mm, $\bar{z} = 0$.

5.104 $\bar{x} = 3.15$ in., $\bar{y} = 6.45$ in., $\bar{z} = 0$.
     SI: $\bar{x} = 80.0$ mm, $\bar{y} = 163.8$ mm, $\bar{z} = 0$.

5.106 $\bar{x} = -\frac{1}{2}r$, $\bar{y} = \bar{z} = 0$.

5.110 $\bar{x} = \frac{5}{8}a$, $\bar{y} = 20a/21\pi$, $\bar{z} = 0$.

5.114   0.125 m below center of tank.

5.116 $\bar{x} = \frac{2}{3}a$, $\bar{y} = \frac{2}{3}b$, $\bar{z} = \frac{2}{3}h$.

5.118 $\bar{x} = 0$, $\bar{y} = \frac{3}{8}h$, $\bar{z} = -\frac{1}{2}a$.

5.120   (a) 20.5×10^3 mm^3.
     (b) 3.07×10^3 mm^3.

5.122 $\bar{x} = 6$ in., $\bar{y} = 1$ in., $\bar{z} = 2.5$ in.
     SI: $\bar{x} = 152.4$ mm, $\bar{y} = 25.4$ mm,
     $\bar{z} = 63.5$ mm.

5.124 $A = 27.0$ kN $\searcharrow 51.6°$; $B = 15.01$ kN $\leftarrow$.

5.126 $A = 81.3$ lb $\uparrow$, $B = 519$ lb $\uparrow$.
     SI: $A = 362$ N $\uparrow$, $B = 2310$ N $\uparrow$.

5.128 $\bar{x} = 6.74$ in., $\bar{y} = -2.14$ in.,
     $\bar{z} = 1.193$ in. SI: $\bar{x} = 171.2$ mm,
     $\bar{y} = -54.4$ mm, $\bar{z} = 30.3$ mm.

5.130   329 mm.

CHAPTER 6

6.2 $F_{AB} = 1600$ lb C; $F_{AC} = 2000$ lb T;
     $F_{BC} = 1709$ lb T. SI: $F_{AB} = 7.12$ kN C;
     $F_{AC} = 8.90$ kN T; $F_{BC} = 7.60$ kN T.

6.4 $F_{AB} = 3900$ N T; $F_{AC} = 4500$ N C;
$F_{BC} = 3600$ N C.

6.6 $F_{AB} = F_{DE} = F_{BG} = F_{DI} = 0$;
$F_{AF} = F_{CH} = F_{EJ} = 400$ N C;
$F_{BC} = F_{CD} = 800$ N C;
$F_{BF} = F_{DJ} = 849$ N C;
$F_{BH} = F_{DH} = 283$ N T;
$F_{FG} = F_{GH} = F_{HI} = F_{IJ} = 600$ N T.

6.8 $F_{AB} = 1800$ lb T; $F_{AC} = 0$;
$F_{BC} = 2250$ lb C; $F_{BD} = 1350$ lb T;
$F_{CD} = 3600$ lb T; $F_{CE} = 1350$ lb C;
$F_{DE} = 4500$ lb C; $F_{DF} = 4050$ lb T.
SI: $F_{AB} = 8.01$ kN T; $F_{AC} = 0$;
$F_{BC} = 10.01$ kN C; $F_{BD} = 6.01$ kN T;
$F_{CD} = 16.01$ kN T; $F_{CE} = 6.01$ kN C;
$F_{DE} = 20.02$ kN C; $F_{DF} = 18.01$ kN T.

6.10 $F_{AB} = F_{DE} = 8$ kN C;
$F_{AF} = F_{FG} = F_{GH} = F_{HE} = 6.93$ kN T;
$F_{BC} = F_{CD} = F_{BG} = F_{DG} = 4$ kN C;
$F_{BF} = F_{DH} = F_{CG} = 4$ kN T.

6.12 $F_{AB} = 26$ kips T; $F_{AD} = 20$ kips T;
$F_{AE} = 26$ kips C; $F_{BC} = 34.7$ kips T;
$F_{BE} = 10$ kips T; $F_{BF} = 10.41$ kips C;
$F_{CF} = 6.67$ kips T; $F_{DE} = 0$;
$F_{EF} = 24$ kips C. SI: $F_{AB} = 115.7$ kN T;
$F_{AD} = 89.0$ kN T; $F_{AE} = 115.7$ kN C;
$F_{BC} = 154.4$ kN T; $F_{BE} = 44.5$ kN T;
$F_{BF} = 46.3$ kN C; $F_{CF} = 29.7$ kN T;
$F_{DE} = 0$; $F_{EF} = 106.8$ kN C.

6.14 $BC, CD, IJ, IL, KL, LM, MN$.

6.16 $AG, AH, BD, DH, BE, EJ, FJ$.

6.18 $F_{AB} = F_{AC} = F_{AD} = P/\sqrt{6}$ comp.;
$F_{BC} = F_{CD} = F_{DB} = P/3\sqrt{6}$ ten.

6.20 (a) $\mathbf{A} = (192$ lb$)\mathbf{i} - (144$ lb$)\mathbf{k}$;
$\mathbf{B} = -(384$ lb$)\mathbf{i} + (576$ lb$)\mathbf{j}$;
$\mathbf{C} = -(576$ lb$)\mathbf{j}$.
(b) $F_{AB} = F_{BC} = 240$ lb C;
$F_{BE} = 576$ lb C; $F_{CE} = 624$ lb T;
other members $= 0$.
SI: $\mathbf{A} = (854$ N$)\mathbf{i} - (641$ N$)\mathbf{k}$;
$\mathbf{B} = -(1708$ N$)\mathbf{i} + (2560$ N$)\mathbf{j}$;
$\mathbf{C} = -(2560$ N$)\mathbf{j}$.
(b) $F_{AB} = F_{BC} = 1068$ N C;
$F_{BE} = 2560$ N C; $F_{CE} = 2780$ N T;
other members $= 0$.

6.22 $F_{DF} = 60$ kips C; $F_{DG} = 15$ kips C.
SI: $F_{DF} = 267$ kN C; $F_{DG} = 66.7$ kN C.

6.24 $F_{CD} = 64.3$ kN T; $F_{CE} = 92.1$ kN C.

6.26 60 kN T.

6.28 $F_{EG} = 45$ kN T; $F_{EF} = 55$ kN C;
$F_{DF} = 45$ kN C.

6.30 $F_{DF} = 8.25$ kips T; $F_{DE} = 3$ kips C;
$F_{CE} = 8$ kips C. SI: $F_{DF} = 36.7$ kN T;
$F_{DE} = 13.34$ kN C; $F_{CE} = 35.6$ kN C.

6.32 $F_{FH} = 2.5$ kips T; $F_{DH} = 18$ kips T.
SI: $F_{FH} = 11.12$ kN T; $F_{DH} = 80.1$ kN T.

6.34 $F_{AB} = 5P/6$ ten.; $F_{KL} = 7P/6$ ten.

6.36 22.5 kN C.

6.38 $F_{AB} = 0$; $F_{EJ} = \frac{2}{3}P$ comp.

6.40 $F_{BD} = 21.3$ kN C; $F_{CE} = 10.67$ kN T.
$F_{CD} = 13.33$ kN T.

6.42 $F_{BE} = 16$ kips C; $F_{CG} = 10$ kips C;
$F_{BG} = 10$ kips T. SI: $F_{BE} = 71.2$ kN C;
$F_{CG} = 44.5$ kN C; $F_{BG} = 44.5$ kN T.

6.44 (a) Completely constrained; det.
(b) Completely constrained; indet.
(c) Completely constrained; det.
(d) Improperly constrained.

6.46 (a) Completely constrained; det.
(b) Improperly constrained.
(c) Completely constrained; indet.
(d) Completely constrained; det.

6.48 $F_{BD} = 300$ lb T; $\mathbf{C}_x = 150$ lb $\leftarrow$,
$\mathbf{C}_y = 180$ lb $\uparrow$. SI: $F_{BD} = 1334$ N T;
$\mathbf{C}_x = 667$ N $\leftarrow$, $\mathbf{C}_y = 801$ N $\uparrow$.

6.50 $\mathbf{A}_x = 560$ N $\rightarrow$, $\mathbf{A}_y = 160$ N $\uparrow$.
$\mathbf{B}_x = 560$ N $\leftarrow$, $\mathbf{B}_y = 280$ N $\downarrow$;
$\mathbf{C} = 160$ N $\downarrow$; $\mathbf{D} = 280$ N $\uparrow$.

6.52 1339 lb C. SI: 5.96 kN C.

6.54 (a) $\mathbf{B}_x = 500$ N $\rightarrow$, $\mathbf{B}_y = 200$ N $\downarrow$.
(b) $\mathbf{B} = 200$ N $\uparrow$.

6.56 $\mathbf{E} = 80$ lb $\downarrow$; $\mathbf{F}_x = 90$ lb $\rightarrow$,
$\mathbf{F}_y = 120$ lb $\uparrow$; $\mathbf{G}_x = 90$ lb $\leftarrow$,
$\mathbf{G}_y = 40$ lb $\downarrow$. SI: $\mathbf{E} = 356$ N $\downarrow$;
$\mathbf{F}_x = 400$ N $\rightarrow$, $\mathbf{F}_y = 534$ N $\uparrow$;
$\mathbf{G}_x = 400$ N $\leftarrow$, $\mathbf{G}_y = 177.9$ N $\downarrow$.

6.58 (a) 683 N. (b) $\mathbf{D} = 1114$ N $\measuredangle 1.6°$.

6.60 $\mathbf{A} = 0$; $\mathbf{M}_A = 2250$ N $\cdot$ m $\downarrow$;
$\mathbf{D} = 2500$ N $\uparrow$.

6.62 (a) 52.7°. (b) $\mathbf{A} = 600$ lb $\uparrow$;
$\mathbf{B} = 457$ lb $\leftarrow$; $\mathbf{D} = 754$ lb $\diagdown 52.7°$.
SI: (b) $\mathbf{A} = 2.67$ kN $\uparrow$; $\mathbf{B} = 2.03$ kN $\leftarrow$;
$\mathbf{D} = 3.35$ kN $\diagdown 52.7°$.

6.64 (a) At each wheel: $\mathbf{A} = 117.5$ kN $\uparrow$;
$\mathbf{B} = 176.9$ kN $\uparrow$. (b) $\mathbf{C} = 8.28$ kN $\rightarrow$;
$\mathbf{D}_x = 8.28$ kN $\leftarrow$, $\mathbf{D}_y = 256$ kN $\downarrow$.

6.66 At each wheel: (a) $\mathbf{A} = 1268$ lb $\uparrow$;
$\mathbf{B} = 713$ lb $\uparrow$; $\mathbf{C} = 994$ lb $\uparrow$.

(b) $\mathbf{B} = 73$ lb $\uparrow$; $\mathbf{C} = 34$ lb $\uparrow$.
SI: *At each wheel:* (a) $\mathbf{A} = 5640$ N $\uparrow$;
$\mathbf{B} = 3170$ N $\uparrow$; $\mathbf{C} = 4420$ N $\uparrow$.
(b) $\mathbf{B} = 324$ N $\uparrow$; $\mathbf{C} = 151$ N $\uparrow$.

6.68 $\mathbf{C}_x = 200$ lb $\leftarrow$, $\mathbf{C}_y = 150$ lb $\uparrow$,
$\mathbf{D} = 300$ lb $\downarrow$, $\mathbf{E}_x = 400$ lb $\rightarrow$,
$\mathbf{E}_y = 300$ lb $\uparrow$, $\mathbf{F}_x = 200$ lb $\leftarrow$,
$\mathbf{F}_y = 150$ lb $\downarrow$. SI: $\mathbf{C}_x = 890$ N $\leftarrow$,
$\mathbf{C}_y = 667$ N $\uparrow$, $\mathbf{D} = 1334$ N $\downarrow$,
$\mathbf{E}_x = 1779$ N $\rightarrow$, $\mathbf{E}_y = 1334$ N $\uparrow$,
$\mathbf{F}_x = 890$ N $\leftarrow$, $\mathbf{F}_y = 667$ N $\downarrow$.

6.70 $a \geqslant 333$ mm.

6.72 $F_{AF} = \frac{1}{4}P$ ten.; $F_{BG} = P/\sqrt{2}$ comp.;
$F_{GD} = P/\sqrt{2}$ ten.; $F_{EH} = \frac{1}{4}P$ comp.

6.74 (a) $\mathbf{C}_x = 12$ kN $\leftarrow$, $\mathbf{C}_y = 6$ kN $\uparrow$.
(b) $\mathbf{B}_x = 12$ kN $\leftarrow$, $\mathbf{B}_y = 6$ kN $\uparrow$.

6.78 (a) Rigid; $\mathbf{A} = 40$ kN $\rightarrow$;
$\mathbf{B} = 56.6$ kN $\searrow 45°$. (b) Not rigid.

6.80 $F_{AB} = F_{BC} = 1600$ lb T; $F_{AE} = 2000$ lb T;
$F_{BF} = 0$; $F_{DE} = 3200$ lb C;
$F_{CF} = 1200$ lb C; $F_{CG} = 2000$ lb T.
SI: $F_{AB} = F_{BC} = 7.12$ kN T;
$F_{AE} = 8.90$ kN T; $F_{BF} = 0$;
$F_{DE} = 14.23$ kN C; $F_{CF} = 5.34$ kN C;
$F_{CG} = 8.90$ kN T.

6.82 $\mathbf{C} = 632$ N $\measuredangle 71.6°$;
$\mathbf{D} = 250$ N $\searrow 36.9°$; $F_{AE} = 750$ N C;
$F_{BF} = 500$ N T.

6.84 $\mathbf{C} = 750$ N $\searrow 36.9°$;
$\mathbf{D} = 671$ N $\measuredangle 26.6°$; $F_{AE} = 1500$ N C;
$F_{BF} = 2250$ N T.

6.86 $\mathbf{C}_x = 2500$ N $\rightarrow$, $\mathbf{C}_y = 1000$ N $\uparrow$;
$\mathbf{D} = 2500$ N $\leftarrow$; $\mathbf{E} = 4500$ N $\leftarrow$;
$\mathbf{F}_x = 4500$ N $\rightarrow$, $\mathbf{F}_y = 1000$ N $\downarrow$.

6.88 $\mathbf{A} = 2.5$ kips $\uparrow$; $\mathbf{B} = 1.5$ kips $\uparrow$;
$\mathbf{C} = 5$ kips $\uparrow$; $\mathbf{D} = 3$ kips $\uparrow$.
SI: $\mathbf{A} = 11.12$ kN $\uparrow$; $\mathbf{B} = 6.67$ kN $\uparrow$;
$\mathbf{C} = 22.2$ kN $\uparrow$; $\mathbf{D} = 13.34$ kN $\uparrow$.

6.90 $\mathbf{A} = P/15 \uparrow$; $\mathbf{D} = 2P/15 \uparrow$; $\mathbf{E} = 8P/15 \uparrow$;
$\mathbf{H} = 4P/15 \uparrow$.

6.92 $\mathbf{D} = 400$ N $\downarrow$; $\mathbf{E} = 800$ N $\uparrow$; $\mathbf{F} = 200$ N $\uparrow$.

6.94 (a) 415 N. (b) 459 N T.

6.96 $\mathbf{C} = 150$ lb $\rightarrow$; $\mathbf{D} = 250$ lb $\nearrow 53.1°$.
SI: $\mathbf{C} = 667$ N $\rightarrow$; $\mathbf{D} = 1112$ N $\nearrow 53.1°$

6.98 14,800 lb. SI: 65.8 kN.

6.100 (a) 140 lb $\cdot$ ft $\rangle$. (b) 60 lb $\cdot$ ft $\rangle$.
SI: (a) 189.8 N $\cdot$ m $\rangle$. (b) 81.3 N $\cdot$ m $\rangle$.

6.102 (a) $\mathbf{M}_C = 50.4$ N $\cdot$ m $\rangle$.
(b) $\mathbf{C}_x = 217$ N $\leftarrow$, $\mathbf{C}_y = 86.9$ N $\uparrow$.

6.104 $\mathbf{A}_x = 4.4$ kN $\leftarrow$, $\mathbf{A}_y = 19.8$ kN $\downarrow$;
$\mathbf{C}_x = 4.9$ kN $\rightarrow$, $\mathbf{C}_y = 19.8$ kN $\uparrow$.

6.106 1689 lb. SI: 7.51 kN.

6.108 (a) 128.0 N. (b) 510 N.

6.110 88.3 kN C.

6.112 $F_{AB} = 18.97$ kips C; $F_{CD} = 4.27$ kips T;
$F_{EF} = 9.61$ kips C. SI: $F_{AB} = 84.4$ kN C;
$F_{CD} = 18.99$ kN T; $F_{EF} = 42.7$ kN C.

6.114 (a) 12.5 kN T. (b) 10 kN $\leftarrow$, 20 kN $\downarrow$.

6.116 $T_1 = 350$ lb; $T_2 = 4150$ lb.
SI: $T_1 = 1.557$ kN; $T_2 = 18.46$ kN.

6.118 (a) $-(21$ N $\cdot$ m$)\mathbf{i}$.
(b) $\mathbf{G} = 0$, $\mathbf{M}_G = (2.70$ N $\cdot$ m$)\mathbf{i}$;
$\mathbf{H} = 0$, $\mathbf{M}_H = -(11.70$ N $\cdot$ m$)\mathbf{i}$.

6.120 (a) 1.5. (b) $4M_A \rangle$.

6.122 (a) $M_A = 43.3$ N $\cdot$ m.
(b) $\mathbf{B} = -(250$ N$)\mathbf{k}$; $\mathbf{D} = (350$ N$)\mathbf{k}$;
$\mathbf{E} = -(100$ N$)\mathbf{k}$.

6.124 5.29 oz. SI: 150.0 g (mass).

6.126 $\mathbf{F} = 484$ lb $\searrow 82.9°$; $F_{AE} = 200$ lb C;
$F_{BD} = 400$ lb T.
SI: $\mathbf{F} = 2150$ N $\searrow 82.9°$; $F_{AE} = 890$ N C;
$F_{BD} = 1779$ N T.

6.128 $F_{CD} = 30$ kN T; $F_{CG} = 0$.

6.130 (a) $\mathbf{B} = 2500$ N $\uparrow$; $\mathbf{C} = 5000$ N $\downarrow$;
$\mathbf{D} = 2500$ N $\uparrow$. (b) 1600 N. (c) 160 mm.

6.132 (a) $\mathbf{A} = 7.85$ kN $\uparrow$, $\mathbf{M}_A = 15.7$ kN $\cdot$ m $\rangle$;
$\mathbf{D} = 22.20$ kN $\nearrow 45°$.
(b) $\mathbf{A} = 3.92$ kN $\uparrow$, $\mathbf{M}_A = 8.44$ kN $\cdot$ m $\rangle$;
$\mathbf{D} = 11.10$ kN $\nearrow 45°$.
(c) $\mathbf{A} = 3.92$ kN $\uparrow$;
$\mathbf{M}_A = 8.44$ kN $\cdot$ m $\rangle$;
$\mathbf{D} = 18.95$ kN $\nearrow 45°$.
(d) $\mathbf{A}_x = 3.92$ kN $\rightarrow$, $\mathbf{A}_y = 3.92$ kN $\uparrow$;
$\mathbf{M}_A = 2.35$ kN $\cdot$ m $\rangle$;
$\mathbf{D} = 11.10$ kN $\nearrow 45°$.

6.134 $F_{DE} = 509$ lb C; $F_{FE} = 7270$ lb T.
SI: $F_{DE} = 2.26$ kN C; $F_{FE} = 32.3$ kN T.

6.136 $\mathbf{B} = 17.07$ lb $\leftarrow$; $\mathbf{E}_x = 36.3$ lb $\rightarrow$,
$\mathbf{E}_y = 32.0$ lb $\downarrow$; $\mathbf{H}_x = 19.20$ lb $\leftarrow$,
$\mathbf{H}_y = 32.0$ lb $\uparrow$. SI: $\mathbf{B} = 75.9$ N $\leftarrow$;
$\mathbf{E}_x = 161.5$ N $\rightarrow$, $\mathbf{E}_y = 142.3$ N $\downarrow$;
$\mathbf{H}_x = 85.4$ N $\leftarrow$, $\mathbf{H}_y = 142.3$ N $\uparrow$.

CHAPTER 7

7.2 (On JG) $\mathbf{F} = 90$ lb $\rightarrow$; $\mathbf{V} = 40$ lb $\uparrow$;
$\mathbf{M} = 60$ lb $\cdot$ ft $\rangle$. SI: $\mathbf{F} = 400$ N $\rightarrow$;
$\mathbf{V} = 177.9$ N $\uparrow$; $\mathbf{M} = 81.3$ N $\cdot$ m $\rangle$.

7.4 (On AJ) $\mathbf{F} = 16.99$ kN $\nearrow 26.6°$;
$\mathbf{V} = 7.15$ kN $\searrow 63.4°$;
$\mathbf{M} = 12.5$ kN $\cdot$ m $\rangle$.

7.6 225 mm.

7.8 (On KD) $\mathbf{F} = 0$; $\mathbf{V} = 30$ lb $\downarrow$;
$\mathbf{M} = 300$ lb $\cdot$ in. $\rangle$. SI: $\mathbf{F} = 0$;
$\mathbf{V} = 133.4$ N $\downarrow$; $\mathbf{M} = 33.9$ N $\cdot$ m $\rangle$.

7.10 (On AJ) $\mathbf{F} = 500$ N $\downarrow$; $\mathbf{V} = 187.5$ N $\leftarrow$;
$\mathbf{M} = 18.75$ N $\cdot$ m $\rangle$.

7.12 (On JC) 54.5 lb $\cdot$ in. $\rangle$. SI: 6.16 N $\cdot$ m $\rangle$.

7.14 (On JC) $\mathbf{F} = (\frac{1}{2}W - W\theta/\pi)\cos\theta \searrow$;
$\mathbf{V} = (\frac{1}{2}W - W\theta/\pi)\sin\theta \nearrow$;
$\mathbf{M} = \frac{1}{2}Wr(1 - \cos\theta)$
$- (Wr/\pi)(\sin\theta - \theta\cos\theta)\rangle$.

7.16 $0.1786W$; $\theta = 49.3°$.

7.18 $M_B = Pab/L$.

7.20 $M = Pa$ between B and C.

7.22 $M = \frac{1}{2}wa^2$ between B and C.

7.24 $M_C = +800$ lb $\cdot$ ft.
SI: $M_C = +1085$ N $\cdot$ m.

7.26 $M_D = +28$ kN $\cdot$ m.

7.28 24 kN.

7.30 $M_C = -4$ kN $\cdot$ m.

7.32 Just to the right of D:
$M = +1440$ lb $\cdot$ in.
SI: $M = +162.7$ N $\cdot$ m.

7.34 $M_C = +2250$ N $\cdot$ m.

7.36 $0.207L$.

7.46 $M_D = +48$ kN $\cdot$ m.

7.48 $M_D = -120$ lb $\cdot$ in.
SI: $M_D = -13.56$ N $\cdot$ m.

7.50 $M = +2000$ lb $\cdot$ ft, 10 ft to the right
of A. SI: $M = +2.71$ kN $\cdot$ m, 3.05 m
to the right of A.

7.52 $M = -2$ kN $\cdot$ m at C;
$M = +0.667$ kN $\cdot$ m, 1.333 m to the left
of B.

7.54 $V = w(\frac{1}{2}L - x)$; $M = \frac{1}{2}w(Lx - x^2)$.

7.56 $V = w_0(L/\pi)\cos(\pi x/L)$;
$M = w_0(L/\pi)^2\sin(\pi x/L)$;
$M_{max} = w_0(L/\pi)^2$, at $x = \frac{1}{2}L$.

7.58 $V_D = +200$ lb; $M_C = -200$ lb $\cdot$ ft.
SI: $V_D = +890$ N; $M_C = -271$ N $\cdot$ m.

7.60 (a) $V = -m$ and $M = 0$ everywhere.
(b) $V = 0$ everywhere; $M_A = -mL$,
$M_B = 0$.

7.62 (a) $\mathbf{E}_x = 1500$ lb $\rightarrow$, $\mathbf{E}_y = 1000$ lb $\uparrow$.
(b) 1803 lb. SI: (a) $\mathbf{E}_x = 6.67$ kN $\rightarrow$,
$\mathbf{E}_y = 4.45$ kN $\uparrow$. (b) 8.02 kN.

7.64 $\mathbf{E}_x = 1600$ N $\rightarrow$, $\mathbf{E}_y = 1100$ N $\uparrow$.

7.66 1 m.

7.68 3575 ft. SI: 1090 m.

7.70 0.833 m.

7.72 (a) $\sqrt{3L\Delta/8}$. (b) 12.25 ft.
SI: (b) 3.73 m.

7.74 8 m to left of B; 14.72 kN.

7.76 $h = 0.273$ ft; $\theta_A = 39.3°$, $\theta_C = 34.3°$.
SI: $h = 83.1$ mm.

7.80 $y = h(1 - \cos\pi x/L)$;
$T_{max} = (w_0L/\pi)[1 + (L/h\pi)^2]^{1/2}$;
$T_0 = w_0L^2/h\pi^2$.

7.84 89.0 ft; 69.0 lb. SI: 27.1 m; 307 N.

7.86 (a) 102.8 m. (b) 10.4 m. (c) 977 N.

7.88 (a) 11.27 m. (b) 29.5 kg/m.

7.90 25.1 ft; 27.7 lb. SI: 7.66 m; 123.4 N.

7.92 (a) 40.6 N. (b) 2.68 m. (c) 106.1 N.

7.94 (a) $L = 1.326T_m/w$. (b) 7.53 mi.
SI: (b) 12.12 km.

7.96 0.1408.

7.98 (a) $1.029(2\pi r)$. (b) $0.657wr$.

7.100 $M_A = -240$ lb $\cdot$ in.; $M_D = +720$ lb $\cdot$ in.
SI: $M_A = -27.1$ N $\cdot$ m;
$M_D = +81.3$ N $\cdot$ m.

7.102 671 N.

7.104 (a) $\mathbf{A}_x = 267$ lb $\leftarrow$, $\mathbf{A}_y = 100$ lb $\uparrow$;
$\mathbf{B}_x = 228$ lb $\rightarrow$, $\mathbf{B}_y = 171$ lb $\uparrow$.
(b) $T_{ACB} = 285$ lb. (c) $T_{DE} = 48.5$ lb.
SI: (a) $\mathbf{A}_x = 1186$ N $\leftarrow$, $\mathbf{A}_y = 445$ N $\uparrow$;
$\mathbf{B}_x = 1013$ N $\rightarrow$, $\mathbf{B}_y = 760$ N $\uparrow$.
(b) $T_{ABC} = 1267$ N. (c) $T_{DE} = 216$ N.

7.106 (On left section) $\mathbf{F} = 1028$ N $\nearrow$;
$\mathbf{V} = 79.1$ N $\nwarrow$; $\mathbf{M} = 30$ N $\cdot$ m $\rangle$.

7.108 (On AJ) $\mathbf{F} = 10.40$ kN $\nwarrow$;
$\mathbf{V} = 0.928$ kN $\nearrow$; $\mathbf{M} = 0.5$ kN $\cdot$ m $\rangle$.

7.110 (On JB) (a) $\mathbf{M}_J = \frac{1}{2}Wr \rangle$.
(b) $\mathbf{M}_{max} = 0.579\,Wr \rangle$, $\theta = 116.3°$.

CHAPTER 8

8.2 (a) 930 N $\leftarrow$. (b) 151.2 N $\leftarrow$.

8.4 (a) 23.7 lb $\measuredangle 36.3°$. (b) 9.47 lb $\measuredangle 13.7°$.
SI: (a) 105.4 N $\measuredangle 36.3°$.
(b) 42.1 N $\measuredangle 13.7°$.

8.6 (a) A and B move.
(b) $\mathbf{F}_A = 8.53$ N $\nwarrow$; $\mathbf{F}_B = 9.48$ N $\nwarrow$.

8.8 5.84 lb. SI: 26.0 N.

8.10 (a) 36 lb. (b) 40 in.
SI: (a) 160.1 N. (b) 1.016 m.

8.12 (a) 98.9 N. (b) Slides.

8.14 45.5 lb $\leftarrow$. SI: 202 N $\leftarrow$.

8.16 (*a*) 54.5 N · m. (*b*) 46.2 N · m.
8.18 36.6°.
8.20 (*b*) Until $b = 2.26$ ft.
SI: Until $b = 0.689$ m.
8.22 371 N (motion impends at *C*).
8.24 0.955 lb. SI: 4.25 N.
8.26 0.1940*L*.
8.28 0.80.
8.30 $\mu_A = 0.222$; $\mu_C = 0.247$.
8.32 (*a*) 8 lb. (*b*) 12 lb.
SI: (*a*) 35.6 N.
(*b*) 53.4 N.
8.34 0.378.
8.36 (*a*) $P = W \tan \phi / \cos \theta$.
(*b*) $M = \frac{1}{2} Wr \sin 2\phi / \cos \theta$.
8.38 60.2 N · m ↻.
8.40 0.520 lb ↗; *A* rolls.
SI: 2.31 N ↗; *A* rolls.
8.42 $\theta \leqslant 45°$.
8.44 54.4°.
8.46 1454 N.
8.48 474 lb. SI: 2.11 kN.
8.50 1793 N.
8.52 80.9°.
8.54 (*a*) **P** = 41.7 kN ←. (*b*) 30 kN →.
8.56 (*a*) 58.5 lb ←. (*b*) Machine will not move. SI: (*a*) 260 N ←.
8.58 832 lb · ft. SI: 1128 N · m.
8.60 16.86 N · m.
8.62 (*a*) 21.3 lb · ft ↓. (*b*) 3.58 lb · ft ↻.
SI: (*a*) 28.9 N · m ↓. (*b*) 4.85 N · m ↻.
8.64 3.60 N · m.
8.66 14.60 lb · in. SI: 1.650 N · m.
8.68 (*a*) 0.24. (*b*) 55.6 lb. SI: (*b*) 247 N.
8.70 0.10.
8.72 $T_{AB} = 310$ N; $T_{CD} = 290$ N;
$T_{EF} = 272$ N.
8.74 5.0 in. SI: 127 mm.
8.76 3.47 lb. SI: 15.44 N.
8.78 577 N.
8.82 2.60 lb. SI: 11.59 N.
8.84 58.9 N.
8.86 12 in. SI: 305 mm.
8.88 (*a*) 0.33. (*b*) 2.66 turns.
8.90 0.116
8.92 46.2 lb · ft. SI: 62.6 N · m.
8.94 88.8 N · m ↻.
8.96 109.7 kg.
8.98 13.17 ft. SI: 4.01 m.
8.100 0.253.

8.102 531 N · m.
8.106 0.27.
8.108 71.5 lb · ft. SI: 97.0 N · m.
8.110 (*a*) 235 N. (*b*) 208 N.
8.112 (*a*) 26.0 N. (*b*) Motion impends at *B*.
8.114 412 lb. SI: 1832 N.
8.116 (*a*) $P = W (\mu^2 \cos^2 \theta - \sin^2 \theta)^{\frac{1}{2}}$.
(*b*) $\tan \beta = (\mu^2 \cot^2 \theta - 1)^{\frac{1}{2}}$, where β is the angle between the direction of motion and the line of greatest slope of the plane. (*c*) 6.52 lb, $\beta = 26.7°$.
SI: (*c*) 29.0 N, $\beta = 26.7°$.

CHAPTER 9

9.2 $2a^3 b / 7$.
9.4 $3a^3 b / 35$.
9.6 $2ab^3 / 15$.
9.8 $3ab^3 / 35$.
9.10 $ab^3 / 9$; $b \sqrt{5/27}$.
9.12 $3a^3 b / 11$; $a \sqrt{5/11}$.
9.14 (*a*) $4a^4 / 3$; $a \sqrt{2/3}$.
(*b*) $17a^4 / 6$; $a \sqrt{17/12}$.
9.16 (*b*) 10.6%; 2.99%; 0.125%.
9.18 $a^3 h / 5$.
9.20 614×10^3 mm⁴; 19.01 mm.
9.22 1281 in⁴; 5.98 in.
SI: 533×10^6 mm⁴; 151.8 mm.
9.24 25 in²; 400 in⁴.
SI: 16.13×10^3 mm²; 166.5×10^6 mm⁴.
9.26 $\bar{I}_x = 1.5 \times 10^6$ mm⁴, $\bar{I}_y = 3 \times 10^6$ mm⁴.
9.28 60.7×10^6 mm⁴.
9.30 (*a*) 159.5×10^6 mm⁴.
(*b*) 31.8×10^6 mm⁴.
9.32 $\bar{I}_x = 313$ in⁴, $\bar{I}_y = 371$ in⁴;
$\bar{k}_x = 3.80$ in., $\bar{k}_y = 4.13$ in.
SI: $\bar{I}_x = 130.3 \times 10^6$ mm⁴,
$\bar{I}_y = 154.4 \times 10^6$ mm⁴;
$\bar{k}_x = 96.5$ mm, $\bar{k}_y = 104.9$ mm.
9.34 4470 in⁴. SI: 1862×10^6 mm⁴.
9.36 $\frac{1}{2} h$.
9.38 $(a + 3b)h / (2a + 4b)$.
9.40 (*a*) $\frac{1}{2} \gamma \pi r^2 h$. (*b*) $\frac{1}{4} \gamma \pi r^4$.
9.44 $b^2 h^2 / 8$.
9.46 $a^4 / 8$.
9.48 -800×10^3 mm⁴.
9.50 -70 in⁴. SI: -29.1×10^6 mm⁴.
9.52 1.627×10^6 mm⁴, 0.881×10^6 mm⁴;
-0.954×10^6 mm⁴.

9.54 $\overline{I}_u = 0.267a^4$, $\overline{I}_v = 0.567a^4$,
$\overline{P}_{uv} = -0.200a^4$.

9.56 $-25.7°$; 2.28×10^6 mm^4,
0.230×10^6 mm^4.

9.58 $+23.2°$; 28.1 in^4, 5.14 in^4.
SI: $+23.2°$; 11.68×10^6 mm^4,
2.14×10^6 mm^4.

9.62 1.627×10^6 mm^4, 0.881×10^6 mm^4;
-0.954×10^6 mm^4.

9.66 $-25.7°$; 2.28×10^6 mm^4,
0.230×10^6 mm^4.

9.72 (a) $\frac{1}{4}ma^2$, $\frac{1}{4}mb^2$. (b) $\frac{1}{4}m(a^2 + b^2)$.

9.74 (a) $I_{AA'} = mb^2/24$; $I_{BB'} = mh^2/18$.
(b) $I_{CC'} = m(3b^2 + 4h^2)/72$.

9.76 $m(3a^2 + L^2)/12$.

9.78 $\frac{1}{3}ma^2$; $a/\sqrt{3}$.

9.80 $5ma^2/18$

9.82 $m(2b^2 + h^2)/10$.

9.84 $2mr^2/3$.

9.86 1.514 kg $\cdot$ m^2; 155.7 mm.

9.88 (a) $md^2/6$. (b) $2md^2/3$. (c) $2md^2/3$.

9.90 (a) 5.14×10^{-3} kg $\cdot$ m^2.
(b) 7.54×10^{-3} kg $\cdot$ m^2.
(c) 3.47×10^{-3} kg $\cdot$ m^2.

9.92 0.0503 lb $\cdot$ ft $\cdot$ s^2; 3.73 in.
SI: 0.0682 kg $\cdot$ m^2; 94.8 mm.

9.94 (a) 20.2 lb $\cdot$ ft $\cdot$ s^2. (b) 42.1 lb $\cdot$ ft $\cdot$ s^2.
(c) 41.3 lb $\cdot$ ft $\cdot$ s^2. SI: (a) 27.4 kg $\cdot$ m^2.
(b) 57.1 kg $\cdot$ m^2. (c) 56.0 kg $\cdot$ m^2.

9.96 $P_{xy} = -0.001199$ kg $\cdot$ m^2, $P_{yz} = P_{zx} = 0$.

9.98 $P_{xy} = 14.05$ lb $\cdot$ ft $\cdot$ s^2, $P_{yz} = P_{zx} = 0$.
SI: $P_{xy} = 19.05$ kg $\cdot$ m^2, $P_{yz} = P_{zx} = 0$.

9.100 (a) $P_{zx} = mca/20$. (b) $P_{xy} = mab/20$;
$P_{yz} = mbc/20$.

9.102 $3ma^2(a^2 + 6h^2)/20(a^2 + h^2)$.

9.104 25.2 lb $\cdot$ ft $\cdot$ s^2. SI: 34.1 kg $\cdot$ m^2.

9.106 (a) $I_x = 2ma^2/3$, $I_y = I_z = 11ma^2/12$,
$P_{xy} = ma^2/4$, $P_{yz} = 0$, $P_{zx} = -ma^2/4$.
(b) $2ma^2/3$.

9.108 $0.315\ ma^2$.

9.110 (a) 2. (b) 0.5.

9.114 $\frac{1}{2}ma^2$.

9.116 (a) $ab^3/30$. (b) $a^3b/6$. (c) $a^2b^2/16$.

9.118 (a) $\frac{1}{2}\rho\pi h(a_2^4 + 2a_2^2a_1^2 - 3a_1^4)$. (b) $a_2/\sqrt{3}$.
(c) $\frac{2}{3}\rho\pi ha_2^4$.

9.120 $\overline{I}_x = 1730$ in^4, $\overline{I}_y = 185.9$ in^4;
$\overline{k}_x = 5.05$ in., $\overline{k}_y = 1.657$ in.

SI: $\overline{I}_x = 720 \times 10^6$ mm^4,
$\overline{I}_y = 77.4 \times 10^6$ mm^4;
$\overline{k}_x = 128.4$ mm, $\overline{k}_y = 42.1$ mm.

9.122 (a) $ma^2/20$. (c) $ma^2/20$, $17ma^2/40$.

9.124 $\overline{I}_x = \overline{I}_z = 0.309$ kg $\cdot$ m^2,
$\overline{I}_y = 0.414$ kg $\cdot$ m^2.

CHAPTER 10

10.2 6.67 lb. SI: 29.7 N.

10.4 400 N.

10.6 $3P \tan \theta$.

10.8 $(Pl/a) \cos \theta \sin^2 \theta$.

10.10 $(Pl/r) \cos^2 \theta$.

10.12 $2Pl \cot \theta$.

10.14 (a) 9.8 lb. (b) 44.8 lb.
SI: (a) 43.6 N. (b) 199.3 N.

10.16 541 N.

10.18 $28.1°$.

10.20 $24.8°$.

10.22 $51.3°$.

10.24 $32.3°$.

10.26 (a) $12.5(HE)$. (b) $40.5°$.

10.28 $78.7°$; $323.8°$; $379.1°$.

10.32 215 lb, 305 lb. SI: 956 N, 1356 N.

10.34 1050 N $\uparrow$.

10.36 $\tan \theta_1 = 2P/3W$, $\tan \theta_2 = 2P/W$.

10.38 0.833 in. $\downarrow$. SI: 21.2 mm $\downarrow$.

10.46 $\theta = 45°$, unstable; $\theta = -135°$, stable.

10.48 $59.0°$, stable.

10.50 $y[1 - a/\sqrt{y^2 + a^2}] = W/k$.

10.52 $10.3°$, stable; $33.1°$, unstable; $90°$, stable.

10.54 (a) $\tan \theta = P/kl$. (b) $\theta = 63.4°$.

10.56 $W \leqslant kr^2/l$.

10.58 $10.9°$; $39.5°$.

10.60 $d > 313$ mm.

10.64 $P < k(l - a)^2/2l$.

10.66 $P < 0.382\ kl$.

10.68 7.50 lb. SI: 33.4 N.

10.70 $11.0°$, stable; $79.0°$, unstable.

10.72 $3.49W$.

10.74 $Q = [P \cos \theta + \frac{1}{2}W \sin \beta \sin \theta]/\cos (\beta - \theta)$.

10.76 $M = 2Wl \sin \theta$.

10.78 $63.4°$.

10.80 $\frac{1}{2}Pl \tan \theta/(1 - \mu \tan \theta)$.

Centroids of Common Shapes of Areas and Lines

Shape		$\bar{x}$	$\bar{y}$	Area
Triangular area			$\dfrac{h}{3}$	$\dfrac{bh}{2}$
Quarter-circular area		$\dfrac{4r}{3\pi}$	$\dfrac{4r}{3\pi}$	$\dfrac{\pi r^2}{4}$
Semicircular area		0	$\dfrac{4r}{3\pi}$	$\dfrac{\pi r^2}{2}$
Semiparabolic area		$\dfrac{3a}{8}$	$\dfrac{3h}{5}$	$\dfrac{2ah}{3}$
Parabolic area		0	$\dfrac{3h}{5}$	$\dfrac{4ah}{3}$
Parabolic spandrel		$\dfrac{3a}{4}$	$\dfrac{3h}{10}$	$\dfrac{ah}{3}$
Circular sector		$\dfrac{2r\sin\alpha}{3\alpha}$	0	αr^2
Quarter-circular arc		$\dfrac{2r}{\pi}$	$\dfrac{2r}{\pi}$	$\dfrac{\pi r}{2}$
Semicircular arc		0	$\dfrac{2r}{\pi}$	πr
Arc of circle		$\dfrac{r\sin\alpha}{\alpha}$	0	$2\alpha r$

Parabolic spandrel: $y = kx^2$